CONDUCTION HEAT TRANSFER

by

Vedat S. Arpacı

University of Michigan

ADDISON-WESLEY PUBLISHING COMPANY
Reading, Massachusetts
Menlo Park, California · London · Amsterdam · Don Mills, Ontario · Sydney

This book is in the
ADDISON-WESLEY SERIES IN
MECHANICS AND THERMODYNAMICS

Copyright © 1966 by Addison-Wesley. All rights reserved. This book, or parts thereof, may not be reproduced in any form without the written permission of the publisher. Printed in the United States of America. Published simultaneously in Canada. Library of Congress Catalog Card No. 66-25602.

ISBN 0-201-00359-7
FGHIJKLM-HA-79

PREFACE

This book is written for engineering students and for engineers working in heat transfer research or on thermal design. A new book on conduction is going to give rise to a number of questions among the members of the latter group: What conceivable reason inspired the author to write another book on conduction? Do we not already have the first and last testaments of conduction by Fourier and by Carslaw and Jaeger? What is the importance of conduction in today's engineering heat transfer studies? And so on. After all, is it not the feeling among engineers that the temperature problems associated with solids are now classical, with many solutions existing in the literature for a number of geometries and boundary conditions? Doubtless, from the viewpoint of mathematics, the foregoing points are all quite valid. One cannot claim, however, that the foundations of engineering conduction are based on mathematics only. And this text is intended, not to serve as an additional catalog of a number of new situations which are not listed in Carslaw and Jaeger, but rather to introduce the reader to engineering conduction.

Problems of engineering heat transfer involve one or a combination of the phenomena called *diffusion, radiation, stability,* and *turbulence.* Among these phenomena diffusion, because of its comparative simplicity, is a logical starting point in the study of heat transfer. In other words, we are not interested in diffusion for its own sake, as was the case for Fourier and for Carslaw and Jaeger. For this reason, the traditional mathematical treatment of the subject is no longer adequate. In order to provide an exposition of applied nature, I have followed the philosophy which I thought would be most suitable to engineering—the combination of physical reasoning with theoretical analysis. I regret that even in a book of this size, I have not been able to do justice to another but equally important aspect of heat transfer, experimental methods. Instead of defending the content of the text, however, I prefer simply to admit that I have chosen to write about only those topics in which I have some confidence of my understanding.

In the planning stage, my intention was to write a text involving the linear diffusion of momentum, mass, electric current, and neutrons as well as heat. This, of course, is a more elegant way to demonstrate the phenomenon of diffusion. Yet, without sacrificing a part of the present text and considerably exceeding the size of a typical text book, there seems to be no way of accomplishing this task. I thus confined myself to writing on the diffusion of heat only. The diffusion of heat in a rigid medium differs from that in a deformable medium, the latter including the diffusion of momentum. From the standpoint of the formulation, the rigid medium is a special case of the deformable medium, and need not be considered separately. From the viewpoint of solutions, however, the techniques applicable to a deformable body are in general not convenient for the linear problems of a rigid body, because of the nonlinear nature of the equations which govern the diffusion of momentum. For this reason, I have devoted the text to diffusion of heat in regid media, the so-called conduction phenomenon only.

Following an introductory chapter, the text is divided into three parts. In Part I, *Formulation*, I have tried hard to break away from the traditional thought that the formulation of conduction problems is merely

$$\frac{dT}{dt} = \boldsymbol{\nabla} \cdot (k\,\boldsymbol{\nabla} T) + \frac{u'''}{\rho c},$$

or another but similar differential equation. I have kept this part somewhat general so it can readily be extended to the case of deformable media. Only the treatment of inertial coordinates, stress tensor, momentum, and moment of momentum are omitted from this discussion. I have devoted Part II, *Solution*, to the simplest and, to a large extent, the general (but not necessarily the most elegant) methods of solution. Thus the potential theory, the source theory, Green's functions, and the transform calculus (with the exception of Laplace transforms) are left untreated. This seems quite adequate for the intended size and level of the text. I have collected topics of advanced or special nature under Part III as *Further Methods of Formulation and Solution*. These include variational calculus, difference and differential-difference formulations, and relaxation, numerical, graphical, and analog solutions.

In general, problems are designed to clarify the physically and/or mathematically important points, and to supplement and extend the text. The classical "handbook" type of material is avoided as much as possible. Aside from the classical problems, the great majority are my own inventions.

With few exceptions, no more engineering background is required of the reader than the customary undergraduate courses in thermodynamics, heat transfer, and advanced calculus. Prior knowledge in fluid mechanics and the fundamentals of vector calculus are helpful but not necessary. The sections involving vectors may be studied without them by using a coordinate system in the usual manner.

The text is the result of a series of revisions of the material originally prepared and mimeographed for use in a senior-graduate course on heat transfer in the Mechanical Engineering Department of The University of Michigan. It may, of course, find application in other fields of endeavor which deal with temperature and associated stress problems in solids.

The following outline and suggestions seem pertinent for a three-credit course. Chapter 1 should be read as a survey based on undergraduate material. Other examples than those employed in this chapter may be utilized, the choice depending on the instructor's taste and the students' background. Chapter 2 is the most important chapter of the text. It may not be possible to master this chapter in one attempt. It is therefore strongly suggested that the chapter be continuously reviewed in the course of study. Elementary parts of Chapter 3 may be eliminated for students who have had a first course on heat transfer. Chapters 4, 5, and 6 are the backbone of solution methods, and should be studied without omission. In general, the time available for a three-credit course does not permit study of all the remaining chapters. For the rest of the course, therefore, one or possibly two chapters out of Chapters 7, 8, 9, and 10 are suggested; again, the choice will depend on the instructor's taste and the students' background.

My first acknowledgment is to my student, friend, and now colleague Professor P. S. Larsen, who read my notes in the course of the writing process and made invaluable suggestions. Thanks are extended to Professor R. J. Schoenhals of Purdue University and Professor J. W. Mitchell of the University of Wisconsin for their constructive criticism of the manuscript. I am grateful to Professor G. J. Van Wylen, then Chairman of the Mechanical Engineering Department and now Dean of Engineering, and to Professors W. Mirsky, H. Merte, and J. R. Cairns of The University of Michigan for reading one chapter of my notes and making some remarks. I am also indebted to Professor G. J. Van Wylen for reducing my teaching load for one semester in the final stage of my writing. Professor J. A. Clark, Professor-in-charge of the Heat Transfer Laboratory, has been a continual source of encouragement and inspiration as a friend and colleague. I am thankful to my students, to Professor C. L. S. Farn of Carnegie Institute of Technology, and to Messrs. L. H. Blake and C. Y. Warner, for helping me in the preparation of some figures.

Last but not least, I must express a word of appreciation to Mrs. B. Ogilvy, whose unusual cooperation often exceeded regular hours in the process of typing my class notes over a period of five years, and to Addison-Wesley Publishing Company, whose competent work made this publication possible.

Ann Arbor, Michigan V. S. A.
June 1966

CONTENTS

Introduction

Chapter 1 Foundations of Heat Transfer 3

    1–1 The place of heat transfer in engineering 3
    1–2 Continuum theory versus molecular theory 9
    1–3 Foundations of continuum heat transfer 10

PART I FORMULATION

Chapter 2 Lumped, Integral, and Differential Formulations 17

    2–1 Definition of concepts 18
    2–2 Statement of general laws 19
    2–3 Lumped formulation of general laws 20
    2–4 Integral formulation of general laws 26
    2–5 Differential formulation of general laws 32
    2–6 Statement of particular laws 37
    2–7 Equation of conduction. Entropy generation due to conductive resistance 44
    2–8 Initial and boundary conditions 46
    2–9 Methods of formulation 59
    2–10 Examples 61

PART II SOLUTION

Chapter 3 Steady One-Dimensional Problems. Bessel Functions . . . 103

    3–1 A general problem 103
    3–2 Composite structures 107
    3–3 Examples 110
    3–4 Principle of superposition 126
    3–5 Heterogeneous solids (variable thermal conductivity) . . 129
    3–6 Power series solutions. Bessel functions 132
    3–7 Properties of Bessel functions 139
    3–8 Extended surfaces (fins, pins, or spines) 144
    3–9 Approximate solutions for extended surfaces 156
    3–10 Higher-order approximations 161

CONTENTS

Chapter 4 Steady Two- and Three-Dimensional Problems. Separation of Variables. Orthogonal Functions 180

- 4–1 Boundary-value problems. Characteristic-value problems . . . 180
- 4–2 Orthogonality of characteristic functions 183
- 4–3 Expansion of arbitrary functions in series of orthogonal functions . 185
- 4–4 Fourier series 186
- 4–5 Separation of variables. Steady two-dimensional cartesian geometry 193
- 4–6 Selection of coordinate axes 214
- 4–7 Nonhomogeneity 217
- 4–8 Steady two-dimensional cylindrical geometry. Solutions by Fourier series 224
- 4–9 Steady two-dimensional cylindrical geometry. Fourier-Bessel series 230
- 4–10 Steady two-dimensional spherical geometry. Legendre polynomials. Fourier-Legendre series 240
- 4–11 Steady three-dimensional geometry 248

Chapter 5 Separation of Variables. Unsteady Problems. Orthogonal Functions 270

- 5–1 Distributed systems having stepwise disturbances 273
- 5–2 Multidimensional problems expressible in terms of one-dimensional ones. Use of one-dimensional charts 290
- 5–3 Time-dependent boundary conditions. Duhamel's superposition integral 307

Chapter 6 Steady Periodic Problems. Complex Temperature 324

Chapter 7 Unsteady Problems. Laplace Transforms 336

- 7–1 Transform calculus 336
- 7–2 An introductory example 337
- 7–3 Properties of Laplace transforms 339
- 7–4 Solutions obtainable by the table of transforms 343
- 7–5 Fourier integrals 365
- 7–6 Inversion theorem for Laplace transforms 372
- 7–7 Functions of a complex variable 374
- 7–8 Evaluation of the inversion theorem in terms of two particular contours 386

7–9	Solutions obtainable by the inversion theorem	390
7–10	Solutions valid for small or large values of time	412

PART III FURTHER METHODS OF FORMULATION AND SOLUTION

Chapter 8 Variational Formulation. Solution by Approximate Profiles . . 435

8–1	Basic problem of variational calculus	435
8–2	Meaning and rules of variational calculus	438
8–3	Steady one-dimensional problems	440
8–4	Ritz method	444
8–5	Steady one-dimensional Ritz profiles	450
8–6	Steady two-dimensional problems	455
8–7	Steady two-dimensional Ritz method	458
8–8	Kantorovich method	460
8–9	Kantorovich method extended	466
8–10	Construction of steady two-dimensional profiles	469
8–11	Unsteady problems	473
8–12	Some definite integrals	478

Chapter 9 Difference Formulation. Numerical and Graphical Solutions . . 483

9–1	Difference formulation of steady problems	483
9–2	Relation between difference and differential formulations	486
9–3	Error in difference formulation	487
9–4	Finer, graded, triangular, and hexagonal networks	489
9–5	Cylindrical and spherical geometries	490
9–6	Irregular boundaries	492
9–7	Solution of steady problems. Relaxation method	493
9–8	Difference formulation of unsteady problems. Stability	503
9–9	Solution of unsteady problems. Step-by-step numerical solution. Binder-Schmidt graphical method	515

Chapter 10 Differential-Difference Formulation. Analog Solution 524

10–1	Analogy between conduction and electricity	524
10–2	Passive circuit elements	527
10–3	Active circuit elements. High-gain DC amplifiers	529
10–4	Examples	533
10–5	Miscellaneous	540

Index	543

The ability to analyze a problem involves a combination of inherent insight and experience. *The former, unfortunately, cannot be learned, but depends on the individual. However, the latter is of equal importance, and can be gained with* patient study.

INTRODUCTION

CHAPTER 1

FOUNDATIONS OF HEAT TRANSFER

The foundations of any engineering science may best be understood by considering the place of that science in relation to other engineering sciences. Therefore, our first concern in this chapter will be to determine the place of heat transfer among the engineering sciences. Next, two modes of heat transfer —diffusion and radiation—will be briefly reviewed. We shall then proceed to a discussion of the *continuum* and the *molecular* approaches to engineering problems, and finally, to a discussion of the foundations of *continuum heat transfer*.

1-1. The Place of Heat Transfer in Engineering

Let us first review four well-known problems taken from the mechanics of rigid and deformable bodies and from thermodynamics. For each problem let us consider two formulations, based on different assumptions. Our concern will be with the nature of the physical laws employed in these formulations. (At this stage, our discussion will necessarily be framed in the conventional terms of existing textbooks; the philosophy of the present text will be set forth in the next chapter.)

Example 1-1. *Free fall of a body.* Consider a body of mass m falling freely under the effect of the gravitational field (Fig. 1-1). We wish to determine the instantaneous location of this body.

Formulation (physics) of the problem: Newton's second law of motion,

$$\mathbf{F} = m\mathbf{a}, \qquad (1\text{-}1)$$

$\mathbf{F}$ being the sum of external forces and $\mathbf{a}$ the acceleration vector, may be reduced to a one-dimensional problem, for if we neglect the resistance R of the surrounding medium, Eq. (1-1) can be written in the form

$$mg = m\frac{d^2x}{dt^2}, \qquad (1\text{-}2)$$

subject to the appropriate initial conditions.

Solution (mathematics) of the problem: Integrating Eq. (1-2) twice with respect to time yields

$$x = \tfrac{1}{2}gt^2 + C_1 t + C_2, \qquad (1\text{-}3)$$

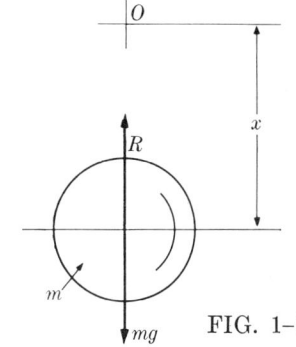

FIG. 1-1

3

where the two integration constants, C_1 and C_2, may be determined from the initial position and velocity of the body.

The formulation and solution of a problem are clearly distinguished in the foregoing trivial case, elaborated for this purpose.

In our second formulation of the problem, let us include the resistance to the motion of the body exerted by the ambient medium. With this consideration we have

$$mg - R = m\frac{d^2x}{dt^2}. \tag{1-4}$$

Equation (1–4) cannot be integrated without further information about the resistance force R. If, for example, this force is assumed to be proportional to the square of the velocity of the body, that is, if

$$\frac{R}{m} = k\left(\frac{dx}{dt}\right)^2, \tag{1-5}$$

then Eq. (1–4) gives

$$g - k\left(\frac{dx}{dt}\right)^2 = \frac{d^2x}{dt^2}, \tag{1-6}$$

where k is a constant. Equation (1–6) is a nonlinear differential equation whose solution is quite involved, and since this solution is unimportant for the present discussion it will not be given here.

Before commenting on the foregoing formulations, let us consider two more problems taken from mechanics.

Example 1–2. *Reaction forces of a beam.* Consider a beam subject to a localized force P. We wish to determine the reaction forces of the beam.

For the first formulation of the problem let us assume that the beam is simply supported (Fig. 1–2). The reaction forces A and B may then be obtained from the conditions

$$\Sigma \text{ Force} = 0, \quad \text{Force balance}, \tag{1-7}$$

$$\Sigma \text{ Moment} = 0, \quad \text{Moment balance}, \tag{1-8}$$

which are the results of Newton's second law of motion as applied to statics problems.

For the second formulation of the problem let us replace one of the simple supports of the beam by a built-in support as shown in Fig. 1–3. The new case can no longer be solved by employing Newton's law only. Because there are three unknowns, the reaction forces A_1, B_1, and the bending moment M, we require one more condition in addition to Eqs. (1–7) and (1–8). This may be obtained by considering the nature of the beam. If, for example, the beam is assumed to be elastic, the additional condition may be derived from *Hooke's law*.

CHAPTER 1

FOUNDATIONS OF HEAT TRANSFER

The foundations of any engineering science may best be understood by considering the place of that science in relation to other engineering sciences. Therefore, our first concern in this chapter will be to determine the place of heat transfer among the engineering sciences. Next, two modes of heat transfer —diffusion and radiation—will be briefly reviewed. We shall then proceed to a discussion of the *continuum* and the *molecular* approaches to engineering problems, and finally, to a discussion of the foundations of *continuum heat transfer*.

1-1. The Place of Heat Transfer in Engineering

Let us first review four well-known problems taken from the mechanics of rigid and deformable bodies and from thermodynamics. For each problem let us consider two formulations, based on different assumptions. Our concern will be with the nature of the physical laws employed in these formulations. (At this stage, our discussion will necessarily be framed in the conventional terms of existing textbooks; the philosophy of the present text will be set forth in the next chapter.)

Example 1-1. *Free fall of a body.* Consider a body of mass m falling freely under the effect of the gravitational field (Fig. 1-1). We wish to determine the instantaneous location of this body.

Formulation (physics) of the problem: Newton's second law of motion,

$$\mathbf{F} = m\mathbf{a}, \tag{1-1}$$

$\mathbf{F}$ being the sum of external forces and $\mathbf{a}$ the acceleration vector, may be reduced to a one-dimensional problem, for if we neglect the resistance R of the surrounding medium, Eq. (1-1) can be written in the form

$$mg = m\frac{d^2x}{dt^2}, \tag{1-2}$$

subject to the appropriate initial conditions.

Solution (mathematics) of the problem: Integrating Eq. (1-2) twice with respect to time yields

$$x = \tfrac{1}{2}gt^2 + C_1 t + C_2, \tag{1-3}$$

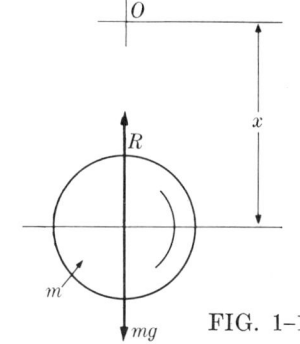

FIG. 1-1

3

where the two integration constants, C_1 and C_2, may be determined from the initial position and velocity of the body.

The formulation and solution of a problem are clearly distinguished in the foregoing trivial case, elaborated for this purpose.

In our second formulation of the problem, let us include the resistance to the motion of the body exerted by the ambient medium. With this consideration we have

$$mg - R = m\frac{d^2x}{dt^2}. \tag{1-4}$$

Equation (1-4) cannot be integrated without further information about the resistance force R. If, for example, this force is assumed to be proportional to the square of the velocity of the body, that is, if

$$\frac{R}{m} = k\left(\frac{dx}{dt}\right)^2, \tag{1-5}$$

then Eq. (1-4) gives

$$g - k\left(\frac{dx}{dt}\right)^2 = \frac{d^2x}{dt^2}, \tag{1-6}$$

where k is a constant. Equation (1-6) is a nonlinear differential equation whose solution is quite involved, and since this solution is unimportant for the present discussion it will not be given here.

Before commenting on the foregoing formulations, let us consider two more problems taken from mechanics.

Example 1-2. *Reaction forces of a beam.* Consider a beam subject to a localized force P. We wish to determine the reaction forces of the beam.

For the first formulation of the problem let us assume that the beam is simply supported (Fig. 1-2). The reaction forces A and B may then be obtained from the conditions

$$\sum \text{Force} = 0, \qquad \text{Force balance}, \tag{1-7}$$

$$\sum \text{Moment} = 0, \qquad \text{Moment balance}, \tag{1-8}$$

which are the results of Newton's second law of motion as applied to statics problems.

For the second formulation of the problem let us replace one of the simple supports of the beam by a built-in support as shown in Fig. 1-3. The new case can no longer be solved by employing Newton's law only. Because there are three unknowns, the reaction forces A_1, B_1, and the bending moment M, we require one more condition in addition to Eqs. (1-7) and (1-8). This may be obtained by considering the nature of the beam. If, for example, the beam is assumed to be elastic, the additional condition may be derived from *Hooke's law*.

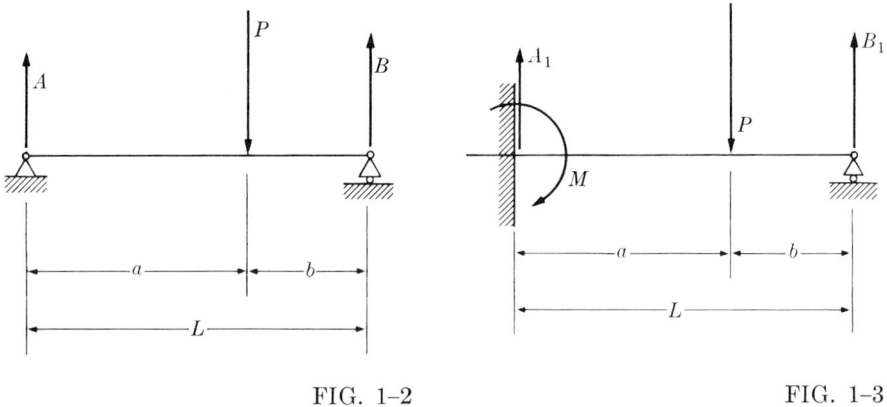

FIG. 1-2   FIG. 1-3

As in Example 1-1, the details of the foregoing formulations and their solutions are not important for the present discussion.

Example 1-3. *Fully developed, steady incompressible flow of a fluid between two parallel plates.* We wish to find the velocity distribution in the flow.

Consider the control volume* shown in Fig. 1-4. Since the acceleration terms are identically zero, when Newton's second law of motion is applied to this control volume it again reduces to a force balance. The normal and tangential components of the *surface forces* per unit area are designated pressure p and shear τ. The *body forces* such as gravity are neglected in this example. The first formulation of the problem will be based on the assumption of an ideal (frictionless) fluid, the second on that of a viscous (Newtonian) fluid.

If the fluid is ideal, the shear stress is zero by definition, and the pressure change in the x-direction is also zero as a consequence of Newton's second law. The fluid may now be replaced by infinitely thin parallel layers with no friction between them. In the absence of any net force, *Newton's first law* states that each layer must either be stationary or move with a uniform velocity. Therefore, if the entrance velocity is constant and uniform, this condition is preserved axially and transversally for all values of the time. In other words, the velocity at any cross section is uniform.

If fluid friction is included, Newton's second law applied to the control volume of Fig. 1-4 yields

$$-\frac{d\tau}{dy} - \frac{dp}{dx} = 0. \tag{1-9}$$

Equation (1-9), however, cannot be solved to give the desired velocity distribution unless an additional condition is specified. One such condition might be a relation between the shear stress and the velocity gradient. If we assume,

---

* The definition of *control volume* will be given in the next chapter.

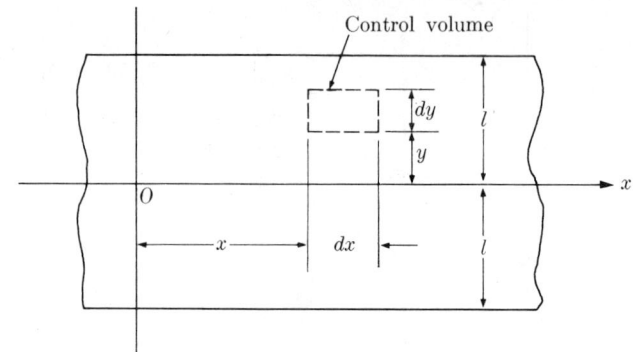

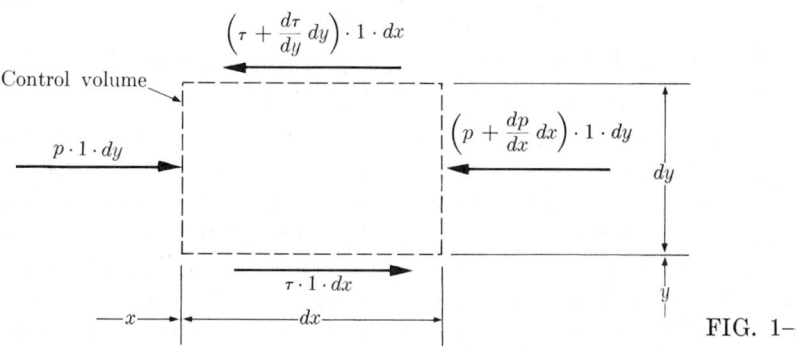

FIG. 1-4

for example, that the fluid is Newtonian, this condition may be stated as

$$\tau = -\mu \frac{du}{dy}, \tag{1-10}$$

where the proportionality constant is the viscosity of the fluid. Introducing Eq. (1–10) into Eq. (1–9) and rearranging gives

$$\frac{d^2 u}{dy^2} = \frac{1}{\mu}\left(\frac{dp}{dx}\right). \tag{1-11}$$

Equation (1–11) and the boundary conditions

$$\frac{du(0)}{dy} = 0, \quad u(l) = 0, \tag{1-12}$$

complete the formulation of the problem. For a constant axial pressure gradient (dp/dx), the solution of Eq. (1–11) satisfying Eq. (1–12) is the well-known parabolic velocity distribution

$$u(y) = \frac{(-dp/dx)l^2}{2\mu}\left[1 - \left(\frac{y}{l}\right)^2\right]. \tag{1-13}$$

Let us now examine the foregoing three examples in the light of the physical laws used in their formulations. As we have just seen, some problems taken from mechanics can be solved by using only Newton's second law of motion, combined sometimes with Newton's first law and/or the conservation of mass; these are called *mechanically determined problems*. The dynamics of rigid bodies in the absence of friction, the statically determined problems of rigid bodies, and the mechanics of ideal fluids provide well-known examples of this class. More specifically, we may place in this category the first formulations of each of the foregoing problems: the free fall of a body without friction, the statically determined beam, and the steady flow of an ideal fluid between two parallel plates.

Some mechanics problems, however, require an extra condition in addition to Newton's laws of motion and the conservation of mass. These are called *mechanically undetermined problems*. The dynamics of rigid bodies with friction and the mechanics of deformable bodies (viscous, elastic, plastic, viscoelastic bodies) provide examples of this group, as illustrated by the second formulations of each of the foregoing examples: the free fall of a body with friction, the statically undetermined beam, and laminar flow between two parallel plates. It is important to note that each of these mechanically undetermined problems employs not only the general laws of mechanics, but also an additional law whose nature depends on the specific problem under consideration. The free fall of a body requires a relation between the resistance force and the velocity, the statically undetermined beam requires a relation between the stress and the strain, and laminar flow between two parallel plates requires a relation between the shear stress and the velocity. Hereafter any such additional law will be called a *particular law*, although the term *constitutive relation* is more frequently used in the literature.

Thermodynamics problems may similarly be divided into two classes. Some thermodynamics problems can be solved by employing the *general* (first and second) *laws of thermodynamics* and, if necessary, the general laws of mechanics; these are called *thermodynamically determined problems*. Others, however, require the use of conditions in addition to the general laws; these are called *thermodynamically undetermined problems*. The following example may be helpful to illustrate the point.

Example 1–4. *Steady one-dimensional isentropic and subsonic flow of an inviscous fluid through an insulated diffuser.* The state of the fluid is given at the inlet; we wish to find the state at the outlet. The notation is shown in Fig. 1–5—the pressure p, the density ρ, the internal energy u, the

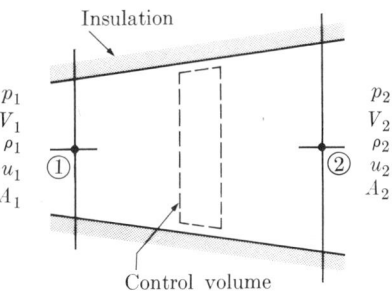

FIG. 1–5

velocity V, and the cross-sectional area A. The inlet properties are identified by the subscript 1 and the outlet properties by the subscript 2.

Let us apply the appropriate general laws to the control volume shown in Fig. 1–5. The law of *conservation of mass* gives

$$d(\rho A V) = 0; \qquad (1\text{–}14)$$

Newton's second law of motion, rearranged by means of Eq. (1–14), results in

$$dp + \rho V \, dV = 0, \qquad (1\text{–}15)$$

and the *first law of thermodynamics*, combined with Eqs. (1–14) and (1–15), yields

$$du + p \, d(1/\rho) = 0. \qquad (1\text{–}16)$$

In the first formulation of the problem we assume that the fluid is incompressible. Then by definition, $\rho_1 = \rho_2$, and the remaining outlet properties V_2, p_2, u_2 are obtained from the integration of Eqs. (1–14), (1–15), and (1–16), respectively, between the inlet and the outlet of the diffuser.

In the second formulation of the problem we consider instead a compressible fluid. Now, in addition to V_2, p_2, and u_2, the outlet density ρ_2 must be determined; therefore, the three conditions given by Eqs. (1–14), (1–15), and (1–16) are no longer sufficient. To complete the formulation, let us recall the manner in which we arrived at the second formulations of Examples 1–1, 1–2, and 1–3. In the present example, the required additional condition may be related to the nature of the fluid. Implicitly, we may write this condition in the form

$$p = p(\rho, u). \qquad (1\text{–}17)$$

In practice, Eq. (1–17) is expressed or tabulated in a number of ways suitable for frequently encountered problems. The most commonly found explicit form of Eq. (1–17) is

$$p = \rho R T, \qquad (1\text{–}18)$$

the so-called *ideal gas law*. Equations (1–17) and (1–18) are forms of a particular law which is usually referred to as the *equation of state*. The outlet state of the fluid may, in principle, be determined from Eqs. (1–14), (1–15), (1–16), and (1–17). However, as before, the details are not important and will not be considered here.

It is now clear that in its first formulation, Example 1–4 is a thermodynamically determined problem and can be solved by using the general laws of thermodynamics, combined with those of mechanics. In its second formulation, however, the problem requires the use of an additional particular law, and therefore it is thermodynamically undetermined.

Gas dynamics and *heat transfer* are the major disciplines which deal with thermodynamically undetermined problems. In addition to the general laws of thermodynamics, gas dynamics depends on the equation of state as a particular law. Heat transfer employs two particular laws, related to the so-called *modes of heat transfer* which we shall now describe.

Diffusion. In diffusion, heat is transferred through a medium or from one to another of two media in contact, if there exists a nonuniform temperature distribution in the medium or between the two media. On the molecular level, the mechanism of diffusion is visualized as the exchange of kinetic energy between the molecules in the regions of high and low temperatures. Particularly, it is attributed to the elastic impacts of molecules in gases, to the motion of free electrons in metals, and to the longitudinal oscillations of atoms in solid insulators of electricity.

Radiation. The true nature of radiation and its transport mechanism have not been completely understood to date. Some of the effects of radiation can be described in terms of electromagnetic waves, and others in terms of quantum mechanics, although neither theory explains all the experimental observations. According to wave theory, for example, during the emission of radiation a body continuously converts part of its internal energy to electromagnetic waves, another form of energy. These waves travel through space with the velocity of light until they strike another body, where part of their energy is absorbed and reconverted into internal energy.

In the foregoing classification we have not considered *convection* to be a mode of heat transfer. Actually, convection is motion of the medium which facilitates heat transfer by diffusion and/or radiation. For customary reasons only, we distinguish between the diffusion of heat in moving or stationary rigid bodies, which we shall call *conduction*, and the diffusion of heat in moving deformable bodies, which we shall call *convection*. Conduction is the subject of this volume; convection and radiation will be treated elsewhere. Nevertheless, examples from convection and radiation are occasionally included in this text when pertinent.

1–2. Continuum Theory Versus Molecular Theory*

In the preceding section the process of heat transfer by diffusion was described in two different ways. From a macroscopic or phenomenological standpoint, heat, as evidenced from experimental observations, is transferred from a region of higher temperature to a region of lower temperature in a medium. From a microscopic or molecular standpoint, the transfer of heat is thought to come about through the exchange of kinetic energy between molecules. This theory, however, is based on hypothesis rather than experiment. The contrast between these two views of heat transfer is reflected in the two alternative approaches to engineering problems in heat transfer.

---

* The continuum theory is also called the *field* or *macroscopic* theory, or *phenomenological* view; the molecular theory is also called the *microscopic* theory.

In the first approach, corresponding to the macroscopic view, the medium is assumed to be a continuum. That is, the mean free path of molecules is small compared with all other dimensions existing in the medium, such that a statistical average (global description) is possible. In other words, the medium fits the definition of the concept of *field*. The properties of a field may be scalar, such as temperature T, or vectorial, such as velocity $\mathbf{V}$. In the second approach, corresponding to the molecular view, either a statistical average of the molecular behavior is not possible or it is possible but not desired. Actually, a very general and logical description of a medium that consists of a spatially distributed molecular structure would be one in which the general laws were written for each separate molecule. Solving the many-particle system in time and space and then relating a required macroscopic concept to molecular behavior would obviously produce the same result as that obtained from the continuum theory.

The reason for not always starting with the molecular approach, apart from the mathematical difficulties and the fact that we actually know little about intermolecular forces, is that the behavior of molecules or small particles of a medium may not be of particular interest. On the contrary, as in most cases of engineering, the problem may be to determine how the medium behaves as a whole—to find, for example, the velocity and/or temperature variations in a medium. Here the convenience of using the field concept becomes clear. Obviously, there are cases in which it is advantageous to use one of the foregoing methods in preference to the other. For example, one would hardly think of solving a conduction problem of a solid body from the particle standpoint, or explaining the behavior of rarefied gases by means of continuum considerations. A large number of problems exists, however, for which both approaches can be used conveniently, the choice depending on previous experience, skill, or taste. From the viewpoint of physical interpretation the only difference is that the averaging process of the molecular structure is undertaken before or after the analysis, depending on the approach used. That is, the statistics either precedes or follows the mechanics (or thermodynamics).

In this text our interest lies not in the individual behavior of the molecules, but rather in their mean effects in space and time. In other words, the problem is to determine how a medium behaves as a whole, or how such parts of it containing a large number of molecules behave. We therefore will be looking at problems of heat transfer from the continuum standpoint.

1–3. Foundations of Continuum Heat Transfer

Any engineering science is based on both theory and experiment. The answer to the question "Why not only experiment or theory rather than both together?" is that each is a tool fundamentally different from the other, and each has its own idealizations and approximations which may not pertain to the other. In the search for reality, both are needed, for reality may be closely approached by cross-checking the results of theoretical and experimental investigations. Therefore, although the present text is directed toward theoretical heat transfer only, let it be clearly understood that this is a matter of the author's area of competence rather than an indication of the importance of theory as opposed to experiment.

Problem-solving in theoretical heat transfer, as in other disciplines of engineering, may be outlined as follows:

A problem posed by reality ←⎤
↓
Formulation by idealization
(Physics)
↓
Solution by approximation
(Mathematics)
↓
Interpretation of physical
meaning of answer ⎦

From this outline we see that two principles are involved in all problems of engineering, the principle of *idealization* and the principle of *approximation*.

In the formulation of a problem, idealizations are necessary even for the definition of concepts and the statement of natural (general and particular) laws. A well-known example of the idealized definition of concepts is the density b* of a physical property B of a medium at a point P. This is defined according to the following idealized limiting process. A small sphere of radius R is considered at P, then the total quantity of the property ΔB contained in the sphere is divided by its volume ΔV, and R (or ΔV) is allowed to approach zero. The foregoing limiting process is actually an unrealistic idealization. As a matter of fact, when the volume of the sphere becomes less than a value $\Delta V_0(R_0)$ the density changes discontinuously, as a result of the molecular structure of the medium. However, extrapolating the density curve from $\Delta V_0(R_0)$ to $\Delta V(0) = 0$ as shown in Fig. 1–6, we substitute for the discontinuous behavior the continuous behavior and, passing from the difference form to the differential form, we have

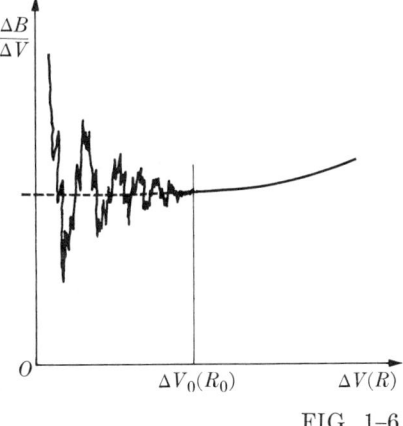

FIG. 1–6

$$b = \lim_{\Delta V \to 0} \frac{\Delta B}{\Delta V} = \frac{dB}{dV}, \quad (1\text{--}19)$$

---

* Here the density is, in the generalized sense, a quantity per unit volume. It may or may not be the mass density.

which is a convenient mathematical definition of density at a point. Well-known examples of applications of this definition are mass density, mass concentration, and electric charge density. The same procedure may be applied to other volume- or mass-dependent properties* of a medium.

The second idealization in the formulation of a problem involves the individual terms appearing in the statement of general laws. This may be illustrated by considering, for example, Newton's second law of motion. The forces described by this law may be categorized as body and surface forces. The body forces acting on a differential volume and the surface forces acting on a differential surface element are both idealized to be identical to a vector by excluding the couples.† Further idealizations may involve the complete omission of certain terms. Without these and many other idealizations, the continuum theory of engineering as used in the mechanics of rigid and deformable bodies, thermodynamics, gas dynamics, heat transfer, or electromagnetics would be impossible to formulate. These fields deal with ideal continua, although the continuous media involved consist of a finite but very large number of discrete individual particles.

Once the natural laws of continuum theory have been established (on the basis of a number of idealizations) we must find ways of formulating the problems posed by reality. The formulation selected depends on our ability to fit problems to the natural laws, a process which often requires further approximations of these laws to make them applicable to the specific problem under consideration.

It should always be kept in mind that a problem may be formulated in a number of approximate ways, and that intuition, insight, and experience are required in order to select the formulation best suited to the problem. Intuition and insight, unfortunately, cannot be taught and depend on the individual. However, the equally important experience can be gained with faithful practice. Similarly, the solution of a formulated problem may be obtained in a number of approximate ways, and again, the selection of the most suitable method of solution requires the intuition, insight, and experience of the individual, although in mathematics rather than in physics. The solution of a well-formulated problem should satisfy the criteria of existence, uniqueness, and stability. Existence and uniqueness, although important, are the concerns of the pure scientist. Stability, on the other hand, is obviously of great importance to the applied scientist.

This text is divided into three parts. Part I deals with the formulation, and Part II with the methods of exact and approximate solutions of conduction problems. Part III is devoted to further formulation and solution methods of an advanced or special nature.

---

* These are sometimes called *extensive* properties.
† These and further assumptions are necessary for the formulation of the Euler and Navier-Stokes equations of fluids.

References

1. L. PRANDTL and O. G. TIETJENS, *Fundamentals of Hydro- and Aerodynamics.* New York: McGraw-Hill, 1934.

2. M. H. SHAMOS and G. M. MURPHY, *Recent Advances in Science.* New York: Interscience Publishers, 1956.

3. A. H. SHAPIRO, *The Dynamics and Thermodynamics of Compressible Flow.* New York: The Ronald Press, 1953.

PART I | **FORMULATION**

CHAPTER 2

LUMPED, INTEGRAL, AND DIFFERENTIAL FORMULATIONS

In Chapter 1 the place of heat transfer among the engineering disciplines was established and the modes of heat transfer—conduction, convection, and radiation—were distinguished. Having this background we now proceed to the general formulation of conduction problems.

The formulation or physics of the analytical phase of an engineering science such as heat transfer is based on *definitions of concepts* and on *statements of natural laws* in terms of these concepts. The natural laws of conduction, like those of other disciplines, can be neither proved nor disproved but are arrived at inductively, on the basis of evidence collected from a wide variety of experiments. As man continues to increase his understanding of the universe, the present statements of natural laws will be refined and generalized. For the time being, however, we shall refer to these statements as the available approximate descriptions of nature, and employ them for the solution of current problems of engineering.

As we saw in Chapter 1, the natural laws may be classified as (1) general laws, and (2) particular laws. *A general law is characterized by the fact that its application is independent of the nature of the medium under consideration.* Examples are the law of conservation of mass, Newton's second law of motion (including momentum and moment of momentum), the first and second laws of thermodynamics, the law of conservation of electric charge, Lorentz's force law, Ampere's circuit law, and Faraday's induction law. The problems of nature which can be formulated completely by using only general laws are called *mechanically, thermodynamically, or electromagnetically determined problems.* On the other hand, the problems which cannot be formulated completely by means of general laws alone are called *mechanically, thermodynamically or electromagnetically undetermined problems.* Each problem of the latter category requires, in addition to the general laws, one or more conditions stated in the form of particular laws. *A particular law is characterized by the fact that its application depends on the nature of the medium under consideration.* Examples are Hooke's law of elasticity, Newton's law of viscosity, the ideal gas law, Fourier's law of conduction, Stefan-Boltzmann's law of radiation, and Ohm's law of electricity.

In this text we shall employ three general laws,

(a) the law of conservation of mass,
(b) the first law of thermodynamics,
(c) the second law of thermodynamics,

and two particular laws,

(d) Fourier's law of conduction,
(e) Stefan-Boltzmann's law of radiation,

each with a different degree of importance. For this reason the first seven sections of this chapter are devoted to a review of these laws and their associated concepts.

2-1. Definition of Concepts

Starting with the hypothesis that the universe is a medium of molecular structure containing energy, let us first define the following concepts.

Continuum (field). A medium in which the smallest volume under consideration contains enough molecules to permit the statistically averaged characteristics to adequately describe the medium. [See, for example, the definition of density b (Eq. 1–19).]

System. A part of a continuum which is separated from the rest of the continuum for convenience in the formulation of a problem. The boundaries of a system may expand or contract, but they are always so assumed that the rest of the continuum does not cross them during any change of the system. The *Lagrangian* method of fluid mechanics is used in the mathematical description of a system.

Control volume. The same as a system, except that the rest of the continuum may cross the fixed or deformable boundaries (control surfaces) of a control volume at one or more places. This is the *only* difference between a control volume and a system. For most problems in this text, control volumes with deformable boundaries are not necessary and, except for simple cases,* are not considered. The *Eulerian* method of fluid mechanics is used in the mathematical description of a control volume.

Property. A macroscopic characteristic of a system or control volume which is ascertained by a statistical averaging procedure. Properties, such as density, velocity, pressure, temperature, internal energy, kinetic energy, potential energy, enthalpy, or entropy, are observed or evaluated quantitatively. The mathematical description of a property B is that between any initial and final conditions the change in B does not depend on the path followed, that is,

$$\int_1^2 dB = B_2 - B_1.$$

State. A condition of a system or control volume which is identified by means of properties. A state may be determined when a sufficient number of independent properties is specified.

Process. A change of any state of a system or control volume.

Cycle. A process whose initial and final states are identical.

Work. A form of energy which is identified as follows. Work is done by a system or control volume on its environment during a process if the system or

---

* See Section 2–3.

the control volume could pass through the same process while the sole effect external to the system or the control volume was the raising of a weight. *Work done by a system or control volume is considered positive; work done on a system or control volume, negative.* The most common forms of work are discussed in the statements of the first law of thermodynamics. [See the developments of Eqs. (2–16) and (2–36).]

Equality of temperature. When any two systems or control volumes are placed in contact with each other, in general they affect each other as evidenced by changes in their properties. The limiting state which the two approach is called the state of equality of temperature. The definition of equality of temperature implies the existence of states of inequality of temperature for this pair.

Heat. A form of energy which is transferred across the boundaries of a system or control volume during a process by virtue of inequality of temperature. *Heat transferred to a system or control volume is considered positive; heat transferred from a system or control volume, negative.*

It can be shown that the work and heat interactions between a system or control volume and the surroundings depend on the path followed by the associated process. Therefore, *work and heat are not properties.*

The natural laws may now be stated in terms of the foregoing concepts. First let us consider the general laws.

2–2. Statement of General Laws

The first step in the statement of a general law is the selection of a system or control volume. Without this step it is meaningless to speak of such concepts as density, velocity, pressure, temperature, heat, work, internal energy, or entropy, which are the terms used in the statements of general laws. Although the well-known, simple forms of the general laws are always written for a system, system analysis becomes inconvenient when dealing with continua in motion, because it is often difficult to identify the boundaries of a moving system for any appreciable length of time. The control-volume approach is therefore generally preferred for continua in motion.

The second step in the statement of a general law is the selection of the form of this law. A general law may be formulated in either of the following forms:

(1) lumped (or averaged);

(2) distributed: (a) integral, (b) differential, (c) variational, (d) difference.

A general law is said to be lumped if its terms are independent of space, and to be distributed if its terms depend on space. This is demonstrated in Fig. 2–1 by a volume property B. At a point $P(\mathbf{r})$, where $\mathbf{r}$ denotes the position vector of P, $B = B(t)$ and $B = B(P, t)$ denote the lumped and distributed values, respectively, of B.

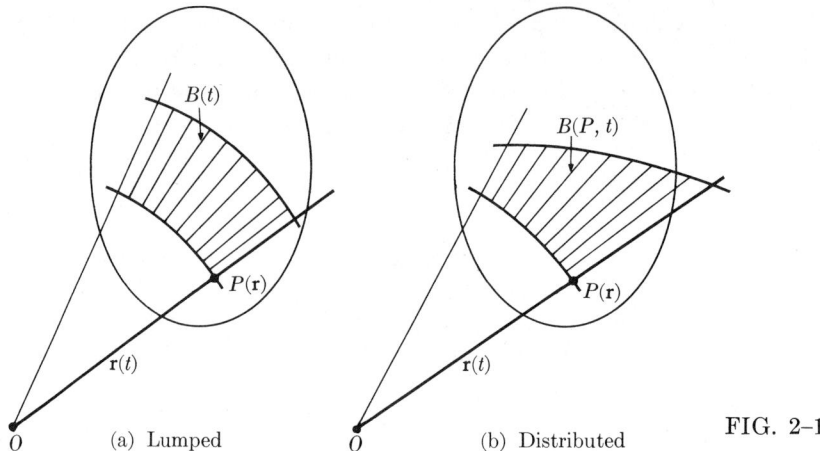

FIG. 2-1

Because of the assumed background of the reader, the development of the variational and difference formulations of the general laws is delayed until Chapters 8, 9, and 10. Here we shall consider only the lumped, integral, and differential formulations of these laws. We shall do this in terms both of a system and of a control volume. It should be noted, however, that there exists a *transformation formula* (Reynold's transport theorem) to convert a general law stated for a system to that for a control volume, thereby eliminating the necessity for separate developments of a general law for both a system and a control volume. It is with mass- or volume-dependent properties that we derive the different forms of this transformation formula suitable to the lumped, integral, and differential formulations of general laws.

2-3. Lumped Formulation of General Laws

First let us develop the form of the transformation formula applicable to the lumped case.

Let B be a mass- or volume-dependent scalar property whose specific value b is

$$b = B/m \quad \text{or} \quad B/V. \tag{2-1}$$

The most frequent examples of B and b are listed in Table 2-1. Here we shall derive the transformation formula in terms of the specific value of the property in relation to the mass, $b = B/m$. The same procedure may be readily extended to the volume-dependent case.

Consider a control volume undergoing a process whose initial and final states are given in Fig. 2-2. During this process the mass Δm_i having the lumped specific value b_i of property B crosses the boundaries of the control volume at the location i. (Here Δ denotes a finite change in any property.) The expansion or contraction of the boundaries of the control volume does not represent any

TABLE 2-1

Property	B	$b = B/m$	$b = B/\mathcal{V}$
Mass	m	1	ρ
Volume	$\mathcal{V} = \begin{cases} mv \\ m/\rho \end{cases}$	v $1/\rho$	1 1
Momentum	$m\mathbf{V}$	$\mathbf{V}$	$\rho\mathbf{V}$
Kinetic energy	$\tfrac{1}{2}mV^2$	$\tfrac{1}{2}V^2$	$\tfrac{1}{2}\rho V^2$
Potential energy	mgz	gz	ρgz
Internal energy	$U = mu$	u	ρu
Total energy	$E = me$	e	ρe
Enthalpy	$H = mh$	h	ρh
Entropy	$S = ms$	s	ρs
Mass concentration	$C = mc$	c	ρc
Electric charge	$q_e = \mathcal{V}\rho_e$	ρ_e/ρ	ρ_e

difficulty and is therefore included in the discussion.* Consider at the same time a system that coincides with the control volume at the final state, but at the initial state occupies the volume $\Delta\mathcal{V}_i$ as well as the control volume. We wish to find the relation between the changes in the property B within the system and within the control volume.

Let B_1, B_2, and B', B'' denote the initial and final values of B for the system and the control volume, respectively. During the process shown in Fig. 2–2

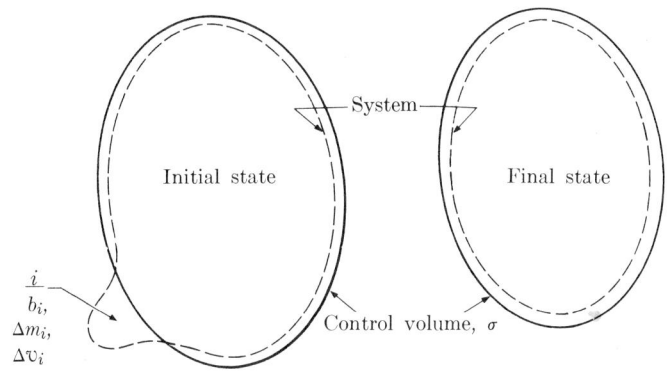

FIG. 2–2

---

* The initial and final states of Fig. 2–2 are shown separately for clarity. While the boundaries are deforming, the whole control volume may or may not be stationary.

the change in B of the system is

$$\Delta B = B_2 - B_1. \qquad (2\text{-}2)$$

Referring again to Fig. 2–2 and expressing B_1 and B_2 in terms of the values for the control volume, we have

$$B_1 = B' + b_i \Delta m_i, \qquad B_2 = B''. \qquad (2\text{-}3)$$

Then, introducing Eq. (2–3) into Eq. (2–2) and denoting the change within the control volume by ΔB_σ, we have

$$\Delta B = \Delta B_\sigma - b_i \Delta m_i. \qquad (2\text{-}4)$$

If the property B crosses the control volume at more than one place, Eq. (2–4) becomes

$$\Delta B = \Delta B_\sigma - \sum_{i=1}^{N} b_i \Delta m_i, \qquad (2\text{-}5)$$

which is the desired transformation formula. Here N is the number of crossings, and positive Δm_i indicates flow to the control volume. Finally, dividing each term of Eq. (2–5) by Δt^* and carrying out the limiting process $\Delta t \to 0$, we find the transformation formula to be

$$\underbrace{\frac{dB}{dt}}_{\text{system}} = \underbrace{\frac{dB_\sigma}{dt} - \sum_{i=1}^{N} b_i w_i}_{\text{control volume}}, \qquad (2\text{-}6)$$

where $w_i = \lim_{\Delta t \to 0} (\Delta m_i / \Delta t)$ is the mass flow rate crossing the control volume at the location i.

Let us now use the transformation formula, Eq. (2–6), to obtain the lumped forms of the general laws for control volumes.

Conservation of mass (lumped formulation). By definition, a system is so constituted that no continuum (mass) may cross its boundaries; therefore, for a system we have

$$\frac{dm}{dt} = 0, \qquad (2\text{-}7)$$

which is the equation for *conservation of mass for lumped systems*. The conservation of mass for the control volume of Fig. 2–2 may now be obtained from Eqs. (2–6) and (2–7). Noting from Table 2–1 that

$$B = m, \qquad b = 1, \qquad (2\text{-}8)$$

---

* Although time is neither a property nor a nonproperty of a system or control volume, hereafter any change in time will be denoted by the same symbol as that used for a change in a property.

and introducing Eq. (2–8) into Eq. (2–6), we have

$$\frac{dm}{dt} = \frac{dm_\sigma}{dt} - \sum_{i=1}^{N} w_i.$$

From this result and Eq. (2–7) it follows that

$$0 = \frac{dm_\sigma}{dt} - \sum_{i=1}^{N} w_i. \tag{2-9}$$

This is the equation for *conservation of mass for lumped control volumes*.

First law of thermodynamics (lumped formulation). Since this law is important from the standpoint of heat transfer it will be developed in detail.

Across the boundaries of a system which completes a cycle the net heat is proportional to the net work:

$$\nabla Q = \nabla W, \tag{2-10}$$

where ∇ denotes the net amount of a nonproperty, Q the heat, and W the work.

If the system undergoes a process, instead, the difference between the net heat and work is equal to the change of the *total energy* (a property) of the system:

$$\Delta E = \nabla Q - \nabla W. \tag{2-11}$$

Here E might be present in a variety of forms, such as internal, kinetic, potential, chemical, and nuclear energy. Thus

$$E = U + \tfrac{1}{2}mV^2 + mgz + U_{\text{chem}} + U_{\text{nucl}}. \tag{2-12}$$

The rate form of Eq. (2–11) gives the *first law of thermodynamics for lumped systems*:

$$\frac{dE}{dt} = q - P, \tag{2-13}$$

where $q = \lim_{\Delta t \to 0} (\nabla Q/\Delta t)$ denotes the rate of heat transfer and $P = \lim_{\Delta t \to 0} (\nabla W/\Delta t)$ the rate of work (power). Note that Eq. (2–13), in terms of our classification of general laws, is the spatially lumped but timewise differential form of the first law of thermodynamics for systems.

Applying the transformation formula of Eq. (2–6) in the usual manner, that is, noting from Table 2–1 that

$$B = E, \quad b = e, \tag{2-14}$$

then introducing Eq. (2–14) into Eq. (2–6) and the result into Eq. (2–13), we find the rate form of the *first law of thermodynamics for lumped control volumes*:

$$\frac{dE_\sigma}{dt} - \sum_{i=1}^{N} e_i w_i = q - P, \tag{2-15}$$

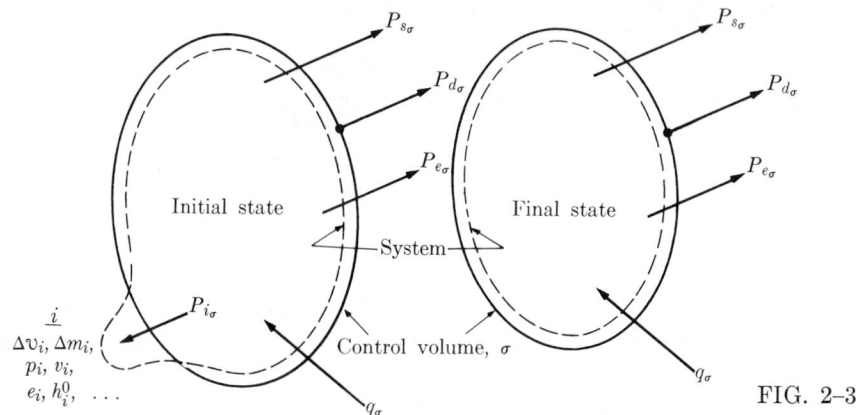

FIG. 2–3

where P is the power leaving and q the rate at which heat is received by the system.

The power P is usually composed of three terms,

$$P = P_d + P_s + P_e,$$

P_d, P_s, and P_e being the displacement, shaft, and electrical powers, respectively. These powers may be conveniently expressed in terms of those leaving the control volume as follows:

$$P_d = (P_d + P_i)_\sigma, \qquad P_s = P_{s_\sigma}, \qquad P_e = P_{e_\sigma} + P_{e_{\Delta v, i}},$$

where P_{d_σ}, P_{i_σ}, P_{s_σ}, and P_{e_σ} denote the displacement, mass flow, shaft, and electrical powers, respectively (Fig. 2–3).

Assume, for example, that the total internal energy is generated at the rate of

$$U' = U'_\sigma + U'_{\Delta v_i}$$

in the system of Fig. 2–3 by means of an external electric circuit (Fig. 2–4a). According to the definition of work, this energy generation is equivalent to the power P_e drawn to the system (Fig. 2–4b). Thus we have

$$P_e = U',$$

where

$$P_e = P_{e_\sigma} + P_{e_{\Delta v, i}}.$$

In the literature the expression "heat generation" is erroneously employed in place of internal energy generation.

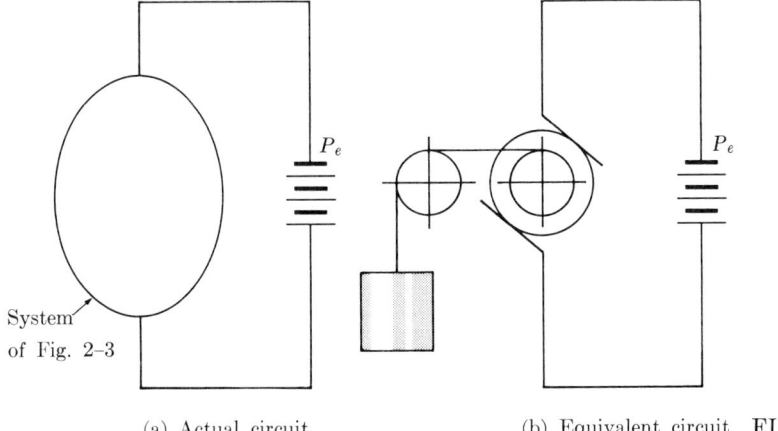

(a) Actual circuit   (b) Equivalent circuit   FIG. 2-4

The electrical power $P_{e\Delta\mathcal{V},i}$, being identical to the electrical internal energy generated in the volume $\Delta\mathcal{V}_i$, must be included in the internal energy flow

$$\sum_{i=1}^{N} e_i w_i$$

across the boundaries of the control volume. If the *friction* is *negligible*, the mass flow power P_{i_σ} reduces to

$$P_{i_\sigma} = -\lim_{\Delta t \to 0} \sum_{i=1}^{N} (p_i \Delta\mathcal{V}_i/\Delta t) = -\sum_{i=1}^{N} p_i v_i w_i,$$

where $\Delta\mathcal{V}_i$ is the volume and v_i the specific volume of the mass Δm_i, and p_i is the pressure on the boundaries of $\Delta\mathcal{V}_i$.

Similarly, the rate of heat q received by the system may be conveniently expressed in terms of q_σ received by the control volume. Thus we have

$$q = q_\sigma + q_{\Delta\mathcal{V}_i}.$$

However, from the standpoint of control-volume analysis $q_{\Delta\mathcal{V}_i}$, like P_{i_σ}, may be considered as an internal energy increase in the volume $\Delta\mathcal{V}_i$ and included in the internal energy flow $\sum_{i=1}^{N} e_i w_i$.

Rearranging Eq. (2-15) according to the foregoing discussion on the power and heat terms, we obtain the explicit form of the *first law of thermodynamics for lumped control volumes*:

$$\frac{dE_\sigma}{dt} = \sum_{i=1}^{N} h_i^0 w_i + q_\sigma - (P_d + P_s + P_e)_\sigma, \tag{2-16}$$

where $h_i^0 = e_i + p_i v_i$ defines the *stagnation enthalpy* per unit mass.

Second law of thermodynamics (lumped formulation). Let us first write the *second law of thermodynamics for lumped systems,*

$$\Delta S \geq \nabla Q / T, \qquad (2\text{-}17)$$

in the rate form

$$dS/dt \geq q/T. \qquad (2\text{-}18)$$

Then, from Table 2–1, inserting

$$B = S, \qquad b = s \qquad (2\text{-}19)$$

into the transformation formula, Eq. (2–6), and the result into Eq. (2–18), we obtain the *second law of thermodynamics for lumped control volumes:*

$$\frac{dS_\sigma}{dt} \geq \sum_{i=1}^{N} s_i w_i + \frac{q}{T}. \qquad (2\text{-}20)$$

Introducing the rate of *entropy generation** S_g, we may write Eq. (2–20) as an equality in the form

$$\frac{dS_\sigma}{dt} = \sum_{i=1}^{N} s_i w_i + \frac{q}{T} + S_g, \qquad (2\text{-}21)$$

which gives the *conservation of entropy for lumped control volumes.* Since S_g is related to the *degree of irreversibility,*† Eq. (2–21) becomes useful for the study of irreversible processes. This point will be elaborated in the differential formulation. [See Eqs. (2–66) through (2–69).]

Having developed the lumped general laws for systems and control volumes, we now proceed to the corresponding integral forms. The general laws stated for control volumes become identical to those for systems as the mass flow across the boundaries approaches zero; therefore, although the starting place will always be the system, we shall develop the integral and differential forms for control volumes only.

2–4. Integral Formulation of General Laws

As with the lumped case, our first step will be the derivation of the appropriate transformation formula. Because of the increased complexity of the integral formulation, however, we shall develop the formula for a *control volume fixed in space.* This proves to be quite satisfactory for most applications of the integral formulations of the general laws.

Consider a control volume which is fixed in space and through which a property B flows (Fig. 2–5). Assume that at time t a system coincides with this control volume. We wish to evaluate the time rate of change of the property within the system.

---

* See the concept of *lost work* in Reference 9, Eq. (7–8).
† As $S_g \to 0$ Eq. (2–21) applies to reversible processes only.

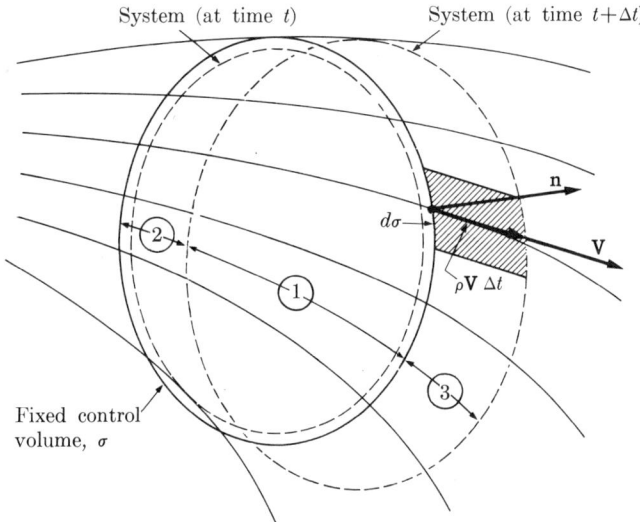

FIG. 2-5

Since the continuum cannot cross its boundaries, this system during a time interval Δt moves with the continuum and at time $t + \Delta t$ occupies another volume in space as shown in Fig. 2-5. Thus the required rate of change may be written in the form

$$\left(\frac{dB}{dt}\right)_{\text{system}} = \lim_{\Delta t \to 0} \frac{[B_1(t + \Delta t) + B_3(t + \Delta t)] - [B_1(t) + B_2(t)]}{\Delta t}, \quad (2\text{-}22)$$

or, by rearranging its terms,

$$\left(\frac{dB}{dt}\right)_{\text{system}} = \lim_{\Delta t \to 0} \left[\underbrace{\frac{B_1(t + \Delta t) - B_1(t)}{\Delta t}}_{\text{first term}} + \underbrace{\frac{B_3(t + \Delta t)}{\Delta t}}_{\text{second term}} - \underbrace{\frac{B_2(t)}{\Delta t}}_{\text{third term}}\right], \quad (2\text{-}23)$$

where B_1, B_2, and B_3 are the instantaneous values of B corresponding to three regions of space, denoted by 1, 2, and 3 in Fig. 2-5.

As $\Delta t \to 0$ the space 1 coincides with the control volume, and the first term on the right of Eq. (2-23) gives the time rate of change of B within the control volume. Because of the fact that B now depends on space as well as time, this rate of change may be conveniently expressed in terms of $B = \int_{\mathcal{V}} b\rho \, d\mathcal{V}$, $\mathcal{V}$ being the fixed volume of the control volume, $d\mathcal{V}$ the volume element, and $b = B/m$. Thus we have for the first term of Eq. (2-23)*

$$\lim_{\Delta t \to 0} \frac{B_1(t + \Delta t) - B_1(t)}{\Delta t} = \left(\frac{dB}{dt}\right)_{\mathcal{V}} = \int_{\mathcal{V}} \frac{\partial(b\rho)}{\partial t} \, d\mathcal{V}. \quad (2\text{-}24)$$

---

* Since the control volume is fixed, $d/dt \equiv \partial/\partial t$, and the order of differentiation and integration is interchangeable.

The second and third terms in Eq. (2–23) respectively give the outgoing and incoming flows of B through the control surface σ. These terms may be evaluated by considering the shaded cylinder shown in Fig. 2–5. The height of this cylinder is $(\mathbf{V}\,\Delta t)\cdot\mathbf{n}$, the volume $(\mathbf{V}\,\Delta t)\cdot\mathbf{n}\,d\sigma$, and the mass $\rho(\mathbf{V}\,\Delta t)\cdot\mathbf{n}\,d\sigma$; $\mathbf{V}$ denotes the velocity of the flow, $\mathbf{n}$ the outward normal vector, and $d\sigma$ the surface element of the control volume. Thus the flow rates of the mass and of the property B through $d\sigma$ are found to be $\rho\mathbf{V}\cdot\mathbf{n}\,d\sigma$ and $b\rho\mathbf{V}\cdot\mathbf{n}\,d\sigma$, respectively. Integrating the latter over the entire control surface, we get

$$\lim_{\Delta t\to 0}\left[\frac{B_3(t+\Delta t)}{\Delta t}-\frac{B_2(t)}{\Delta t}\right]=\int_\sigma b\rho\mathbf{V}\cdot\mathbf{n}\,d\sigma. \qquad (2\text{--}25)$$

The explicit form of Eq. (2–25) for outgoing and incoming flows is illustrated in Fig. 2–6.

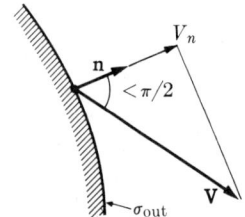

$$\lim_{\Delta t\to 0}\frac{B_3(t+\Delta t)}{\Delta t}=\int_\sigma b\rho\mathbf{V}\cdot\mathbf{n}\,d\sigma$$

$$=\int_{\sigma_{\text{out}}} b\rho V_n\,d\sigma_{\text{out}},$$

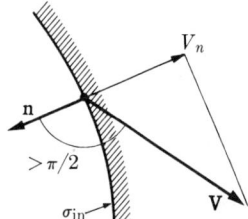

$$\lim_{\Delta t\to 0}\frac{B_2(t)}{\Delta t}=\int_\sigma b\rho\mathbf{V}\cdot\mathbf{n}\,d\sigma$$

$$=-\int_{\sigma_{\text{in}}} b\rho V_n\,d\sigma_{\text{in}}$$

FIG. 2–6

Finally, introducing Eqs. (2–24) and (2–25) into Eq. (2–23), we find the desired transformation formula to be*

$$\left(\frac{dB}{dt}\right)_{\text{system}}=\underbrace{\int_v \frac{\partial(b\rho)}{\partial t}\,dv+\int_\sigma b\rho\mathbf{V}\cdot\mathbf{n}\,d\sigma}_{\text{control volume}}. \qquad (2\text{--}26)$$

It is worth noting that the foregoing transformation formula, as just derived, is based on the physical interpretation of the definitions of system and control volume. The same result might also be obtained from the mathematical de-

---

* Note that by convention the signs for the flow terms of the lumped and integral forms of the transformation formula are opposite.

scription of these, namely, the Lagrangian and Eulerian representations of the continuum. This mathematical description will not be given here.*

Let us now use Eq. (2–26) to establish the integral form of the general laws for control volumes, starting from the statement of these laws as applied to systems.

Conservation of mass (integral formulation). Introducing the previously used appropriate values

$$B = m, \quad b = 1 \tag{2-8}$$

into Eq. (2–26) gives

$$\left(\frac{dm}{dt}\right)_{\text{system}} = \int_v \frac{\partial \rho}{\partial t}\, dv + \int_\sigma \rho \mathbf{V} \cdot \mathbf{n}\, d\sigma. \tag{2-27}$$

Since the left-hand side of Eq. (2–27) is equal to zero, we have

$$0 = \int_v \frac{\partial \rho}{\partial t}\, dv + \int_\sigma \rho \mathbf{V} \cdot \mathbf{n}\, d\sigma, \tag{2-28}$$

the equation for *conservation of mass for integral control volumes*.

First law of thermodynamics (integral formulation). Let us develop the integral form of the first law of thermodynamics for the control volume of Fig. 2–5, starting with the system shown in the figure. The rate form of the first law for this system is

$$(dE/dt)_{\text{system}} = (\delta Q/dt)_{\text{system}} - (\delta W/dt)_{\text{system}}, \tag{2-29}$$

where d and δ denote the differential change in a property and the amount of a nonproperty, respectively.†

Referring to

$$B = E, \quad b = e \tag{2-14}$$

and the transformation formula given by Eq. (2–26), we may rearrange Eq. (2–29) as follows:

$$\left(\frac{dE}{dt}\right)_{\text{system}} = \int_v \frac{\partial (e\rho)}{\partial t}\, dv + \int_\sigma e\rho \mathbf{V} \cdot \mathbf{n}\, d\sigma. \tag{2-30}$$

Since at time t the system occupies the control volume, the rate of heat transfer $(\delta Q/dt)_{\text{system}}$ and the rate of work $(\delta W/dt)_{\text{system}} = P$ (power) across the boundaries of the system may be conveniently expressed in terms of the control volume.

---

* See, for example, Reference 12, p. 84.
† The order of the amount δ is the same as that of the differential change d.

Thus, introducing the *heat flux** vector **q** (Fig. 2–7) and integrating the rate of heat transfer $\mathbf{q}\cdot\mathbf{n}\,d\sigma$ from the surface element $d\sigma$ over the entire control surface, we have

$$(\delta Q/dt)_{\text{system}} = -\int_\sigma \mathbf{q}\cdot\mathbf{n}\,d\sigma, \qquad (2\text{--}31)$$

where the negative sign is in accordance with the sign convention for heat.

The rate of work $(\delta W/dt)_{\text{system}}$ may be conveniently discussed, as it was in the lumped case, by considering its components. These components are the rates of work done by

(a) the system on the surrounding pressure,

$$(\delta W/dt)_d = P_d,$$

(b) the system on the surrounding viscous stresses,

$$(\delta W/dt)_\tau = P_\tau, \qquad (2\text{--}32)$$

(c) the shaft of the system on the surroundings,

$$(\delta W/dt)_s = P_s, \qquad (2\text{--}33)$$

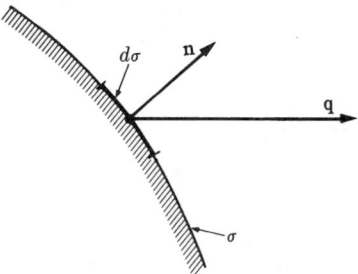

FIG. 2–7

(d) the electrical power of the system on the surroundings,

$$(\delta W/dt)_e = P_e.$$

During the time interval Δt the work done by the system against the pressure on the surface element $d\sigma$ is $p\,d\sigma(\mathbf{V}\,\Delta t)\cdot\mathbf{n}$, where $(\mathbf{V}\,\Delta t)\cdot\mathbf{n}$, the height of the cylinder, is the distance moved normal to $d\sigma$. The rate of this work, $p\mathbf{V}\cdot\mathbf{n}\,d\sigma$, integrated over the entire control surface gives

$$P_p = \int_\sigma p\mathbf{V}\cdot\mathbf{n}\,d\sigma = \int_\sigma (p/\rho)\rho\mathbf{V}\cdot\mathbf{n}\,d\sigma. \qquad (2\text{--}34)$$

The second integral of Eq. (2–34), obtained by multiplying and dividing the integrand of the first integral by ρ, becomes more convenient when the definition of enthalpy is used. [See Eqs. (2–36) and (2–37).]

---

* The definitions of *flow* and *flux* are clearly distinguished in this text. *Flow* is employed to indicate the convection (motion) of a mass- or volume-dependent property across a surface. Examples are mass, momentum, energy, enthalpy, and entropy transfer by convection, and convective electric current. *Flux*, on the other hand, is used for the rate of diffusion per unit area of a property (or nonproperty) through a surface due to the motion of molecules. This includes mass, momentum, and heat transfer by diffusion, and conductive electric current. The notation employed for conduction heat transfer terms is as follows: heat transfer, Q (Btu); rate of heat transfer, q (Btu/hr); and rate of heat transfer per unit area, q with a subscript or superscript, such as q'', q_n, q_x, ..., or $\mathbf{q}$ (Btu/ft$^2\cdot$hr).

For the case in which P_e denotes the power drawn to the system from an external electric circuit, we have

$$P_e = -\int_v u''' \, dv, \qquad (2\text{-}35)$$

where u''' is the *rate of local internal energy per unit volume*, generated electrically in the system.

Introducing Eqs. (2–30), (2–31), (2–32), (2–33), (2–34), and (2–35) into Eq. (2–29) gives

$$\int_v \frac{\partial(e\rho)}{\partial t} \, dv + \int_\sigma e\rho \mathbf{V} \cdot \mathbf{n} \, d\sigma = -\int_\sigma \mathbf{q} \cdot \mathbf{n} \, d\sigma$$
$$-\int_\sigma \frac{p}{\rho} \rho \mathbf{V} \cdot \mathbf{n} \, d\sigma - P_\tau - P_s + \int_v u''' \, dv. \qquad (2\text{-}36)$$

Furthermore, by employing the definition of *stagnation enthalpy* per unit mass, $h^0 = e + pv$, Eq. (2–36) may be rearranged to yield the *first law of thermodynamics for integral control volumes*:

$$\int_v \frac{\partial(e\rho)}{\partial t} \, dv + \int_\sigma h^0 \rho \mathbf{V} \cdot \mathbf{n} \, d\sigma = -\int_\sigma \mathbf{q} \cdot \mathbf{n} \, d\sigma - P_\tau - P_s + \int_v u''' \, dv. \qquad (2\text{-}37)$$

Second law of thermodynamics (integral formulation). The rate form of this law written for the system of Fig. 2–5 is

$$dS/dt \geq q/T. \qquad (2\text{-}18)$$

The right-hand side of Eq. (2–18) may be expressed in terms of the heat flux vector $\mathbf{q}$ as

$$q/T = -\int_\sigma (\mathbf{q}/T) \cdot \mathbf{n} \, d\sigma. \qquad (2\text{-}38)$$

Inserting
$$B = S, \qquad b = s \qquad (2\text{-}19)$$

into Eq. (2–26), and then the result, together with Eq. (2–38), into Eq. (2–18) yields the *second law of thermodynamics for integral control volumes*:

$$\int_v \frac{\partial(s\rho)}{\partial t} \, dv + \int_\sigma s\rho \mathbf{V} \cdot \mathbf{n} \, d\sigma \geq -\int_\sigma \left(\frac{\mathbf{q}}{T}\right) \cdot \mathbf{n} \, d\sigma. \qquad (2\text{-}39)$$

As with Eq. (2–21) of the lumped case, by adding the total entropy generation

$$S_g = \int_v s''' \, dv$$

to the right-hand side of Eq. (2–39) we obtain the equation for *conservation of entropy for integral control volumes*:

$$\int_v \frac{\partial(s\rho)}{\partial t}\,d\mathcal{V} + \int_\sigma s\rho \mathbf{V}\cdot\mathbf{n}\,d\sigma = -\int_\sigma \left(\frac{\mathbf{q}}{T}\right)\cdot\mathbf{n}\,d\sigma + \int_v s'''\,d\mathcal{V}, \quad (2\text{–}40)$$

where s''' is the *rate of local entropy generation per unit volume*.

Having finished the study of the integral forms of the general laws, we now proceed to the differential forms of these laws.

2–5. Differential Formulation of General Laws

There are two ways of obtaining the differential forms of the general laws. One of these starts with the integral form and, using the divergence (Green's) theorem, converts the surface integrals of this form to the volume integrals; then omitting the volume-integral operation results in the differential form. The second method directly establishes the differential forms in terms of an appropriately chosen differential control volume. The former method, being shorter for the present discussion, is preferred here.

We first note the divergence theorem, which states that for a volume $\mathcal{V}$ enclosed by a piecewise smooth surface σ,

$$\int_\sigma \mathbf{a}\cdot\mathbf{n}\,d\sigma = \int_v \boldsymbol{\nabla}\cdot\mathbf{a}\,d\mathcal{V}, \quad (2\text{–}41)$$

where $\mathbf{a}$ is any continuously differentiable vector. Then by employing Eq. (2–41), the differential forms of the general laws may be obtained from their integral forms.

Conservation of mass (differential formulation). The integral form of this law, Eq. (2–28), after its surface integral is converted into a volume integral by using Eq. (2–41) with $\mathbf{a} = \rho\mathbf{V}$, may be rearranged in the form

$$0 = \int_v \left[\frac{\partial\rho}{\partial t} + \boldsymbol{\nabla}\cdot(\rho\mathbf{V})\right]d\mathcal{V}. \quad (2\text{–}42)$$

Since Eq. (2–42) is true for an arbitrary control volume, the integrand itself must vanish everywhere, thus yielding the *conservation of mass for differential control volumes*:

$$\frac{\partial\rho}{\partial t} + \boldsymbol{\nabla}\cdot(\rho\mathbf{V}) = 0. \quad (2\text{–}43)$$

Noting the well-known vectorial identity

$$\boldsymbol{\nabla}\cdot(\alpha\boldsymbol{\beta}) = \alpha\boldsymbol{\nabla}\cdot\boldsymbol{\beta} + \boldsymbol{\beta}\cdot\boldsymbol{\nabla}\alpha \quad (2\text{–}44)$$

(where α is a scalar, $\boldsymbol{\beta}$ is a vector) and the definition of *derivative following the motion*,*

$$\frac{d\alpha}{dt} = \frac{\partial \alpha}{\partial t} + \mathbf{V} \cdot \boldsymbol{\nabla}\alpha, \tag{2-45}$$

we find that an alternative form of Eq. (2-43) is

$$\frac{d\rho}{dt} + \rho \boldsymbol{\nabla} \cdot \mathbf{V} = 0. \tag{2-46}$$

For solids and incompressible fluids $\rho = $ const, and Eq. (2-46) reduces to

$$\boldsymbol{\nabla} \cdot \mathbf{V} = 0. \tag{2-47}$$

Before considering the other general laws, let us rearrange the integral form of the transformation formula, Eq. (2-26), in the light of the differential form of the law of conservation of mass. Converting the surface integral of Eq. (2-26) into a volume integral, and employing Eqs. (2-44) and (2-43), respectively, we have

$$\left(\frac{dB}{dt}\right)_{\text{system}} = \int_v \rho \frac{db}{dt} \, dv. \tag{2-48}$$

Hereafter this form of the transformation formula will be used for the differential formulation of general laws.

First law of thermodynamics (differential formulation). The left-hand side, Eq. (2-30), of the integral form of the first law of thermodynamics, Eq. (2-36), may readily be written by transforming the left-hand side of Eq. (2-30) by Eq. (2-48). Thus we have

$$\int_v \rho \frac{de}{dt} \, dv = -\int_\sigma \mathbf{q} \cdot \mathbf{n} \, d\sigma - \int_\sigma p\mathbf{V} \cdot \mathbf{n} \, d\sigma$$

$$- P_\tau - P_s + \int_v u''' \, dv. \tag{2-49}$$

Since our interest lies in solids and in frictionless incompressible fluids, $P_\tau = 0$. Furthermore, considering the special case where $P_s = 0$, and converting the surface integrals related to the heat flux and the rate of pressure work to volume integrals, then eliminating the volume-integral operation, we get

$$\rho \frac{de}{dt} + \boldsymbol{\nabla} \cdot (p\mathbf{V} + \mathbf{q}) = u'''. \tag{2-50}$$

---

* This derivative is often denoted by D/Dt, and is commonly referred to as the *material derivative, substantial derivative,* or *convective derivative.*

Expanding $\nabla \cdot (p\mathbf{V})$ by Eq. (2–44), and for $\rho = $ const, using Eq. (2–47), we get

$$\nabla \cdot (p\mathbf{V}) = \mathbf{V} \cdot \nabla p. \tag{2–51}$$

Inserting Eq. (2–51) into Eq. (2–50) gives

$$\rho \frac{de}{dt} + \mathbf{V} \cdot \nabla p + \nabla \cdot \mathbf{q} = u''', \tag{2–52}$$

which is the *first law of thermodynamics for differential control volumes*. Since Eq. (2–52) is a statement of the law of *conservation of total energy*, we may deduce a number of results from it.

In the absence of thermal, chemical, and nuclear effects, Eq. (2–52) reduces to the law of *conservation of mechanical energy for differential control volumes*:

$$\rho \frac{d}{dt} (\tfrac{1}{2} V^2 + gz) + \mathbf{V} \cdot \nabla p = 0. \tag{2–53}$$

Equation (2–53), being directly available from Newton's second law of motion, applies equally to processes involving thermal, chemical, and/or nuclear effects.

For a steady process, the derivative of p following the motion is

$$\frac{dp}{dt} = \mathbf{V} \cdot \nabla p. \tag{2–54}$$

Introducing Eq. (2–54) into Eq. (2–53) and rearranging terms yields

$$\frac{d}{dt} \left(\frac{p}{\rho} + \frac{1}{2} V^2 + gz \right) = 0, \tag{2–55}$$

which integrates to *Bernoulli's equation* along a streamline:

$$\frac{p}{\rho} + \frac{1}{2} V^2 + gz = \text{const.} \tag{2–56}$$

The difference between Eqs. (2–53) and (2–52) gives the law of *conservation of thermal energy for differential control volumes*:

$$\rho \frac{du}{dt} + \nabla \cdot \mathbf{q} = u'''. \tag{2–57}$$

Equation (2–57) may further be rearranged in the light of thermodynamic considerations. A state of any property of a continuum which is homogeneous and invariable in composition (but which may be a mixture of two phases) is completely determined by two independent properties.* Thus, considering $u(v, T)$ and $h(p, T)$, and introducing the definitions of *specific heat at constant volume*

---

* This is the pure substance of thermodynamics. See, for example, Reference 9, p. 30.

and *specific heat at constant pressure*,

$$c_v = \left(\frac{\partial u}{\partial T}\right)_v, \qquad c_p = \left(\frac{\partial h}{\partial T}\right)_p,$$

we have

$$du = \left(\frac{\partial u}{\partial v}\right)_T dv + c_v\, dT, \tag{2-58}$$

$$dh = \left(\frac{\partial h}{\partial p}\right)_T dp + c_p\, dT. \tag{2-59}$$

For solids and incompressible fluids $v = $ const, and Eq. (2-58) reduces to

$$du = c_v\, dT. \tag{2-60}$$

Furthermore, if $p = $ const, Eq. (2-59) becomes

$$dh = c_p\, dT, \tag{2-61}$$

and from the definition of *enthalpy* per unit mass, $h = v + pv$, we get

$$dh = du, \qquad p = \text{const}, \quad v = \text{const}. \tag{2-62}$$

Thus combining Eqs. (2-60), (2-61), and (2-62) gives

$$c_p = c_v = c. \tag{2-63}$$

If p varies, $c_p - c_v$ is negligibly small for solids and incompressible fluids,* and Eq. (2-63) still holds, although only approximately.

Introducing Eq. (2-63) into Eq. (2-60) and the result into Eq. (2-57), we have finally

$$\rho c \frac{dT}{dt} + \nabla \cdot \mathbf{q} = u''', \tag{2-64}$$

as an alternative statement of the law of *conservation of thermal energy for differential control volumes*. In Section 2-7, Eq. (2-64) will be the starting point in deriving the differential form of the equation of conduction.

Second law of thermodynamics (differential formulation). Converting the left-hand side of the integral form of this law, Eq. (2-39), by the transformation formula, Eq. (2-48), and the right-hand side by the divergence theorem, Eq. (2-41), and eliminating the volume-integral operation, we get

$$\rho \frac{ds}{dt} + \nabla \cdot \left(\frac{\mathbf{q}}{T}\right) \geq 0, \tag{2-65}$$

which is the *second law of thermodynamics for differential control volumes*.

---

* See, for example, Reference 9, p. 413, Example 14-4.

Similarly, the integral form of the conservation of entropy, Eq. (2–40), readily yields the *conservation of entropy for differential control volumes:*

$$\rho \frac{ds}{dt} + \nabla \cdot \left(\frac{\mathbf{q}}{T}\right) = s'''. \tag{2-66}$$

Equation (2–66) becomes useful in the evaluation of s''' in terms of $\mathbf{q}$ and T alone. First, from the thermodynamics relation

$$du = T\,ds - p\,dv,$$

we have its rate form for $v = \text{const}$

$$\frac{du}{dt} = T\frac{ds}{dt}. \tag{2-67}$$

Now the elimination of du/dt and ds/dt among Eqs. (2–57), (2–66), and (2–67) gives

$$s''' = \nabla \cdot \left(\frac{\mathbf{q}}{T}\right) - \frac{1}{T}(\nabla \cdot \mathbf{q}) + \frac{u'''}{T}. \tag{2-68}$$

After we expand $\nabla \cdot (\mathbf{q}/T)$ by Eq. (2–44), Eq. (2–68) may be reduced to

$$s''' = -\frac{1}{T^2}(\mathbf{q} \cdot \nabla T) + \frac{u'''}{T}. \tag{2-69}$$

Equation (2–69) gives, for a differential control volume, the *rate of entropy generation per unit volume* in conduction problems. [When the effect of friction is included in the incompressible fluid, the dissipation of friction causes an increase in entropy generation which is reflected by an additional term in Eq. (2–69). This term, being of little interest for the present study, is not given here.]

We have thus completed the lumped, integral, and differential formulations of the general laws. As preparation for the next section, it will be helpful to focus our attention for a moment on the objective of the study of conduction, which is the design of thermal devices, for example, heat exchangers. This objective cannot be accomplished without evaluating the *temperature* of and *heat transfer* to or from such a device. The temperature is important to the *mechanical design*, since it enables us to calculate *thermal deformation and stress*. The rate of heat transfer is important to the *thermal design*, since it helps us to determine the size of the device, say the heat transfer area of heat exchangers. Thus a heat transfer problem, in general, requires the simultaneous evaluation of $\mathbf{q}$ and T. Keeping this in mind, let us reconsider, for example, the conservation of thermal energy, Eq. (2–64). This equation gives a relation between but is not sufficient for the evaluation of $\mathbf{q}$ and T. Therefore, we are forced to find other equations relating $\mathbf{q}$ to T. Experimental observations show that the additional relations needed are dependent on the continuum under consideration; hence the equations expressing such relations are statements of particular laws, which will now be discussed.

2-6. Statement of Particular Laws

Two particular laws, *Fourier's law of conduction* and *Stefan-Boltzmann's law of radiation*, are considered in this section. The former, being directly related to conduction, is emphasized. The latter becomes useful only when radiation governs the heat transfer from the boundaries of the continuum under consideration.

Fourier's law of conduction. Microscopic theories such as the kinetic theory of gases and the free-electron theory of metals have been developed to the point where they can be used to predict conduction through media. However, the macroscopic or continuum theory of conduction, which is the subject matter of this text, disregards the molecular structure of continua. Thus conduction is taken to be phenomenological and its effects are determined by experiment as follows.

Consider a solid flat plate of thickness L (Fig. 2-8). Part of this plate is assumed to be bounded by an imaginary cylinder of small cross section A and whose axis is normal to the surfaces of the plate. This cylinder is supposed to be so far from the ends of the plate that no heat crosses its peripheral surface; that is, the transfer of heat is one-dimensional along the axis of the cylinder. Let the temperatures of the surfaces of the plate be T_1, T_2 and let us assume, for example, that $T_1 > T_2$. According to the first law of thermodynamics, under steady conditions there must be a constant rate of heat q through any cross section of the cylinder parallel to the surfaces of the plate. From the second law of thermodynamics we know that the direction of this heat is from the higher temperature to the lower.

Experimental observations of different solids lead us to the fact that for sufficiently small values of the temperature difference between the surfaces of the plate,

$$q = kA \frac{T_1 - T_2}{L}, \qquad (2\text{-}70)$$

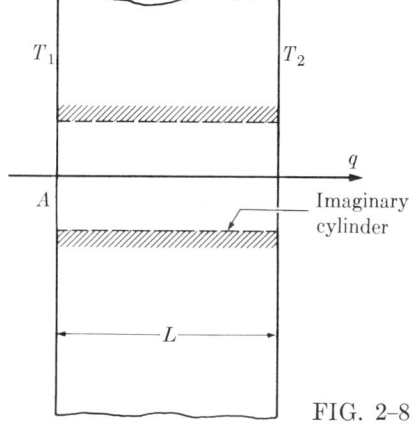

FIG. 2-8

where k is a constant, the so-called *thermal conductivity* of the material of the plate. Thus the heat flux due to conduction is found to be

$$q_n = k \frac{T_1 - T_2}{L}, \qquad (2\text{-}71)$$

where the subscript of q_n indicates the direction of this flux. Equation 2-71 gives *Fourier's law for homogeneous isotropic continua*.

Equation (2-70) may also be used for a fluid (liquid or gas) placed between two plates a distance L apart, provided that suitable precautions are taken to eliminate convection and radiation. Equation (2–70) therefore describes the conduction of heat in fluids as well as in solids.

Let us now note that continua may be classified according to variations in thermal conductivity. A continuum is said to be *homogeneous* if its conductivity does not vary from point to point within the continuum, and *heterogeneous* if there is such variation. Furthermore, continua in which the conductivity is the same in all directions are said to be *isotropic*, whereas those in which there exists directional variation of conductivity are said to be *anisotropic*.* Some materials consisting of a fibrous structure exhibit anisotropic character, for example, wood and asbestos. Materials having a porous structure, such as wool or cork, are examples of heterogeneous continua. In this text, except where explicitly stated otherwise, we shall be studying only the problems of isotropic continua. Because of the symmetry in the conduction of heat in isotropic continua, the flux of heat at a point must be normal to the *isothermal surface*† through this point.

Suppose now that the plate of Fig. 2–8 is isotropic but heterogeneous. Let the temperatures of two isothermal surfaces corresponding to the locations x and $x + \Delta x$ be T and $T + \Delta T$, respectively (Fig. 2–9). Since this plate may be assumed to be locally homogeneous, Eq. (2–71) can be used for a layer of the plate having the thickness Δx as $\Delta x \to 0$. Thus it becomes possible to state the differential form of Fourier's law of conduction, giving the heat flux at x in the direction of increasing x, as follows:

$$q_x = -k \lim_{\Delta x \to 0} \left(\frac{\Delta T}{\Delta x} \right) = -k \frac{\partial T}{\partial x}, \qquad (2\text{–}72)$$

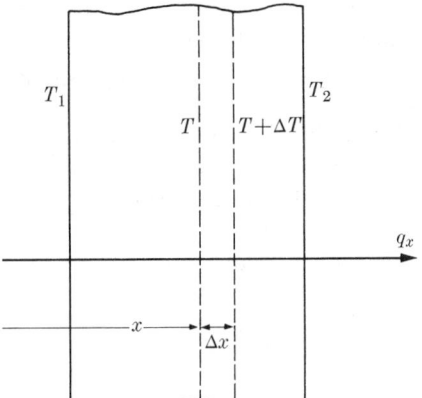

FIG. 2–9

(*Fourier's law for heterogeneous isotropic continua*). In Eq. (2–72), by introducing a minus sign we have made q_x positive in the direction of increasing x. It is important to note that this equation is independent of the temperature distribution. Thus, for example, in Fig. 2–10(a) $\partial T/\partial x < 0$ and $q_x > 0$, whereas in Fig. 2–10(b) $\partial T/\partial x > 0$ and $q_x < 0$. Both results agree with the second law of thermodynamics in that the heat diffuses from higher to lower temperatures.

---

* Once we have classified continua according to their conductivity, it becomes clear that the solids used in the experiments which suggest Fourier's law of conduction must necessarily be homogeneous and isotropic.

† A surface described instantaneously in a continuum such that at every point upon it the temperature is the same.

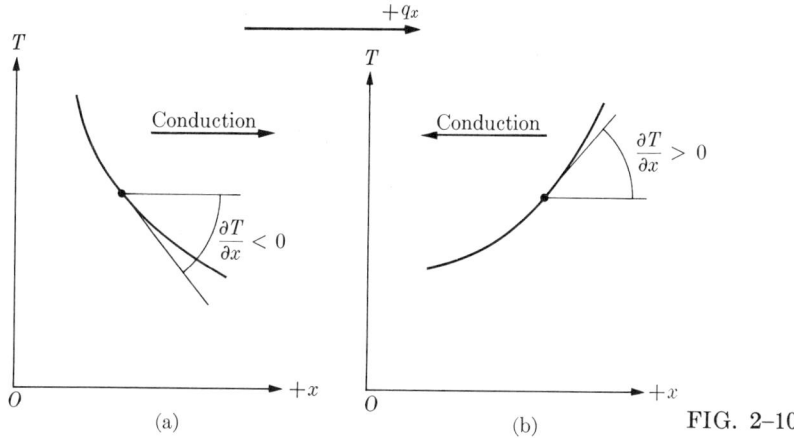

FIG. 2-10

Equation (2-72) may be readily extended to any isothermal surface if we state that the heat flux across an isothermal surface is

$$q_n = -k\frac{\partial T}{\partial n}, \qquad (2\text{-}73)$$

where $\partial/\partial n$ represents differentiation along the normal to the surface. Denoting this flux by a vector $\mathbf{q}$ coinciding with n, we have

$$q_n = |\mathbf{q}|. \qquad (2\text{-}74)$$

Here $\mathbf{q}$ may be expressed in terms of a coordinate system.* This yields

$$\mathbf{q} = -k\,\nabla T, \qquad (2\text{-}75)$$

which is the *vectorial form of Fourier's law for heterogeneous isotropic continua.*

The heat flux at a point P across any nonisothermal surface is now determined by the heat flux across the isothermal surface through the same point (Fig. 2-11). If at P the normal vector $\mathbf{m}$ to a nonisothermal surface has direction cosines (α, β, γ) relative to a coordinate system, the magnitude of the heat flux across this surface is

$$q_m = \mathbf{q}\cdot\mathbf{m}. \qquad (2\text{-}76)$$

---

* In terms of cartesian coordinates, for example,

$$\mathbf{q} = q_x\mathbf{i} + q_y\mathbf{j} + q_z\mathbf{k},$$

where $\mathbf{i}, \mathbf{j}, \mathbf{k}$ are the unit vectors in the x-, y-, and z-directions, respectively. Noting from Eq. (2-72) that $q_x = -k(\partial T/\partial x)$, and similarly that $q_y = -k(\partial T/\partial y)$, $q_z = -k(\partial T/\partial z)$, we have

$$\mathbf{q} = -k\left(\mathbf{i}\frac{\partial}{\partial x} + \mathbf{j}\frac{\partial}{\partial y} + \mathbf{k}\frac{\partial}{\partial z}\right)T,$$

which, by definition, is identical to Eq. (2-75).

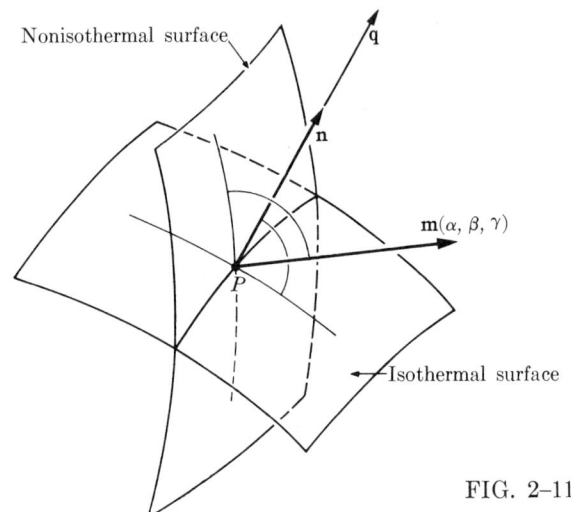

FIG. 2-11

Introducing Eq. (2-75) into Eq. (2-76) yields

$$q_m = -k(\nabla T \cdot \mathbf{m}). \tag{2-77}$$

Since, according to vector calculus,

$$\frac{\partial T}{\partial m} = \nabla T \cdot \mathbf{m},$$

Eq. (2-77) may also be written in the form

$$q_m = -k\frac{\partial T}{\partial m}, \tag{2-78}$$

which gives the magnitude of the heat flux across any surface; here $\partial/\partial m$ represents differentiation in the direction of the normal.

Thus far we have considered Fourier's law for isotropic continua only. In practice, anisotropic continua are also important. The most frequent examples of these are crystals, wood, and laminated materials such as transformer cores. For such continua the direction of the heat flux vector at a point is no longer normal to the isothermal surface through the point. Generalizing Fourier's law for isotropic continua, we may assume each component of the heat flux vector to be linearly dependent on all components of the temperature gradient at the point. Thus, for example, the cartesian form of *Fourier's law* for heterogeneous*

---

* The vectorial form of this law is

$$\mathbf{q} = -\boldsymbol{\kappa} \cdot \nabla T,$$

where $\boldsymbol{\kappa}$ is the *conductivity tensor;* the components of this tensor are called the *conductivity coefficients.*

anisotropic continua becomes

$$q_x = -\left(k_{11}\frac{\partial T}{\partial x} + k_{12}\frac{\partial T}{\partial y} + k_{13}\frac{\partial T}{\partial z}\right),$$

$$q_y = -\left(k_{21}\frac{\partial T}{\partial x} + k_{22}\frac{\partial T}{\partial y} + k_{23}\frac{\partial T}{\partial z}\right), \quad (2\text{-}79)$$

$$q_z = -\left(k_{31}\frac{\partial T}{\partial x} + k_{32}\frac{\partial T}{\partial y} + k_{33}\frac{\partial T}{\partial z}\right).$$

The physical dimensions of the property k in British thermal units are $[k] \equiv \text{Btu/ft·hr·°F}$. The numerical value of k for different continua varies from practically zero for gases under extremely low pressures to about 7000 Btu/ft·hr·°F for a natural copper crystal at very low temperatures. The value of k for a continuum depends in general on the chemical composition, the physical state, and the structure, temperature, and pressure.

In solids the pressure dependency, being very small, is always neglected. For narrow temperature intervals the temperature dependency may also be negligible. Otherwise a linear relation is assumed in the form

$$k = k_0(1 + \beta T), \quad (2\text{-}80)$$

where β is small and negative for most solids.

To illustrate the numerical values, the thermal conductivities of some gases, liquids, and solids are given in Fig. 2–12* as functions of the temperature. The experimental methods for determining the thermal conductivity of continua are many and varied. These, however, have been treated quite extensively in the literature, and will not be given in this text:†

Stefan-Boltzmann's law of radiation. Before the statement of this law is given, a brief review of a number of concepts seems appropriate.

From the viewpoint of electromagnetics, radiant heat transfer, like radio waves, light, cosmic rays, etc., is energy in the form of electromagnetic waves differing only in wavelength from other radiations. When radiant energy impinges on a surface, one fraction of it, α, is absorbed; another fraction, ρ, is reflected; and the remainder, τ, is transmitted. Thus

$$\alpha + \rho + \tau = 1, \quad (2\text{-}81)$$

where α, ρ, and τ are respectively called the *absorptivity*, *reflectivity*, and *transmissivity* of the surface. Equation (2–81) reduces to

$$\alpha + \rho = 1 \quad (2\text{-}82)$$

---

* From M. Jakob and G. A. Hawkins, *Elements of Heat Transfer*. New York: John Wiley & Sons, 1957. Reproduced by permission.
† See, for example, Chapter 9 of Reference 14.

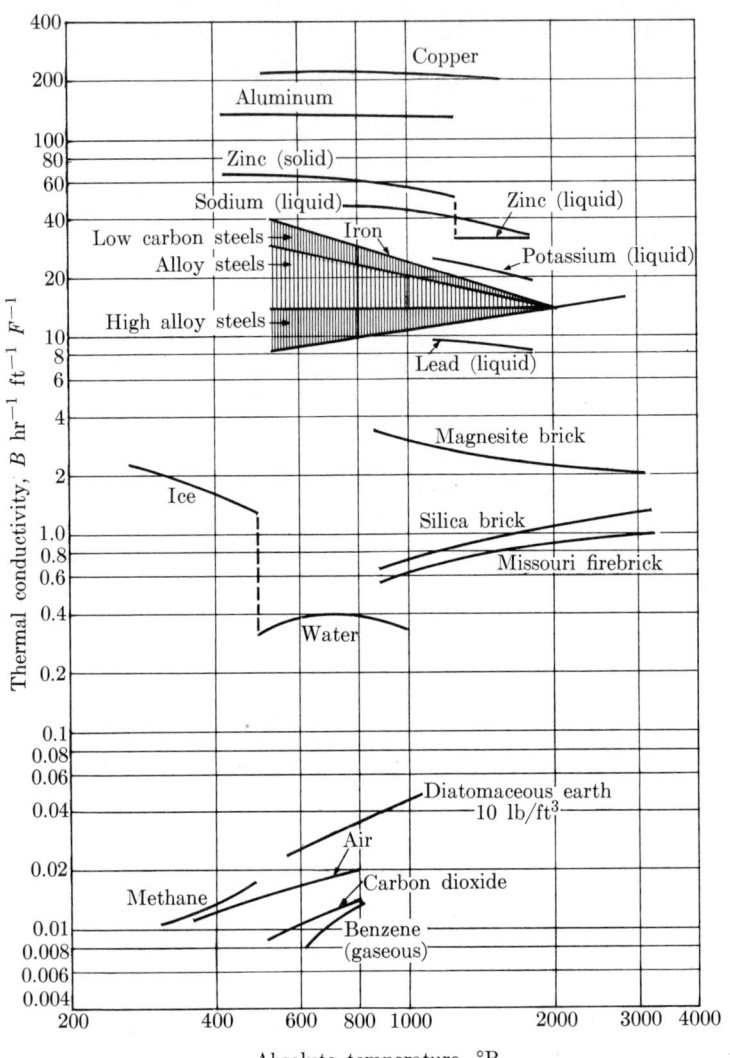

FIG. 2–12

for *opaque* continua, since $\tau = 0$, and to

$$\alpha + \tau = 1 \tag{2–83}$$

for *transparent* continua, since $\rho = 0$.

A surface which absorbs all radiation incident upon it ($\alpha = 1$) or at a specified temperature emits the maximum possible radiation is called a *black surface*.

The emissivity of a surface is defined as

$$\epsilon = q/q_b, \tag{2–84}$$

where q and q_b are the radiant heat fluxes from this surface and from a black surface, respectively, at the same temperature. Thus $\epsilon = 1$ for a black surface.

Under thermal equilibrium $\alpha = \epsilon$ for any surface.*

Consider now two isothermal surfaces A_1 and A_2 having the emissivities ϵ_1, ϵ_2 and the *absolute* temperatures T_1, T_2. These surfaces, together with a third insulated surface, complete an enclosure (Fig. 2–13). It has been shown experimentally by Stefan and later proved thermodynamically by Boltzmann that under steady conditions and in the presence of a nonabsorbing continuum or vacuum, the radiant heat flux q_{12} between the surfaces A_1 and A_2 is governed by *Stefan-Boltzmann's law of radiation* as follows:

$$q_{12} = \sigma \bar{\mathcal{F}}_{12}(T_1^4 - T_2^4), \tag{2-85}$$

where σ is the *Stefan-Boltzmann constant*

$$0.17 \times 10^{-8} \text{ Btu/ft}^2 \cdot \text{hr} \cdot {}^\circ\text{R}^4.$$

$\bar{\mathcal{F}}_{12}$ is a factor which, depending on the emissivity and the relative position of two surfaces, has the form

$$\frac{1}{\bar{\mathcal{F}}_{12}} = \left(\frac{1}{\epsilon_1} - 1\right) + \frac{1}{\bar{F}_{12}} + \frac{A_1}{A_2}\left(\frac{1}{\epsilon_2} - 1\right), \tag{2-86}$$

where

$$\bar{F}_{12} = \frac{A_2 - A_1 F_{12}^2}{A_1 + A_2 - 2A_1 F_{12}}. \tag{2-87}$$

Here F_{12} is the so-called *geometric view factor*. Physically F_{12} represents the fraction of the total radiation from the surface A_1 which is intercepted by the surface A_2. This factor becomes unity for a surface which is enclosed by another surface or for two parallel plates having negligible radiation losses from the ends. As the insulated surface approaches zero, $\bar{F}_{12} \to F_{12}$.

For configurations including more than three surfaces, the evaluation of the radiant heat flux becomes involved. Furthermore, the determination of the geometric view factors for any but simple geometries are often complex; hence these will not be given here.

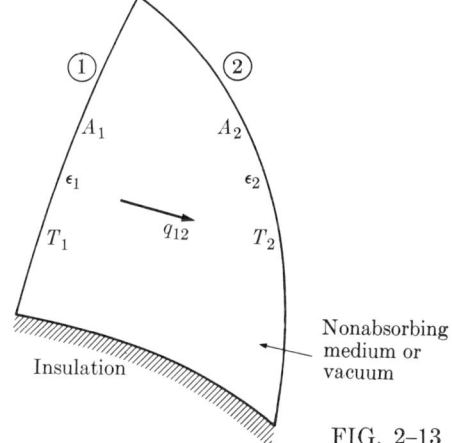

FIG. 2–13

---

* A result of Kirchhoff's law. See, for example, Reference 14, Section 4–2.

2-7. Equation of Conduction.
Entropy Generation Due to Conductive Resistance

When we introduce Fourier's law, Eqs. (2–75) or (2–79), into the law of conservation of thermal energy, Eq. (2–64), the differential form of the equation of heat conduction may be obtained in terms of the temperature alone.

First let us consider an isotropic continuum. Inserting Eq. (2–75) into Eq. (2–64), we have the *equation of conduction for heterogeneous isotropic solids and frictionless incompressible fluids*:

$$\rho c \frac{dT}{dt} = \nabla \cdot (k \nabla T) + u'''. \tag{2-88}$$

Equation (2–88) may be rearranged by means of Eq. (2–44) to give

$$\rho c \frac{dT}{dt} = \nabla k \cdot \nabla T + k \nabla^2 T + u''', \tag{2-89}$$

where ∇^2 denotes the well-known *Laplacian operator*. If k is a function of space only, Eq. (2–89) is linear. On the other hand, when k depends on temperature alone, by use of the vectorial identity $\nabla k = (dk/dT)\nabla T$ Eq. (2–89) may be modified as

$$\rho c \frac{dT}{dt} = \frac{dk}{dT}(\nabla T)^2 + k\nabla^2 T + u''', \tag{2-90}$$

which is nonlinear. [Which term makes Eq. (2–90) nonlinear?] For homogeneous isotropic continua k is constant, and Eq. (2–89) reduces to the *equation of conduction for homogeneous isotropic solids and frictionless incompressible fluids*:

$$\frac{dT}{dt} = a \nabla^2 T + \frac{u'''}{\rho c}, \tag{2-91}$$

where

$$a = k/\rho c \tag{2-92}$$

is the so-called *thermal diffusivity*.

Next suppose that the continuum is anisotropic.* In terms of cartesian coordinates, for example, introducing Eq. (2–79) into Eq. (2–64) gives

$$\rho c \frac{dT}{dt} = \frac{\partial}{\partial x}\left(k_{11}\frac{\partial T}{\partial x} + k_{12}\frac{\partial T}{\partial y} + k_{13}\frac{\partial T}{\partial z}\right) + \frac{\partial}{\partial y}\left(k_{21}\frac{\partial T}{\partial x} + k_{22}\frac{\partial T}{\partial y} + k_{23}\frac{\partial T}{\partial z}\right)$$

$$+ \frac{\partial}{\partial z}\left(k_{31}\frac{\partial T}{\partial x} + k_{32}\frac{\partial T}{\partial y} + k_{33}\frac{\partial T}{\partial z}\right) + u''', \tag{2-93}$$

---

* This case does not have any physical significance for fluids.

the *equation of conduction for heterogeneous anisotropic solids*.* If the conductivity coefficients, though different from each other, remain constant in space, Eq. (2–93) reduces to

$$\rho c \frac{dT}{dt} = k_{11} \frac{\partial^2 T}{\partial x^2} + k_{22} \frac{\partial^2 T}{\partial y^2} + k_{33} \frac{\partial^2 T}{\partial z^2}$$

$$+ (k_{12} + k_{21}) \frac{\partial^2 T}{\partial x\, \partial y} + (k_{23} + k_{32}) \frac{\partial^2 T}{\partial y\, \partial z} \qquad (2\text{--}94)$$

$$+ (k_{31} + k_{13}) \frac{\partial^2 T}{\partial z\, \partial x} + u''',$$

the *equation of conduction for homogeneous anisotropic solids*. The intended scope of this text prevents any further discussion of anisotropic continua.†

Once the temperature variation of any continuum undergoing a desired process is obtained, the entropy generation of this process, related to the temperature by the use of Fourier's law, may be readily evaluated. Hence in terms of isotropic continua, for example, introducing Eq. (2–75) into Eq. (2–69), we have

$$s''' = k \frac{(\boldsymbol{\nabla} T)^2}{T^2} + \frac{u'''}{T}. \qquad (2\text{--}95)$$

So far we have been able to derive expressions, namely, the equations of conduction given by Eqs. (2–88), (2–91), (2–93), and (2–94), which satisfy a partial differential equation in terms of the unknown temperature, rather than an algebraic equation. Since the solution of a differential equation involves a number of integration constants, the completion of the formulation requires that we state an equal number of appropriate conditions in space and time to determine these constants. This is the concern of the next section.

---

* In terms of the conductivity tensor $\boldsymbol{\kappa}$ the general representation of Eq. (2–93) is

$$\rho c \frac{dT}{dt} = \boldsymbol{\nabla} \cdot (\boldsymbol{\kappa} \cdot \boldsymbol{\nabla} T) + u''',$$

which, by use of the tensor calculus, may be rearranged in the form

$$\rho c \frac{dT}{dt} = (\boldsymbol{\nabla} \cdot \boldsymbol{\kappa}) \cdot (\boldsymbol{\nabla} T) + \boldsymbol{\kappa} : \boldsymbol{\nabla}(\boldsymbol{\nabla} T) + u'''.$$

Similarly, the general form of Eq. (2–94) is

$$\rho c \frac{dT}{dt} = \boldsymbol{\kappa} : \boldsymbol{\nabla}(\boldsymbol{\nabla} T) + u'''.$$

† Interested readers may refer to Sections 1–17, 1–18, 1–19, and 1–20 of Reference 2, and the literature cited in the same reference.

2-8. Initial and Boundary Conditions

These conditions are the mathematical descriptions of experimental observations. *Their number in the direction of each independent variable of a problem is equal to the order of the highest derivative of the governing differential equation in the same direction.* An example taken from conduction may illustrate this statement. Consider the equation of conduction written, for example, in cartesian coordinates for a homogeneous isotropic solid moving with velocity **V**,

$$\frac{\partial T}{\partial t} + v_x \frac{\partial T}{\partial x} + v_y \frac{\partial T}{\partial y} + v_z \frac{\partial T}{\partial z} = a\left(\frac{\partial^2 T}{\partial x^2} + \frac{\partial^2 T}{\partial y^2} + \frac{\partial^2 T}{\partial z^2}\right) + \frac{u'''}{\rho c}, \qquad (2\text{-}96)$$

where v_x, v_y, v_z are the components of **V**. The solution of Eq. (2-96), regardless of the mathematical method employed, requires a single integration in time and a double integration in each of the three space variables involved. Thus, referring to the condition in time as the *initial condition* and the conditions in space as the *boundary conditions*, we say that Eq. (2-96), together with one initial and six boundary conditions, completes the differential formulation of the problem.* Let us now consider in detail the initial and boundary conditions appropriate for heat conduction problems.

Initial (volume) condition. For an unsteady problem the temperature of a continuum under consideration must be known at some instant of time. In many cases this instant is most conveniently taken to be the beginning of the problem. Mathematically speaking, if the initial condition is given by $T_0(\mathbf{r})$, the solution of this problem, $T(\mathbf{r}, t)$, must be such that at all points of the continuum

$$\lim_{t \to 0} T(\mathbf{r}, t) = T_0(\mathbf{r}). \qquad (2\text{-}97)$$

Boundary (surface) conditions. The most frequently encountered boundary conditions in conduction are as follows.

(1) *Prescribed temperature.* The surface temperature of the boundaries is specified to be a constant or a function of space and/or time. This is the easiest boundary condition from the viewpoint of mathematics, yet a difficult one to materialize physically, except for the limiting case $h \to \infty$ described below, under (4).

(2) *Prescribed heat flux.* The heat flux across the boundaries is specified to be a constant or a function of space and/or time. The mathematical description of this condition may be given in the light of Kirchhoff's current law; that is, the algebraic sum of heat fluxes at a boundary must be equal to zero. Hereafter the sign is to be assumed positive for the heat flux to the boundary and negative for that from the boundary. Thus, remembering that the statement

---

* Clearly, the velocity involved in Eq. (2-96), although affecting the solution of this equation, does not change the number of initial and boundary conditions needed.

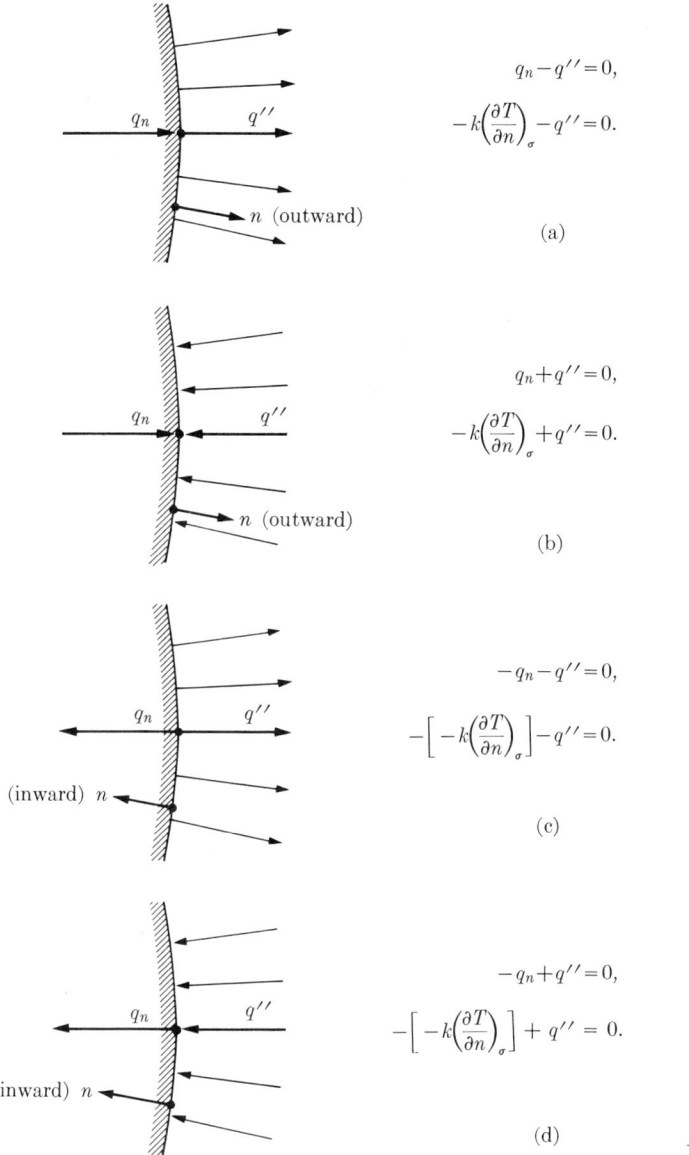

FIG. 2–14

of Fourier's law, $q_n = -k(\partial T/\partial n)$, is independent of the actual temperature distribution, and selecting the direction of q_n conveniently such that it becomes positive, we have from Fig. 2–14

$$\pm k \left(\frac{\partial T}{\partial n}\right)_\sigma = \pm q'', \qquad (2\text{--}98)$$

where $\partial/\partial n$ denotes differentiation along the normal of the boundary. The plus and minus signs of the left-hand side of Eq. (2–98) correspond to the differentiations along the inward and outward normals, respectively, and the plus and minus signs of the right-hand side correspond to the heat flux from and to the boundary, respectively.

A practical example of this case is encountered in the experimental evaluation of the forced-convection heat transfer coefficient in tubes as follows. Consider a constant internal energy as being generated electrically in the walls of an externally insulated tube through which a fluid flows in a prescribed manner. Under steady conditions and with the assumption that the electric resistivity and the thermal conductivity of the tube walls are constant, the fluid becomes subjected to a constant peripheral heat flux.*

(3) *No heat flux (insulation).* This, prescribed

$$\left(\frac{\partial T}{\partial n}\right)_\sigma = 0, \tag{2–99}$$

is a special form of the previous case, obtained by inserting $q'' = 0$ into Eq. (2–98). The illustrative example considered below indicates the practical importance of this boundary condition.

We wish to transfer heat from one surface of a flat electric heater plate through a solid plate, say plate 1, for a specific purpose (Fig. 2–15). Any heat transfer from the other surface of the heater is considered to be a heat loss, and is not desired. In practice the heat loss is reduced by placing another flat plate, plate 2 (insulator), next to the second surface of the heater. We are interested in the geometric and thermal properties of the insulator.

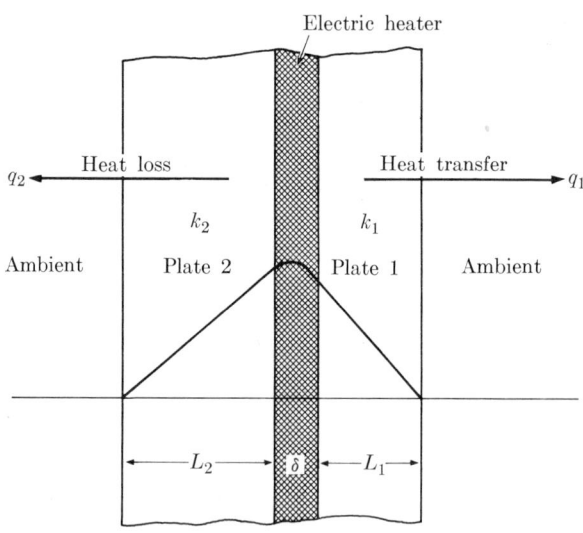

FIG. 2–15

---

* The same physical situation is reconsidered in (4). See also Problem 3–8.

Let us assume, for the sake of simplicity, that the left and right ambients have negligible thermal resistances.* Denoting the thickness of the heater by δ, the thermal conductivity and the thickness of the plates by k_1, k_2 and L_1, L_2, and the rate of internal energy generation per unit volume by u''', we have under steady conditions

$$\delta u''' = q_1 + q_2, \qquad q_1 \frac{L_1}{k_1} = q_2 \frac{L_2}{k_2},$$

and from these,

$$q_1 = \frac{\delta u'''}{1 + (L_1/k_1)/(L_2/k_2)}, \qquad q_2 = \frac{\delta u'''}{1 + (L_2/k_2)/(L_1/k_1)}.$$

The desired condition that $q_2 \to 0$ may be readily obtained by letting $(L_2/k_2)/(L_1/k_1) \to \infty$, or

$$\frac{L_2}{k_2} \gg \frac{L_1}{k_1}. \tag{2-100}$$

We learn from Eq. (2-100) that only the thickness and the thermal conductivity of plates are important for the conduction of heat through these plates.† Hence the heat loss through plate 2 can be eliminated by letting either $L_2 \to \infty$ or $k_2 \to 0$ compared with L_1 and k_1, respectively. Since a plate of $L_2 = \infty$ or $k_2 = 0$ is physically impossible, the foregoing insulation may never be accomplished in the absolute sense. The larger the thickness or the smaller the thermal conductivity, however, the better the insulation will be.

If the heat loss through plate 2 is to be completely eliminated, the use of another heater becomes necessary (Fig. 2-16). Then, by properly adjusting the

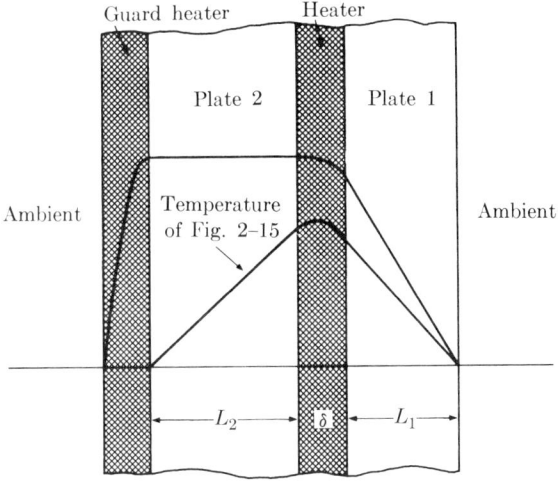

FIG. 2-16

---

* The case of finite resistance of the ambient is introduced in (4).
† The temperature drop (or temperature gradient) across a plate is immaterial for the conductive character of the plate.

power supply to the second heater, all internal energy generated in the original heater may be transferred through plate 1. The second heater, often referred to as the *guard heater*, is an important experimental tool for the control of heat transfer, since it permits accurate thermal conductivity measurements.

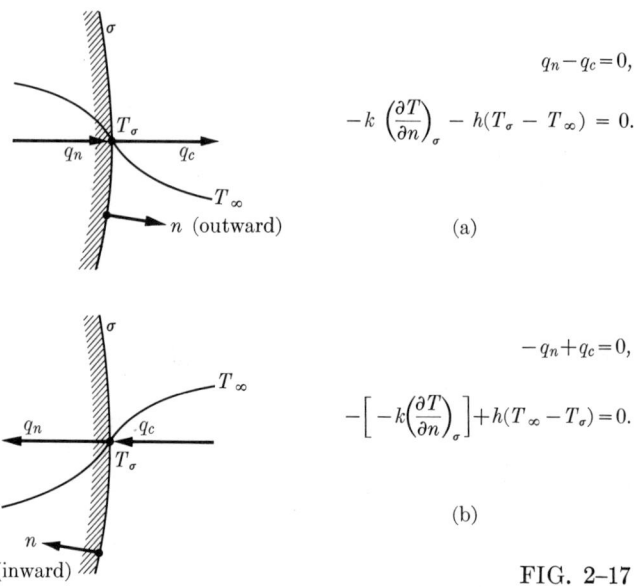

FIG. 2-17

(4) *Heat transfer to the ambient by convection.* When the heat transfer across the boundaries of a continuum cannot be prescribed, it may be assumed to be proportional to the temperature difference between the boundaries and the ambient. Thus we have

$$q_c = h(T_\sigma - T_\infty), \qquad (2\text{-}101)$$

where T_σ is the temperature of the solid boundaries, T_∞ is the temperature of the ambient at a distance far from the boundaries, and h, the proportionality constant,* is the so-called *heat transfer coefficient*. Equation (2-101) is *Newton's cooling law*. It is important, however, to note that this relation is not phenomenological like Fourier's law of conduction and Stefan-Boltzmann's law of radiation. Since it is based on an assumption only, it cannot be considered a natural (particular) law; it will therefore be referred to as the definition of the heat transfer coefficient. Despite its weak foundation, Eq. (2-101), being the only relation available for expressing the unspecified heat transfer to the ambient, plays a significant role in conduction problems.

---

* h, by definition, is assumed to be positive.

TABLE 2-2

Condition		h (Btu/ft^2 · hr · °F)
Free convection	Gases	1–5
	Water	20–150
Forced convection	Gases	2–50
	Water	50–2000
	Viscous oils	10–300
	Liquid metals	1000–20,000
Phase change	Boiling liquids	500–10,000
	Condensing vapors	1000–20,000

Thus with the consideration that the sum of heat fluxes at the boundary must be equal to zero, and in the light of Eqs. (2–73) and (2–101), the required boundary condition may be stated in the form

$$\pm k \left(\frac{\partial T}{\partial n}\right)_\sigma = h(T_\sigma - T_\infty), \tag{2-102}$$

where $\partial/\partial n$ denotes the differentiation along the normal. The plus and minus signs of the left member of Eq. (2–102) correspond to the differentiations along the inward and outward normals, respectively (Fig. 2–17). It should be kept in mind that q_n shown in Fig. 2–17 is a positive quantity, obtained by arbitrarily selecting it in the direction of the normal. Actually, Eq. (2–102) is independent of the temperature distribution and the direction of the heat transfer.

The range of values of heat transfer coefficients occurring under various conditions will now be presented, to give the reader a feeling for the order of magnitudes involved. It should be remembered that h, similar to but more strongly than k, depends on certain variables. These may include the space, time, geometry, flow conditions, and physical properties. The spacewise averaged, steady values of commonly encountered heat transfer coefficients are given in Table 2–2.

The wide variation in the values of heat transfer coefficients suggests further investigations of the boundary condition under study for the limiting values of h. This will be done in terms of a frequently encountered practical situation as follows. Consider a tube of inner and outer radii R_i, R_o through which a fluid flows under specified steady conditions (Fig. 2–18). The steady and uniform internal energy per unit volume is generated at the rate of u''' in the tube walls. The temperature of the surroundings and the bulk temperature of the

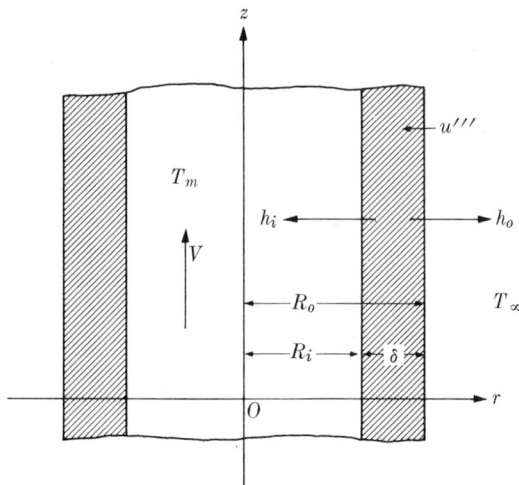

FIG. 2–18

fluid* are T_∞ and T_m, respectively, and the inside and outside heat transfer coefficients are h_i, h_o. The radial boundary conditions for the tube may be written in the form

$$q_i = +k\frac{\partial T(R_i, z)}{\partial r} = h_i[T(R_i, z) - T_m],$$

$$q_o = -k\frac{\partial T(R_o, z)}{\partial r} = h_o[T(R_o, z) - T_\infty].$$

Let us first consider the case of an insulated tube, where $q_o = 0$. In this case, under steady conditions all the internal energy generated in the tube walls is transferred to the inside fluid. Thus $q_i = u'''\delta$. Since q_i is constant for a given u''' and δ, the larger the inside heat transfer coefficient h_i gets, the smaller the temperature difference $T(R_i, z) - T_m$ becomes. In the limit as $h_i \to \infty$, $T(R_i, z)$ tends to T_m, so that the present boundary condition is reduced to that of prescribed surface temperature given under (1). The boiling of liquids in insulated tubes is an example of this case. For a constant q_i, the smallest h_i gives the largest temperature difference between $T(R_i, z)$ and T_m.

Next consider the case of a bare tube. Now $u'''\delta$ is transferred to the surroundings as well as to the inside fluid. Assume that T_m and T_∞ are of the same order, and that the tube wall is thin enough that the difference between

---

* The bulk temperature of a fluid is

$$T_m = \frac{1}{\pi R_i^2 \rho_m c_{p_m} V} \int_0^{R_i} 2\pi r v(r) \rho c_p T(r)\, dr,$$

where ρ_m, c_{p_m}, and V are the density, specific heat at constant pressure, and mean velocity of the fluid, all evaluated at the bulk temperature T_m.

$T(R_i, z)$ and $T(R_o, z)$ may be neglected. Then the heat transfer to the inside fluid and to the surroundings becomes approximately proportional to the inside and outside heat transfer coefficients, respectively. If, for example, $h_o \ll h_i$, the heat transfer to the surroundings may be neglected compared with that to the inside fluid. Thus the outer surface of the tube may be assumed to be insulated, the boundary condition described under (3). An example of this case is water flowing through a tube (forced convection to liquids) surrounded by the stationary atmosphere (free convection to gases). The temperature sketches for three cases, $h_i \sim h_o$, $h_o \ll h_i$, either one of these and large h_i, are left to the reader. From the foregoing discussion we learn the important fact that *the magnitude of the heat transfer coefficient is decisive for the type of boundary condition to be used in the formulation of a problem.*

The study of the magnitude of h suggests a similar investigation in regard to the magnitude of k. For this purpose let us return to the case of an insulated tube. For a given q_i, the larger the thermal conductivity, the smaller the temperature gradient $\partial T(R_i, z)/\partial r$. In the limit as $k \to \infty$, $\partial T(R_i, z)/\partial r$ approaches zero, which implies that the radial temperature distribution in the tube wall can be neglected. This, since it leads to a radially *lumped analysis*, may considerably simplify the formulation of the problem. On the other hand, small or moderate values of k require a radially *distributed analysis*. (Lumped and distributed analyses will be considered in the formulation of the five illustrative examples given in Section 2–10.) Thus *the magnitude of the thermal conductivity plays an important role in the formulation of the conduction equation of a problem.*

Finally, the dimensionless form of the boundary condition under consideration, written in the form

$$\pm \left[\frac{\partial T}{\partial (r/R)} \right]_\sigma = \frac{hR}{k}(T_\sigma - T_\infty), \tag{2-103}$$

indicates that the simultaneous effects of h and k may be investigated in terms of a single dimensionless number, the *Biot modulus*,

$$hR/k = \text{Bi},$$

where R is a characteristic length. Rearranged in the form of $(R/k)/(1/h)$, the Biot modulus may physically be interpreted as the ratio of the internal and external resistances of a problem in the direction for which Eq. (2–103) applies.

(5) *Heat transfer to the ambient by radiation.* Let us reconsider Fig. 2–13, and find, for example, the boundary condition prescribing heat transfer by radiation from the boundaries of continuum 1.

When T_1 is uniform but unspecified, to express the heat flux across the surfaces of 1 by conduction and radiation the required boundary condition may be written in the form

$$\pm k \left(\frac{\partial T}{\partial n} \right)_\sigma = \sigma \bar{\mathcal{F}}_{12}(T_1^4 - T_2^4), \tag{2-104}$$

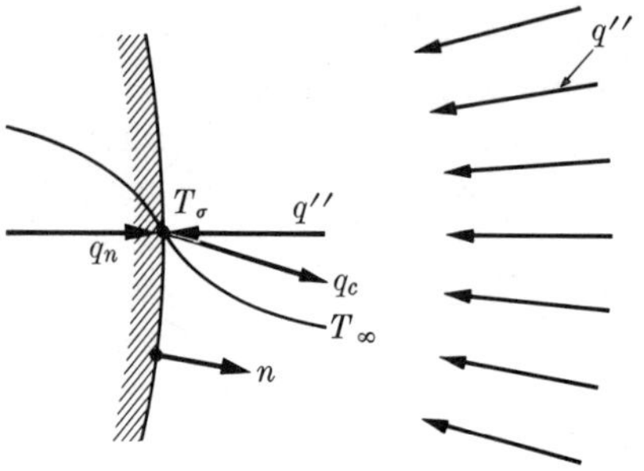

FIG. 2-19

$$q_n + q'' - q_c = 0,$$

$$-k\left(\frac{\partial T}{\partial n}\right)_\sigma + q''$$

$$-h(T_\sigma - T_\infty) = 0.$$

where, as before, the plus and minus signs of the conduction term correspond to differentiation in the direction of inward and outward normals, respectively. Equation (2–104) is independent of the actual temperature distribution. Since it involves the fourth power of the dependent variable, it is a nonlinear boundary condition.

The combination of Eqs. (2–102) and (2–104) gives the simultaneous heat transfer by convection and radiation from the boundaries of the continua. In practice, such simultaneous transfer is the actual case. The importance of radiation relative to convection depends, to a large extent, on the temperature level; radiation increases rapidly with increasing temperature. Even at room temperatures, however, for low rates of convection, say free convection to air, radiation may contribute up to fifty percent of the total heat transfer.

(6) *Prescribed heat flux acting at a distance.* Consider a continuum that transfers heat to the ambient by convection while receiving the net radiant heat flux q'' from a distant source (Fig. 2–19).* The heat transfer coefficient is h, and the ambient temperature T_∞.

This boundary condition may be readily obtained as

$$\pm k\left(\frac{\partial T}{\partial n}\right)_\sigma + q'' = h(T_\sigma - T_\infty), \qquad (2\text{–}105)$$

where the signs of the conduction term depend on the direction of normal in the usual manner. Equation (2–105), like Eq. (2–104), is independent of the actual temperature distribution. Any body surrounded by the atmosphere, capable of receiving radiant heat, and near a radiant source (a light bulb or a sun lamp) or exposed to the sun exemplifies the foregoing boundary condition.

(7) *Interface of two continua of different conductivities k_1 and k_2.* When two continua have a common boundary (Fig. 2–20), the heat flux across this boundary evaluated from both continua, regardless of the direction of normal, gives

$$k_1\left(\frac{\partial T_1}{\partial n}\right)_\sigma = k_2\left(\frac{\partial T_2}{\partial n}\right)_\sigma. \qquad (2\text{–}106)$$

---

* Hereafter figures illustrating the statement of boundary conditions are drawn for one direction of the boundary normal.

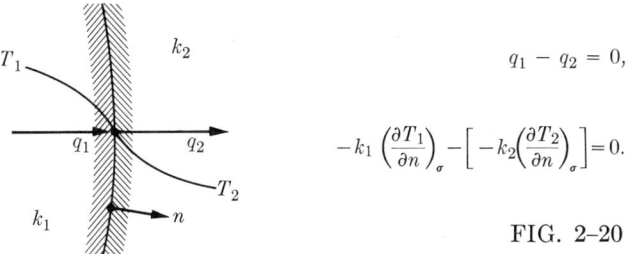

FIG. 2-20

$$q_1 - q_2 = 0,$$

$$-k_1\left(\frac{\partial T_1}{\partial n}\right)_\sigma - \left[-k_2\left(\frac{\partial T_2}{\partial n}\right)_\sigma\right] = 0.$$

Furthermore, a second condition may be specified along this boundary relating the temperatures of the two continua. If the continua are assumed, for example, to be solid and in intimate contact, mathematical idealization suggests the equality of temperatures

$$(T_1)_\sigma = (T_2)_\sigma. \tag{2-107}$$

However, Eq. (2-107) is a difficult condition to satisfy in practice. Even for perfectly smooth surfaces pressed together, heat transfer takes place between the two continua across the so-called *contact resistance*.* This resistance, which is difficult to measure or specify, causes a temperature difference between the continua along the interface. Despite this fact, Eq. (2-107) necessarily finds extensive use in the formulation of conduction problems. Composite walls and insulated tubes are well-known examples of this case.

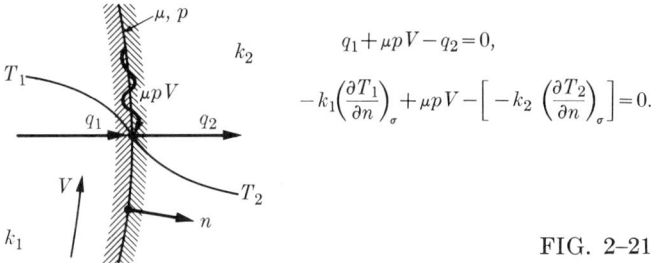

FIG. 2-21

$$q_1 + \mu p V - q_2 = 0,$$

$$-k_1\left(\frac{\partial T_1}{\partial n}\right)_\sigma + \mu p V - \left[-k_2\left(\frac{\partial T_2}{\partial n}\right)_\sigma\right] = 0.$$

(8) *Interface of two continua in relative motion.* Consider two solid continua in contact, one moving relative to the other (Fig. 2-21). The local pressure on the common boundary is p, the coefficient of dry friction μ, and the relative velocity V.

Noting that the heat transfer to both continua by conduction is equal to the work done by friction, we have

$$\pm k_1\left(\frac{\partial T}{\partial n}\right)_\sigma + \mu p V = \pm k_2\left(\frac{\partial T_2}{\partial n}\right)_\sigma, \tag{2-108}$$

where the minus signs of the conduction terms correspond to the normal shown

---

* See References 15 and 16.

in Fig. 2–21. Again, for idealized intimate contact we may assume that the temperatures of the two continua are the same on the boundary, as expressed previously by Eq. (2–107).

The friction brake is an important practical case of the foregoing boundary condition. However, wear and high temperatures make this boundary condition impractical for continuous operation. The obvious remedy, lubrication, is beyond the scope of the text and is not considered here.

(9) *Moving interface of two continua (change of phase).* When part of a continuum has temperatures below the temperature at which the continuum changes from one phase to another by virtue of the liberation or absorption of heat, there exists a moving boundary between the two phases. For problems in this category, the way in which the boundary moves has to be determined together with the temperature variation in the continuum.

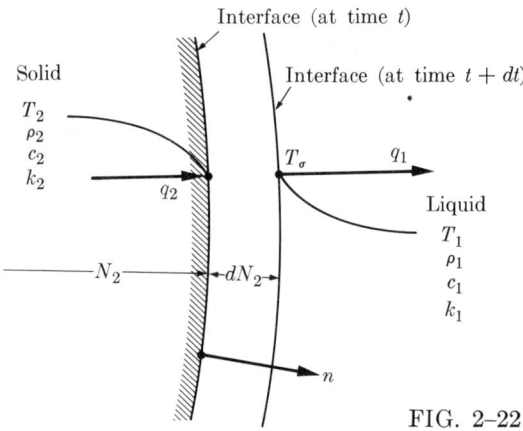

FIG. 2–22

Consider, for example, the solidification of a liquid. Here our concern is the boundary condition on the moving interface $N_2(t)$ (Fig. 2–22). The thermal properties of the liquid and solid are distinguished by the subscripts 1 and 2, respectively. Since the densities of the two phases are not the same, in the time interval dt the solid of thickness dN_2 is formed from the liquid of thickness dN_1. Applying the first law of thermodynamics to the system shown in Fig. 2–23, whose initial state is the liquid of thickness dN_1 and whose final state is the solid of thickness dN_2, we have

$$\rho_2 u_2\, dN_2 - \rho_1 u_1\, dN_1 = q_2\, dt - q_1\, dt - p(dN_2 - dN_1), \quad (2\text{--}109)$$

where p is the pressure of the continuum.

Noting the continuity

$$\rho_2\, dN_2 = \rho_1\, dN_1, \quad (2\text{--}110)$$

and the latent heat of fusion

$$h_{sl} = h_1 - h_2, \quad (2\text{--}111)$$

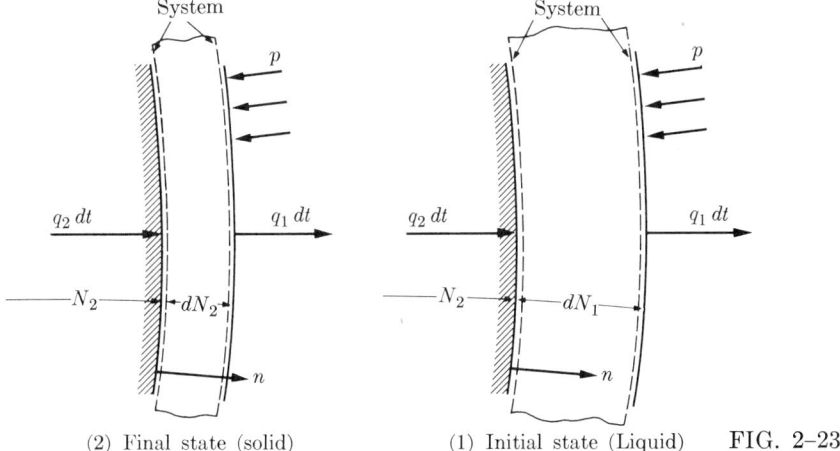

FIG. 2-23

we may write the rate form of Eq. (2-109) as

$$-\rho_2 h_{sl} \frac{dN_2}{dt} = q_2 - q_1. \quad (2\text{-}112)$$

Expressing q_1 and q_2 by Fourier's law, we get finally

$$-\rho_2 h_{sl} \frac{dN_2}{dt} = \pm k_2 \left(\frac{\partial T_2}{\partial n}\right)_\sigma - \left[\pm k_1 \left(\frac{\partial T_1}{\partial n}\right)_\sigma\right], \quad (2\text{-}113)$$

where the plus and minus signs of the conduction terms correspond to the differentiation $\partial/\partial n$ along the inward and outward normals, respectively, of the solid phase.

The foregoing boundary condition may also be obtained by an alternative procedure* as follows. Since $\rho_2 \neq \rho_1$, say $\rho_2 > \rho_1$, the process of solidification gives rise to a velocity V_1 in the fluid proportional to the rate of the difference between the volumes of the two phases (Fig. 2-24a). Thus we have

$$V_1 = \frac{dN_1 - dN_2}{dt},$$

which may be rearranged by continuity, Eq. (2-110), as

$$V_1 = \left(\frac{\rho_2}{\rho_1} - 1\right) \frac{dN_2}{dt}. \quad (2\text{-}114)$$

To simplify the analysis, let us imagine that an observer is traveling with the

---

* A similar method is employed to evaluate the velocity of propagation of a plane pressure pulse in a stationary compressible fluid filling a pipe of uniform cross section. See, for example, Reference 5, Chapter 3, p. 44.

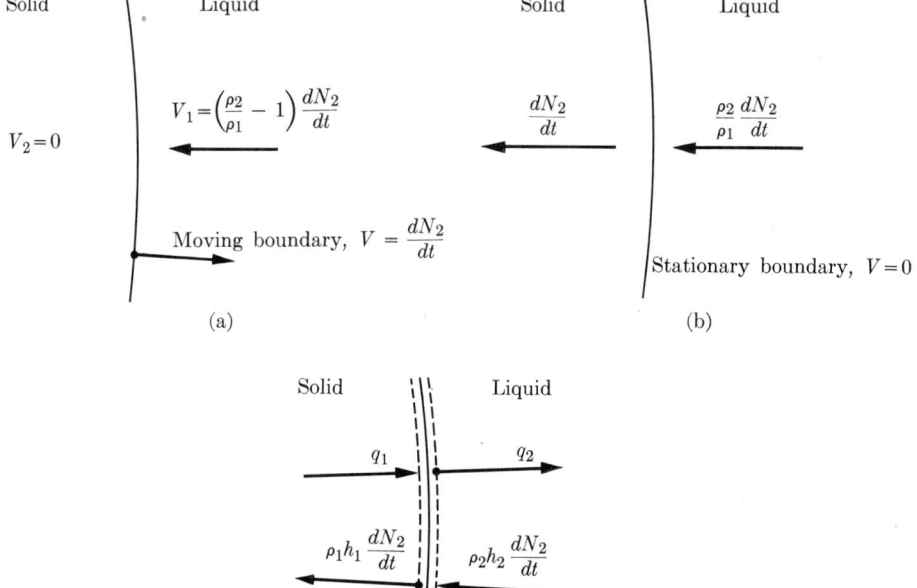

FIG. 2-24

moving boundary. Figure 2-24(b) shows the appearance of the solidification process to such an observer. Then the first law of thermodynamics applied to a control volume surrounding the stationary boundary of Fig. 2-24(c) yields

$$0 = \rho_2(h_2 - h_1)\frac{dN_2}{dt} + q_2 - q_1. \qquad (2\text{-}115)$$

Given Eq. (2-111), Eq. (2-115) reduces to Eq. (2-112).

The problem of solidification may be simplified considerably when the temperature variation in the liquid is not of interest. In this case, if we express q_1 in terms of a heat transfer coefficient h, Eq. (2-113) may be rearranged as

$$-\rho_2 h_{sl}\frac{dN_2}{dt} = \pm k_2 \left(\frac{\partial T_2}{\partial n}\right)_\sigma - h(T_\sigma - T_\infty), \qquad (2\text{-}116)$$

where T_σ is the temperature of solidification and T_∞ the temperature of the liquid far from the moving boundary. Problems involving change of phase are of great practical importance. Ice formation both in geophysics and ice manufacturing, the solidification of metals in casting, and the condensation and evaporation of fluids are typical examples.

2-9. Methods of Formulation

In the preceding sections of this chapter we have established the general formulation of conduction phenomena. We might now expect to obtain the formulation of any specific problem from the general formulation. This, of course, is possible, but it is not always convenient, especially if the problem under consideration is to be lumped in one or more directions. (This point will be clarified by the problem of Fig. 2-27.) Furthermore, the application of the general formulation to a specific problem is a mathematical process which eliminates the physics of the formulation, an important aspect of practical problems. By contrast, the physical approach which will be stressed in this text treats each problem individually from the start of its formulation by bringing the physics into each phase of the formulation. To illustrate this statement let us compare the two methods in the light of three problems requiring the one-dimensional formulation of the first law of thermodynamics.

The first problem is that of the one-dimensional cartesian system shown in Fig. 2-25. When we equate the time rate of change of internal energy to the net heat transfer across the boundaries of the system, the physical approach yields

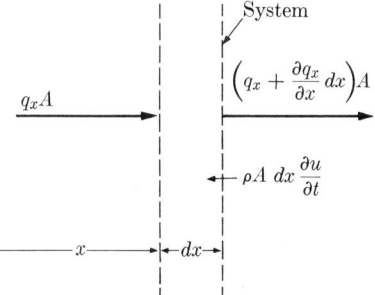

$$\rho \frac{\partial u}{\partial t} + \frac{\partial q_x}{\partial x} = 0. \qquad (2\text{-}117)$$

The general formulation, reduced to the one-dimensional cartesian form of Eq. (2-57), gives the same result.

FIG. 2-25

Next, let us consider an insulated solid rod of radius R, cross section A, and periphery P (Fig. 2-26). By either method, the first law of thermodynamics stated for the one-dimensional system shown in Fig. 2-26 yields the result of the previous problem, Eq. (2-117).

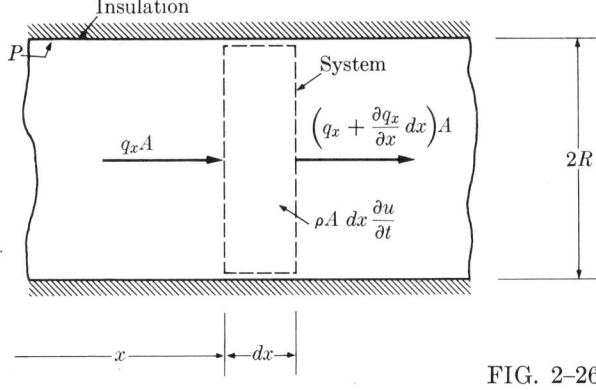

FIG. 2-26

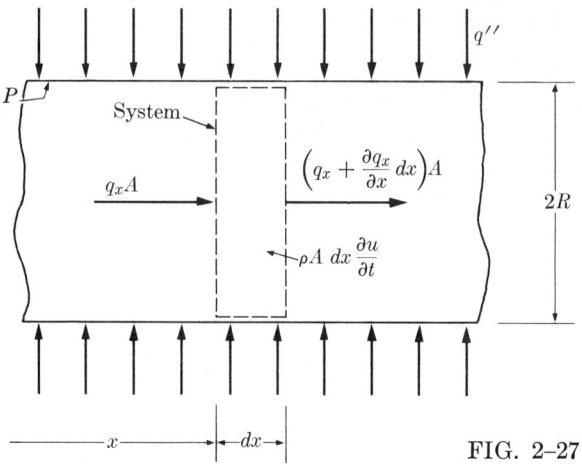

FIG. 2-27

Finally, consider the solid rod of the previous problem now subjected to the uniform peripheral heat flux q'' (Fig. 2-27). The physical approach, in which we apply the first law to the one-dimensional system shown in Fig. 2-27, results in

$$\rho \frac{\partial u}{\partial t} + \frac{\partial q_x}{\partial x} = \frac{q''P}{A}. \qquad (2\text{-}118)$$

By contrast, the one-dimensional form of the general formulation, leading again to Eq. (2-117), does not include the effect of the peripheral heat flux. This difficulty, however, may be circumvented by considering instead the two-dimensional form of Eq. (2-57),

$$\rho \frac{\partial u^*}{\partial t} + \frac{\partial q_x^*}{\partial x} + \frac{1}{r}\frac{\partial}{\partial r}(rq_r^*) = 0, \qquad (2\text{-}119)$$

where u^*, q_x^*, and q_r^* now depend on r as well as x and t. Next, averaging Eq. (2-119) radially, that is, multiplying each term by $2\pi r\, dr$ and integrating the result over the interval $(0, R)$, yields

$$A\rho \frac{\partial u}{\partial t} + A\frac{\partial q_x}{\partial x} + 2\pi r q_r \Big|_0^R = 0, \qquad (2\text{-}120)$$

which is identical to Eq. (2-118). Here the radially averaged value of a dependent variable, say u, is defined as

$$u(x, t) = \frac{2\pi}{A} \int_0^R r u^*(r, x, t)\, dr.$$

The discussion on the foregoing three examples may be generalized as follows. A given problem may be formulated either by considering the appropriate specific case of the general formulation or by following, from the start, an individual formulation suitable to the problem. Whenever the general formulation is available the former method may be used, but this requires the mathematical interpretation of the general formulation in the light of the problem under consideration. The latter method, on the other hand, involves following certain steps in a basic procedure for individual formulation, given below. For one- or multidimensional problems which are formulated to include all dimensions of the problem (such as the first two of the foregoing examples), the general or mathematical approach proves to be slightly shorter than the individual or physical approach. However, for multidimensional problems which we wish to formulate in fewer dimensions, that is, which we wish to lump in one or more directions (such as the third example), the mathematical approach, requiring an averaging process, becomes lengthy and inconvenient.

The foregoing argument and the emphasis, in this text, on practical applications of the study of heat transfer thus suggest that the physical approach be the preferred method of formulation. For convenience and later reference, this method of formulation is summarized in the following five steps:

(i) *Define an appropriate system or control volume.* This step includes the selection of (a) a coordinate system, (b) a lumped or distributed formulation, and (c) a system or control volume in terms of (a) and (b).

(ii) *State the general laws for* (i). The general laws, except in their lumped forms, are written in terms of a coordinate system. The differential forms of these laws depend on the direction but not the origin of coordinates, whereas the integral forms depend on the origin as well as the direction of coordinates. Although the differential forms apply locally, the lumped and integral forms are stated for the entire system or control volume.

(iii) *State the particular laws for* (ii). The particular law describing the diffusion of heat (or momentum, mass, or electricity) is differential, applies locally, and depends on the direction but not the origin of coordinates.

(iv) *Obtain the governing equation from* (ii) *and* (iii). This, such as the equation of conduction, may be an algebraic, differential, or other equation involving the desired dependent variable, say the temperature, as the only unknown. The governing equation (except for its flow terms) is independent of the origin and direction of coordinates.

(v) *Specify the initial and/or boundary conditions pertinent to* (iv). These conditions depend on the origin as well as the direction of coordinates.

2-10. Examples

In this section the emphasis is placed on formulation; however, for those problems whose formulation leads to an ordinary differential equation of first order or to one of second order with constant coefficients we shall also give the solution.

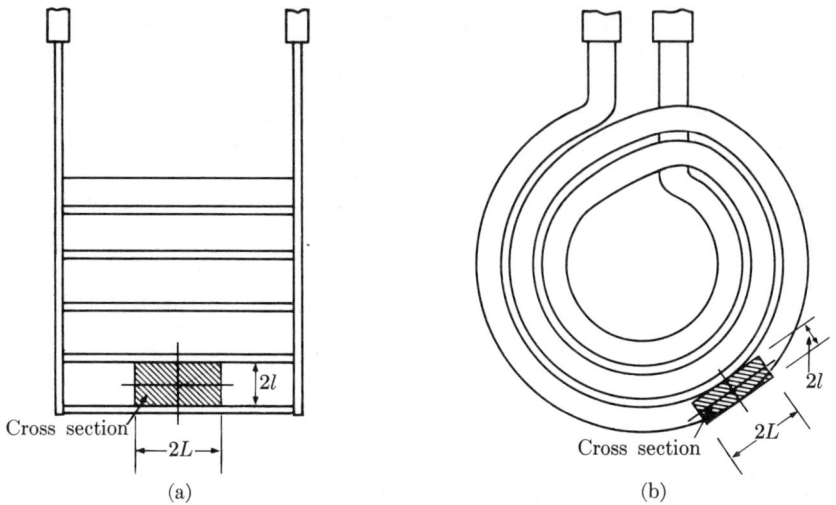

FIG. 2–28

Example 2–1. Consider an electric heater made from a solid rod of rectangular cross section ($2L \times 2l$) and designed according to one of the forms shown in Fig. 2–28. The temperature variation along the rod can be neglected. In addition, the effect of curvature is negligibly small for the coil type of heater of Fig. 2–28(b). The internal energy generation u''' in the heater is uniform. The heat transfer coefficient is denoted by h and the ambient temperature by T_∞. We wish to formulate the steady conduction problem suitable to this heater.

Proceeding according to the five basic steps mentioned in the previous section, we shall now give the lumped, differential, and integral formulations of the problem.

I. *Lumped formulation.*

(i) *System or control volume.* The lumped system covers the entire cross section of the heater (Fig. 2–29). Since the problem is assumed to be two-dimensional, the length of the rod(s) does not affect the formulation; for purposes of illustration, however, let us consider a unit length.

(ii) *General law.* The first law of thermodynamics, Eq. (2–16), applied to Fig. 2–29 yields

$$0 = -2q_1(2L \cdot 1) - 2q_2(2l \cdot 1) + u'''(2L \cdot 2l \cdot 1). \qquad (2\text{–}121)$$

(iii) *Particular law.* The formulation, being lumped, does not require any particular law.

(iv) *Governing equation.* The absence of a particular law makes the governing equation identical to the general law.

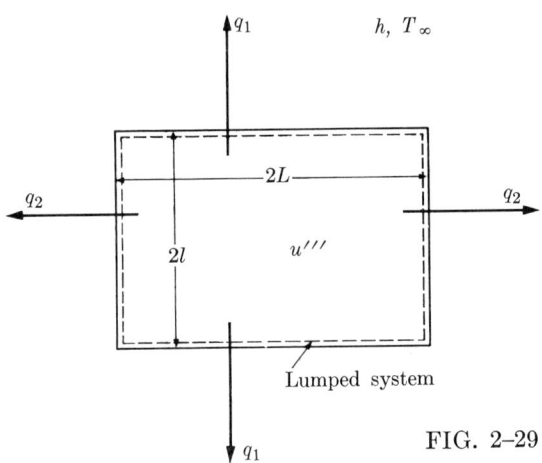

FIG. 2-29

(v) *Initial and boundary conditions.* The requirement of a steady formulation eliminates the need of any initial condition. The definition of h gives the single boundary condition

$$q_1 \text{ (and } q_2) = h(T - T_\infty). \tag{2-122}$$

Equations (2-121) and (2-122) complete the lumped formulation of the problem. If we introduce the latter into the former, this formulation may be written in terms of the unknown temperature T as follows:

$$0 = -2h(2L \cdot 1)(T - T_\infty) - 2h(2l \cdot 1)(T - T_\infty) + u'''(2L \cdot 2l \cdot 1). \tag{2-123}$$

The simplicity of Eq. (2-123) readily allows the lumped temperature of the heater rod to be obtained as

$$T = T_\infty + \frac{u'''(2L \cdot 2l)}{2h(2L + 2l)}. \tag{2-124}$$

In the limit as $h \to \infty$ the temperature of the heater approaches the ambient temperature T_∞. [See the discussion in Section 2-8 regarding the boundary condition of type (4).]

II. *Differential formulation*

(i) *System or control volume.* Consider the two-dimensional differential system shown in Fig. 2-30. The horizontal direction is arbitrarily denoted by x and the vertical one by y. The direction and origin of the coordinates need not be specified yet. To fix ideas, however, we designate the rightward x and the upward y as positive.

64 LUMPED, INTEGRAL, DIFFERENTIAL FORMULATIONS [2–10]

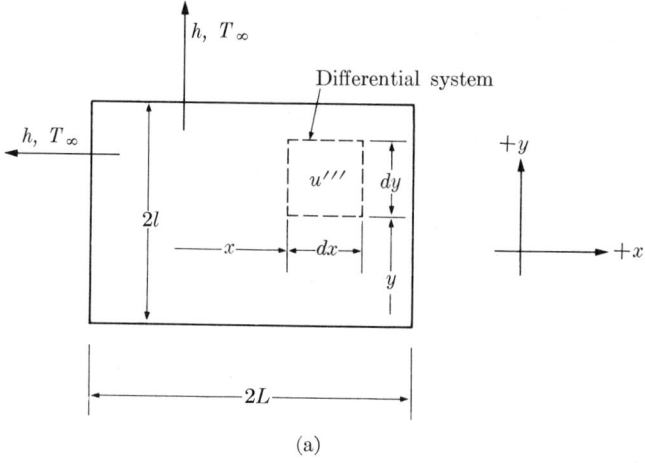

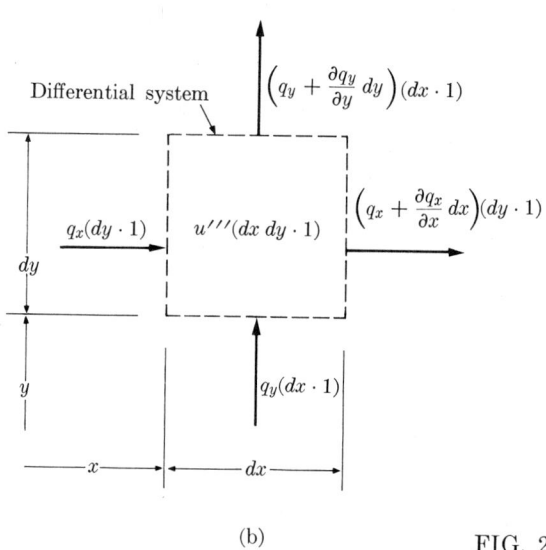

(b) FIG. 2–30

(ii) *General law.* The first law of thermodynamics applied to the differential system of Fig. 2–30(a) and interpreted in terms of Fig. 2–30(b) gives

$$0 = +q_x(dy \cdot 1) - \left(q_x + \frac{\partial q_x}{\partial x} dx\right)(dy \cdot 1)$$

$$+ q_y(dx \cdot 1) - \left(q_y + \frac{\partial q_y}{\partial y} dy\right)(dx \cdot 1) + u'''(dx \cdot dy \cdot 1),$$

which may be simplified to

$$-\frac{\partial q_x}{\partial x} - \frac{\partial q_y}{\partial y} + u''' = 0. \tag{2-125}$$

(iii) *Particular law.* The two cartesian components of the vectorial form of Fourier's law to be used for isotropic continua are

$$q_x = -k\frac{\partial T}{\partial x}, \qquad q_y = -k\frac{\partial T}{\partial y}. \tag{2-126}$$

(iv) *Governing equation.* Introducing Eq. (2–126) into Eq. (2–125) gives

$$\frac{\partial}{\partial x}\left(k\frac{\partial T}{\partial x}\right) + \frac{\partial}{\partial y}\left(k\frac{\partial T}{\partial y}\right) + u''' = 0, \tag{2-127}$$

which for constant k reduces to

$$\frac{\partial^2 T}{\partial x^2} + \frac{\partial^2 T}{\partial y^2} + \frac{u'''}{k} = 0. \tag{2-128}$$

Equation (2–127) or Eq. (2–128) is the governing equation (of conduction) for the problem under study. It is clear that these equations may readily be obtained from the general vectorial forms given by Eqs. (2–88) and (2–91) by considering their steady two-dimensional cartesian forms.

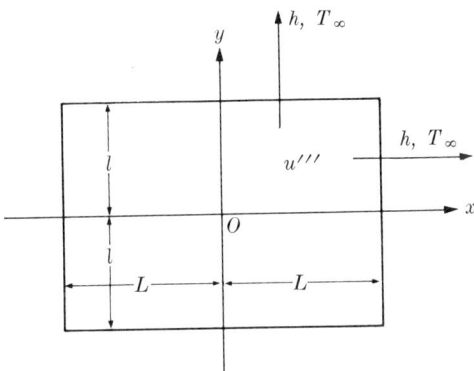

FIG. 2-31

(v) *Initial and boundary conditions.* As with the lumped formulation, no initial condition is needed. The order of highest x- and y-derivatives of Eq. (2–128), on the other hand, requires that two boundary conditions be specified in each direction. Before these can be stated, of course, the origin and direction of coordinates must be chosen. Noting the thermal as well as the geometric symmetry of the problem, we select the coordinate system shown in Fig. 2–31.

Thus the boundary conditions may be written in the form

$$\frac{\partial T(0, y)}{\partial x} = 0, \quad -k\frac{\partial T(L, y)}{\partial x} = h[T(L, y) - T_\infty],$$

$$\frac{\partial T(x, 0)}{\partial y} = 0, \quad -k\frac{\partial T(x, l)}{\partial y} = h[T(x, l) - T_\infty]. \quad (2\text{-}129)$$

[Restate these conditions in the coordinate system whose origin is at one of the corners of the heater. Defend Eq. (2-129) compared with the new form of the boundary conditions.]

Equation (2-127) or Eq. (2-128), together with Eq. (2-129), completes the differential formulation of the problem. The solution of this problem requires further mathematical background and is deferred to Chapter 4. (See Example 4-10, which deals with the limiting case $h \to \infty$.)

III. *Integral formulation*

(i) *System or control volume.* This formulation, as demonstrated below, requires the simultaneous use of the systems of the lumped and differential formulations (Fig. 2-32).

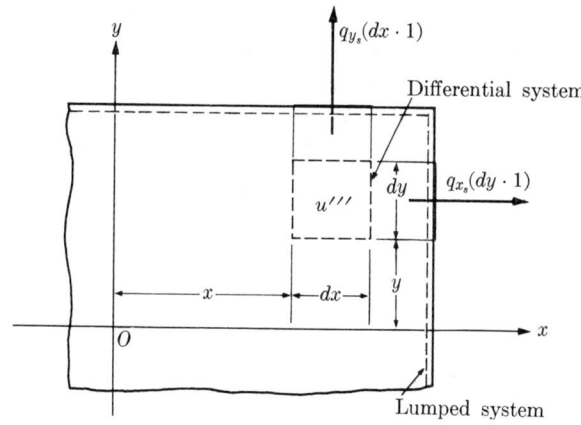

FIG. 2-32

(ii) *General law.* The first law of thermodynamics applied to the lumped system of Fig. 2-32, but with its terms interpreted by the differential system of the same figure, results in

$$-\int_0^l (q_x)_s \, dy - \int_0^L (q_y)_s \, dx + u''' L l = 0. \quad (2\text{-}130)$$

The same result may also be obtained following the mathematical approach instead, by integrating the differential form of the first law of thermodynamics, Eq. (2-125), over the cross section of the heater. This yields

$$\int_0^L \int_0^l \left(-\frac{\partial q_x}{\partial x} - \frac{\partial q_y}{\partial y} + u'''\right) dx \, dy = 0. \quad (2\text{-}131)$$

The equality of Eqs. (2–130) and (2–131) may be readily shown by carrying out the appropriate integrations in Eq. (2–131).

(iii) *Particular law.* Since q_x and q_y apply locally, Fourier's law given by Eq. (2–126) is equally valid for the present case.

(iv) *Governing equation.* Introducing Eq. (2–126) into Eqs. (2–130) and (2–131) gives the integral form of the equation of conduction as

$$\int_0^l \left(\frac{\partial T}{\partial x}\right)_s dy + \int_0^L \left(\frac{\partial T}{\partial y}\right)_s dx + \frac{u'''}{k} Ll = 0 \qquad (2\text{–}132)$$

or

$$\int_0^L \int_0^l \left(\frac{\partial^2 T}{\partial x^2} + \frac{\partial^2 T}{\partial y^2} + \frac{u'''}{k}\right) dx\, dy = 0. \qquad (2\text{–}133)$$

Thus we have two integral forms corresponding to the differential formulation of the equation of conduction, one obtained by physical considerations, the other by the mathematical approach, which involves the integration of the suitable differential form over the cross section of the heater. It is clear that Eq. (2–133) is easier to establish than Eq. (2–132) when the corresponding differential form is available. The two equations are, of course, identical.

(v) *Initial and boundary conditions.* Since the terms of Eqs. (2–132) and (2–133) apply locally, the initial and boundary conditions of the differential formulation are also valid for the present integral formulation. Hence Eq. (2–132) or Eq. (2–133), together with Eq. (2–129), completes the integral formulation of the problem.

The integral formulation is useful for obtaining approximate solutions, which are convenient for problems whose exact solution is rather involved algebraically, and indispensable for complex problems having no exact solution. Solution of the integral formulation requires no further mathematics than that which we assume the reader to have; hence we shall give the method here. This method is based on the selection of an approximate profile for the unknown (dependent) variable, say the temperature. The profile, containing an unknown *parameter* to be determined, is assumed to be composed of the product of simple (polynomial, circular, etc.) functions.* Each function in this product depends on only one of the independent variables entering the problem, and is chosen such that the boundary conditions are satisfied.† When this product form is intro-

---

* This, however, is an assumption only, and may not lead to a solution. The existence of a solution implies the validity of the assumption.
† Although not common in conduction heat transfer, the integral method is extensively used under the name of Karman-Polhausen procedure for approximate solutions of the velocity and temperature boundary-layer problems of fluid mechanics and convection heat transfer.

duced into the integral formulation, the result of integration specifies the unknown parameter,* and when, in turn, this value of the parameter is inserted into the product form, an approximate solution is obtained for the problem under consideration.

Let us now apply this general procedure to the present problem. For the temperature of the heater suppose that a product solution exists in the form

$$T(x, y) - T_\infty = X(x)Y(y),$$

where X and Y are functions of x and y, respectively. Restricting ourselves to the case of large h and assuming, for example, parabolic profiles in both directions such that they satisfy the boundary conditions, we may write a first approximation of the heater temperature as

$$T(x, y) - T_\infty = (L^2 - x^2)(l^2 - y^2)a_0, \qquad (2\text{--}134)$$

where a_0 is the unknown parameter to be determined. Equation (2–134) is the first-order polynomial *Ritz profile*, from the well-known *Ritz method* of the variational calculus, which will be discussed in Chapter 8.

Inserting Eq. (2–134) into Eq. (2–132) or Eq. (2–133) yields

$$a_0 = \frac{3}{4} \frac{u'''/k}{L^2 + l^2}. \qquad (2\text{--}135)$$

Combining Eqs. (2–134) and (2–135) and rearranging gives the first-order polynomial Ritz profile of the desired temperature distribution in the form

$$\frac{T(x, y) - T_\infty}{u''' l^2/k} = \frac{3}{4} \frac{[1 - (x/L)^2][1 - (y/l)^2]}{1 + (l/L)^2}. \qquad (2\text{--}136)$$

Let us now comment on the accuracy of this approximate solution. Since the boundary conditions are exactly satisfied, the maximum error is anticipated at the location farthest from the boundaries, namely, at the origin of the coordinate system. And in fact, when we insert $x/L = y/l = 0$ and a specific value of l/L, say $l/L = 1$, into Eq. (2–136), we find the error to be 27.3%, an appreciable amount.† However, this error may be reduced greatly by the second-order approximation, which will be considered later. (See Example 4–11.)

We may also use an alternative procedure, the so-called *Kantorovich method*, for the selection of approximate profiles. This method is based on a generalization of the Ritz procedure. Again assume that a product solution exists composed of functions depending on only one independent variable. One of these functions, the *parameter function*,‡ is left unspecified. The new profile satisfies

---

* Product forms of higher orders depend on more than one parameter.
† This error is evaluated by comparing Eq. (2–136) with the exact solution, Eq. (4–133), obtained by solving the differential formulation of the problem.
‡ In the Ritz method the term "parameter" refers to a constant parameter.

the boundary conditions of the problem only in the direction of specified functions, and when substituted into the integral formulation it yields a differential equation in terms of the parameter function. The integration constants of the solution of this differential equation are determined according to the boundary conditions in the direction of the parameter function. As we shall see in Examples 2–2 and 2–3, the Kantorovich method is especially convenient for unsteady problems.

If we return now to the specific problem under study and leave, for example, the x-direction of Eq. (2–134) unspecified,* the first-order polynomial *Kantorovich profile* becomes

$$T(x, y) - T_\infty = (l^2 - y^2)X(x), \qquad (2\text{–}137)$$

which satisfies the boundary conditions only in the y-direction.

Introducing Eq. (2–137) into Eq. (2–133)† and integrating the latter with respect to y yields

$$\int_0^L (\tfrac{2}{3}l^2 X'' - 2X + u'''/k)\, dx = 0. \qquad (2\text{–}138)$$

Since Eq. (2–138) is true for an arbitrary length L, the integrand itself must vanish everywhere in the interval $(0, L)$. Thus the parameter function $X(x)$ satisfies the differential equation

$$X'' - (3/l^2)X = 3u'''/2kl^2, \qquad (2\text{–}139)$$

subject to the boundary conditions in x, which have not been employed so far. These conditions may be determined as follows. Consider, for example, the boundary condition at $x = L$, $T(L, y) = T_\infty$. In terms of the product solution, this condition may be written in the form

$$T(L, y) - T_\infty = 0 = X(L)Y(y). \qquad (2\text{–}140)$$

However, Eq. (2–140) cannot be valid for all values of $Y(y)$ unless $X(L) = 0$. Similarly, the other condition, resulting from the symmetry of temperature, is found to be $dX(0)/dx = 0$. Thus the boundary conditions in x are

$$dX(0)/dx = 0, \qquad X(L) = 0. \qquad (2\text{–}141)$$

The solution of Eq. (2–139) satisfying Eq. (2–141) is

$$X(x) = \frac{u'''}{2k}\left(1 - \frac{\cosh(\sqrt{3}/l)x}{\cosh(\sqrt{3}/l)L}\right). \qquad (2\text{–}142)$$

---

* As will be explained in Chapter 8, this choice cannot be made arbitrarily, for the direction to be left unspecified affects the accuracy of the procedure.
† Equation (2–133) is more suitable to the Kantorovich method than Eq. (2–132).

Finally, inserting Eq. (2-142) into Eq. (2-137) and rearranging gives the first-order polynomial Kantorovich profile for the desired temperature distribution of the heater in the form

$$\frac{T(x,y) - T_\infty}{u''' l^2 / k} = \frac{1}{2}\left[1 - \left(\frac{y}{l}\right)^2\right]\left(1 - \frac{\cosh (\sqrt{3}/l)x}{\cosh (\sqrt{3}/l)L}\right). \quad (2\text{-}143)$$

For a square plate the temperature at the origin of the coordinate system, leading to the maximum error, deviates 11.5% from that of the exact solution. As expected, this result is more accurate than that of the Ritz procedure because less arbitrary restrictions are imposed on the Kantorovich profile. The accuracy of the first-order Kantorovich profile, like that of the first-order Ritz profile, may be considerably improved by the second-order profile. (See Example 4-11.)

It is worth noting that the first-order Ritz and Kantorovich profiles evaluated by the *variational form* of the same problem yield more accurate results. Thus the aforementioned error of 27.3% under the integral-Ritz procedure may be reduced to 6.05% by use of the variational-Ritz. Similarly, the 11.5% of the integral-Kantorovich becomes 2.68% by use of the variational-Kantorovich. This justifies the study of the *variational calculus* which is the concern of Chapter 8. The reason why the calculus of variations gives more accurate results compared with those of the integral method is also explained in that chapter. (See the discussion following Example 8-6.)

Example 2-2. Consider a pool reactor (Fig. 2-33) whose core is constructed from a number of vertical fuel plates of thickness $2L$. Initially the system has the uniform temperature T_∞; then assume that the constant nuclear internal energy u''' is uniformly generated in these plates. The heat transfer coefficient between the plates and the coolant is h. The temperature of the coolant re-

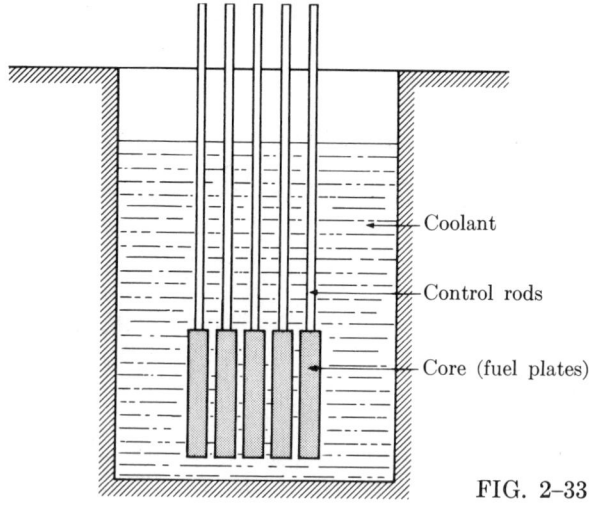

FIG. 2-33

mains constant, and the thickness of the plates is small compared with other dimensions. Thus, if the end effects are neglected, the heat transfer may be taken to be one-dimensional. We wish to formulate the unsteady temperature problem of the reactor.

The lumped, differential, and integral forms of the problem will be formulated for one of the plates of the core, again by following the five basic steps of formulation, although these steps will no longer be elaborated.

I. *Lumped formulation.* The whole plate is taken to be the system (Fig. 2–34). The lumped first law of thermodynamics, Eq. (2–16), applied to this system reduces to

$$\frac{dE}{dt} = -2Aq_n, \qquad (2\text{--}144)$$

where A denotes the surface area of one side of the plate. Note that the internal energy generation u''' can no longer be identified as a power input to the system by an external electrical source. It consists, rather, of continuous changes in the composition of the nuclear fuel of which the plates are composed, as fissionable material is turned into internal energy.* These composition changes are generally small enough that thermal properties may be assumed constant.

Thus when we make use of the definition of specific heat, the left-hand side of Eq. (2–144) becomes

$$\frac{dE}{dt} = \rho(A \cdot 2L)c\frac{dT}{dt} - (A \cdot 2L)u'''. \qquad (2\text{--}145)$$

Inserting Eq. (2–145) into Eq. (2–144), we get the lumped form of the first law of thermodynamics:

$$\rho cL\frac{dT}{dt} = -q_n + u'''L. \qquad (2\text{--}146)$$

Since there is no need for the particular law, Eq. (2–146) is also the governing equation of the problem.

The initial and boundary conditions, respectively, are

$$T(0) = T_\infty, \qquad (2\text{--}147)$$

$$q_n = h(T - T_\infty). \qquad (2\text{--}148)$$

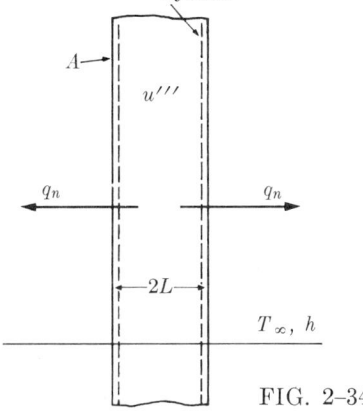

FIG. 2–34

---

* A similar argument pertains to the case of distributed internal energy sources and sinks resulting from exothermic and endothermic chemical reactions, respectively. See Eq. (2–12).

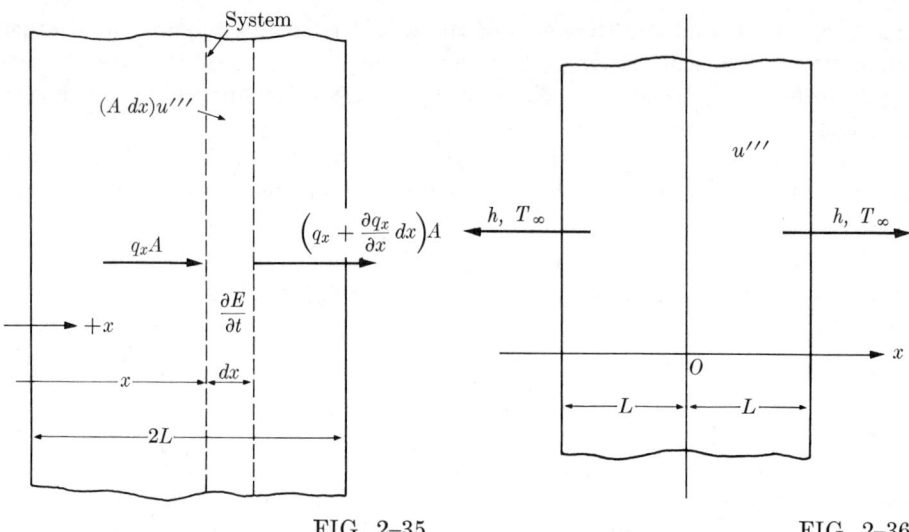

FIG. 2–35   FIG. 2–36

Thus Eqs. (2–146), (2–147), and (2–148) completely describe the lumped formulation of the problem. The trivial solution of this formulation may be readily obtained by solving the combination of Eqs. (2–146) and (2–148),

$$\rho c L \frac{dT}{dt} = -h(T - T_\infty) + u'''L, \qquad (2\text{–}149)$$

subject to Eq. (2–147). The result is

$$\frac{T(t) - T_\infty}{u'''L/h} = 1 - e^{-mt}, \qquad (2\text{–}150)$$

where $m = h/\rho c L$.

II. *Differential formulation.* Consider the one-dimensional differential system shown in Fig. 2–35. The rightward x is assumed to be positive. The first law of thermodynamics written for Fig. 2–35 gives

$$\frac{\partial E}{\partial t} = -A \frac{\partial q_x}{\partial x} dx. \qquad (2\text{–}151)$$

Here the time rate of change of total internal energy may be evaluated in a manner similar to that of the lumped formulation. Hence

$$\frac{\partial E}{\partial t} = \rho c (A\,dx) \frac{\partial T}{\partial t} - (A\,dx) u'''. \qquad (2\text{–}152)$$

Introducing Eq. (2–152) into Eq. (2–151) and rearranging yields the appropriate form of the general law, the first law of thermodynamics, as follows:

$$\rho c \frac{\partial T}{\partial t} = -\frac{\partial q_x}{\partial x} + u'''. \qquad (2\text{–}153)$$

Finally, considering the particular law, the x-component of Fourier's law for isotropic continua,

$$q_x = -k\frac{\partial T}{\partial x}, \tag{2-154}$$

and inserting Eq. (2-154) into Eq. (2-153), we find that the governing equation of the problem is

$$\rho c\frac{\partial T}{\partial t} = \frac{\partial}{\partial x}\left(k\frac{\partial T}{\partial x}\right) + u''', \tag{2-155}$$

which for constant k reduces to

$$\frac{\partial T}{\partial t} = a\frac{\partial^2 T}{\partial x^2} + \frac{u'''}{\rho c}. \tag{2-156}$$

Equations (2-155) and (2-156) are the one-dimensional cartesian forms of the differential conduction equations, Eqs. (2-88) and (2-91), respectively.

The origin of the coordinate axis must now be determined, before we can state the initial and boundary conditions. The thermal and geometric symmetry of the problem suggests the middle plane of the plate to be the origin of x (Fig. 2-36). In terms of this coordinate the appropriate initial and boundary conditions are:

Initial: $\quad T(x, 0) = T_\infty,$

Boundary: $\quad \dfrac{\partial T(0, t)}{\partial x} = 0, \quad -k\dfrac{\partial T(L, t)}{\partial x} = h[T(L, t) - T_\infty],$ \hfill (2-157)

Thus Eq. (2-155) or Eq. (2-156), subject to Eq. (2-157), completes the differential formulation of the problem. The solution of this formulation requires further mathematics and is left to Chapter 5. (See Example 5-4 and Problem 5-8.)

III. *Integral formulation.* Let us reconsider the systems employed for the lumped and differential formulations, and specify the origin of the coordinate axis at this step of the formulation. The first law of thermodynamics, when applied to the lumped system of Fig. 2-37, results in the previously obtained relation

$$\frac{dE}{dt} = -2Aq_n. \tag{2-144}$$

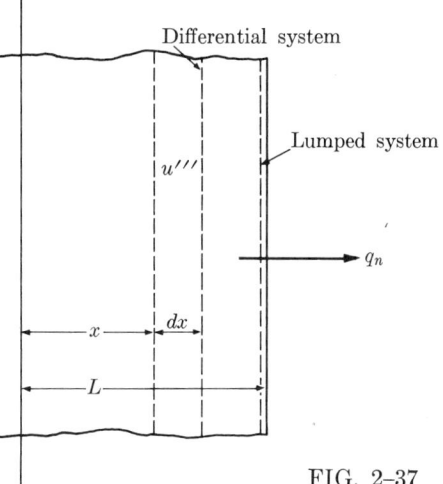

FIG. 2-37

Evaluating dE/dt in terms of the differential system integrated over the thickness of the plate, and noting the symmetry, we find that the integral form of the first law of thermodynamics is

$$\frac{d}{dt}\int_0^L \rho c T\, dx - u'''L = -q_n. \tag{2-158}$$

The particular law of the differential formulation, Eq. (2-154), being also valid for the integral formulation, gives

$$q_n = q_{x=L} = -k\left(\frac{\partial T}{\partial x}\right)_{x=L}. \tag{2-159}$$

Thus, inserting Eq. (2-159) into Eq. (2-158), we obtain the governing equation:

$$\frac{d}{dt}\int_0^L \rho c T\, dx - u'''L = k\left(\frac{\partial T}{\partial x}\right)_{x=L}. \tag{2-160}$$

The initial and boundary conditions are identical to those of the differential formulation. Hence Eq. (2-160), together with Eq. (2-157), completes the integral formulation of the problem.

An approximate solution of the foregoing formulation may now be obtained. Let us first consider the simple case $h \to \infty$. Although the spacewise temperature distribution of an unsteady problem may be easily approximated, its timewise variation is often difficult to guess. For these problems, the Kantorovich profile, being expressible in terms of an unspecified parameter function in time, becomes important. More specifically, if an unsteady problem asymptotically tends to a steady solution, and if the unsteady profile within a scale factor resembles the steady distribution,* the Kantorovich profile of the problem may be conveniently constructed from the steady solution. Let us illustrate the point in terms of the problem under consideration. Leaving the details of the trivial steady solution of the problem to the reader, we consider its result in terms of Fig. 2-36:

$$\frac{T(x) - T_\infty}{u'''L^2/k} = \frac{1}{2}\left[1 - \left(\frac{x}{L}\right)^2\right]. \tag{2-161}$$

An unsteady first-order Kantorovich profile may now be assumed in the form

$$\frac{T(x,t) - T_\infty}{u'''L^2/k} = \frac{1}{2}\left[1 - \left(\frac{x}{L}\right)^2\right]\tau_0(t). \tag{2-162}$$

---

* The Kantorovich profile for problems whose unsteady temperature variations do not resemble their steady distributions will be explained in Example 2-3.

Furthermore, the unsteady scale factor $\tau_0(t)$ may conveniently be taken as the unspecified parameter function to be determined. Thus introducing Eq. (2–162) into Eq. (2–160), integrating the latter with the assumption of constant properties, and rearranging gives the differential equation

$$\frac{d\tau_0}{dt} + \frac{3a}{L^2}(\tau_0 - 1) = 0, \qquad (2\text{–}163)$$

subject to the condition

$$\tau_0(0) = 0. \qquad (2\text{–}164)$$

The solution of Eq. (2–163) satisfying Eq. (2–164) is

$$\tau_0(t) = 1 - \exp(-3at/L^2). \qquad (2\text{–}165)$$

Inserting Eq. (2–165) into Eq. (2–162), we obtain the first-order Kantorovich profile for the temperature variation of the plate in the form

$$\frac{T(x,t) - T_\infty}{u'''L^2/k} = \frac{1}{2}\left[1 - \left(\frac{x}{L}\right)^2\right]\left[1 - \exp\left(-3\frac{at}{L^2}\right)\right]. \qquad (2\text{–}166)$$

As $t \to \infty$ Eq. (2–166) approaches the steady solution of the problem, Eq. (2–161).

The case of finite h can be treated in the same way. Again multiplying the steady solution of the problem by the unspecified parameter function $\tau_0(t)$, we have

$$\frac{T(x,t) - T_\infty}{u'''L^2/k} = \frac{1}{2}\left[1 - \left(\frac{x}{L}\right)^2 + \frac{2}{\text{Bi}}\right]\tau_0(t), \qquad (2\text{–}167)$$

where $\text{Bi} = hL/k$. [Does $\tau_0(t)$ of Eq. (2–167) have any physical significance?] Inserting Eq. (2–167) into Eq. (2–160), and integrating the latter, we obtain the differential equation

$$\frac{d\tau_0}{dt} + \frac{3a}{L^2}\left(\frac{\text{Bi}}{\text{Bi}+3}\right)(\tau_0 - 1) = 0, \qquad (2\text{–}168)$$

subject to the condition

$$\tau_0(0) = 0. \qquad (2\text{–}164)$$

The solution of Eq. (2–168), first satisfied by Eq. (2–164), then introduced into Eq. (2–167), gives the first-order Kantorovich profile of our problem. Thus we have

$$\frac{T(x,t) - T_\infty}{u'''L^2/k} = \frac{1}{2}\left[1 - \left(\frac{x}{L}\right)^2 + \frac{2}{\text{Bi}}\right]\left[1 - \exp\left(-\frac{3a}{L^2}\frac{\text{Bi}}{\text{Bi}+3}t\right)\right]. \qquad (2\text{–}169)$$

[What is the limiting form of Eq. (2–169) as $\text{Bi} \to \infty$?]

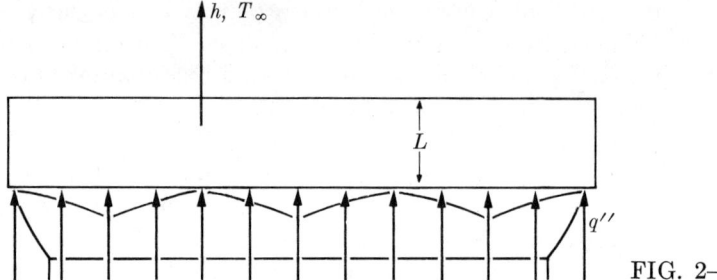

FIG. 2-38

The following alternative procedure for the selection of approximate profiles is suggested to the reader for further exercise. Write a Kantorovich profile by assuming the middle plane and surface temperatures of the plate as unspecified parameter functions to be determined. Then write the temperature of the plate corresponding to the case of finite h, in terms of a two-parameter profile. Next, utilizing the surface boundary condition, eliminate one of the parameter functions of this profile. Compare the result with Eq. (2-167).

In the next example, our primary interest lies, not in the lumped and differential formulations of the problem, but rather in the integral formulation, which requires that we define a new concept, the *penetration depth*.

Example 2-3. A hot plate of thickness L (Fig. 2-38) initially assumes the ambient temperature T_∞. From this condition, the bottom of the plate is subjected to the uniform heat flux q''. The upward heat transfer coefficient is h. The thickness L of the plate is small compared with its other dimensions, such that the heat loss from the sides may be neglected. We wish to formulate the unsteady temperature problem.

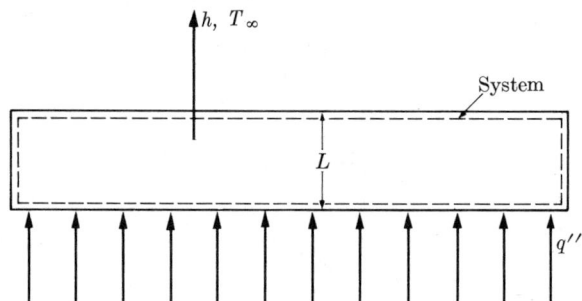

FIG. 2-39

I. *Lumped formulation.* Applying the first law of thermodynamics, combined with the definition of heat transfer coefficient, to the lumped system of Fig. 2-39 results in the lumped formulation of the problem as follows:

$$\rho c L \frac{dT}{dt} = q'' - h(T - T_\infty) \qquad (2\text{-}170)$$

and

$$T(0) = T_\infty. \qquad (2\text{-}171)$$

The solution of Eq. (2–170) satisfying Eq. (2–171) is

$$\frac{T(t) - T_\infty}{q''/h} = 1 - e^{-mt}, \qquad (2\text{–}172)$$

where $m = h/\rho cL$.

II. *Differential formulation.* The first law of thermodynamics and Fourier's law of conduction, combined for the differential system of Fig. 2–40, result in

$$\rho c \frac{\partial T}{\partial t} = \frac{\partial}{\partial x}\left(k \frac{\partial T}{\partial x}\right), \qquad (2\text{–}173)$$

which for constant k reduces to

$$\frac{\partial T}{\partial t} = a \frac{\partial^2 T}{\partial x^2}. \qquad (2\text{–}174)$$

The initial and boundary conditions in x measured from the bottom* upward are

$$T(x, 0) = T_\infty; \qquad -k \frac{\partial T(0, t)}{\partial x} = q'',$$

$$-k \frac{\partial T(L, t)}{\partial x} = h[T(L, t) - T_\infty]. \qquad (2\text{–}175)$$

These, combined with Eq. (2–173) or Eq. (2–174), complete the differential formulation of the problem. The solution of this formulation is left to Chapter 5. [See Eq. (5–34).]

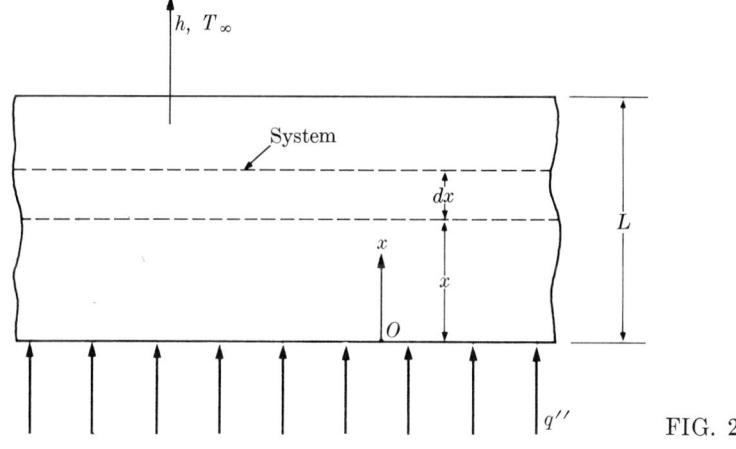

FIG. 2–40

---

* This is the first problem for which the appropriate coordinate axis is not obvious. The selection of the most suitable reference frame is important in view of the complexity of solution. This question will be clarified in Chapters 3 and 4.

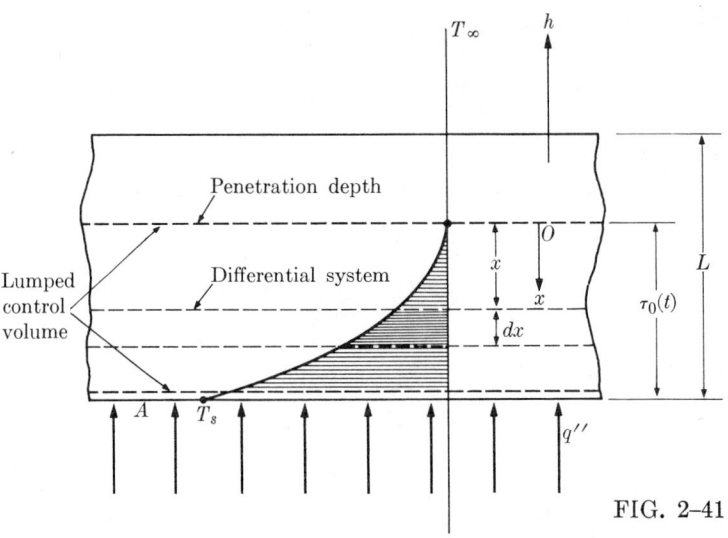

FIG. 2–41

III. *Integral formulation.* Let us first divide the problem into two domains with respect to time, such that the first time domain $0 \leq t \leq t_0$ terminates when the effect of applied heat flux reaches the upper surface; the second domain $t \geq t_0$ is valid for the remaining part of the transient while the temperature of the upper surface rises to its steady value.

The first domain of the problem can be best described by the *penetration depth* of the temperature,[*] $\tau_0(t)$, shown in Fig. 2–41. In the same figure the appropriate *lumped control volume* and the differential system are also indicated. The coordinate axis measured from the moving penetration depth downward proves to be convenient for the construction of the temperature profile. [See Eq. (2–181).]

The first law of thermodynamics applied to the lumped control volume of Fig. 2–41 yields

$$\frac{dE}{dt} = \rho c A T_\infty \frac{d\tau_0}{dt} + q'' A, \qquad (2\text{--}176)$$

where A is the surface area of one side of the plate. Expressing the total internal energy E of the lumped control volume in terms of the internal energy of the differential system gives the appropriate integral formulation of the general law:

$$\frac{d}{dt}\int_0^{\tau_0(t)} \rho c T\, dx = \rho c T_\infty \frac{d\tau_0}{dt} + q''. \qquad (2\text{--}177)$$

Although the problem is distributed, none of the terms of Eq. (2–177) requires

---

[*] This is analogous to the concept of velocity and temperature boundary-layer thicknesses of the boundary-layer theory.

the use of any particular law. However, the x-component of Fourier's particular law will be used later in connection with boundary conditions. [See Eq. (2–179).] Thus the governing equation of the first domain is identical to Eq. (2–177).

The uniform initial temperature of the plate gives the initial condition

$$T(x, 0) = T_\infty. \tag{2-178}$$

The boundary condition on the lower surface of the plate, relating the constant heat flux q'' to the temperature by the particular law, may be written in the form

$$-k\left(\frac{\partial T}{\partial x}\right)_{x=\tau_0(t)} = -q''. \tag{2-179}$$

Since in this time domain the penetration depth is less than the thickness of the plate, the boundary condition expressing the heat transfer from the upper surface is not valid. Instead, zero heat flux across the plane surface of the penetration depth

$$\left(\frac{\partial T}{\partial x}\right)_{x=0} = 0 \tag{2-180}$$

should be used. Thus Eqs. (2–177), (2–178), (2–179), and (2–180) complete the integral formulation of the problem in the first time domain. Before proceeding with the formulation of the second time domain, however, let us find an approximate solution for the first time domain.

Since the unsteady temperature variation in this domain approaches, not a known steady temperature, but the unknown initial temperature of the second domain, the unsteady Kantorovich profile cannot be constructed from the steady solution. (Note the Kantorovich profile of Example 2–2.) This suggests a second approach in constructing the unsteady Kantorovich profiles, one which requires the selection of functions (polynomial, circular, etc.) satisfying the boundary conditions. Although the order of the polynomial (or the type of circular function) is to a large extent arbitrary, simplicity may be used as a guide for the construction of these profiles. Thus, for example, the polynomial of least order satisfying the boundary conditions is generally the best approximation of the actual profile. In the present case we may assume the parabola

$$T(x, t) - T_\infty = \left(\frac{q''}{2k}\right)\frac{x^2}{\tau_0(t)}, \tag{2-181}$$

which satisfies the boundary conditions given by Eqs. (2–179) and (2–180). Introducing Eq. (2–181) into Eq. (2–177), and integrating the latter with the assumption of constant properties, we obtain the trivial but *nonlinear* differential equation

$$d\tau_0^2 = 6a\, dt, \tag{2-182}$$

subject to the initial condition

$$\tau_0(0) = 0. \tag{2-183}$$

The solution of Eq. (2-182) which satisfies Eq. (2-183) is

$$\tau_0(t) = (6at)^{1/2}. \tag{2-184}$$

Thus the first-order Kantorovich profile giving the temperature variation in the first time domain of the problem is found to be

$$T(x, t) - T_\infty = \left(\frac{q''}{2k}\right)\frac{x^2}{(6at)^{1/2}}, \quad \text{when} \quad 0 \le t \le t_0. \tag{2-185}$$

Furthermore, inserting $\tau_0(t_0) = L$ into Eq. (2-184), we get the penetration time of the applied heat flux q'' to the upper surface of the plate as

$$t_0 = L^2/6a. \tag{2-186}$$

For a comparison among the following three different solutions of the problem, let us consider, for example, the temperature variation at the bottom of the plate. From the solution under study, introducing $x = \tau_0(t) = (6at)^{1/2}$ into Eq. (2-185), we obtain

$$T_s(t) - T_\infty = \left(\frac{3}{2}\right)^{1/2}\left(\frac{q''}{k}\right)(at)^{1/2}. \tag{2-187}$$

The exact solution of the problem, which will be seen in Chapter 7*, gives

$$T_s(t) - T_\infty = \left(\frac{4}{\pi}\right)^{1/2}\left(\frac{q''}{k}\right)(at)^{1/2}. \tag{2-188}$$

Hence the error involved in Eq. (2-187) is 8.4%. If the approximate profile given by Eq. (2-181) were introduced into the variational form of the formulation (see Example 8-8), the result would be

$$T_s(t) - T_\infty = \left(\frac{5}{4}\right)^{1/2}\left(\frac{q''}{k}\right)(at)^{1/2}. \tag{2-189}$$

Equation (2-189) leads to an error of approximately 0.89%. The foregoing comparison thus indicates once more the importance of the variational calculus as an approximate method.

Having finished the study of the first time domain, we now return to the formulation of the problem for the second time domain. Consider the lumped and differential systems shown in Fig. 2-42. The coordinate axis is measured

---

* The special case of Problem 7-1 corresponding to $h = 0$.

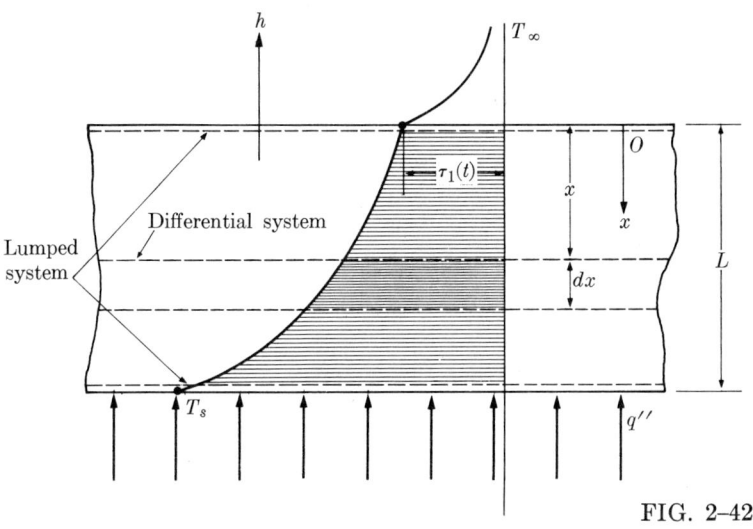

FIG. 2-42

from the upper surface downward. The first law of thermodynamics and Fourier's law of conduction result in the appropriate integral form of the governing equation:

$$\frac{d}{dt}\int_0^L \rho c T\, dx = q'' - k\left(\frac{\partial T}{\partial x}\right)_{x=0}. \qquad (2\text{-}190)$$

The initial condition of this domain is the final condition of the first domain. Also, the boundary condition on the lower surface should be modified according to the new coordinate axis as

$$-k\left(\frac{\partial T}{\partial x}\right)_{x=L} = -q''. \qquad (2\text{-}191)$$

Furthermore, the boundary at the upper surface, now transferring heat to the ambient, satisfies the condition

$$+k\left(\frac{\partial T}{\partial x}\right)_{x=0} = h(T_{x=0} - T_\infty). \qquad (2\text{-}192)$$

Thus Eqs. (2-190), (2-185), for $t = t_0$, (2-191), and (2-192) describe the integral formulation of the problem for the second time domain.

Finally, let us find an approximate temperature variation for the second time domain. Although this problem, like that of Example 2-2, approaches a steady solution at the end of the second time domain, an unsteady Kantorovich profile cannot be constructed by scaling the exact steady solution of the problem in terms of an unspecified parameter function. In this case a procedure analogous to that of the first time domain is followed. The simplest possible profile is constructed from the combination of an unspecified parameter function in time and appropriately selected space functions satisfying the boundary conditions of the problem. It is important to note that this profile must approach that of the first time domain as $t \to t_0$, and of the steady solution as $t \to \infty$.

Let us assume the parabolic profile

$$T(x,t) - T_\infty = Ax^2 + Bx + C.$$

If we take, for example, the upper surface temperature $\tau_1(t)$ as the unspecified parameter function to be determined, this profile may be written in the form

$$T(x,t) - T_\infty = \left(\frac{q''}{h} - \tau_1\right)\frac{hx^2}{2kL} + \left(1 + \frac{hx}{k}\right)\tau_1. \quad (2\text{--}193)$$

As $t \to \infty$ Eq. (2–193) approaches the steady solution of the problem,

$$T(x) - T_\infty = \left(1 + \frac{hx}{k}\right)\frac{q''}{h},$$

and it becomes identical to the solution of the first time domain as $t \to t_0$.

Inserting Eq. (2–193) into Eq. (2–190) and integrating the latter with the assumption of constant properties, we obtain the *linear* differential equation

$$\frac{d\tau_1}{dt} + \left(\frac{m}{1 + \text{Bi}/3}\right)\left(\tau_1 - \frac{n}{m}\right) = 0, \quad (2\text{--}194)$$

subject to the initial condition

$$\tau_1(t_0) = 0. \quad (2\text{--}195)$$

Here $m = h/\rho cL$, $n = q''/\rho cL$, and $\text{Bi} = hL/k$.

The solution of Eq. (2–194) satisfying Eq. (2–195) is

$$\tau_1(t) = \frac{n}{m}\left\{1 - \exp\left[-\frac{m(t - t_0)}{1 + \text{Bi}/3}\right]\right\}. \quad (2\text{--}196)$$

Finally, combining Eqs. (2–196) and (2–193) and rearranging gives the first-order Kantorovich profile for the second time domain of the problem in the form

$$\frac{T(x,t) - T_\infty}{q''L/k} = \frac{1}{2}\left(\frac{x}{L}\right)^2 + \left[\frac{1}{\text{Bi}} + \left(1 - \frac{x}{2L}\right)\frac{x}{L}\right]\left\{1 - \exp\left[-\frac{m(t - t_0)}{1 + \text{Bi}/3}\right]\right\}, \quad (2\text{--}197)$$

when $t \geq t_0$.

So far we have considered three simple problems that were so stated as to include the complete information for the formulation. Our main objective has been to develop the ability to formulate problems in the lumped, differential, and integral forms. In a given physical situation, however, the formulation of a problem often requires that we make a number of assumptions as part of the formulation. For this reason, in the next two examples only the physics of the problem is described. The necessary information for formulation is then given in the course of formulation.

Example 2-4. Two rigid circular disks are coaxially pressed together by the external load P as shown in Fig. 2–43. Initially the system is stationary and at the ambient temperature T_∞. Then the upper plate suddenly assumes the constant angular velocity ω. The coefficient of dry friction between the disks is μ. How should we formulate the problem?

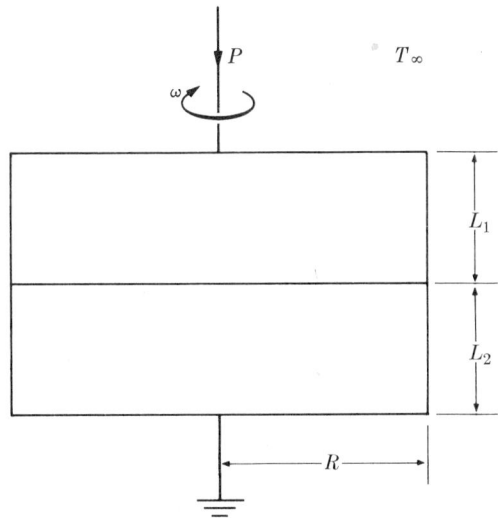

FIG. 2–43

Note first that the coefficients of upward, downward, and horizontal heat transfer resulting from gravitational free convection are different. Second, the heat transfer from the upper disk is increased many times by centrifugal free convection. Thus we have four different heat transfer coefficients, the upward h_1 and horizontal h_3 for the rotating disk, the downward h_2 and horizontal h_4 for the stationary disk. To simplify the formulation, the disks are assumed to be homogeneous and isotropic.

I. *Differential formulation.* The properties of the upper and lower disks are distinguished by the subscripts 1 and 2. The first law of thermodynamics and Fourier's law of conduction written for the two-dimensional cylindrical systems of Fig. 2–44 give the differential formulation of the problem as follows:

$$\frac{\partial T_1}{\partial t} = a_1 \left[\frac{1}{r} \frac{\partial}{\partial r}\left(r \frac{\partial T_1}{\partial r} \right) + \frac{\partial^2 T_1}{\partial z^2} \right], \qquad (2\text{-}198)$$

$$\frac{\partial T_2}{\partial t} = a_2 \left[\frac{1}{r} \frac{\partial}{\partial r}\left(r \frac{\partial T_2}{\partial r} \right) + \frac{\partial^2 T_2}{\partial z^2} \right], \qquad (2\text{-}199)$$

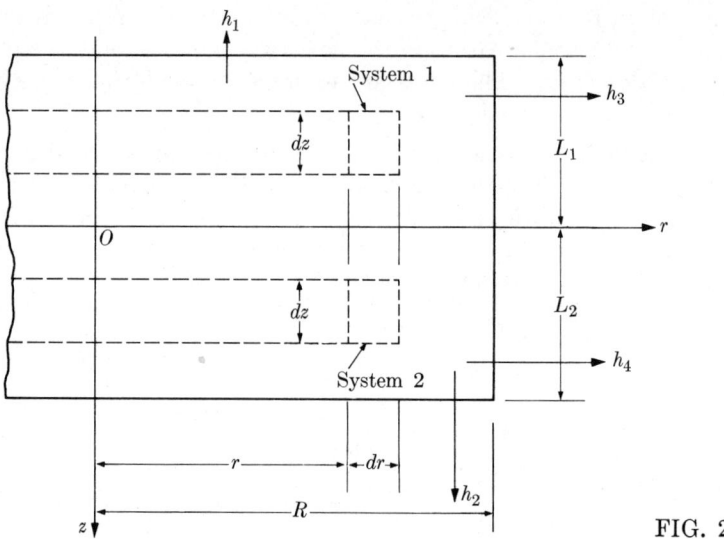

FIG. 2-44

with two initial and eight boundary conditions:*

$$T_1(r, z, 0) = T_\infty, \tag{2-200}$$

$$T_2(r, z, 0) = T_\infty, \tag{2-201}$$

$$+k_1 \frac{\partial T_1(r, -L_1, t)}{\partial z} = h_1[T_1(r, -L_1, t) - T_\infty], \tag{2-202}$$

$$T_1(r, 0, t) = T_2(r, 0, t), \tag{2-203}$$

$$-k_1 \frac{\partial T_1(r, 0, t)}{\partial z} + \mu p(r)\omega r = -k_2 \frac{\partial T_2(r, 0, t)}{\partial z}, \tag{2-204}$$

$$-k_2 \frac{\partial T_2(r, L_2, t)}{\partial z} = h_2[T_2(r, L_2, t) - T_\infty], \tag{2-205}$$

$$\frac{\partial T_1(0, z, t)}{\partial r} = 0, \tag{2-206}$$

$$-k_1 \frac{\partial T_1(R, z, t)}{\partial r} = h_3[T_1(R, z, t) - T_\infty], \tag{2-207}$$

$$\frac{\partial T_2(0, z, t)}{\partial r} = 0, \tag{2-208}$$

$$-k_2 \frac{\partial T_2(R, z, t)}{\partial r} = h_4[T_2(R, z, t) - T_\infty], \tag{2-209}$$

where $p(r)$ is the local pressure between the disks.

---

* As with the second time domain of Example 2-3, the selection of axial origin is irrelevant for the present formulation and will not be discussed here.

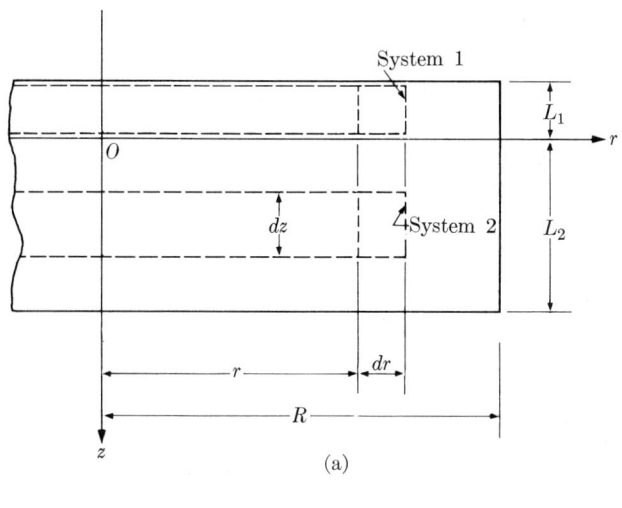

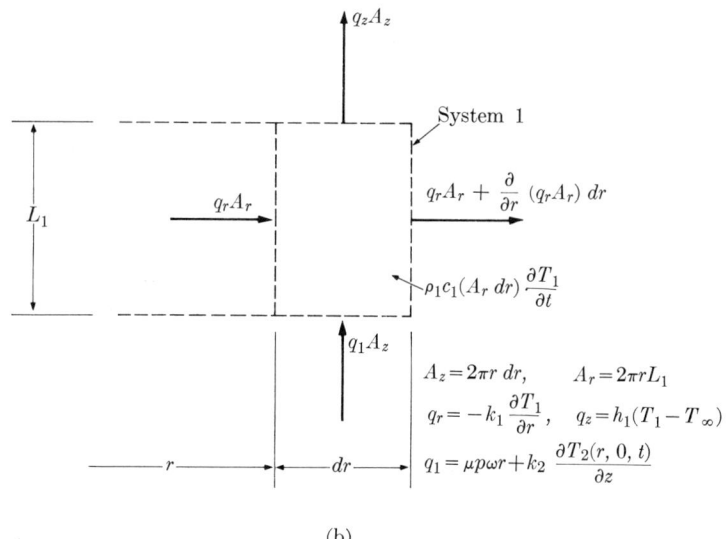

FIG. 2-45

The foregoing differential formulation, being classical, is easy to establish, but it is difficult or even impossible to solve. On the other hand, whenever physics permits there may be simpler formulations of the same problem, possibly leading to a solution. Let us demonstrate now how certain simplifications may be brought into the differential formulation.

(a) *Axially lumped upper disk.* When $k_1 \gg k_2$ or $L_1 \ll L_2$, the axial temperature variation of the upper disk may be lumped. Thus the first law of thermodynamics, Fourier's law of conduction, and the definition of heat transfer coefficient applied to system 1, Fig. 2-45, yield the equation of conduction for the upper disk as follows:

$$\frac{\partial T_1}{\partial t} = \frac{a_1}{r}\frac{\partial}{\partial r}\left(r\frac{\partial T_1}{\partial r}\right) - \frac{h_1}{\rho_1 c_1 L_1}(T_1 - T_\infty) + \frac{\mu p(r)\omega r}{\rho_1 c_1 L_1} - \left[-k_2\frac{\partial T_2(r,0,t)}{\partial z}\right]. \tag{2-210}$$

However, noting the equality of interface temperatures,

$$T_1(r,t) = T_2(r,0,t), \tag{2-211}$$

we may transform Eq. (2-210) to a boundary condition for the lower disk. [See Eq. (2-212).] The equation of conduction for the lower disk remains unchanged. Thus the formulation of the problem for the present case becomes

$$\frac{\partial T_2}{\partial t} = a_2\left[\frac{1}{r}\frac{\partial}{\partial r}\left(r\frac{\partial T_2}{\partial r}\right) + \frac{\partial^2 T_2}{\partial z^2}\right], \tag{2-199}$$

subject to

$$T_2(r,z,0) = T_\infty, \tag{2-201}$$

$$\frac{\partial T_2(r,0,t)}{\partial t} = \frac{a_1}{r}\frac{\partial}{\partial r}\left[r\frac{\partial T_2(r,0,t)}{\partial r}\right] - \frac{h_1}{\rho_1 c_1 L_1}[T_2(r,0,t) - T_\infty]$$

$$+ \frac{\mu p \omega r}{\rho_1 c_1 L_1} - \frac{1}{\rho_1 c_1 L_1}\left[-k_2\frac{\partial T_2(r,0,t)}{\partial z}\right], \tag{2-212}$$

$$-k_2\frac{\partial T_2(r,L_2,t)}{\partial z} = h_2[T_2(r,L_2,t) - T_\infty], \tag{2-205}$$

$$\frac{\partial T_2(0,z,t)}{\partial r} = 0, \tag{2-208}$$

$$-k_4\frac{\partial T_2(R,z,t)}{\partial r} = h_4[T_2(R,z,t) - T_\infty]. \tag{2-209}$$

Note that the heat transfer coefficient h_3 has disappeared from the formulation of the problem. (What boundary condition is then satisfied at the peripheral surface of the upper disk? Is it possible to assume insulation on this surface?)

For the limiting case $L_1 \to 0$, the present formulation remains unchanged except for the boundary condition given by Eq. (2-212), which simplifies to

$$0 = -h_1[T_2(r,0,t) - T_\infty] + \mu p \omega r - \left[-k_2\frac{\partial T_2(r,0,t)}{\partial z}\right]. \tag{2-213}$$

We thus learn that by neglecting axial temperature variation in the upper disk only, the formulation of the problem is reduced from two partial differential equations and two initial plus eight boundary conditions to one partial differential equation and one initial plus four boundary conditions.

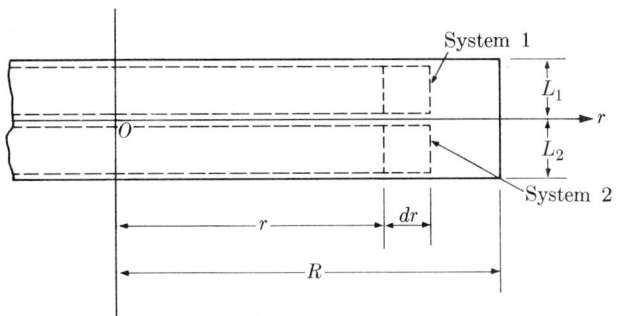

FIG. 2–46

(b) *Both disks axially lumped.* When both disks are thin such that L_1 and $L_2 \ll R$, or when k_1 and k_2 are large, the axial temperature variation in both disks may be neglected.

Thus, considering the axially lumped and radially differential systems shown in Fig. 2–46, we may obtain a single equation of conduction in the form

$$\frac{\partial T}{\partial t} = \frac{\bar{a}}{r}\frac{\partial}{\partial r}\left(r\frac{\partial T}{\partial r}\right) - \frac{\bar{h}}{\overline{\rho c L}}(T - T_\infty) + \frac{\mu p \omega r}{2\overline{\rho c L}}, \quad (2\text{–}214)$$

where T is the common temperature of the disks, $\bar{a} = \overline{kL}/\overline{\rho cL}$, $\overline{kL} = (k_1 L_1 + k_2 L_2)/2$, $\overline{\rho cL} = (\rho_1 c_1 L_1 + \rho_2 c_2 L_2)/2$, and $\bar{h} = (h_1 + h_2)/2$.

The initial and boundary conditions to be imposed on Eq. (2–214) are

$$T(r, 0) = T_\infty, \quad (2\text{–}215)$$

$$\frac{\partial T(0, t)}{\partial r} = 0, \quad (2\text{–}216)$$

$$-\bar{k}\frac{\partial T(R, t)}{\partial r} = \bar{h}[T(R, t) - T_\infty], \quad (2\text{–}217)$$

where $\bar{k} = (k_1 + k_2)/2$. It is, of course, possible to replace Eq. (2–217) with the approximate condition

$$\frac{\partial T(R, t)}{\partial r} \cong 0. \quad (2\text{–}218)$$

II. *Integral formulation.* A two-dimensional integral formulation of the problem, involving a penetration surface which is no longer plane, is beyond the scope of the present discussion. The one-dimensional integral formulation corresponding to case (b), however, is trivial and is left to the reader as an exercise.

The solutions of the foregoing four formulations, which do not concern us here, require that the pressure distribution $p(r)$ between the disks be specified. This determines the work done by friction. Two commonly assumed cases are

(i) constant pressure, $p = \text{const}$,
(ii) constant wear, $pr = \text{const}$.

The latter of these is physically more realistic, and also turns out to be more convenient for obtaining a solution.*

Example 2–5. An insulated thin-walled vessel initially contains superheated water vapor at temperature T_v (Fig. 2–47). Then the external surface of the bottom of the vessel is exposed to the surrounding air at temperature $T_\infty (T_\infty < T_v)$. We wish to formulate the problem in terms of the instantaneous thickness $X(t)$ of the condensate and the other variables involved.

The following assumptions and facts may be used in the formulation of the problem:

(a) The condensate-vapor interface is at the saturation temperature T_s.

(b) The problem is one-dimensional because of the peripheral insulation of the vessel.

(c) The temperature drop across the bottom thickness of the vessel may be neglected.

(d) The properties of the vapor and condensate are constant, and are distinguished by the subscripts 1 and 2, respectively.

(e) The constant density difference between the vapor and condensate causes a downward flow of vapor at the uniform velocity†

$$V = \left(\frac{\rho_2}{\rho_1} - 1\right)\frac{dX}{dt}.$$

Hence the two-domain one-dimensional differential formulation of the problem in x, measured from the bottom of the vessel upward, may be summarized as follows:

Vapor:

$$\frac{\partial T_1}{\partial t} - \left(\frac{\rho_2}{\rho_1} - 1\right)\frac{dX}{dt}\frac{\partial T_1}{\partial x} = a_1 \frac{\partial^2 T_1}{\partial x^2},$$

$$T_1(x, 0) = T_v,$$

$$T_1(\infty, t) = T_v,$$

Condensate:

$$\frac{\partial T_2}{\partial t} = a_2 \frac{\partial^2 T_2}{\partial x^2},$$

$$T_2(x, 0) = T_s,$$

$$+k_2 \frac{\partial T_2(0, t)}{\partial x} = h[T_2(0, t) - T_\infty],$$

---

* Compare the two solutions of Problem 3–30 corresponding to the cases of constant pressure and wear.
† See Eq. (2–114).

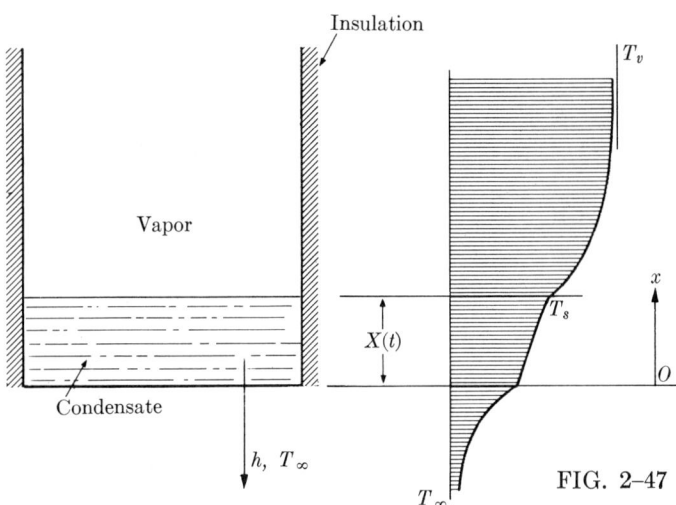

FIG. 2-47

coupled along the condensate-vapor interface by the boundary conditions

$$T_1(X, t) = T_2(X, t) = T_s,$$

$$-k_1 \frac{\partial T_1(X, t)}{\partial x} = -k_2 \frac{\partial T_2(X, t)}{\partial x} + \rho_2 h_{vc} \frac{dX}{dt}.$$

(How many boundary conditions are needed for the foregoing formulation and how many have been stated?)

The differential formulations of two-domain problems are often difficult to solve, especially when the domains are finite. With appropriate physical reasoning, however, these may be occasionally reduced to simpler problems. Let us first consider that the vapor is at the saturation temperature, or assume that the superheat is small enough that it can be neglected. Thus $T_1(x, t) = T_s$, and the formulation is reduced to that of the condensate, for which (leaving out the subscript) we have

$$\frac{\partial T}{\partial t} = a \frac{\partial^2 T}{\partial x^2},$$

$$T(x, 0) = T_s,$$

$$+k \frac{\partial T(0, t)}{\partial x} = h[T(0, t) - T_\infty], \qquad (2\text{-}219)$$

$$T(X, t) = T_s,$$

$$-k \frac{\partial T(X, t)}{\partial x} + \rho h_{vc} \frac{dX}{dt} = 0.$$

Even the solution of this simplified formulation involves mathematical difficulties.

Let us next consider a quasi-steady formulation as a further simplification. Given a small rate of condensation, which is often the case, the time rate of change of internal energy of the condensate may be neglected. Thus the previous formulation is reduced to

$$0 = \frac{\partial^2 T}{\partial x^2}, \tag{2-220}$$

$$+k\frac{\partial T(0, t)}{\partial x} = h[T(0, t) - T_\infty], \tag{2-221}$$

$$T(X, t) = T_s, \tag{2-222}$$

$$-k\frac{\partial T(X, t)}{\partial x} + \rho h_{vc}\frac{dX}{dt} = 0. \tag{2-223}$$

Note that the unsteadiness of this formulation is associated only with the boundary condition given by Eq. (2-223).

The solution of the foregoing case may be readily obtained. Solving for T from Eqs. (2-220), (2-221), and (2-222), then inserting the result into Eq. (2-223) yields

$$\rho h_{vc}\frac{dX}{dt} = \frac{T_s - T_\infty}{X/k + 1/h}, \tag{2-224}$$

or

$$\frac{1}{k} X \, dX + \frac{1}{h} dX = \left(\frac{T_s - T_\infty}{\rho h_{vc}}\right) dt, \tag{2-225}$$

subject to

$$X(0) = 0. \tag{2-226}$$

The solution of Eq. (2-225) which satisfies Eq. (2-226) is

$$\frac{X(t)}{k/h} = -1 + \left[1 + \frac{2h^2(T_s - T_\infty)t}{k\rho h_{vc}}\right]^{1/2}. \tag{2-227}$$

Our final simplification is to eliminate the temperature drop across the condensate. This is a valid assumption when k is large or t is small. Hence, leaving out the thermal resistance X/k of the condensate in Eq. (2-224), we have

$$\rho h_{vc}\frac{dX}{dt} = \frac{(T_s - T_\infty)}{1/h}, \tag{2-228}$$

which satisfies Eq. (2-226) and integrates to

$$X(t) = \frac{h(T_s - T_\infty)}{\rho h_{vc}} t. \tag{2-229}$$

[What is the limiting form of Eq. (2-227) as $k \to \infty$ or $t \to 0$?]

References

1. L. PRANDTL and O. G. TIETJENS, *Fundamentals of Hydro- and Aerodynamics.* New York: McGraw-Hill, 1934.
2. H. S. CARSLAW and J. C. JAEGER, *Conduction of Heat in Solids.* Oxford: Clarendon Press, 1959.
3. R. B. BIRD, W. E. STEWART, and E. N. LIGHTFOOT, *Transport Phenomena.* New York: Wiley, 1960.
4. A. H. SHAPIRO, Class Notes on "Advanced Fluid Mechanics." MIT, 1956.
5. A. H. SHAPIRO, *The Dynamics and Thermodynamics of Compressible Flow.* New York: The Ronald Press, 1953.
6. J. H. KEENAN, Class Notes on "Advanced Engineering Thermodynamics." MIT, 1955.
7. J. H. KEENAN, *Thermodynamics.* New York: Wiley, 1941.
8. M. W. ZEMANSKY, *Heat and Thermodynamics.* New York: McGraw-Hill, 1957.
9. G. J. VAN WYLEN, *Thermodynamics.* New York: Wiley, 1959.
10. J. C. HUNSAKER and B. G. RIGHTMIRE, *Engineering Applications of Fluid Mechanics.* New York: McGraw-Hill, 1947.
11. I. H. SHAMES, *Mechanics of Fluids.* New York: McGraw-Hill, 1962.
12. R. ARIS, *Vectors, Tensors, and the Basic Equations of Fluid Mechanics.* Englewood Cliffs: Prentice-Hall, 1962.
13. M. JAKOB and G. A. HAWKINS, *Elements of Heat Transfer.* New York: Wiley, 1957.
14. M. JAKOB, *Heat Transfer I.* New York: Wiley, 1949.
15. T. N. CETINKALE and M. FISHENDEN, "Thermal Conductance of Metal Surfaces in Contact," *General Discussion on Heat Transfer.* IME, London and ASME, New York, 271 (1951).
16. H. FENECH and W. M. ROHSENOW, "Prediction of Thermal Conductance of Metallic Surfaces in Contact." *Trans. ASME, C, Journal of Heat Transfer,* **85,** 15 (1963).
17. A. L. LONDON and R. A. SEBAN, "Rate of Ice Formation." *Trans. ASME,* **65,** 771 (1943).
18. M. A. BIOT, "New Methods in Heat Flow Analysis with Application to Flight Structures." *Journal of Aeronautical Sciences,* **24,** 857 (1957).

Problems

2-1. A chamber contains liquid and vapor water in equilibrium at a temperature a little less than the critical temperature. One pound of liquid is withdrawn through a valve at the bottom of the chamber while the chamber is surrounded by a constant-temperature bath which prevents the temperature of the contents from changing from its initial value (Fig. 2-48). The specific volume of the liquid is not negligible as compared with that of the vapor. (a) Find an expression for the increase in volume of

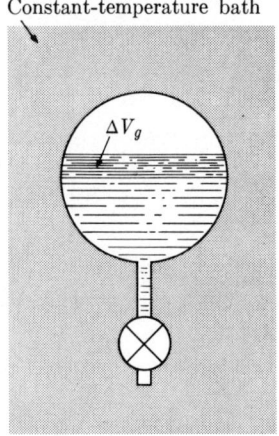

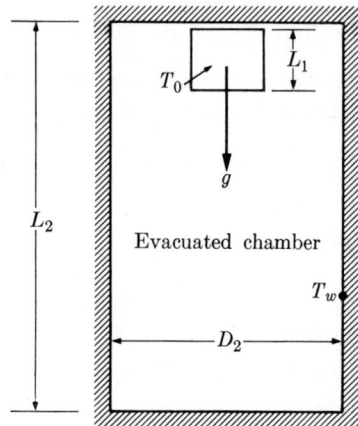

FIG. 2-48   FIG. 2-49

the vapor phase in the chamber, in terms only of saturation properties of liquid and vapor water. (b) Find an expression for the heat transferred from the bath to the contents of the chamber as a result of the withdrawal of the liquid, in terms only of saturation properties.

2-2. A square copper plate of thickness δ having the initial temperature T_0 is dropped into an evacuated vertical chamber whose walls are maintained at the constant temperature $T_w(\gg T_0)$ (Fig. 2-49). Using the data given below, compute the temperature of the plate when it reaches the bottom of the chamber.

$$
\begin{aligned}
&L_1 = 6 \text{ in.} & &\delta = \tfrac{1}{4} \text{ in.} \\
&L_2 = 40 \text{ ft} & &D_2 = 10 \text{ ft} \\
\text{copper} &\begin{cases} \rho_1 = 500 \text{ lbm/ft}^3 \\ c_1 = 0.1 \text{ Btu/lbm} \cdot {}^\circ\text{F} \end{cases} & &T_w = 2000{}^\circ\text{R} \\
& & &T_0 = 60{}^\circ\text{F} \\
&\epsilon_1 = 0.8 & &g = 32.2 \text{ ft}^2/\text{sec} \\
&\epsilon_2 = 0.4 & &
\end{aligned}
$$

2-3. A solid rod moving through a tube melts as a result of the uniform heat flux q'' applied peripherally to the tube, and assumes a parabolic velocity profile (Fig. 2-50). The density ρ of the solid is approximately equal to the density of the fluid. The velocity of the solid is V, the temperature of the solid at the inlet is T_0, the latent heat of melting is h_{fs}, and the specific heat of the solid and fluid are c_s, c_f, respectively. The friction between the solid rod and the tube may be neglected. The axial conduction is negligible compared to the enthalpy flow in both the solid and the fluid. Assuming a parabolic radial temperature distribution for the fluid, find this distribution at the distance L from the inlet of the tube.

2-4. Consider the steady one-dimensional flows of a frictionless incompressible fluid through a constant-area tube and a diffuser of the same length (Fig. 2-51). The tube and diffuser are subjected peripherally to the same uniform heat flux q''. The

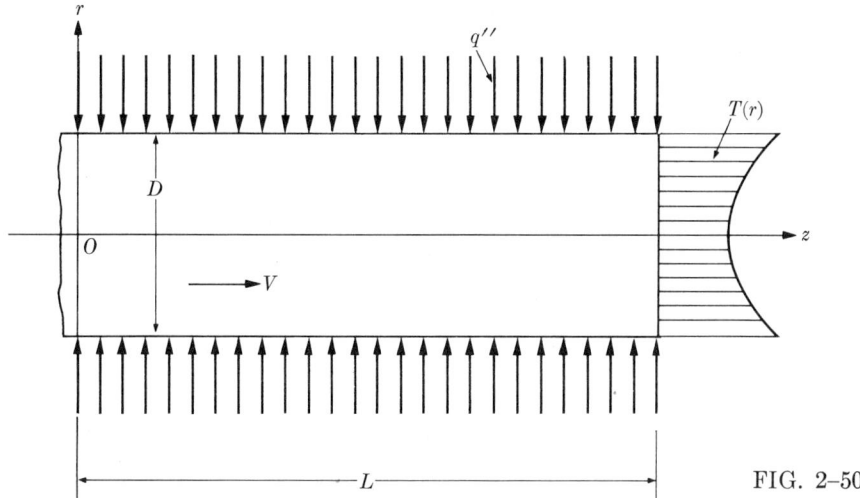

FIG. 2-50

inlet diameter and inlet velocity of the diffuser are identical to those of the tube. (a) Is the exit temperature of the diffuser higher or lower than that of the tube? Base your statement on physical reasoning rather than mathematics. (b) Support your argument with a simple analysis in which the axial conduction may be neglected.

2-5. Reconsider Example 2-5. Assume that the vapor is at the saturation temperature and that the temperature of the condensate may be lumped. Apply the first law of thermodynamics directly to the condensate. Express the heat loss from the bottom of the condensate in terms of a heat transfer coefficient. Compare the result with Eq. (2-228).

2-6. Consider a three-dimensional, differential control volume in cartesian, cylindrical, and spherical coordinates. (a) Obtain the special forms of ∇T and $\nabla^2 T$ in these coordinates. (Here T denotes any scalar property.) (b) Derive the equation of conduction for a homogeneous, isotropic, frictionless incompressible fluid by following the five basic steps of formulation. (c) Utilizing (a), write the cartesian, cylindrical, and spherical forms of Eq. (2-91). (d) Compare the results of (b) and (c).

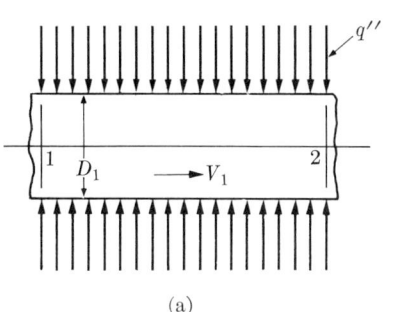
(a)

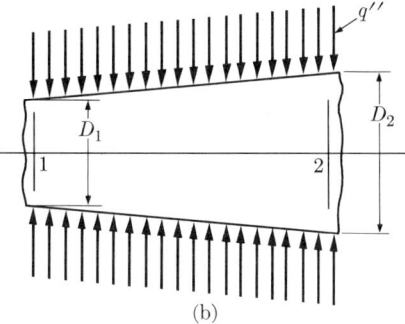
(b)

FIG. 2-51

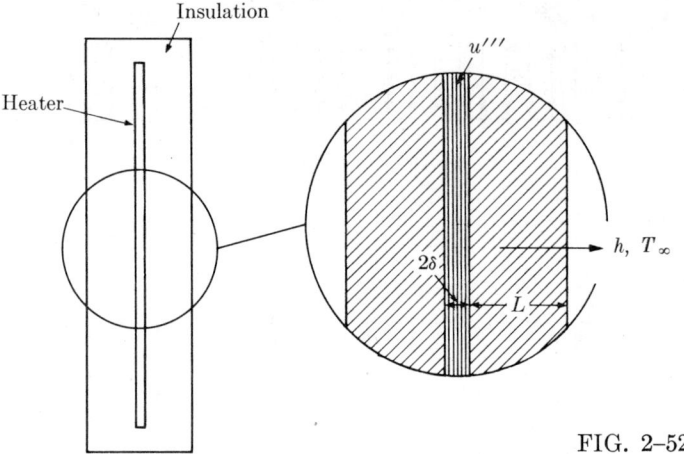

FIG. 2-52

2-7. Reconsider Example 2-1. Using the integral form of the equation of conduction and assuming (a) circular (trigonometric) functions in both directions, (b) a circular function for the shorter side but an unspecified parameter function for the longer side, find the steady temperature of the heater.

2-8. Derive the lumped and integral formulations of Example 2-3 starting from its differential formulation.

Hint: For the integral formulation suitable to the first time domain, integrate the differential formulation over the interval $(0, \tau_0)$; rearrange the result by *Leibnitz's integral formula,*

$$\frac{d}{dx}\int_{a(x)}^{b(x)} F(x, \xi)\, d\xi = \int_{a(x)}^{b(x)} \frac{\partial F(x, \xi)}{\partial x}\, d\xi + F(x, b)\frac{db}{dx} - F(x, a)\frac{da}{dx}.$$

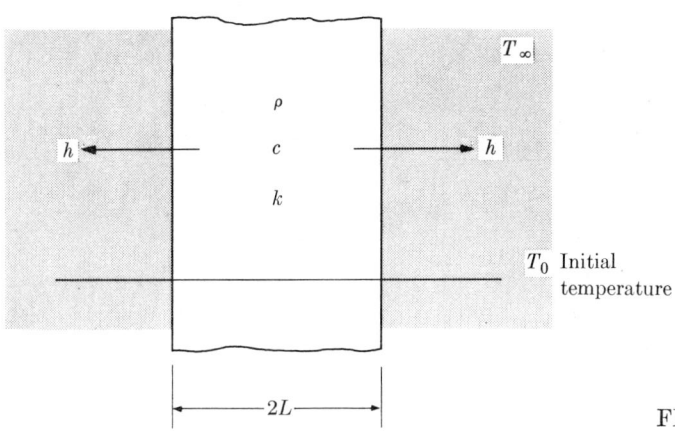

FIG. 2-53

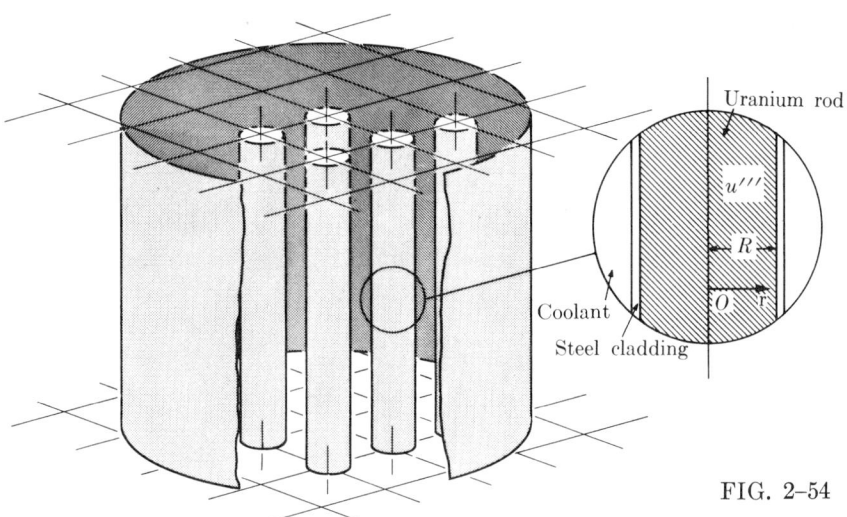

FIG. 2-54

2-9. Internal energy is to be generated electrically in a flat metal plate of thickness 2δ for heating purposes. To obtain low surface temperatures, and for electrical insulation, this plate is covered by an electrical insulator of thickness L which is also a poor thermal conductor (Fig. 2-52). The heater is at the ambient temperature T_∞ initially; then the internal energy u''' is suddenly generated in the heater. Make a one-dimensional analysis.

2-10. A flat plate of thickness $2L$ having the uniform initial temperature T_0 is plunged suddenly into a bath at constant temperature T_∞ (Fig. 2-53). The heat transfer coefficient is h. The density, heat capacity, and thermal conductivity of the plate are ρ, c and k, respectively. Formulate the problem.

2-11. The core of a pool reactor is made of cylindrical fuel elements, each composed of a uranium rod of radius R and a stainless steel cladding of negligible thickness (Fig. 2-54). The reactor has the uniform temperature T_∞ initially; then internal energy is assumed suddenly to be generated in the uranium rods as

$$u'''/u_0''' = 1 - (r/R)^2,$$

where u_0''' is the internal energy generation at the center line. The temperature of the coolant is held constant at T_∞. The heat transfer coefficient is large. Formulate the problem.

2-12. A solid sphere of radius R having the initial temperature T_0 is dropped into boiling water at temperature T_∞. Formulate the problem.

2-13. Reformulate the unsteady problem associated with Example 2-1 for a large heat transfer coefficient. Assume that the heater is at the ambient temperature T_∞ initially and that the uniform internal energy u''' is suddenly generated in the heater.

2-14. Consider the lumped formulation of Examples 2-2 and 2-3. Evaluate the total entropy generation in these examples.

96 LUMPED, INTEGRAL, DIFFERENTIAL FORMULATIONS

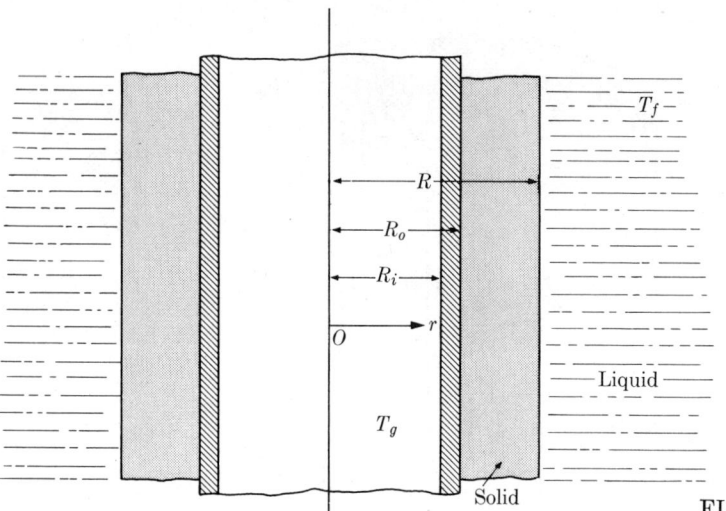

FIG. 2-55

2-15. A liquid at the freezing temperature T_f is being frozen around a tube cooled by an internal flow of gas at temperature T_g (Fig. 2-55). Formulate the problem in terms of the instantaneous location R of the cylindrical solid-liquid interface and the other parameters involved.*

2-16. Reconsider the differential formulation of Example 2-5 given by Eq. (2-219). (a) Find the corresponding integral formulation by following the physical approach. (b) Obtain an approximate solution of the problem in terms of polynomials.

2-17. Consider the unsteady one-dimensional flow of a frictionless incompressible fluid through a diffuser. The diffuser is subjected to the uniform peripheral heat flux q'' (Fig. 2-56). Find the local and total entropy generation in this flow, assuming that the axial conduction is (a) negligible, and (b) not negligible.

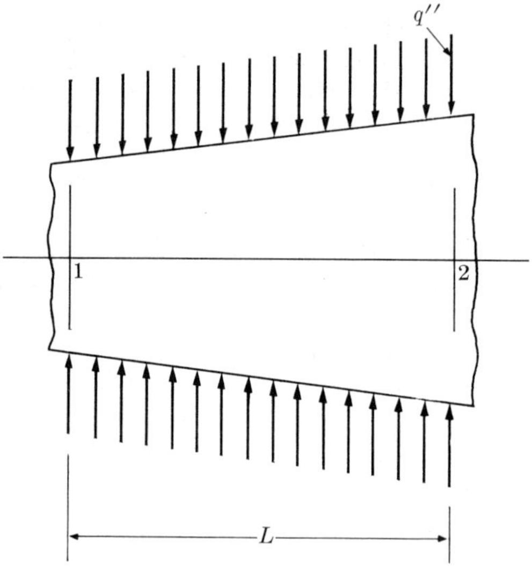

2-18. Consider a two-dimensional homogeneous anisotropic body, say a flat plate of laminated material as shown in Fig. 2-57. The laminations of the plate make an angle α with the normal n to the surfaces. Assume that the maximum and minimum values of the thermal conductivity, say k_ξ and k_η, are in the directions of ξ and η, respectively. (These directions and the corresponding values of the conductivity are called the *principal axes* and *the principal values*, respectively. It may be shown that the values of the conductivity in other directions vary as an

FIG. 2-56

---

* See Reference 17.

ellipse whose axes are the principal values of the conductivity.) The components of the heat flux vector in ξ and η are

$$q_\xi = -k_\xi \frac{\partial T}{\partial \xi} \quad \text{and} \quad q_\eta = -k_\eta \frac{\partial T}{\partial \eta}.$$

(a) Show that the components of the heat flux vector in x and y are

$$q_x = -k_\xi \sin\alpha \frac{\partial T}{\partial \xi} + k_\eta \cos\alpha \frac{\partial T}{\partial \eta},$$

$$q_y = -k_\xi \cos\alpha \frac{\partial T}{\partial \xi} - k_\eta \sin\alpha \frac{\partial T}{\partial \eta}.$$

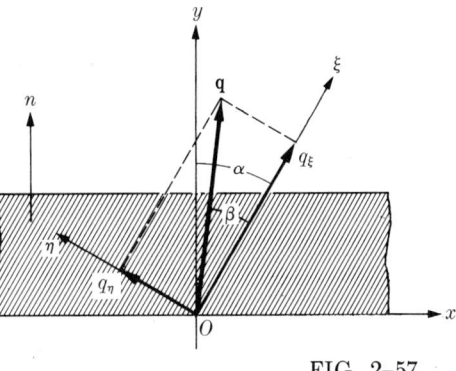

FIG. 2-57

(b) Express q_x and q_y in terms of $\partial T/\partial x$ and $\partial T/\partial y$, and rearrange the result according to the geometric considerations so that

$$q_x = -(k_\xi \sin^2\alpha + k_\eta \cos^2\alpha)\frac{\partial T}{\partial x} - (k_\xi - k_\eta)\sin\alpha\cos\alpha\frac{\partial T}{\partial y},$$

$$q_y = -(k_\xi - k_\eta)\cos\alpha\sin\alpha\frac{\partial T}{\partial x} - (k_\xi \cos^2\alpha + k_\eta \sin^2\alpha)\frac{\partial T}{\partial y}.$$

(c) Show that the two-dimensional conduction equation for homogeneous anisotropic materials is

$$\rho c \frac{\partial T}{\partial t} = (k_\xi \sin^2\alpha + k_\eta \cos^2\alpha)\frac{\partial^2 T}{\partial x^2} + (k_\xi \cos^2\alpha + k_\eta \sin^2\alpha)\frac{\partial^2 T}{\partial y^2}$$

$$+ (k_\xi - k_\eta)\cos 2\alpha \frac{\partial^2 T}{\partial x\, \partial y}.$$

(d) What is the special form of the foregoing equation corresponding to homogeneous isotropic materials? (e) What value for the conductivity would be obtained if the plate were held between the isothermal surfaces of a conductivity test apparatus? (f) Show that under the conditions of part (e),

$$\tan\beta = \frac{q_\eta}{q_\xi} = \frac{k_\eta(\partial T/\partial \eta)}{k_\xi(\partial T/\partial \xi)} = \frac{k_\eta[\sin\alpha(\partial T/\partial y) - \cos\alpha(\partial T/\partial x)]}{k_\xi[\cos\alpha(\partial T/\partial y) + \sin\alpha(\partial T/\partial x)]};$$

show also that given $\partial T/\partial x = 0$,

$$\tan\beta = \frac{k_\eta}{k_\xi}\tan\alpha,$$

which implies $\beta < \alpha$, and that the heat flux vector is no longer normal to isothermal surfaces as it would be if the material were isotropic.

2-19. A frictionless incompressible fluid having the velocity U flows steadily through a tube (Fig. 2-58). The inlet temperature of the fluid is T_0. The initial temperatures of the tube and fluid are the same and equal to the inlet temperature of the fluid. The uniform heat flux q'' is suddenly applied to the outer surface of the tube.

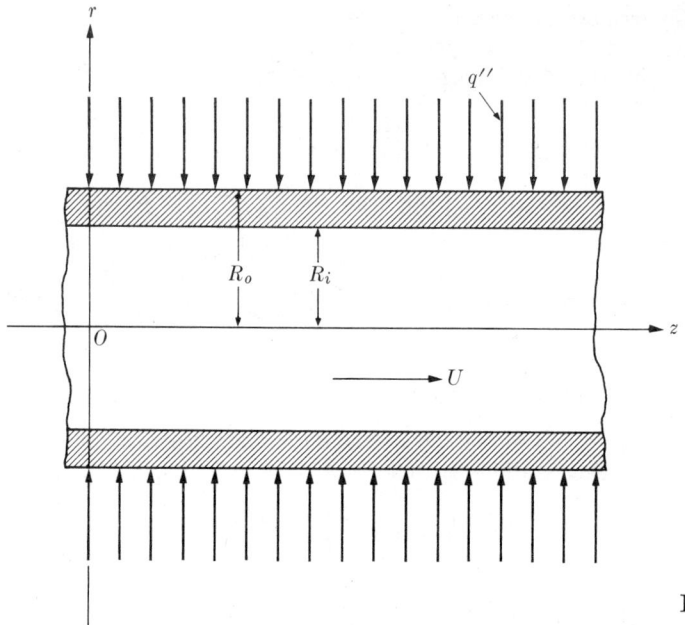

FIG. 2-58

The axial conduction is negligible. Give the differential formulation of the problem, employing only one of the following assumptions at a time: (i) None. (ii) The thickness of the tube is negligible. (iii) The thickness of the tube is small. (iv) The fluid is radially lumped, and the heat transfer coefficient is (a) large, or (b) moderate. (v) The fluid and tube are radially lumped.*

2–20. A frictionless incompressible fluid having the velocity U_∞ flows steadily over a semi-infinite plate of negligible thickness (Fig. 2–59). The temperatures of the fluid and the wall are T_∞ and T_w, respectively. Find the thickness of the steady penetration depth (boundary layer) of the temperature depending on x and the other variables involved.

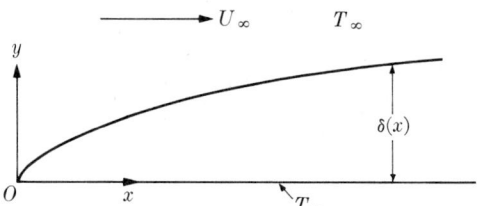

FIG. 2-59

---

* Note that a formulation based on assumption (i), although it is more exact, is difficult to solve. Formulations based on (ii) and (iii), without any or with only partial effect of the tube wall, are models for fluid flow problems. On the other hand, formulations using (iv-a) and (iv-b), allowing radial variation in the tube but not in the fluid, become convenient for thermal stress problems. The formulation based on (v), being completely lumped in the radial direction, is the simplest model. This may be utilized to evaluate the bulk temperature of the fluid at a desired location.

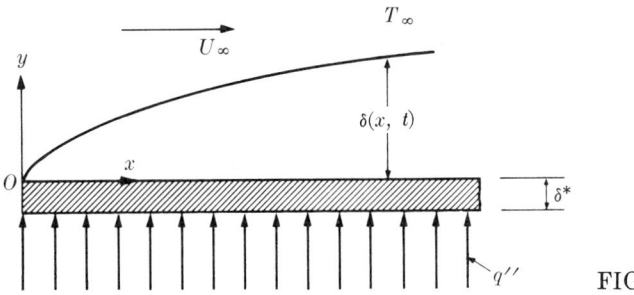

FIG. 2-60

2-21. A frictionless incompressible fluid having the velocity U_∞ flows steadily over a thin semi-infinite plate of thickness δ^* (Fig. 2-60). The initial temperatures of the fluid and the wall are the same, T_∞. Then the uniform heat flux q'' is suddenly applied to the bottom of the plate. Find the unsteady penetration of the heat flux into the fluid.

PART II | SOLUTION

CHAPTER 3

STEADY ONE-DIMENSIONAL PROBLEMS. BESSEL FUNCTIONS

The lumped and integral formulations of steady one-dimensional problems involve the solving of algebraic equations only. The differential formulation of the same problems, on the other hand, involves the solving of first- or second-order differential equations. As we mentioned in Chapter 2, methods for solving linear differential equations of first order, and of second order with constant coefficients, are assumed to be known to the reader. Hence for steady one-dimensional problems resulting in such equations we shall give the solution. For problems of temperature-dependent thermal conductivity the differential formulation yields a nonlinear differential equation of the second order, and for problems of space-dependent thermal conductivity, as well as for those of extended surfaces with variable cross sections, the differential formulation results in a linear differential equation of second order with variable coefficients. In Sections 3–5 and 3–6, we shall introduce the methods appropriate to solution of both these types of equations. First, however, we shall consider a general problem in which a number of important concepts are defined.

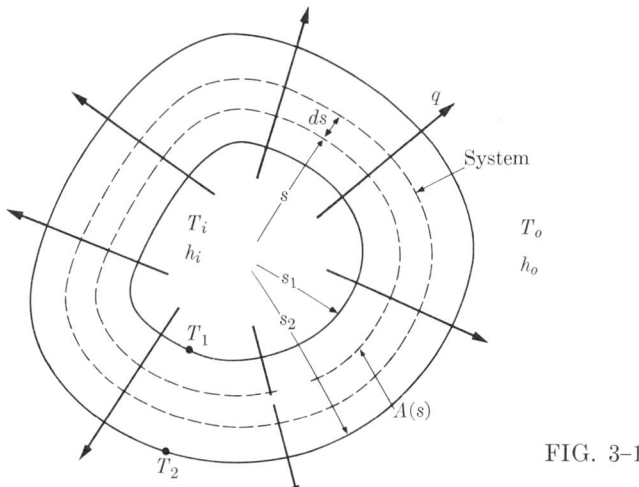

FIG. 3–1

3–1. A General Problem

Consider a long, hollow cylinder or a thick-walled, closed shell of constant wall thickness whose cross section is shown in Fig. 3–1. This cylinder or shell contains a fluid at temperature T_i and is surrounded by an ambient at temperature T_o. Let us suppose that $T_i > T_o$. The inside and outside heat transfer coefficients are h_i and h_o, respectively. We wish to know the temperature distribution of and the heat transfer through this cylinder or shell.

The method of solution employed here is convenient for one-dimensional problems in which $q = $ const at every cross section. The first two steps of the formulation (see Section 2–9) result in

$$d[q_s A(s)] = 0, \qquad (3\text{--}1)$$

which readily gives

$$q = q_s A(s) = \text{const}, \qquad (3\text{--}2)$$

where s denotes the space variable and $A(s)$ the corresponding heat transfer area. The third step of the formulation is

$$q_s = -k \frac{dT}{ds}. \qquad (3\text{--}3)$$

The fourth step is the introduction of Eq. (3–3) into Eq. (3–2). Rearranging and integrating the result between the inner and outer surfaces with the assumption that k is constant, we get

$$q = \frac{T_1 - T_2}{(1/k)\int_{s_1}^{s_2} ds/A(s)} = \frac{T_1 - T_2}{R}, \qquad (3\text{--}4)$$

where

$$R = \frac{1}{k} \int_{s_1}^{s_2} \frac{ds}{A(s)} \qquad (3\text{--}5)$$

is the so-called *conductive resistance*, and the subscripts 1 and 2 refer to the inner and outer surfaces, respectively. Equation (3–4) is analogous to Ohm's law,

$$i = \frac{E_1 - E_2}{R_e},$$

for steady electric current; the conduction heat transfer q corresponds to the electric current i, the temperature drop $T_1 - T_2$ to the potential drop $E_1 - E_2$, and the conductive resistance R to the electrical resistance R_e. (What is the differential form of Ohm's law?)

Since ambient temperatures are more easily measured than surface temperatures, it is convenient to express q in terms of the ambient temperature. From the last step of the formulation, we have

$$q = \frac{T_i - T_1}{1/h_i A(s_1)} = \frac{T_i - T_1}{R_i}, \qquad (3\text{--}6)$$

where

$$R_i = \frac{1}{h_i A(s_1)} \qquad (3\text{--}7)$$

is the *convective resistance* between the inside fluid and the inner surface of the wall; similarly, we have

$$q = \frac{T_2 - T_o}{1/h_o A(s_2)} = \frac{T_2 - T_o}{R_o}, \qquad (3\text{--}8)$$

where
$$R_o = \frac{1}{h_o A(s_2)} \qquad (3\text{-}9)$$

is the outside convective resistance. Solving Eqs. (3–4), (3–6), and (3–8) for their respective temperature differences, then adding the results side by side eliminates T_1 and T_2, and yields

$$q = \frac{T_i - T_o}{R_i + R + R_o}. \qquad (3\text{-}10)$$

The same result may be readily obtained by considering the analogy between the diffusion of heat and electric current. Thus the problem becomes analogous to the evaluation of an electric current through three resistors connected in series (Fig. 3–2).

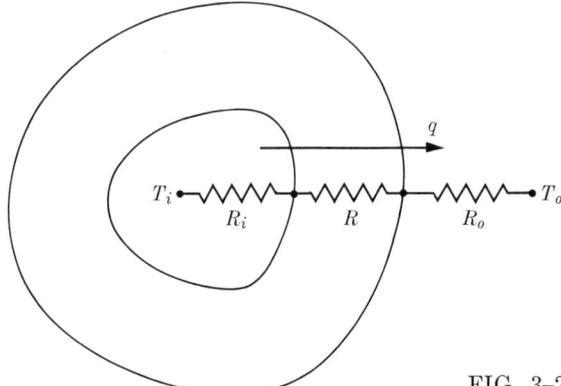

FIG. 3–2

It is sometimes convenient to simplify Eq. (3–10) by writing it in terms of the so-called *over-all coefficient of heat transfer* U, which is defined according to

$$q = UA(T_i - T_o), \qquad (3\text{-}11)$$

and

$$\frac{1}{UA} = R_i + R + R_o. \qquad (3\text{-}12)$$

Since U depends on A, the statement of U is ambiguous until an area is chosen. Noting that

$$UA = U_i A(s_1) = U_o A(s_2),$$

where U_i and U_o denote the over-all heat transfer coefficients based on the inner and outer surface areas, respectively, we may write the outside coefficient U_o, for example, as

$$\frac{1}{U_o} = \frac{A(s_2)/A(s_1)}{h_i} + \frac{A(s_2)}{k}\int_{s_1}^{s_2}\frac{ds}{A(s)} + \frac{1}{h_o}. \qquad (3\text{-}13)$$

To determine the temperature distribution of the problem we reconsider Eq. (3–4) now integrated from an arbitrary location s to the outer surface* in the form

$$q = \frac{T - T_2}{(1/k)\int_s^{s_2} ds/A(s)}, \qquad (3\text{–}14)$$

where T is the temperature of the location s. Elimination of T_2 between Eqs. (3–8) and (3–14) in the usual manner yields

$$q = \frac{T - T_o}{(1/k)\int_s^{s_2} ds/A(s) + 1/h_o A(s)}. \qquad (3\text{–}15)$$

Finally, equating Eq. (3–11) to Eq. (3–13) and rearranging the result in terms of U_o, we obtain the desired temperature distribution:

$$\frac{T - T_o}{T_i - T_o} = U_o \left[\frac{A(s_2)}{k}\int_s^{s_2}\frac{ds}{A(s)} + \frac{1}{h_o}\right]. \qquad (3\text{–}16)$$

An alternative method of solution which leads first to the temperature distribution and then to the heat transfer may be obtained from the formal formulation of the problem. Thus the governing equation is found, from the first four steps of the formulation, to be

$$\frac{d}{ds}\left[A(s)\frac{dT}{ds}\right] = 0, \qquad (3\text{–}17)$$

subject to the boundary conditions

$$\begin{aligned}+k\frac{dT(s_1)}{ds} &= h_i[T(s_1) - T_i], \\ -k\frac{dT(s_2)}{ds} &= h_o[T(s_2) - T_o],\end{aligned} \qquad (3\text{–}18)$$

obtained from the last step of the formulation. Substituting the solution of Eq. (3–17),

$$T = C\int^s \frac{ds}{A(s)} + D, \qquad (3\text{–}19)$$

into Eq. (3–18) results in

$$\begin{aligned}\frac{kC}{A(s_1)} &= h_i\left[C\int^{s_1}\frac{ds}{A(s)} + D - T_i\right], \\ -\frac{kC}{A(s_2)} &= h_o\left[C\int^{s_2}\frac{ds}{A(s)} + D - T_o\right].\end{aligned} \qquad (3\text{–}20)$$

---

*An alternative expression for the temperature distribution is obtained by carrying this integration out from the inner surface to the arbitrary location s.

The values of C and D from Eq. (3-20), introduced into Eq. (3-19), give the previously obtained temperature distribution, Eq. (3-16). The heat transfer may now be found by inserting Eq. (3-16) into the combination of Eqs. (3-2) and (3-3); this yields the previous result, Eq. (3-10).

For convenience, we shall apply the procedure of the foregoing problem to three important cases, the cartesian, cylindrical, and spherical geometries. The over-all heat transfer coefficients based on the outer surface area and the temperature distributions of these geometries are:

$$\frac{1}{U_o} = \frac{1}{U} = \frac{1}{h_i} + \frac{L}{k} + \frac{1}{h_o}, \qquad \text{cartesian,} \qquad (3\text{-}21)$$

$$\frac{1}{U_o} = \frac{(R_2/R_1)}{h_i} + \frac{R_2}{k}\ln\left(\frac{R_2}{R_1}\right) + \frac{1}{h_o}, \qquad \text{cylindrical,} \qquad (3\text{-}22)$$

$$\frac{1}{U_o} = \frac{(R_2/R_1)^2}{h_i} + \frac{R_2}{k}\left(\frac{R_2}{R_1} - 1\right) + \frac{1}{h_o}, \qquad \text{spherical,} \qquad (3\text{-}23)$$

$$\frac{T - T_o}{T_i - T_o} = U_o\left(\frac{x_2 - x}{L} + \frac{1}{h_o}\right), \qquad \text{cartesian,} \qquad (3\text{-}24)$$

$$\frac{T - T_o}{T_i - T_o} = U_o\left[\frac{R_2}{k}\ln\left(\frac{R_2}{r}\right) + \frac{1}{h_o}\right], \qquad \text{cylindrical,} \qquad (3\text{-}25)$$

$$\frac{T - T_o}{T_i - T_o} = U_o\left[\frac{R_2}{k}\left(\frac{R_2}{r} - 1\right) + \frac{1}{h_o}\right], \qquad \text{spherical.} \qquad (3\text{-}26)$$

3-2. Composite Structures

Assume that the hollow cylinder or the thick-walled, closed shell of Fig. 3-1 is composed of N layers of materials having different thicknesses and thermal conductivities (Fig. 3-3). The contact resistance between the layers is negligible. We wish to find the heat transfer from the inner fluid to the surrounding ambient, and the temperature distribution of the structure.

Extending the analogy between the diffusion of heat and electric current to the present case, we readily obtain

$$\frac{1}{UA} = R_i + \sum_{n=1}^{N} R_n + R_o. \qquad (3\text{-}27)$$

The explicit form of U based on the outside surface is

$$\frac{1}{U_o} = \frac{A(s_{N+1})/A(s_1)}{h_i} + A(s_{N+1}) \sum_{n=1}^{N} \frac{1}{k_n} \int_{s_n}^{s_{n+1}} \frac{ds}{A(s)} + \frac{1}{h_o}. \qquad (3\text{-}28)$$

Equation (3-28) reduces to Eq. (3-13) for $N = 1$.

To obtain the temperature distribution in the structure we first express q in terms of the temperature difference $T - T_o$ and the corresponding resistances (from the series R_n, R_{n+1}, R_{n+2}, ..., R_N). The result is

$$q = \frac{T - T_o}{(1/k_n)\int_s^{s_{n+1}} ds/A(s) + \sum_{m=n+1}^{N} (1/k_m)\int_{s_m}^{s_{m+1}} ds/A(s) + 1/h_o A(s_{N+1})}, \quad (3\text{--}29)$$

where T denotes the temperature of the location s (Fig. 3–3). Then eliminating q between Eqs. (3–11) and (3–29), we find that the desired temperature distribution in terms of U_o is

$$\frac{T - T_o}{T_i - T_o} = U_o \left[\frac{A(s_{N+1})}{k_n} \int_s^{s_{n+1}} \frac{ds}{A(s)} + A(s_{N+1}) \sum_{m=n+1}^{N} \frac{1}{k_m} \int_{s_m}^{s_{m+1}} \frac{ds}{A(s)} + \frac{1}{h_o} \right]. \quad (3\text{--}30)$$

Equation (3–30) reduces to Eq. (3–16) for $N = 1$.

The cartesian, cylindrical, and spherical forms of Eqs. (3–28) and (3–30) are listed below:

$$\frac{1}{U_o} = \frac{1}{U} = \frac{1}{h_i} + \sum_{n=1}^{N} \frac{L_n}{k_n} + \frac{1}{h_o}, \quad \text{cartesian,} \quad (3\text{--}31)$$

$$\frac{1}{U_o} = \frac{(R_{N+1}/R_1)}{h_i} + R_{N+1} \sum_{n=1}^{N} \frac{1}{k_n} \ln\left(\frac{R_{n+1}}{R_n}\right) + \frac{1}{h_o}, \quad \text{cylindrical,} \quad (3\text{--}32)$$

$$\frac{1}{U_o} = \frac{(R_{N+1}/R_1)^2}{h_i} + R_{N+1}^2 \sum_{n=1}^{N} \frac{1}{k_n}\left(\frac{1}{R_n} - \frac{1}{R_{n+1}}\right) + \frac{1}{h_o}, \quad \text{spherical,} \quad (3\text{--}33)$$

$$\frac{T - T_o}{T_i - T_o} = U_o \left[\frac{x_{n+1} - x}{k_n} + \sum_{m=n+1}^{N} \frac{x_{m+1} - x_m}{k_m} + \frac{1}{h_o} \right], \quad \text{cartesian,} \quad (3\text{--}34)$$

$$\frac{T - T_o}{T_i - T_o} = U_o \left[\frac{R_{N+1}}{k_n} \ln\left(\frac{R_{n+1}}{r}\right) \right.$$
$$\left. + R_{N+1} \sum_{m=n+1}^{N} \frac{1}{k_m} \ln\left(\frac{R_{m+1}}{R_m}\right) + \frac{1}{h_o} \right], \quad \text{cylindrical,} \quad (3\text{--}35)$$

$$\frac{T - T_o}{T_i - T_o} = U_o \left[\frac{R_{N+1}}{k_n}\left(\frac{R_{N+1}}{r} - \frac{R_{N+1}}{R_{n+1}}\right) \right.$$
$$\left. + R_{N+1}^2 \sum_{m=n+1}^{N} \frac{1}{k_m}\left(\frac{1}{R_m} - \frac{1}{R_{m+1}}\right) + \frac{1}{h_o} \right], \quad \text{spherical.} \quad (3\text{--}36)$$

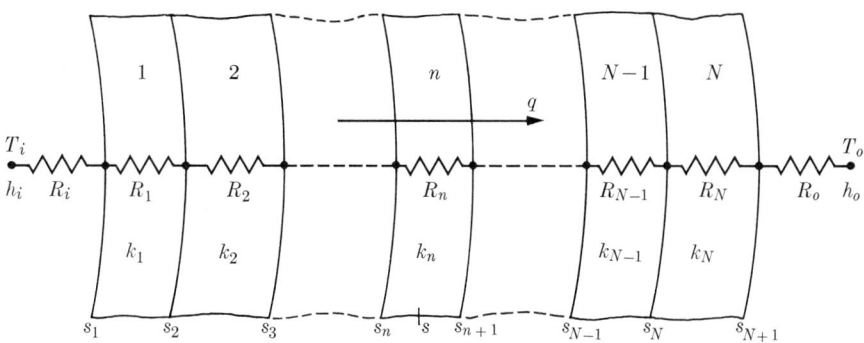

FIG. 3-3

In practice, the combination of series- and parallel-connected structures is also important, especially in cartesian geometry. For example, consider a wall composed of hollow concrete blocks, such as is used in building construction (Fig. 3-4). Actually, the heat transfer through this type of wall is not one-dimensional. However, a one-dimensional analysis gives satisfactory results for practical problems.

Again employing the electrical analogy, we readily obtain

$$\frac{1}{UA} = R_i + R_1 + \frac{1}{1/R_3 + 1/R_4 + 1/R_5} + R_2 + R_o.$$

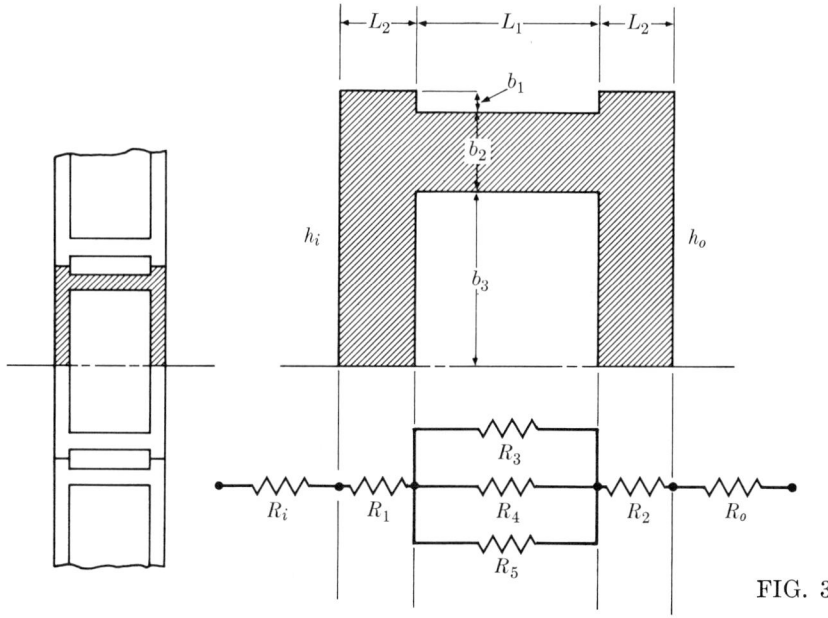

FIG. 3-4

Thus we have, per unit width of the wall perpendicular to the cross section shown in Fig. 3–4,

$$\frac{1}{h_i(b_1+b_2+b_3)} + \frac{L_2}{k_2(b_1+b_2+b_3)} + \frac{1}{k_1b_1/L_1 + k_2b_2/L_1 + k_1b_3/L_1}$$
$$+ \frac{L_2}{k_2(b_1+b_2+b_3)} + \frac{1}{h_o(b_1+b_2+b_3)},$$

and hence per unit area of the wall

$$\frac{1}{U} = \frac{1}{h_i} + \frac{L_2}{k_2} + \frac{L_1}{\epsilon_1 k_1 + \epsilon_2 k_2} + \frac{L_2}{k_2} + \frac{1}{h_o},$$

where

$$\epsilon_1 = (b_1+b_3)/(b_1+b_2+b_3) \quad \text{and} \quad \epsilon_2 = b_2/(b_1+b_2+b_3).$$

3–3. Examples

In this section a number of physical and mathematical facts are demonstrated in terms of examples selected from cartesian, cylindrical, and spherical geometries.

Example 3–1. A flat plate of thickness L separates two media having temperatures T_i and T_o. The heat transfer coefficients are h_i and h_o (Fig. 3–5). We wish to eliminate the heat loss from the ambient of higher temperature.

We learned in Section 3–2 that the total resistance between two ambients can be increased by adding insulation to one or both surfaces of the plate. Regardless of the thickness and material of the insulation, however, the heat loss from the ambient of higher temperature cannot be completely eliminated

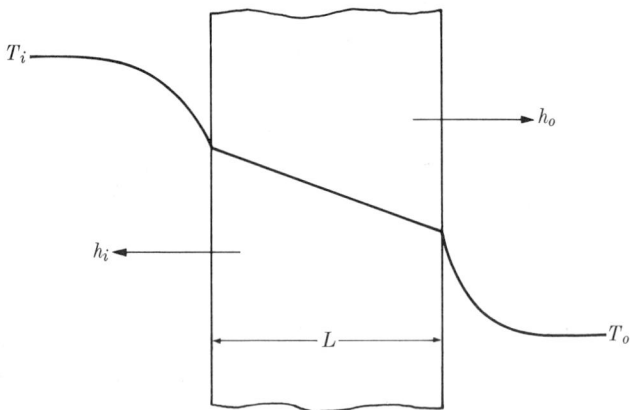

FIG. 3–5

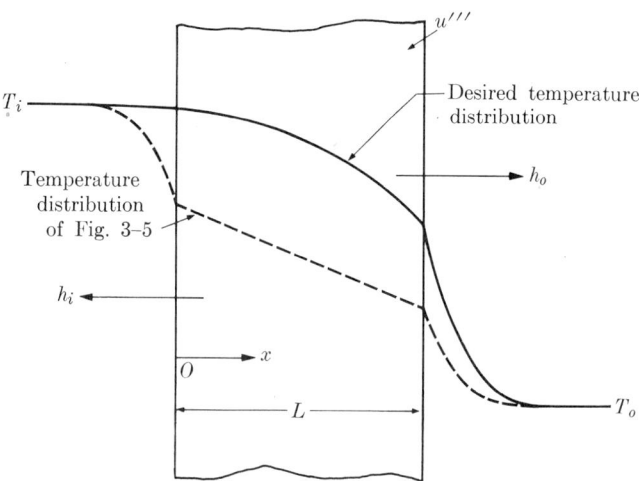

FIG. 3-6

by such insulation. On the other hand, a proper amount of uniform internal energy u''' generated electrically in the plate* may reduce the heat loss to zero (Fig. 3-6). The problem is thus reduced to finding the appropriate value of u'''.

The formulation of the problem in x measured from the left surface of the plate rightward is

$$\frac{d^2T}{dx^2} + \frac{u'''}{k} = 0, \qquad (3\text{-}37)$$

$$T(0) = T_i, \qquad (3\text{-}38)$$

$$\frac{dT(0)}{dx} = 0, \qquad (3\text{-}39)$$

$$-k\frac{dT(L)}{dx} = h_o[T(L) - T_o]. \qquad (3\text{-}40)$$

[Why three boundary conditions rather than two? What are the unknowns to be determined? Describe the physics of the problem as given by Eqs. (3-37), (3-38), (3-40) and as given by Eqs. (3-37), (3-39), (3-40).]

Introducing the solution of Eq. (3-37),

$$T = -\frac{u'''x^2}{2k} + Ax + B, \qquad (3\text{-}41)$$

---

* If the plate is not an electrical conductor or if electric current through the plate is undesirable, another electrically heated plate (guard heater), electrically insulated from but in good thermal contact with the original plate, can be utilized. See also the problem related to Fig. 2-16.

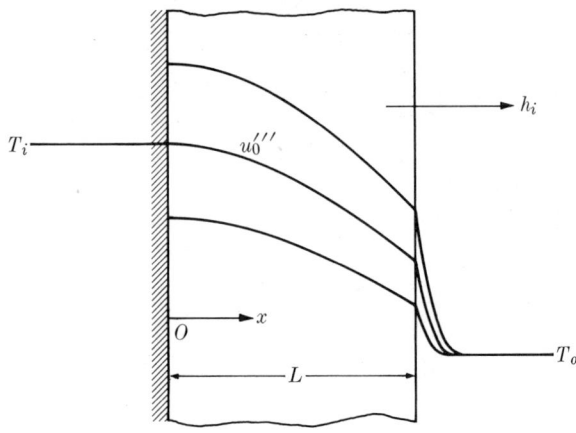

FIG. 3-7

into Eqs. (3-39) and (3-40), we obtain

$$0 = A, \quad u'''L = h_o(-u'''L^2/2k + B - T_o).$$

Thus, by inserting the values of A and B into Eq. (3-41), we find that the temperature distribution corresponding to an arbitrary value of u''' is

$$\frac{T - T_o}{u'''L^2/2k} = 1 - \left(\frac{x}{L}\right)^2 + 2\left(\frac{k}{h_o L}\right). \tag{3-42}$$

A sketch of Eq. (3-42) is given in Fig. 3-7 for various values of u'''. Among these temperature profiles, however, there is only one which satisfies the remaining boundary condition, Eq. (3-38), and which therefore corresponds to the desired value u_o''' of u'''. Thus, combining Eqs. (3-38) and (3-42), we have

$$\frac{u_o'''L^2}{2k} = \frac{T_i - T_o}{1 + 2(k/h_o L)}.$$

[Re-solve this example by considering Eq. (3-39) or Eq. (3-40) as the last boundary condition to be used.]

Example 3-2. The fuel element of a pool reactor is composed of flat plates of thickness $2L_1$ and cladding material of thickness $(L_2 - L_1)$ bonded to the surfaces of these plates (Fig. 3-8). Uniform (nuclear) internal energy u''' is assumed to be generated in the plates only. The heat transfer coefficient is h, the temperature of the coolant T_∞. We need to know the temperature distribution of the fuel element.

This is an example of a *multidomain* problem. The formulation of such problems involves more than one governing equation. Because of the geometric and thermal symmetries, x is measured from the middle plane of the fuel element. Denoting properties of the plate and of the cladding by the subscripts 1 and 2, respectively, we obtain

$$\frac{d^2 T_1}{dx^2} + \frac{u'''}{k_1} = 0, \qquad 0 \leq x \leq L_1, \qquad (3\text{--}43)$$

$$\frac{d^2 T_2}{dx^2} = 0, \qquad L_1 \leq x \leq L_2, \qquad (3\text{--}44)$$

$$\frac{dT_1(0)}{dx} = 0,$$

$$T_1(L_1) = T_2(L_1) = T_{12}, \qquad k_1 \frac{dT_1(L_1)}{dx} = k_2 \frac{dT_2(L_1)}{dx}, \qquad (3\text{--}45)$$

$$-k_2 \frac{dT_2(L_2)}{dx} = h[T_2(L_2) - T_\infty].$$

Introducing the solution of Eq. (3–43),

$$T_1 = -\frac{u''' x^2}{2k_1} + Ax + B, \qquad (3\text{--}46)$$

and that of Eq. (3–44),

$$T_2 = Cx + D, \qquad (3\text{--}47)$$

into the boundary conditions, Eq. (3–45), we obtain the following four simultaneous algebraic equations:

$$A = 0, \qquad -u''' L_1^2 / 2k_1 + B = CL_1 + D,$$
$$-u''' L_1 = k_2 C, \qquad -k_2 C = h(CL_2 + D - T_\infty).$$

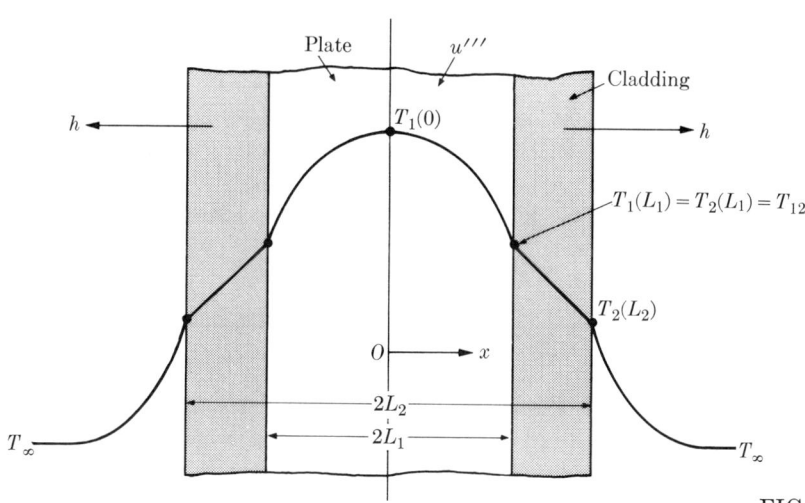

FIG. 3–8

Solving these in terms of A, B, C, and D, then inserting the result into Eqs. (3–46) and (3–47) gives the temperature distribution of the fuel element as

$$\frac{T_1 - T_\infty}{u'''L_1^2/2k_1} = 1 - \left(\frac{x}{L_1}\right)^2 - 2\left(\frac{k_1}{k_2}\right) - 2\left(\frac{L_2}{L_1}\right)\left(\frac{k_1}{k_2}\right)\left(1 + \frac{k_2}{hL_2}\right) \qquad (3\text{–}48)$$

for $0 \le x \le L_1$, and

$$\frac{T_2 - T_\infty}{u'''L_1^2/k_2} = -\left(\frac{x}{L}\right) + \left(\frac{L_2}{L_1}\right)\left(1 + \frac{k_2}{hL_2}\right) \qquad (3\text{–}49)$$

for $L_1 \le x \le L_2$.

When we wish to know only the temperature at some specific locations of the fuel element, say $T_1(0)$, $T_1(L_1) = T_2(L_1) = T_{12}$, and $T_2(L_2)$, the electrical analogy may be employed in place of the foregoing procedure. Thus we have

$$T_2(L_2) - T_\infty = \frac{1}{h}(u'''L_1),$$

$$T_{12} - T_\infty = \left(\frac{1}{h} + \frac{L_2 - L_1}{k}\right)(u'''L_1),$$

where $(u'''L_1)$ is the heat flux across each surface of the fuel element. Furthermore, noting from Eq. (3–27) that

$$T_1(0) - T_{12} = -\frac{u'''L_1^2}{2k_1},$$

we obtain

$$T_1(0) - T_\infty = \left(\frac{L_1}{2k_1} + \frac{1}{h} + \frac{L_2 - L_1}{k_2}\right)(u'''L_1).$$

Example 3–3. A constant-property inviscous liquid having the far upstream temperature T_0 and velocity V flows steadily through an infinitely long tube of cross-sectional area A and periphery P. The wall thickness of the tube is negligible. The downstream half of the tube is subjected to the constant heat flux q''; the upstream half is insulated (Fig. 3–9). We wish to know the axial temperature distribution of the liquid, based on a radially lumped analysis.

This problem will be investigated in detail because it finds considerable application in reactor technology and the computation of heat transfer coefficients in tubes. Let us first clarify an important fact which is not obvious from the statement of the problem. The applied heat flux q'' is going to be axially conducted through the fluid from the heated to the unheated half of the tube, thus giving rise to a temperature distribution in the unheated half. Since the axial enthalpy flow is in the direction opposite to that of the conduction, it has a reverse effect on the temperature rise. Thus the temperature distribution

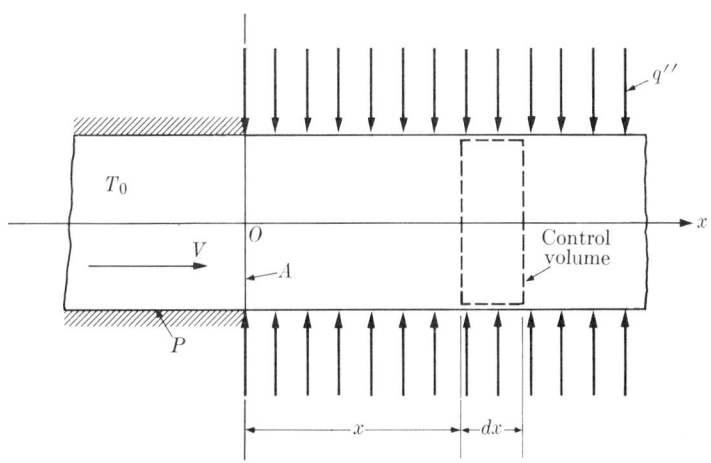

FIG. 3-9

in the tube depends on the relative importance of axial conduction and axial enthalpy flow. Although the fluid constitutes a single domain, the change in the peripheral boundary condition from insulation to constant heat flux suggests that we treat this as a two-domain problem for mathematical convenience. The formulation of the unheated section may be readily obtained by substituting $q'' = 0$ into that of the heated section; therefore, details of the formulation are given only for the heated section.

Consider the radially lumped, axially differential control volume shown in Fig. 3-9. Let us apply the general laws to this control volume as follows.

Conservation of mass:
$$V = \text{const}. \tag{3-50}$$

Momentum (mechanical energy or Bernoulli's equation):*
$$\frac{d}{dx}\left(\frac{p}{\rho} + gz\right) = 0. \tag{3-51}$$

First law of thermodynamics (total energy), in terms of Fig. 3-10:
$$-A\frac{dq_x}{dx} - \rho A V \frac{dh_0}{dx} + q''P = 0. \tag{3-52}$$

It follows from the definition of h_0, Eqs. (3-50) and (3-51), that
$$dh_0/dx = du/dx. \tag{3-53}$$

Neglecting $c_p - c_v$ for liquids, we have, in terms of a common specific heat c,
$$du = c\,dT. \tag{3-54}$$

---

* See the development of Eq. (2-56).

Introducing Eq. (3–54) into Eq. (3–53) and the result into Eq. (3–52), we get for the thermal energy

$$-A\frac{dq_x}{dx} - \rho cAV\frac{dT}{dx} + q''P = 0. \tag{3–55}$$

Finally, Eq. (3–55), rearranged with the x-component of Fourier's law,

$$q_x = -k\frac{dT}{dx},$$

gives the governing equation of the heated section in the following form:

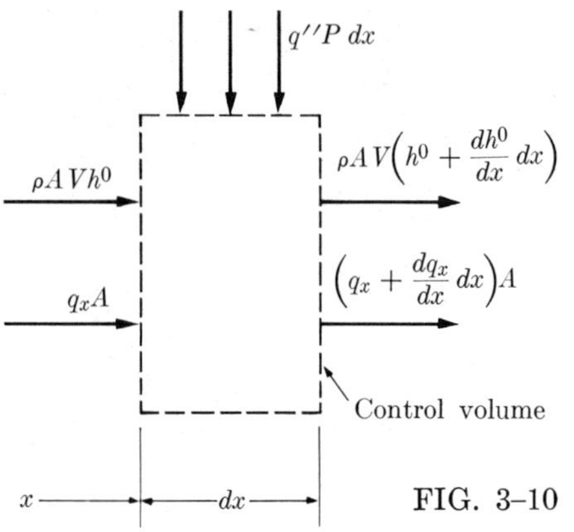

FIG. 3-10

$$\frac{d^2T}{dx^2} - \frac{\rho cV}{k}\frac{dT}{dx} + \frac{q''P}{kA} = 0. \tag{3–56}$$

[Modify Eq. (3–56) assuming that the tube of Fig. 3–9 is vertical. Reformulate the problem for an ideal gas.] With $q'' = 0$, Eq. (3–56) becomes the governing equation of the unheated section.

Distinguishing the temperatures of the unheated and heated sections by the subscripts 1 and 2, respectively, we thus have the formulation of the problem as follows:

$$\begin{aligned} &\frac{d^2T_1}{dx^2} - \frac{\rho cV}{k}\frac{dT_1}{dx} = 0, \qquad -\infty < x \le 0, \\ &\frac{d^2T_2}{dx^2} - \frac{\rho cV}{k}\frac{dT_2}{dx} + \frac{q''P}{kA} = 0, \qquad 0 \le x < +\infty, \end{aligned} \tag{3–57}$$

subject to

$$\begin{gathered} T_1(-\infty) = T_0, \\ T_1(0) = T_2(0), \qquad \frac{dT_1(0)}{dx} = \frac{dT_2(0)}{dx}, \\ T_2(\infty) \sim x, \end{gathered} \tag{3–58}$$

where the last boundary condition indicates a linear temperature distribution in the liquid for large values of x. (If this point is not clear to the reader, he should first consider the general solution of the heated section, then ask himself whether an exponential temperature increase in this section is possible for a constant q''.)

The solution of Eq. (3–57) that is satisfied by Eq. (3–58) gives the temperature distribution of the liquid as

$$\frac{T_1(x) - T_0}{q''/\rho c V} = \left(\frac{a}{V\lambda}\right) e^{(V\lambda/a)(x/\lambda)}, \quad -\infty < x \leq 0,$$

$$\frac{T_2(x) - T_0}{q''/\rho c V} = \left(\frac{a}{V\lambda}\right)\left[1 + \left(\frac{V\lambda}{a}\right)\frac{x}{\lambda}\right], \quad 0 \leq x < +\infty,$$

(3–59)

where $a = k/\rho c$ and $\lambda = A/P$. The dimensionless number $a/V\lambda = k/\rho c V\lambda$ expresses the ratio between axial conduction and axial enthalpy flow. The inverse of this number is the so-called *Péclèt modulus*. [What is the limiting form of Eq. (3–59) as $k \to 0$?] The effect of axial conduction is shown in Fig. 3–11 for various values of $a/V\lambda$. Since the temperature at $x = 0$ is inversely proportional to the Péclèt modulus, when, say, Pé ≥ 100, the temperature distribution in the insulated section and the effect of axial conduction in the tube may be neglected.*

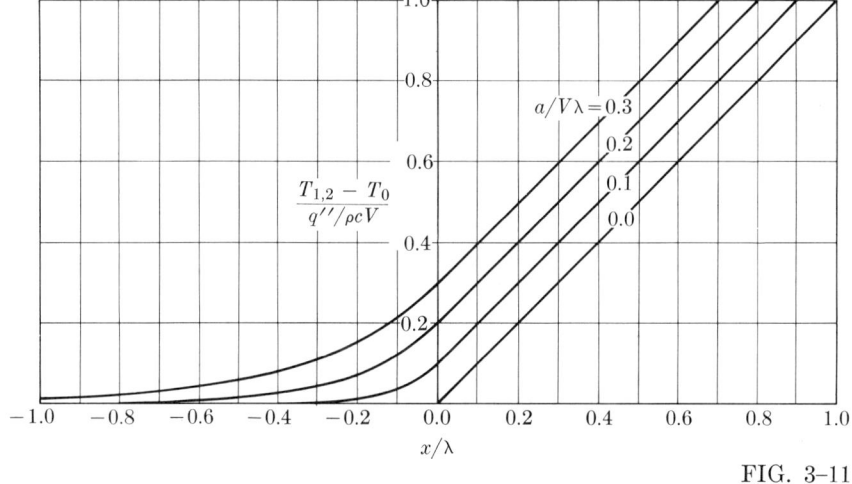

FIG. 3–11

Example 3–4. A length L rather than one-half of the tube of Example 3–3 is subjected to the constant peripheral heat flux q'' (Fig. 3–12). Again we wish to find the radially lumped and axially distributed temperature of the liquid.

The problem now becomes a three-domain problem. Introducing the subscript 3 for the temperature of the insulated downstream region, we proceed

---

* The effect of axial conduction will further be explored in two-dimensional problems. See Example (4–7).

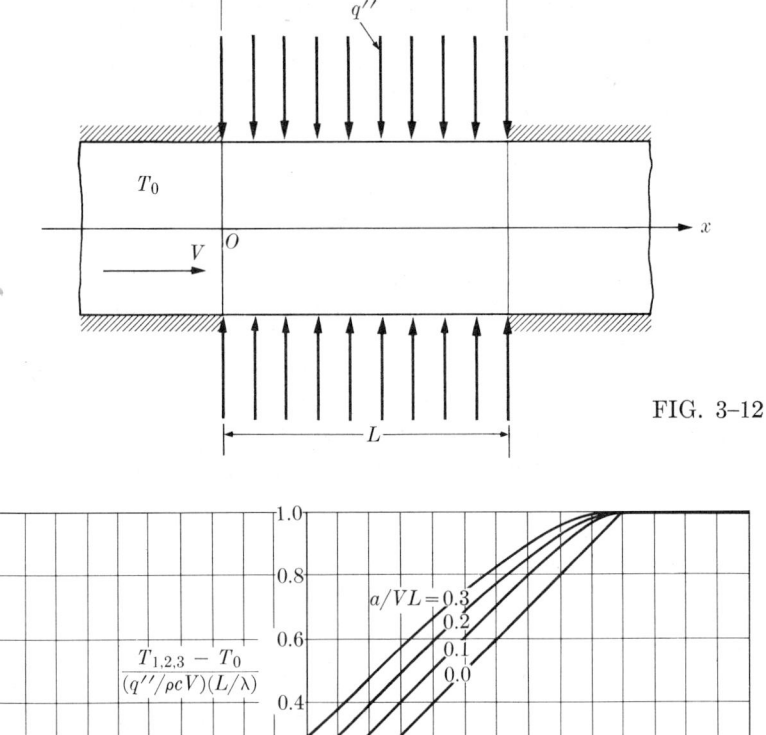

FIG. 3-12

FIG. 3-13

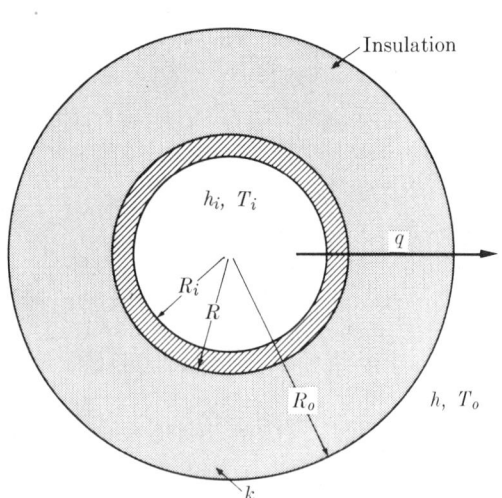

FIG. 3-14

as in Example 3–3. The governing equations are found to be

$$\frac{d^2 T_1}{dx^2} - \frac{\rho c V}{k}\frac{dT_1}{dx} = 0, \qquad -\infty < x \le 0,$$

$$\frac{d^2 T_2}{dx^2} - \frac{\rho c V}{k}\frac{dT_2}{dx} + \frac{q''P}{kA} = 0, \qquad 0 \le x \le L, \qquad (3\text{–}60)$$

$$\frac{d^2 T_3}{dx^2} - \frac{\rho c V}{k}\frac{dT_3}{dx} = 0, \qquad L \le x < \infty,$$

subject to the boundary conditions

$$T_1(-\infty) = T_0,$$

$$T_1(0) = T_2(0), \qquad \frac{dT_1(0)}{dx} = \frac{dT_2(0)}{dx},$$

$$T_2(L) = T_3(L), \qquad \frac{dT_2(L)}{dx} = \frac{dT_3(L)}{dx}, \qquad (3\text{–}61)$$

$$T_3(+\infty) = \text{finite}.$$

The statement of the last boundary condition may be clearly understood by considering the general solution of the third domain.

The solution of Eq. (3–60) that satisfies Eq. (3–61) gives the temperature distribution in the liquid as

$$\frac{T_1(x) - T_0}{(q''/\rho c V)(L/\lambda)} = \left(\frac{a}{VL}\right)(1 - e^{-VL/a})e^{(VL/a)(x/L)}, \qquad -\infty < x \le 0,$$

$$\frac{T_2(x) - T_0}{(q''/\rho c V)(L/\lambda)} = \frac{x}{L} + \left(\frac{a}{VL}\right)[1 - e^{-(VL/a)(1-x/L)}], \qquad 0 \le x \le L, \qquad (3\text{–}62)$$

$$\frac{T_3(x) - T_0}{(q''/\rho c V)(L/\lambda)} = 1, \qquad L \le x < +\infty,$$

where $\lambda = A/P$, as in Example 3–3. [What are the limiting forms of Eq. (3–62) as $L \to \infty$ or $k \to 0$?] In Fig. 3–13 the temperature of the liquid, Eq. (3–62), is plotted against x/L for various values of a/VL.

Example 3–5. A fluid having the temperature T_i flows through a pipe of inner and outer radii R_i, R. The inside and outside heat transfer coefficients are h_i and h_o, respectively. The temperature of the outside fluid is T_o. Let us assume that $T_o < T_i$. We wish to decrease the heat loss from the inside fluid by insulating the pipe (Fig. 3–14). The thermal conductivity of the pipe wall compared with that of the insulation, and the inside heat transfer coefficient compared with the outside coefficient may be assumed to be large. (Condensing steam in a pipe is a typical example of this case.) Find the heat loss from the inner fluid as a function of the thickness of insulation.

Because of the relative magnitude of the resistances involved, the inside convective resistance and the conductive resistance of the pipe may be neglected. Hence the inner surface of the insulation assumes the inside fluid temperature T_i, approximately. It follows from appropriate rearrangement of Eq. (3-32) that the approximate over-all heat transfer coefficient, based on the outer area of the insulation, is

$$\frac{1}{U_o} = \frac{R_o}{k} \ln\left(\frac{R_o}{R}\right) + \frac{1}{h_o}.$$

Furthermore, since $A_o = 2\pi R_o L$, the heat loss from the inside fluid per length L of the pipe, $q = U_o A_o (T_i - T_o)$, may be written

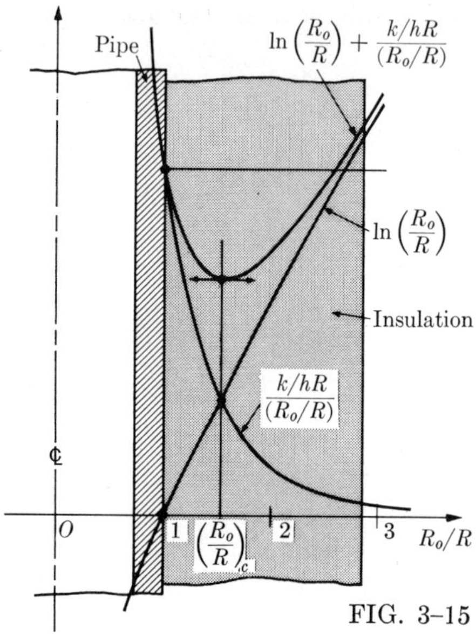

$$\frac{q}{2\pi k L (T_i - T_o)} = \frac{1}{\ln(R_o/R) + (k/hR)/(R_o/R)}. \qquad (3\text{-}63)$$

FIG. 3-15

Inspection of Eq. (3-63) reveals the important fact that when the thickness of the insulation is varied, the first and second terms of the denominator of the right-hand side of Eq. (3-63) vary inversely. This suggests the possibility of an extremum for the heat loss from the inside fluid. The existence of this extremum may be readily shown by equating to zero the first derivative of Eq. (3-63) with respect to R_o/R. The result is

$$\frac{dq}{d(R_o/R)} = -2\pi k L (T_i - T_o) \frac{[1/(R_o/R) - (k/hR)/(R_o/R)^2]}{[\ln(R_o/R) + (k/hR)/(R_o/R)]^2} = 0. \qquad (3\text{-}64)$$

The zero of Eq. (3-64) that is of practical importance gives

$$(R_o/R)_c = k/hR \quad \text{or} \quad (R_o)_c = k/h, \qquad (3\text{-}65)$$

where $(R_o)_c$ is the so-called *critical radius* of the insulation. Introducing Eq. (3-65) into Eq. (3-63), we find that the extremum value of the heat loss from the pipe is

$$\frac{q}{2\pi k L (T_i - T_o)} = \frac{1}{1 + \ln(k/hR)}. \qquad (3\text{-}66)$$

Furthermore, the second derivative of Eq. (3-63) evaluated at $(R_o/R)_c = k/hR$ yields

$$\left.\frac{d^2 q}{d(R_o/R)^2}\right|_{R_o/R = k/hR} = -\frac{2\pi L h^2 (T_i - T_o)}{[1 + \ln(k/hR)]^2} < 0, \qquad (3\text{-}67)$$

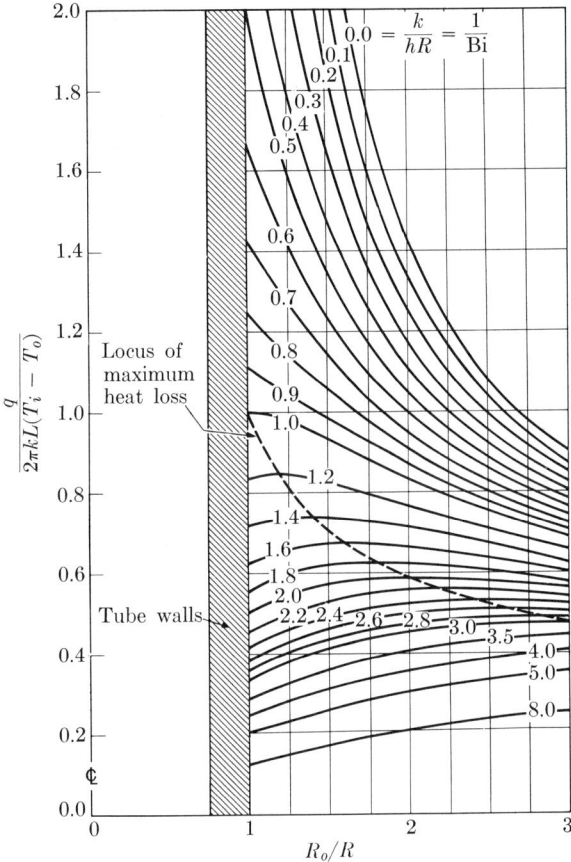

FIG. 3-16

which indicates that the extremum value of the heat loss given by Eq. (3-66) corresponds to a maximum! This surprising result, that is, that *the heat loss from pipes can be increased by insulation*, may be further clarified by sketching the denominator of the right-hand side of Eq. (3-63),

$$\ln\left(\frac{R_o}{R}\right) + \frac{k/hR}{R_o/R}, \qquad (3\text{-}68)$$

as shown in Fig. 3-15. The variation in the first term of Eq. (3-68), which is related to that in the conductive resistance of the insulation, is *logarithmic*, while the variation in the second term, which is proportional to that in the outside convective resistance, is *hyperbolic*. Thus the sum of the two terms, as we see from Fig. 3-15, assumes a minimum value at the critical radius. This in turn gives the maximum value of the heat loss. In Fig. 3-16 the heat loss from tubes is plotted against R_o/R for various values of $1/\text{Bi} = k/hR$.

Example 3-6. An electric wire of radius R is uniformly insulated with plastic to produce an outer radius R_o (Fig. 3–17). The electrical resistance and thermal conductivity of this wire are ρ_e (ohms × length) and k_w, respectively. The thermal conductivity of the insulation is k, the heat transfer coefficient h, and the ambient temperature T_∞. We wish to determine the maximum current that this wire can carry without heating the plastic above its allowable operating temperature $T_{\max}$.

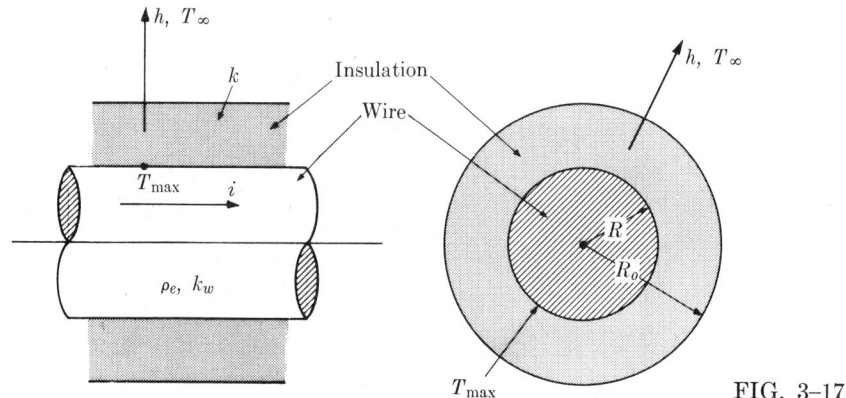

FIG. 3-17

Since axial conduction is negligible, the internal energy

$$i^2(\rho_e/\pi R^2) \tag{3-69}$$

generated electrically per unit length of the wire is removed radially through the conductive resistance of the insulation and the convective resistance of the ambient. Note that the effective total resistance of Example 3–5 is identical to that of the present example. Thus rearranging Eq. (3–63) according to the notation of Fig. 3–17 and equating the result to Eq. (3–69) gives

$$\frac{i_{\max}^2 \rho_e}{\pi R^2} = \frac{2\pi k(T_{\max} - T_\infty)}{\ln(R_o/R) + (k/hR)/(R_o/R)} \tag{3-70}$$

or

$$i_{\max} = \left[\frac{2\pi^2 R^2 (k/\rho_e)(T_{\max} - T_\infty)}{\ln(R_o/R) + (k/hR)/(R_o/R)}\right]^{1/2}. \tag{3-71}$$

If it is possible and permissible to change the thickness of the insulation and hence the critical radius, we get

$$i_{\max_{\max}} = \left[\frac{2\pi^2 R^2 (k/\rho_e)(T_{\max} - T_\infty)}{1 + \ln(k/hR)}\right]^{1/2}. \tag{3-72}$$

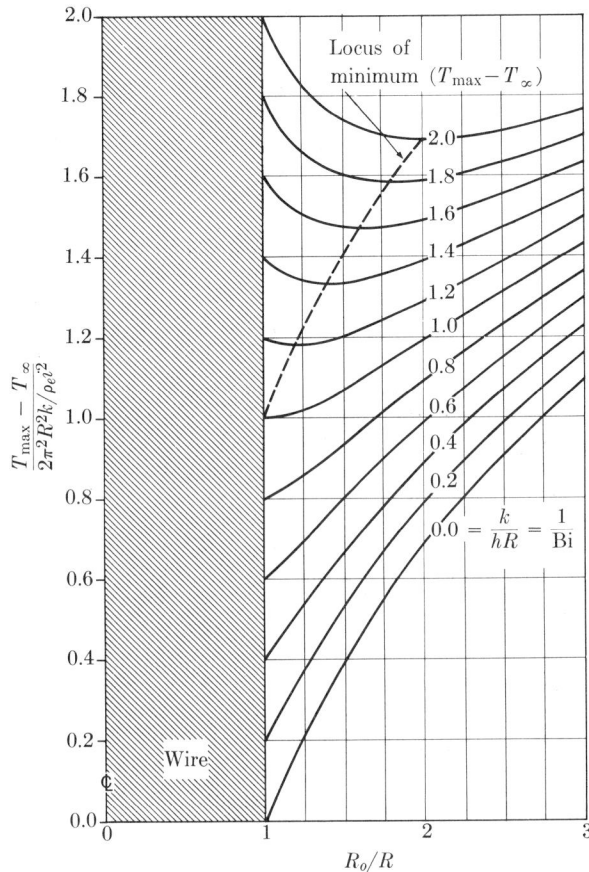

FIG. 3-18

A slightly different version of the same problem is the variation in the interface temperature as a function of the thickness of the plastic insulation for a specified electric current. This temperature may be readily obtained from Eq. (3–70) in the form

$$\frac{T_{\max} - T_\infty}{2\pi^2 R^2 k/\rho_e i^2} = \ln\left(\frac{R_o}{R}\right) + \frac{k/hR}{R_o/R}. \tag{3-73}$$

The behavior of $\ln(R_o/R) + (k/hR)/(R_o/R)$ was sketched in Fig. 3–15. Equation (3–73) is now plotted against R_o/R for various values of k/hR (Fig. 3–18). The variation here is the inverse of that shown in Fig. 3–16. [Compare Eq. (3–63) with Eq. (3–73).]

Note the difference between our interests in Examples 3–5 and 3–6. In Example 3–5 we wished to decrease q for constant inner-fluid and ambient temperatures; in Example 3–6 we wished to produce a low wire-temperature distribution for a constant i (or q) and ambient temperature.

Example 3–7. The fuel element of a reactor consists of a sphere of fissionable material with radius R, surrounded by a spherical shell of cladding with outer radius R_o (Fig. 3–19). The temperature of the coolant is T_∞, and the heat transfer coefficient is h. The nuclear internal energy generated in the sphere can be approximated by a parabola as

$$u'''(r) = u_0'''[1 - (r/R)^2],$$

where u_0''' is the nuclear energy generation at the center of the sphere. We wish to know the temperature distribution in the fuel element.

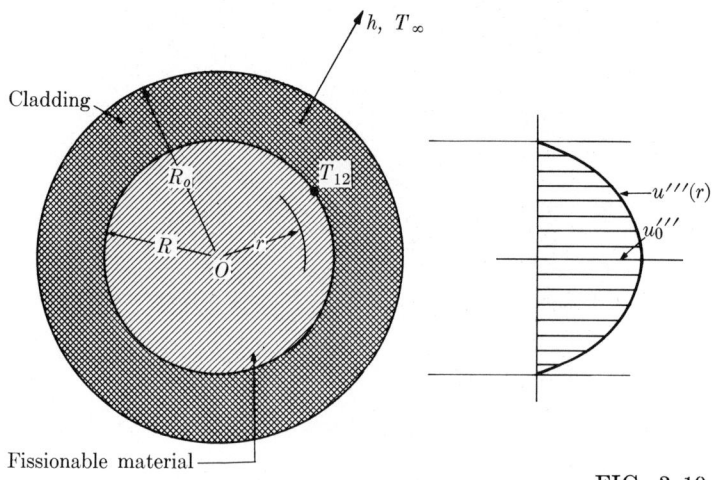

FIG. 3–19

The formal procedure of the formulation leads to a two-domain problem. However, leaving this and its solution to the reader as an exercise, we instead treat the problem as follows. The total internal energy generated in the sphere is

$$4\pi u_0''' \int_0^R [1 - (r/R)^2] r^2 \, dr = \tfrac{8}{15}\pi R^3 u_0'''.$$

Under steady conditions this energy, in the form of heat q, is transferred to the coolant through the conductive resistance of the cladding and the outside convective resistance. Distinguishing the properties of the sphere and the cladding by the subscripts 1 and 2, respectively, and denoting the interface temperature by T_{12}, we have

$$\tfrac{8}{15}\pi R^3 u_0''' = q = U_o A_o (T_{12} - T_\infty), \tag{3–74}$$

where $A_o = 4\pi R_o^2$, and

$$\frac{1}{U_o} = \frac{R_o}{k_2}\left(\frac{R_o}{R} - 1\right) + \frac{1}{h},$$

obtained from Eq. (3–23). Using the values of A_o and U_o, we may rearrange Eq. (3–74) in the form

$$\tfrac{8}{15}\pi R^3 u_0''' = \frac{4\pi R_o^2(T_{12} - T_\infty)}{(R_o/k_2)(R_o/R - 1) + 1/h}. \tag{3–75}$$

Equation (3–75) readily gives the interface temperature T_{12} as

$$T_{12} - T_\infty = \frac{2}{15}\frac{u_0''' R^2}{k_2}\left[1 - \frac{1}{R_o/R} + \frac{k_2/hR}{(R_o/R)^2}\right]. \tag{3–76}$$

By means of an appropriate arrangement of Eq. (3–26), the temperature of the cladding may now be obtained in terms of $T_{12} - T_\infty$. The result is

$$\frac{T_2(r) - T_\infty}{T_{12} - T_\infty} = U_o\left[\frac{R_o}{k_2}\left(\frac{R_o}{r} - 1\right) + \frac{1}{h}\right], \qquad R \le r \le R_o. \tag{3–77}$$

Finally, the temperature problem of the sphere may be formulated in the form

$$\frac{d}{dr}\left(r^2\frac{dT_1}{dr}\right) + \frac{u_0'''}{k_1}\left[1 - \left(\frac{r}{R}\right)^2\right]r^2 = 0,$$

$$T_1(0) = \text{finite}, \qquad T_1(R) = T_{12}. \tag{3–78}$$

The solution of Eq. (3–78) gives the temperature of the sphere with respect to the interface temperature as

$$\frac{T_1(r) - T_{12}}{u_0''' R^2/k_1} = \frac{7}{60} - \left(\frac{r}{R}\right)^2\left[\frac{1}{6} - \frac{1}{20}\left(\frac{r}{R}\right)^2\right], \qquad 0 \le r \le R. \tag{3–79}$$

Note the procedure followed in determining the temperature in Examples 3–6 and 3–7. Although both examples formally are two-domain problems, the temperatures involved may be obtained without following a two-domain formulation, because in both cases radial heat flux is available from the internal energy generation. Thus, multiplying this flux by the sum of the appropriate conductive and convective resistances, we obtain the temperature of the plastic or the cladding relative to the known ambient or coolant temperature. We then evaluate the temperature of the wire or fuel element in terms of the interface temperature of the plastic or the cladding by following the steps of formal procedure. This method was also employed in the alternative solution of the problem of Example 3–2.

Another aspect of Example 3–7 is safe operation of the reactor, which is generally determined by a maximum allowable fuel temperature. For specified values of u''' and R, the influence of cladding thickness on this temperature may be investigated much as it would for cylindrical geometry.* Rearranging Eq. (3–75) for this purpose, we get

$$\theta = \frac{T_{12} - T_\infty}{2u_0''' R^2/15k_2} = 1 - \frac{1}{R_o/R} + \frac{k_2/hR}{(R_o/R)^2}.\qquad(3\text{–}80)$$

The only physically significant extremum of θ, obtained from

$$\frac{d\theta}{d(R_o/R)} = \frac{1}{(R_o/R)^2} - \frac{2k_2/hR}{(R_o/R)^3},$$

gives the critical radius of the cladding as

$$(R_o/R)_c = 2(k_2/hR) \qquad \text{or} \qquad (R_o)_c = 2k_2/h.\qquad(3\text{–}81)$$

It can be shown that $d^2\theta/d(R_o/R)^2 > 0$ for $R_o = (R_o)_c$. Thus T_{12} assumes its minimum value for a specified u''' and R when the outer radius of the cladding corresponds to the critical radius. It is clear that this minimum, if attainable, implies the minimum temperature distribution in the fuel element.

3–4. Principle of Superposition

The solution of linear problems, such as the problems of conduction with constant properties, may often be reduced to the solution of a number of simpler problems by employing the *principle of superposition*. Given their simplicity, one-dimensional problems do not require the use of superposition and cannot fully show its importance. We make a start here, however, merely to familiarize the reader with the procedure, which may be only occasionally convenient for steady one-dimensional problems, but which may be indispensable for multi-dimensional or unsteady problems. (This will be clarified in Section 4–7.)

First let us introduce a necessary mathematical concept. A linear differential equation or a linear boundary condition is said to be *homogeneous* if, when satisfied by a function $y(x)$, it is also satisfied by $Cy(x)$, where C is an arbitrary constant. In other words, a linear differential equation is homogeneous when *all* its terms include either the unknown function or one of its derivatives. Similarly, a boundary condition is homogeneous when an unknown function or its derivatives, or any linear combination of this function and its derivatives, vanishes at the boundary. Thus, for example,

$$\frac{d^2y}{dx^2} + f_1(x)\frac{dy}{dx} + f_2(x)y = f_3(x)\qquad(3\text{–}82)$$

---

* Actually, the thickness of cladding is determined by nuclear considerations rather than the fuel temperature. The part of the problem under study, although unrealistic for this reason, illustrates the determination of the critical radius for a spherical geometry.

is a nonhomogeneous linear differential equation of second order. The equation

$$\frac{d^2y}{dx^2} + f_1(x)\frac{dy}{dx} + f_2(x)y = 0 \tag{3-83}$$

is a homogeneous linear differential equation of second order. At the boundary $x = a$,

$$y(a) = \alpha, \qquad y'(a) = \beta, \qquad \text{or} \qquad y'(a) + \gamma y(a) = \delta$$

denotes a nonhomogeneous linear boundary condition, whereas

$$y(a) = 0, \qquad y'(a) = 0, \qquad \text{or} \qquad y'(a) + \gamma y(a) = 0$$

represents a homogeneous linear boundary condition.

We know from mathematics that the general solution of Eq. (3–82) may be written by the superposition of three particular solutions. Thus

$$y = y_0 + C_1 y_1 + C_2 y_2, \tag{3-84}$$

where the first right-hand term corresponds to a particular solution of the nonhomogeneous equation, and the others to particular solutions of the homogeneous equation, Eq. (3–83). The constants C_1 and C_2 are to be determined according to the boundary conditions of the problem. The proof that Eq. (3–84) is the solution of Eq. (3–82) may be readily shown by direct substitution.

The superposition principle of interest to us concerns the formulation of problems rather than their solutions. Thus the formulation of a problem under consideration is written as the sum of the formulations of a number of simpler problems. *The number of these problems is equal to the number of nonhomogeneities involved in the formulation of the actual problem.** A simple problem may illustrate the point. Reconsider Example 3–1, a flat plate of thickness L separating two media at temperatures T_i and T_o, the heat transfer coefficients being h_i and h_o (Fig. 3–20). Instead of finding the appropriate value u_0''' of the internal

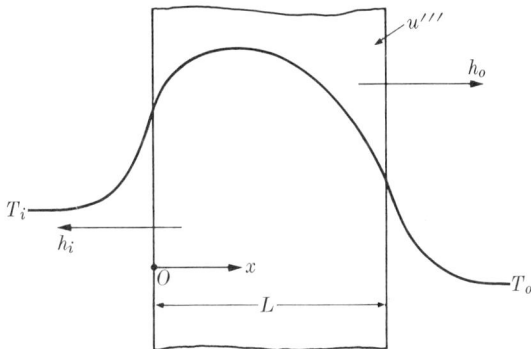

FIG. 3–20

---

* In physically significant problems, *potential* is used rather than nonhomogeneity. For example, thermal potential, electric potential, etc.

energy generation which eliminates the heat loss from the ambient of higher temperature, let us find the temperature of the plate corresponding to an arbitrary u'''.

The formulation of the problem results in

$$\frac{d^2T}{dx^2} + \frac{u'''}{k} = 0, \quad +k\frac{dT(0)}{dx} = h_i[T(0) - T_i], \quad -k\frac{dT(L)}{dx} = h_o[T(L) - T_o]. \tag{3-85}$$

Equation (3–85) has three potentials, u''', T_i, T_o, giving rise to the temperature distribution in the plate. Thus the problem can be separated into three problems (Fig. 3–21),

$$T = T_1 + T_2 + T_3, \tag{3-86}$$

where T_1, T_2, and T_3 satisfy the following equations:

$$\frac{d^2T_1}{dx^2} + \frac{u'''}{k} = 0, \quad +k\frac{dT_1(0)}{dx} = h_iT_1(0), \quad -k\frac{dT_1(L)}{dx} = h_oT_1(L); \tag{3-87}$$

$$\frac{d^2T_2}{dx^2} = 0, \quad +k\frac{dT_2(0)}{dx} = h_i[T_2(0) - T_i], \quad -k\frac{dT_2(L)}{dx} = h_oT_2(L); \tag{3-88}$$

and

$$\frac{d^2T_3}{dx^2} = 0, \quad +k\frac{dT_3(0)}{dx} = h_iT_3(0), \quad -k\frac{dT_3(L)}{dx} = h_o[T_3(L) - T_o]. \tag{3-89}$$

The validity of this superposition may be readily shown by adding Eqs. (3–87), (3–88), and (3–89) side by side as indicated by Eq. (3–86). The result is Eq. (3–85).

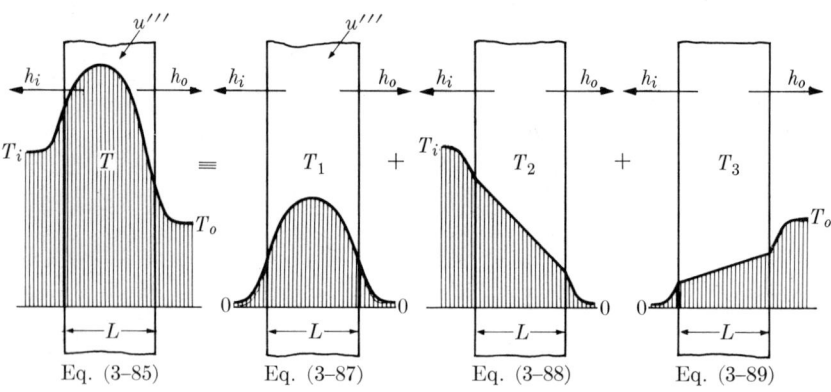

FIG. 3–21

Inspection of Eqs. (3–85), (3–87), (3–88), and (3–89) reveals two important facts: (i) *superposition affects the homogeneity but not the type of differential equation or boundary conditions;* (ii) *each problem involved in a superposition has the same geometry.*

Aside from the foregoing, the principle of superposition is also frequently used in the selection of a reference temperature. This selection is completely arbitrary. In the preceding problem, for instance, we used zero. It is more convenient, however, to measure the temperature above any one of the ambient temperatures. Thus the superposition

$$T(x) = T_0 + \theta(x) \quad \text{or} \quad T(x) = T_i + \psi(x)$$

may be employed. When the former is used, for example, the formulation of the problem becomes

$$\frac{d^2\theta}{dx^2} + \frac{u'''}{k} = 0, \quad +k\frac{d\theta(0)}{dx} = h_i[\theta(0) - \theta_i], \quad -k\frac{d\theta(L)}{dx} = h_o\theta(L),$$
(3–90)

where $\theta_i = T_i - T_o$. The new formulation, Eq. (3–90), has only two nonhomogeneities. Therefore it can be separated into two problems rather than three.

3–5. Heterogeneous Solids (Variable Thermal Conductivity)

Heterogeneous solids are becoming increasingly important because of the large ranges of temperature involved in current problems of technology, as in reactor fuel elements, space vehicle components, solidification of castings, etc. In this section, a general method of solution will be developed for unsteady three-dimensional temperature problems of heterogeneous solids.

Let us reconsider the equation of heat conduction for heterogeneous solids,

$$\rho c \frac{dT}{dt} = \nabla \cdot (k \nabla T) + u'''. \tag{2-88}$$

If k, c, and u''' are functions of space only, Eq. (2–88) becomes a linear differential equation with variable coefficients. The solution of such an equation requires no additional mathematics (see Problem 3–37). If k and c are dependent on temperature but independent of space, however, Eq. (2–88) becomes nonlinear and difficult to solve. (Does temperature dependence of u''' bring any complication into the formulation?) Usually, numerical methods have to be employed. A number of analytical methods are also available. One of these, *Kirchhoff's method*, is to a large extent general and is discussed below.

Equation (2–88) may be reduced to a linear differential equation by introducing a new temperature θ related to the temperature T of the problem by the

Kirchhoff transformation,

$$\theta = \frac{1}{k_R} \int_{T_R}^{T} k(T)\, dT, \tag{3-91}$$

where T_R denotes a convenient reference temperature, and $k_R = k(T_R)$. T_R and k_R are introduced merely to give θ the dimensions of temperature and a definite value. It follows from Eq. (3-91) that

$$\frac{d\theta}{dt} = \frac{k}{k_R} \frac{dT}{dt} \tag{3-92}$$

and

$$\nabla \theta = \frac{k}{k_R} \nabla T. \tag{3-93}$$

Inserting Eqs. (3-92) and (3-93) into Eq. (2-88), we have

$$\frac{d\theta}{dt} = a \nabla^2 \theta + \left(\frac{a}{k_R}\right) u''', \tag{3-94}$$

where a and u''' are expressed as functions of the new variable θ. For many solids, however, the temperature dependence of a can be neglected compared to that of k. In such cases, if u''' is independent of T, Eq. (3-94) becomes identical to Eq. (2-91) except for the different but constant coefficient of u'''. Thus the solutions obtained for homogeneous solids may be readily utilized for heterogeneous solids by replacing T by θ and ρc by k_R/a, provided that the boundary conditions prescribe T or $\partial T/\partial n$. This remark does not hold if the boundary conditions involve the convective term $h(T_\sigma - T_\infty)$. The following one-dimensional example illustrates the use of the method.

Example 3-8. A liquid is boiled by a flat electric heater plate of thickness $2L$. The internal energy u''' generated electrically may be assumed to be uniform. The boiling temperature of the liquid, corresponding to a specified pressure, is T_∞ (Fig. 3-22). We wish to find the steady temperature of the plate for (i) $k = k(T)$; (ii) $k = k_R(1 + \beta T)$.

The formulation of the problem is

$$\frac{d}{dx}\left(k \frac{dT}{dx}\right) + u''' = 0,$$

$$\frac{dT(0)}{dx} = 0, \tag{3-95}$$

$$T(L) = T_\infty.$$

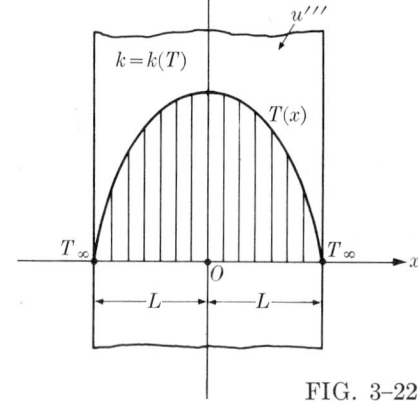

FIG. 3-22

Employing the one-dimensional form of Eq. (3–93),

$$\frac{d\theta}{dx} = \frac{k}{k_R}\frac{dT}{dx},$$

we may transform Eq. (3–95) to

$$\frac{d^2\theta}{dx^2} + \frac{u'''}{k_R} = 0, \qquad \frac{d\theta(0)}{dx} = 0, \qquad \theta(L) = \theta_\infty, \tag{3-96}$$

where, according to Eq. (3–91),

$$\theta_\infty = \frac{1}{k_R}\int_{T_R}^{T_\infty} k(T)\,dT. \tag{3-97}$$

The solution of Eq. (3–96) is

$$\frac{\theta(x) - \theta_\infty}{u'''L^2/2k_R} = 1 - \left(\frac{x}{L}\right)^2. \tag{3-98}$$

Introducing Eqs. (3–91) and (3–97) into Eq. (3–98), we obtain the temperature of the plate in terms of T as follows:

$$\frac{(1/k_R)\int_{T_\infty}^{T} k(T)\,dT}{u'''L^2/2k_R} = 1 - \left(\frac{x}{L}\right)^2. \tag{3-99}$$

For the special case $k = k_R(1 + \beta T)$, Eq. (3–99) becomes

$$\frac{[T(x) - T_\infty] + (\beta/2)[T^2(x) - T_\infty^2]}{u'''L^2/2k_R} = 1 - \left(\frac{x}{L}\right)^2. \tag{3-100}$$

Solving Eq. (3–100) for T and disregarding the physically meaningless root, we find that the temperature of the plate is

$$\frac{T(x) - T_\infty}{u'''L^2/2k_R} = \left(\frac{1 + \beta T_\infty}{\beta u'''L^2/2k_R}\right)$$
$$\times \left[-1 + \sqrt{1 + \left(\frac{2}{1 + \beta T_\infty}\right)\left(\frac{\beta u'''L^2/2k_R}{1 + \beta T_\infty}\right)\left[1 - \left(\frac{x}{L}\right)^2\right]}\,\right]. \tag{3-101}$$

As $\beta \to 0$, Eq. (3–101) tends to the temperature of constant k.

If the reference temperature T_R of the Kirchhoff transformation is taken to be the ambient temperature T_∞, Eq. (3–101) may be reduced to

$$\frac{T(x) - T_\infty}{u'''L^2/2k_\infty} = \left(\frac{1}{\beta u'''L^2/2k_\infty}\right)\left[-1 + \sqrt{1 + 2\left(\frac{\beta u'''L^2}{2k_\infty}\right)\left[1 - \left(\frac{x}{L}\right)^2\right]}\,\right], \tag{3-102}$$

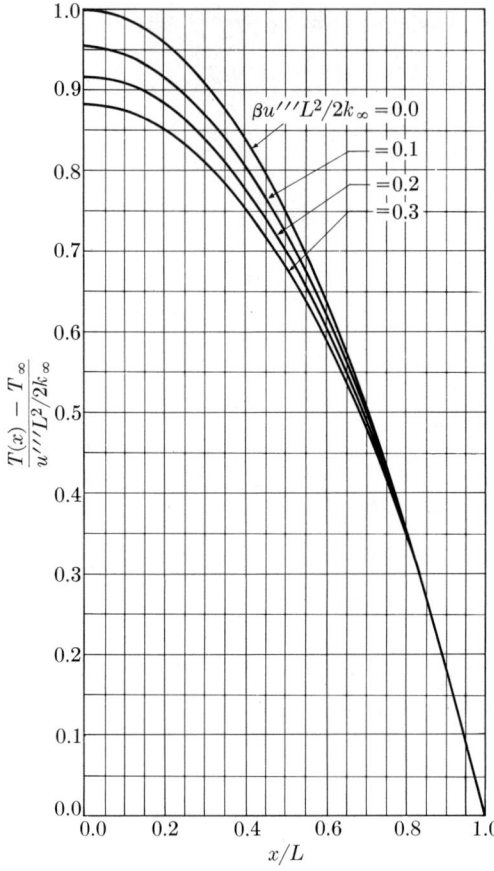

FIG. 3-23

obtained by substituting $\beta T_\infty = 0$ into Eq. (3-101) and noting that $k_R = k_\infty$ for this case. In Fig. 3-23 the effect of linear conductivity on the temperature of the plate, calculated from Eq. (3-102) for various values of $\beta u'''L^2/2k_\infty$, is shown as a function of x/L.

3-6. Power Series Solutions. Bessel Functions

In Section 3-7 we shall discuss a class of one-dimensional problems associated with extended surfaces (fins, pins, or spines). When the cross section of an extended surface is variable, the formulation of the problem results in a second-order linear differential equation with variable coefficients. This differential equation is a form of *Bessel's equation*, except in a special case which leads to the so-called *equidimensional equation*. The solution methods suitable to second-order linear differential equations with constant coefficients are not suitable to those with variable coefficients. We may, however, recall that equations with variable coefficients possess solutions expressible, over an appropriate interval, in terms of power series. This section is therefore devoted to a brief review of the power series solution of Bessel's equation and the properties of *Bessel functions*. Before this review, let us first recall the definition of power series.

An infinite series in the form

$$y(x) = a_0 + a_1(x - x_0) + a_2(x - x_0)^2 + \cdots = \sum_{k=0}^{\infty} a_k(x - x_0)^k$$

is called a *power series* expansion of the function $y(x)$ in the neighborhood of $x = x_0$, and is defined as

$$y(x) = \lim_{K \to \infty} \sum_{k=0}^{K} a_k(x - x_0)^k.$$

For an interval of x in which the foregoing limit exists, the series is said to *converge* in this interval. The reader may refer to any text on differential equations for the convergence criteria of power series.

We may now consider the method of power series solutions. Since the method is applicable to linear differential equations with constant as well as variable coefficients, it may be illustrated in terms of the following simple differential equation with constant coefficients:

$$\frac{d^2 y}{dx^2} + y = 0. \tag{3-103}$$

Let us assume a power series in the form

$$y(x) = a_0 + a_1 x + a_2 x^2 + \cdots = \sum_{k=0}^{\infty} a_k x^k, \tag{3-104}$$

which converges in an interval including $x = 0$. Inserting Eq. (3-104) into Eq. (3-103) and rearranging gives

$$(a_0 + 2a_2) + (a_1 + 6a_3)x + (a_2 + 12a_4)x^2 + \cdots = 0. \tag{3-105}$$

Equation (3-105) is valid over an interval of x provided the coefficients of all powers of x vanish independently in this interval. It then follows that

$$a_2 = -\tfrac{1}{2} a_0,$$
$$a_3 = -\tfrac{1}{6} a_1,$$
$$a_4 = -\tfrac{1}{12} a_2 = \tfrac{1}{24} a_0,$$
$$a_5 = -\tfrac{1}{20} a_3 = \tfrac{1}{120} a_1,$$
$$\vdots$$

Introducing these values into Eq. (3-104), we obtain the solution of Eq. (3-103) in the form

$$y(x) = a_0(1 - x^2/2 + x^4/24 - \cdots) + a_1(x - x^3/6 + x^5/120 - \cdots),$$

which may also be expressed as

$$y(x) = a_0 \sum_{k=0}^{\infty} (-1)^k \frac{x^{2k}}{(2k)!} + a_1 \sum_{k=0}^{\infty} (-1)^k \frac{x^{2k+1}}{(2k+1)!}. \tag{3-106}$$

The two series appearing in this equation are the Maclaurin expansions of cos x and sin x, respectively. Hence Eq. (3-106) is equivalent to

$$y(x) = a_0 \cos x + a_1 \sin x. \tag{3-107}$$

Clearly, Eq. (3-107) may also be obtained by the classical method which suggests solutions in the form $y = e^{rx}$, where r is to be determined from the characteristic equation that is obtained by the introduction of $y = e^{rx}$ into the differential equation (3-103).

We next consider the second-order linear differential equation with variable coefficients, Bessel's equation,

$$x \frac{d}{dx}\left(x \frac{dy}{dx}\right) + (m^2 x^2 - \nu^2) y = 0, \tag{3-108}$$

where m is a parameter, and ν may be zero, a fractional number, or an integer.

The solution of Eq. (3-108) may be obtained by the use of power series in a manner analogous to, but somewhat more involved than, the solution of Eq. (3-103). The result is

$$y(x) = a_0 J_\nu(mx) + a_1 Y_\nu(mx). \tag{3-109}$$

In Eq. (3-109),

$$J_\nu(mx) = \sum_{k=0}^{\infty} (-1)^k \frac{(mx/2)^{2k+\nu}}{k!\,\Gamma(k+\nu+1)} \tag{3-110}$$

and

$$Y_\nu(mx) = \frac{(\cos \nu\pi) J_\nu(mx) - J_{-\nu}(mx)}{\sin \nu\pi}, \tag{3-111}$$

where

$$J_{-\nu}(mx) = \sum_{k=0}^{\infty} (-1)^k \frac{(mx/2)^{2k-\nu}}{k!\,\Gamma(k-\nu+1)}. \tag{3-112}$$

The function appearing in the denominator of Eqs. (3-110) and (3-112), known as the *Gamma function*, is defined in the form

$$\Gamma(n+1) = n\Gamma(n) = n!, \qquad \Gamma(1) = 0! = 1$$

for integers, and in the form

$$\Gamma(\nu)\Gamma(\nu - 1) = \pi/\sin \pi\nu, \qquad \Gamma(\tfrac{1}{2}) = \pi^{1/2}$$

for fractional numbers.

If ν is not an integer, $J_\nu(mx)$ and $J_{-\nu}(mx)$ are independent solutions of Eq. (3–108), but if ν is an integer, say n, then

$$J_n(mx) = (-1)^n J_{-n}(mx). \tag{3–113}$$

To obtain a second solution of Eq. (3–108) which is valid for all values of ν, Eq. (3–111) is so defined that

$$\lim_{\nu \to n} Y_\nu(mx) \to Y_n(mx),$$

where

$$\pi Y_n(mx) = 2\left(\ln \frac{mx}{2} + \gamma\right) J_n(mx) - \sum_{k=0}^{n-1} \frac{(n-k-1)!}{k!} \left(\frac{mx}{2}\right)^{2k-n}$$

$$+ \sum_{k=0}^{\infty} (-1)^{k+1} [\varphi(k) + \varphi(k+n)] \frac{(mx/2)^{2k+n}}{k!(n+k)!}, \tag{3–114}$$

in which

$$\varphi(k) = \sum_{m=1}^{k} \frac{1}{m}, \quad \varphi(0) = 0, \quad \text{and} \quad \gamma = 0.5772\ldots$$

The function $J_\nu(mx)$ is known as the *Bessel function of the first kind, of order ν*, and the function $Y_\nu(mx)$ as the *Bessel function of the second kind, of order ν*.

An equation related to Eq. (3–108) is the *modified Bessel equation*

$$x \frac{d}{dx}\left(x \frac{dy}{dx}\right) - (m^2 x^2 + \nu^2) y = 0. \tag{3–115}$$

Inspection reveals that the replacement of x by ix reduces Eq. (3–115) to Eq. (3–108). Hence the solution of Eq. (3–115) may readily be obtained by replacing x by ix in Eq. (3–109). We have then

$$y(x) = a_0 J_\nu(imx) + a_1 Y_\nu(imx). \tag{3–116}$$

It follows from Eq. (3–110) that

$$J_\nu(imx) = \sum_{k=0}^{\infty} (-1)^k i^{2k+\nu} \frac{(mx/2)^{2k+\nu}}{k!\Gamma(k+\nu+1)} = i^\nu \sum_{k=0}^{\infty} \frac{(mx/2)^{2k+\nu}}{k!\Gamma(k+\nu+1)}. \tag{3–117}$$

However, rather than using Eq. (3–116) as the general solution of Eq. (3–115), it is customary to replace $J_\nu(imx)$ by $I_\nu(mx)$ as a first particular solution defined in the form

$$I_\nu(mx) = \sum_{k=0}^{\infty} \frac{(mx/2)^{2k+\nu}}{k!\Gamma(k+\nu+1)}. \tag{3–118}$$

The comparison of Eqs. (3–117) and (3–118) gives

$$J_\nu(imx) = i^\nu I_\nu(mx). \tag{3–119}$$

If ν is not an integer, $I_{-\nu}(mx)$ is independent of $I_\nu(mx)$ and is therefore another particular solution of Eq. (3–115); the complete solution may then be written as a linear combination of $I_\nu(mx)$ and $I_{-\nu}(mx)$. However, to obtain a second particular solution suitable to all values of ν, we define*

$$K_\nu(mx) = \frac{\pi}{2} \frac{I_{-\nu}(mx) - I_\nu(mx)}{\sin \nu \pi}, \qquad (3\text{–}120)$$

so that

$$\lim_{\nu \to n} K_\nu(mx) \to K_n(mx),$$

where n is an integer. With this definition

$$K_n(mx) = (-1)^{n+1} \left(\ln \frac{mx}{2} + \gamma\right) I_n(mx)$$

$$+ \frac{1}{2} \sum_{k=0}^{n-1} (-1)^k \frac{(n-k-1)!}{k!} \left(\frac{mx}{2}\right)^{2k-n}$$

$$+ \tfrac{1}{2}(-1)^n \sum_{k=0}^{\infty} [\varphi(k) + \varphi(k+n)] \frac{(mx/2)^{2k+n}}{k!(n+k)!}, \qquad (3\text{–}121)$$

where the definitions of $\varphi(k)$ and γ are identical to those given above in conjunction with Eq. (3–114).

Now we may write the general solution of Eq. (3–115) in the alternative form

$$y(x) = a_0 I_\nu(mx) + a_1 K_\nu(mx). \qquad (3\text{–}122)$$

The function $I_\nu(mx)$ is known as the *modified Bessel function of the first kind, of order ν*, and the function $K_\nu(mx)$ as the *modified Bessel function of the second kind, of order ν*.

Many tables of the Bessel functions and the modified Bessel functions have been compiled. The reader is referred to such tables for numerical computations. (Because of the size of the literature no specific reference is cited here.)

We next demonstrate that the solution of differential equations in the general form

$$x^2 \frac{d^2 y}{dx^2} + ax \frac{dy}{dx} + b^2 x^2 y = 0 \qquad (3\text{–}123)$$

can be expressed in terms of the Bessel functions. First, we introduce the dependent variable change $y = x^\nu z$ and rearrange Eq. (3–123) to give

$$x^\nu \frac{d^2 z}{dx^2} + (a + 2\nu) x^{\nu-1} \frac{dz}{dx} + \{b^2 x^\nu + [(a-1)\nu + \nu^2] x^{\nu-2}\} z = 0.$$

Now, adjusting ν so that $a + 2\nu = 1$ and dividing each term by $x^{\nu-2}$, we get

$$x \frac{d}{dx}\left(x \frac{dz}{dx}\right) + (b^2 x^2 - \nu^2) z = 0, \qquad (3\text{–}124)$$

---

* Note that $K_\nu(mx)$ is not defined through $Y_\nu(mx)$.

which is identical to Eq. (3–108). Therefore, if the solution of Eq. (3–124) is $Z_\nu(bx)$, then that of Eq. (3–123) is

$$y(x) = x^\nu Z_\nu(bx), \tag{3-125}$$

where Z_ν is used for the general representation of the Bessel functions of order ν, and $\nu = (1 - a)/2$.

Finally, we consider differential equations in the general form

$$\frac{d}{dx}\left(x^\alpha \frac{dy}{dx}\right) + \gamma^2 x^\beta y = 0, \tag{3-126}$$

where α and β are positive, and γ may be real or imaginary. The formulation of a problem related to an extended surface with variable cross section frequently leads to the general form given in Eq. (3–126). We now show, by means of a variable change, that Eq. (3–126) is another form of Bessel's equation. Introducing the independent variable change $x = t^\mu$, we rearrange Eq. (3–126) to give

$$t^2 \frac{d^2y}{dt^2} + t[\mu(\alpha - 1) + 1]\frac{dy}{dt} + \gamma^2 \mu^2 t^{\mu(\beta-\alpha+2)} y = 0.$$

Adjusting μ so that $\mu(\beta - \alpha + 2) = 2$, we have

$$t^2 \frac{d^2y}{dt^2} + t[\mu(\alpha - 1) + 1]\frac{dy}{dt} + \gamma^2 \mu^2 t^2 y = 0, \tag{3-127}$$

which has the form of Eq. (3–123). Thus the solution of Eq. (3–126) may be written from Eq. (3–125) by appropriately relating the parameters involved in Eq. (3–127) to those in Eq. (3–123). We have then

$$y(t) = t^\nu Z_\nu(\gamma\mu t), \tag{3-128}$$

where $\nu = \mu(1 - \alpha)/2 = (1 - \alpha)/(\beta - \alpha + 2)$. If we return to the original independent variable and insert $x^{1/\mu}$ in place of t, Eq. (3–128) becomes

$$y(x) = x^{\nu/\mu} Z_\nu(\gamma\mu x^{1/\mu}) \tag{3-129}$$

provided $\beta - \alpha + 2 \neq 0$. The parameters involved in Eq. (3–129) are related to those in Eq. (3–126) by $\nu = (1 - \alpha)/(\beta - \alpha + 2)$, $1/\mu = (\beta - \alpha + 2)/2$, and $\nu/\mu = (1 - \alpha)/2$.

If the sign of the second term of Eq. (3–126) is negative, then by replacing γ by $i\gamma$ in Eq. (3–129), we may express the solution in terms of the Bessel functions of the first and second kinds with imaginary argument, or, equivalently, in terms of the modified Bessel functions of the first and second kinds with real argument.

The special case of Eq. (3–126) for $\beta - \alpha + 2 = 0$ may be found by expanding the first term of this equation and dividing the result by $x^{\alpha-2}$.

TABLE 3-1

SOLUTION OF $\dfrac{d}{dx}\left(x^\alpha \dfrac{dy}{dx}\right) + \gamma^2 x^\beta y = 0$

Case (i): $\beta - \alpha + 2 \neq 0$. The general solution is

$$y(x) = x^{\nu/\mu} Z_\nu(|\gamma|\mu x^{1/\mu}),$$

where

$$\nu = (1 - \alpha)/(\beta - \alpha + 2), \qquad \mu = 2/(\beta - \alpha + 2), \qquad \nu/\mu = (1 - \alpha)/2;$$

two particular solutions, corresponding to Z_ν and to be selected according to γ and ν, are shown below.

γ	ν	Particular solutions	
Real	Fractional	J_ν	$J_{-\nu}$ (or Y_ν)
	Zero or integer	J_n	Y_n
Imaginary	Fractional	I_ν	$I_{-\nu}$ (or K_ν)
	Zero or integer	I_n	K_n

Case (ii): $\beta - \alpha + 2 = 0$. The general solution is

$$y(x) = x^r;$$

two particular solutions, to be determined according to the roots of

$$r^2 + (\alpha - 1)r + \gamma^2 = 0,$$

are shown below.

$(\alpha - 1)^2 - 4\gamma^2$	Particular solutions	
Positive	x^{r_1}	x^{r_2}
Zero	x^δ	$x^\delta \ln x$
Negative	$x^\delta \cos(\epsilon \ln x)$	$x^\delta \sin(\epsilon \ln x)$

Here

$$r_{1,2} = \tfrac{1}{2}\{(1 - \alpha) \pm [(\alpha - 1)^2 - 4\gamma^2]^{1/2}\},$$
$$\delta = \tfrac{1}{2}(1 - \alpha), \qquad \epsilon = \tfrac{1}{2}[4\gamma^2 - (\alpha - 1)^2]^{1/2}.$$

Thus we obtain

$$x^2 \frac{d^2 y}{dx^2} + \alpha x \frac{dy}{dx} + \gamma^2 y = 0, \tag{3-130}$$

which is known as the *equidimensional equation* (also called *Euler's equation* or *Cauchy's equation*). It may easily be shown that Eq. (3-130) is reduced to an equation with constant coefficients by the transformation $x = e^u$. The result is

$$\frac{d^2 y}{du^2} + \alpha \frac{dy}{du} + \gamma^2 y = 0. \tag{3-131}$$

The general solution of Eq. (3-131), $y = e^{ru}$, readily gives that of Eq. (3-130) in the form

$$y(x) = (e^u)^r = x^r. \tag{3-132}$$

Inserting Eq. (3-132) into Eq. (3-130), we obtain the characteristic equation

$$r^2 + (\alpha - 1)r + \gamma^2 = 0. \tag{3-133}$$

Introducing the roots of Eq. (3-133) into Eq. (3-132) yields two particular solutions of Eq. (3-130).

For convenience in the solution of problems related to extended surfaces with variable cross sections, the particular solutions of Eq. (3-126) are summarized in Table 3-1.

3-7. Properties of Bessel Functions

In the properties considered below, Z_ν denotes any Bessel function of order ν, and x a complex number unless otherwise specified.

1. *Bessel functions of the third kind, or Hankel functions of the first and second kinds, of order ν are defined to be*

$$H_\nu^{(1),(2)}(x) = J_\nu(x) \pm Y_\nu(x). \tag{3-134}$$

2. *Derivatives of Bessel functions:*

$$\frac{d}{dx}[x^\nu Z_\nu(mx)] = \begin{cases} mx^\nu Z_{\nu-1}(mx), & Z = J, Y, I, H^{(1)}, H^{(2)} \\ -mx^\nu Z_{\nu-1}(mx), & Z = K \end{cases} \tag{3-135}$$

$$\frac{d}{dx}[x^{-\nu} Z_\nu(mx)] = \begin{cases} -mx^{-\nu} Z_{\nu+1}(mx), & Z = J, Y, K, H^{(1)}, H^{(2)} \\ mx^{-\nu} Z_{\nu+1}(mx), & Z = I. \end{cases} \tag{3-136}$$

A special case of Eq. (3-136) corresponding to $\nu = 0$ is

$$\frac{d}{dx}[Z_0(mx)] = \begin{cases} -mZ_1(mx), & Z = J, Y, K, H^{(1)}, H^{(2)} \\ mZ_1(mx), & Z = I \end{cases} \tag{3-137}$$

$$\frac{d}{dx}[Z_\nu(mx)] = \begin{cases} mZ_{\nu-1}(mx) - (\nu/x)Z_\nu(mx), & Z = J, Y, I, H^{(1)}, H^{(2)} \\ -mZ_{\nu-1}(mx) - (\nu/x)Z_\nu(mx), & Z = K \end{cases} \tag{3-138}$$

$$\frac{d}{dx}[Z_\nu(mx)] = \begin{cases} -mZ_{\nu+1}(mx) + (\nu/x)Z_\nu(mx), & Z = J, Y, K, H^{(1)}, H^{(2)} \\ mZ_{\nu+1}(mx) + (\nu/x)Z_\nu(mx), & Z = I. \end{cases} \tag{3-139}$$

3. *Relations between some Bessel and circular (trigonometric) or hyperbolic functions:*

$$J_{1/2}(x) = (2/\pi x)^{1/2} \sin x,$$
$$J_{-1/2}(x) = (2/\pi x)^{1/2} \cos x, \tag{3-140}$$

$$I_{1/2}(x) = (2/\pi x)^{1/2} \sinh x,$$
$$I_{-1/2}(x) = (2/\pi x)^{1/2} \cosh x. \tag{3-141}$$

4. *Relations between some Bessel functions:*

$$J_\nu(xe^{im\pi}) = e^{im\nu\pi} J_\nu(x), \tag{3-142}$$

$$Y_\nu(xe^{im\pi}) = e^{-im\nu\pi} Y_\nu(x) + 2i \sin m\nu\pi \cot \nu\pi J_\nu(x), \tag{3-143}$$

$$I_\nu(xe^{\pm i\pi/2}) = e^{\pm i\nu\pi/2} J_\nu(x), \tag{3-144}$$

$$K_\nu(xe^{\pm i\pi/2}) = \pm i \frac{\pi}{2} e^{\mp i\nu\pi/2} [-J_\nu(x) \pm i Y_\nu(x)]$$
$$= \mp i \frac{\pi}{2} e^{\mp i\nu\pi/2} H_\nu^{(2),(1)}(x), \tag{3-145}$$

$$J_\nu(xe^{\pm i3\pi/4}) = \text{ber}_\nu x \pm i \text{bei}_\nu x, \tag{3-146}$$

where x is real, and ber and bei stand for Bessel-real and Bessel-imaginary, respectively;

$$K_\nu(xe^{\pm i\pi/4}) = e^{\pm i\nu\pi/2}(\text{ker}_\nu x \pm i \text{kei}_\nu x), \tag{3-147}$$

where x is real, and ker and kei stand for Kelvin-real and Kelvin-imaginary, respectively.

5. *Behavior of Bessel functions for small arguments.* It follows from the form of the power series solutions discussed in this section that the series representations of the Bessel functions considered so far converge rapidly for small arguments. Retaining the first few terms of these series, we may obtain the behavior of Bessel functions for small arguments. For example, from Eq. (3–110),

$$J_0(x) = 1 - \frac{(x/2)^2}{(1!)^2} + \frac{(x/2)^4}{(2!)^2} - \cdots, \tag{3-148}$$

$$J_1(x) = \frac{x}{2} - \frac{(x/2)^3}{1!2!} + \frac{(x/2)^5}{2!3!} - \cdots, \tag{3-149}$$

$$\vdots$$

$$J_\nu(x) = \frac{(x/2)^\nu}{\Gamma(\nu+1)} \left\{ 1 - \frac{(x/2)^2}{1!(\nu+1)} + \frac{(x/2)^4}{2!(\nu+1)(\nu+2)} - \cdots \right\}, \tag{3-150}$$

and from Eq. (3–118),

$$I_0(x) = 1 + \frac{(x/2)^2}{(1!)^2} + \frac{(x/2)^4}{(2!)^2} + \cdots, \qquad (3\text{–}151)$$

$$I_1(x) = \frac{x}{2} - \frac{(x/2)^3}{1!2!} + \frac{(x/2)^5}{2!3!} + \cdots, \qquad (3\text{–}152)$$

$$\vdots$$

$$I_\nu(x) = \frac{(x/2)^\nu}{\Gamma(\nu+1)}\left\{1 + \frac{(x/2)^2}{1!(\nu+1)} + \frac{(x/2)^4}{2!(\nu+1)(\nu+2)} + \cdots\right\}. \qquad (3\text{–}153)$$

The small-argument expansion of other Bessel functions may be written in a similar way.

6. *Asymptotic (large-argument) behavior of Bessel functions.* Asymptotic expansions require a different treatment than that of power series. This will not be given here. Only the results corresponding to a number of frequently encountered cases are listed below:

$$J_\nu(x) \sim \left(\frac{2}{\pi x}\right)^{1/2}\left\{\left[1 - \frac{(4\nu^2-1^2)(4\nu^2-3^2)}{2!(8x)^2} + \cdots\right]\cos\left(x - \frac{\pi}{4} - \frac{\nu\pi}{2}\right)\right.$$

$$\left. - \left[\frac{4\nu^2-1^2}{1!8x} - \cdots\right]\sin\left(x - \frac{\pi}{4} - \frac{\nu\pi}{2}\right)\right\}, \qquad (3\text{–}154)$$

$$Y_\nu(x) \sim \left(\frac{2}{\pi x}\right)^{1/2}\left\{\left[1 - \frac{(4\nu^2-1^2)(4\nu^2-3^2)}{2!(8x)^2} + \cdots\right]\sin\left(x - \frac{\pi}{4} - \frac{\nu\pi}{2}\right)\right.$$

$$\left. + \left[\frac{4\nu^2-1^2}{1!8x} - \cdots\right]\cos\left(x - \frac{\pi}{4} - \frac{\nu\pi}{2}\right)\right\}, \qquad (3\text{–}155)$$

$$I_\nu(x) \sim \frac{e^x}{(2\pi x)^{1/2}}\left\{1 - \frac{4\nu^2-1^2}{1!8x} + \frac{(4\nu^2-1^2)(4\nu^2-3^2)}{2!(8x)^2} - \cdots\right\}, \qquad (3\text{–}156)$$

$$K_\nu(x) \sim \left(\frac{\pi}{2x}\right)^{1/2} e^{-x}\left\{1 + \frac{4\nu^2-1^2}{1!8x} + \frac{(4\nu^2-1^2)(4\nu^2-3^2)}{2!(8x)^2} + \cdots\right\}. \qquad (3\text{–}157)$$

7. *Graphical representation of the general behavior of Bessel functions.* Graphs of the general behavior of Bessel functions are shown in Fig. 3–24.

Having thus completed our review of Bessel functions we may now proceed to demonstrate the use of these functions in the solution of problems related to extended surfaces.

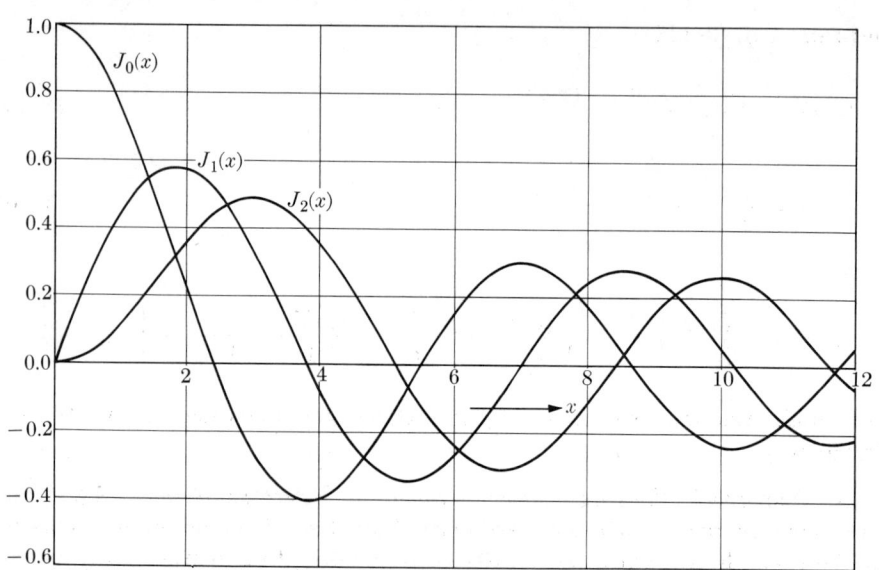

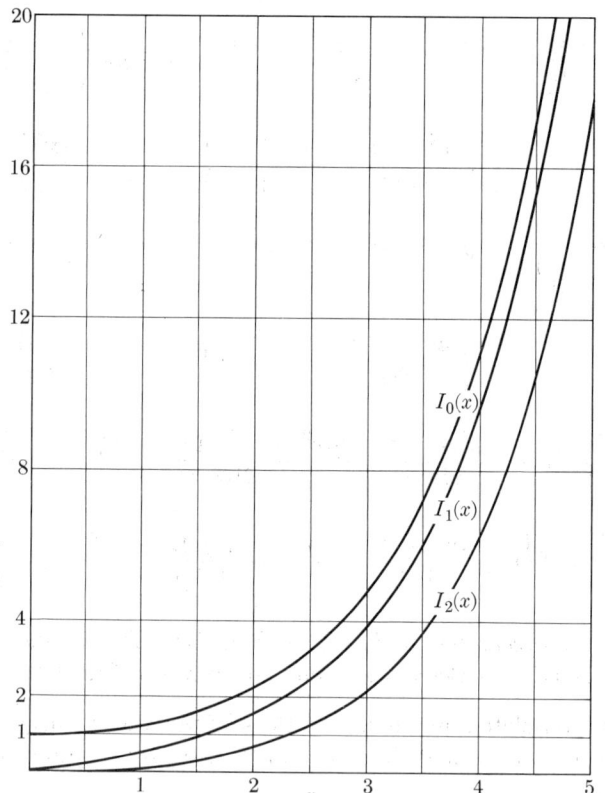

FIG. 3-24 (a)

FIG. 3-24 (c)

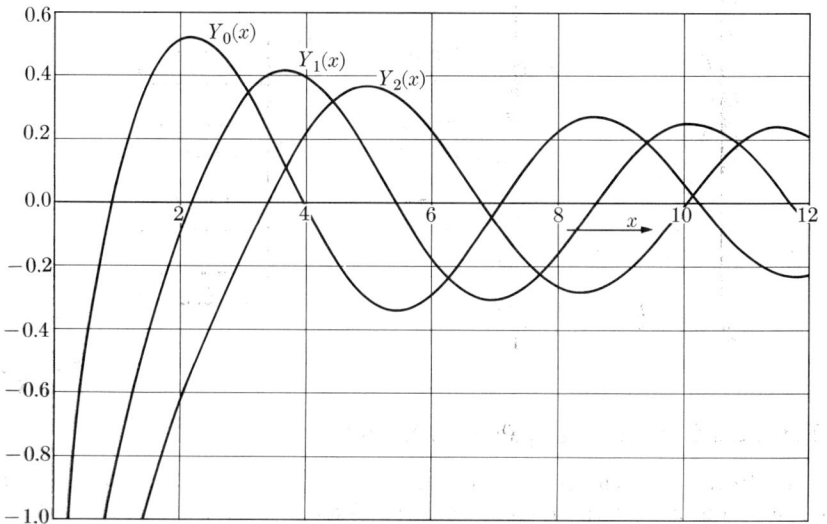

FIG. 3-24 (b)

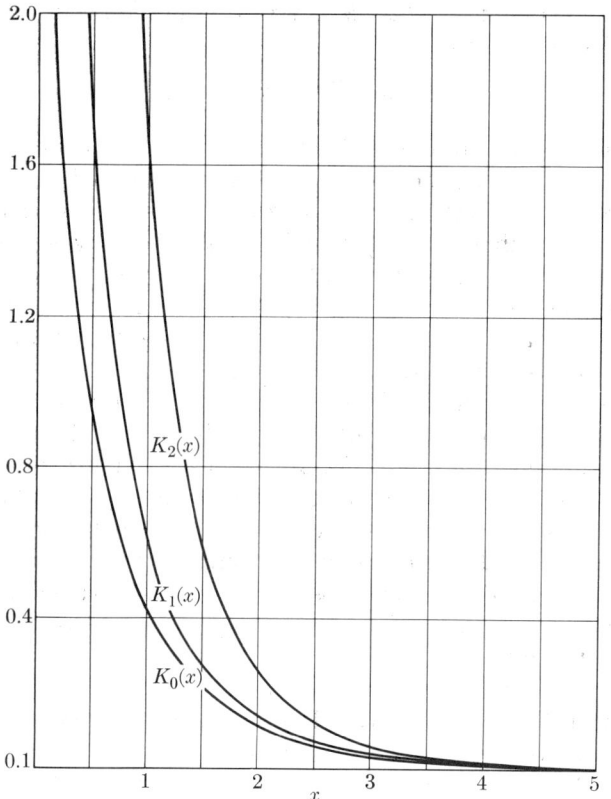

FIG. 3-24 (d)

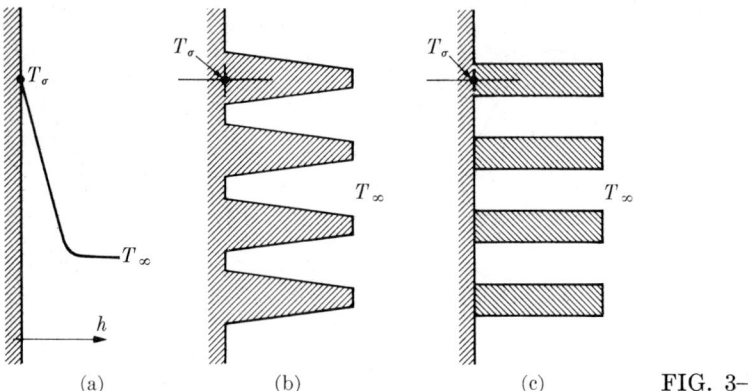

FIG. 3-25

3-8. Extended Surfaces (Fins, Pins, or Spines)

Before we formulate heat transfer problems associated with extended surfaces, let us briefly discuss what we mean by extended surfaces and why they deserve special attention. For this purpose, consider first a wall at temperature T_σ transferring heat by convection to an ambient at temperature T_∞ (Fig. 3-25a). The rate of heat transfer from this wall may be evaluated in terms of a heat transfer coefficient in the form

$$q = hA(T_\sigma - T_\infty). \tag{3-158}$$

One of the prime objectives of the study of heat transfer is to find ways of controlling this q. For example, the design of a heat exchanger is often based on achieving the smallest possible heat transfer area (for lightness or compactness) or else the largest possible amount of heat transfer for any given size heat exchanger. Clearly, q of Eq. (3-158) may be increased by increasing (i) the temperature difference between the wall and the ambient, (ii) the heat transfer coefficient, or (iii) the heat transfer area (keeping the projected area unchanged). The first case needs no explanation; the second case is the subject matter of texts on convective heat transfer; the third case is the concern of this section.

The surface area of a wall may, in principle, be increased in two ways as shown in Figs. 3-25(b) and (c). In Fig. 3-25(b) the extended surfaces are integral parts of the base material, obtained by a casting or extruding process. In Fig. 3-25(c) the extended surfaces, which may or may not be made from the base material, are attached to the base by pressing, soldering, or welding. The same geometry is obtained, though less frequently, by machining the base material. In practice, manufacturing technology and cost dictate the selection of the most desirable form.

Applications of extended surfaces are numerous, particularly in heat transfer to gaseous media. Since in this case the corresponding heat transfer coefficient is low, a small, compact heat exchanger may be achieved only by the use of extended surfaces. Well-known examples of the use of extended surfaces are

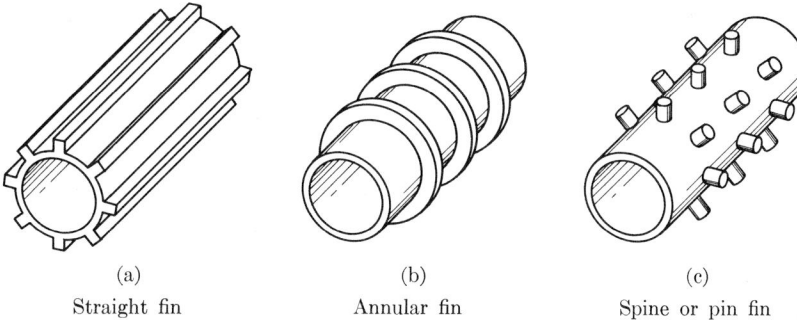

(a) Straight fin   (b) Annular fin   (c) Spine or pin fin

FIG. 3–26

the fins on car radiators and in heating units, liquid-to-gas or gas-to-gas heat exchangers, boilers, and air-cooled engines. For later reference, the most common extended surfaces may be classified as *straight fins, annular fins,* and *spines* or *pin fins* (Fig. 3–26).

Let us now return to our objective, the study of heat transfer through extended surfaces. Since the temperature of an extended surface does not remain constant along its length, because of transversal heat transfer by convection to the surroundings, the heat transfer from extended surfaces cannot be evaluated from Eq. (3–158). Thus we are forced to evaluate first the temperature distribution in extended surfaces, and then the heat transfer in terms of this temperature distribution.

Let us consider an extended surface with variable cross section (Fig. 3–27). We make the following assumptions. (i) The transversal characteristic length, say the thickness of the fin, is small compared to the axial length. Thus the transversal temperature distribution is negligible compared to the axial temperature distribution. (ii) The thermal conductivity is constant (that is, we

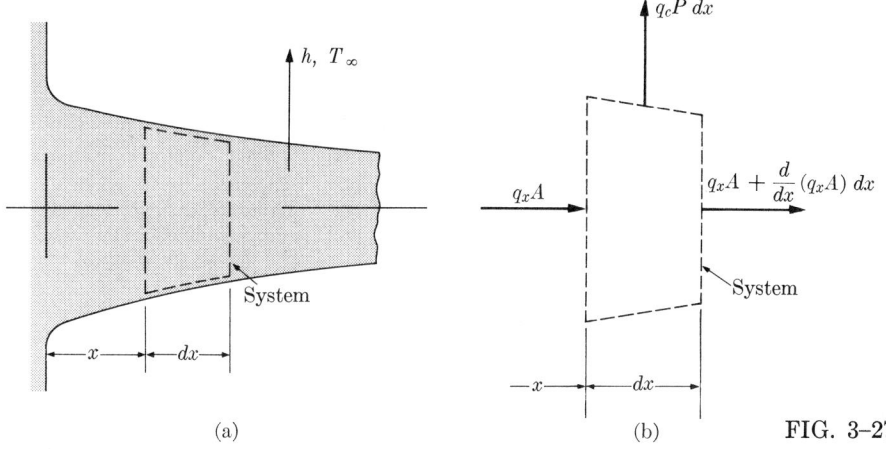

FIG. 3–27

are neglecting its dependence on temperature). (iii) The heat transfer coefficient to be used is the radially and axially averaged value of the actual heat transfer coefficient.

Under steady conditions and using assumption (i), from the first four steps of the formulation we readily obtain

$$\frac{d}{dx}\left(kA\frac{dT}{dx}\right) - q_c P = 0, \tag{3-159}$$

where T is the local temperature, x the axial distance, k the thermal conductivity, A the variable cross-sectional area, q_c the transversal heat flux, and P the periphery. Equation (3–159) is to be satisfied by the transversal boundary condition,

$$q_c = h(T - T_\infty), \tag{3-160}$$

and two boundary conditions in x, one related to the base, the other to the tip of the extended surface, all constituting the fifth and last step of the formulation.

From the standpoint of solution it is convenient to insert Eq. (3–160) into Eq. (3–159). We have then

$$\frac{d}{dx}\left(kA\frac{dT}{dx}\right) - hP(T - T_\infty) = 0, \tag{3-161}$$

to be supplemented only by the boundary conditions in x. Considering a constant k according to assumption (ii), and measuring temperatures above the ambient, we may further rearrange Eq. (3–161) to give

$$\frac{d}{dx}\left(A\frac{d\theta}{dx}\right) - \frac{hP}{k}\theta = 0, \tag{3-162}$$

where $\theta = T - T_\infty$. Note that we have used only the first two of our assumptions in the formulation of the problem. Assumption (iii), constant h, will be employed later to simplify the integration of the formulation.

Since we have already discussed in detail the most frequently encountered boundary conditions in heat transfer problems (see Section 2–8), the boundary conditions to be imposed on Eq. (3–162) need no specific attention. We therefore proceed to a number of illustrative examples, considering first extended surfaces with constant cross sections, and then those with variable cross sections.

Extended surfaces with constant cross sections. When the cross-sectional area is constant, Eq. (3–163) reduces to

$$\frac{d^2\theta}{dx^2} - m^2\theta = 0, \tag{3-163}$$

where $m^2 = hP/kA$.

The general solution of Eq. (3–163) can be written in the form

$$\theta(x) = C_1 e^{mx} + C_2 e^{-mx} \tag{3-164}$$

or

$$\theta(x) = C_3 \cosh mx + C_4 \sinh mx. \tag{3-165}$$

As will be seen in the use of boundary conditions, Eqs. (3–164) and (3–165) are suitable to problems related to infinitely long and finite extended surfaces, respectively.

Example 3–9. Consider an infinitely long fin (Fig. 3–28) whose base temperature T_0 is specified. We wish to find the temperature distribution in and the heat transfer from the fin.

Selecting the base of the fin as the origin of x, and noting that the temperature of the fin approaches that of the ambient as $x \to \infty$, we may write the boundary conditions of the problem as

$$\theta(0) = \theta_0, \tag{3-166}$$

$$\lim_{x \to \infty} \theta(x) \to 0, \tag{3-167}$$

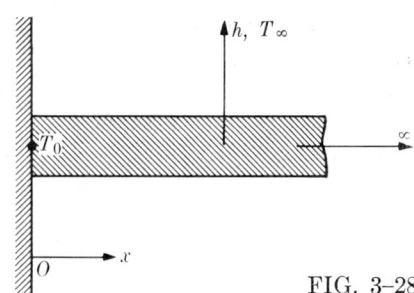

FIG. 3–28

where $\theta_0 = T_0 - T_\infty$.

Equation (3–167) implies that C_1 of Eq. (3–164) must be identically zero. Equation (3–166) readily gives $C_2 = \theta_0$, and the desired temperature distribution is found to be

$$\frac{\theta(x)}{\theta_0} = e^{-mx}. \tag{3-168}$$

The heat transfer from the fin may now be evaluated in terms of this temperature distribution by simply integrating the local convection along the fin. Thus we have

$$q = \int_0^\infty hP\theta \, dx = hP\theta_0 \int_0^\infty e^{-mx} \, dx = \theta_0 (hPkA)^{1/2}. \tag{3-169}$$

Noting that the total heat transfer from the fin by convection must be supplied to the base of the fin by conduction, we may get the same result in terms of conduction as

$$q = -kA \left(\frac{d\theta}{dx}\right)_{x=0} = -kA\theta_0 \frac{d}{dx}(e^{-mx})|_{x=0}$$
$$= \theta_0 (hPkA)^{1/2}.$$

The latter way of evaluating heat transfer, involving a differentiation rather than an integration, is a more convenient one, especially for complicated problems.

Example 3–10. Consider a fin of finite length L. The base temperature T_0 of the fin is specified, and the tip of the fin is insulated (Fig. 3–29). We wish to find the temperature distribution in and the heat transfer from the fin.

For this problem the tip of the fin is more convenient as the origin of x. The boundary conditions are then

$$\frac{d\theta(0)}{dx} = 0, \qquad (3\text{-}170)$$

$$\theta(L) = \theta_0, \qquad (3\text{-}171)$$

where, as before, $\theta_0 = T_0 - T_\infty$.

Since the fin is finite in length, we refer to the general solution given by Eq. (3–165). The use of Eq. (3–170), or, equivalently, the fact that the temperature distribution is symmetric with respect to x, hence is composed of even functions only, yields $C_4 = 0$. Next, the consideration of Eq. (3–171) gives $C_3 = \theta_0/\cosh mL$. Therefore,

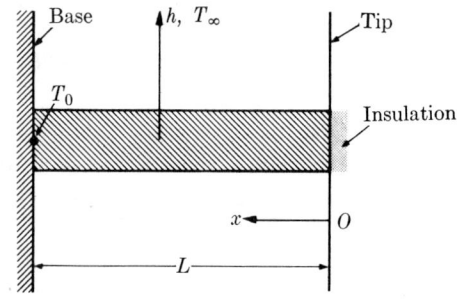

FIG. 3–29

$$\frac{\theta(x)}{\theta_0} = \frac{\cosh mx}{\cosh mL}. \qquad (3\text{-}172)$$

[Re-solve the same problem assuming that the origin of x is at the base of the fin and starting from the general solution given by Eq. (3–164). Compare the algebraic complexity of the two procedures.]

The total heat transfer from the fin, evaluated from the conduction at the base of the fin, is

$$q = -\left[-kA\left(\frac{d\theta}{dx}\right)_{x=L}\right]$$

$$= \frac{kA\theta_0}{\cosh mL} \frac{d}{dx}(\cosh mx)\bigg|_{x=L}$$

$$= \theta_0(hPkA)^{1/2} \tanh mL. \qquad (3\text{-}173)$$

Since $\lim_{x\to\infty} \tanh x \to 1$, Eq. (3–173) approaches Eq. (3–169) as $mL \to \infty$. In other words, heat transfer from a finite fin approaches that from an infinite fin as the length of the finite fin increases indefinitely. This statement is independent of the boundary condition employed at the tip of the fin, since the effect of the tip diminishes as $L \to \infty$. The condition $mL \to \infty$ may also be interpreted as $m \to \infty$ for a given L. This case, as readily seen from the definition of m, $(hP/kA)^{1/2}$, corresponds to $h \to \infty$ or $k \to 0$.

Before proceeding to extended surfaces of variable cross section it may be useful to introduce a basis for the evaluation and comparison of extended surfaces. Such a basis is usually described in terms of an *extended-surface efficiency*.

There are two customary definitions for this efficiency as the ratio of the actual to a hypothetical heat transfer:

$$\eta_b = \frac{\text{Actual heat transfer from extended surface}}{\text{Heat transfer from wall without fin}}, \tag{3-174}$$

$$\eta = \frac{\text{Actual heat transfer from extended surface}}{\text{Heat transfer from extended surface at base temperature}}. \tag{3-175}$$

The denominator of Eq. (3–174) denotes the heat transfer from an area of the wall equivalent to the base area of the extended surface; the heat transfer to be evaluated by numerator and denominator together is based on the same temperature difference, base minus ambient. Since the temperature of a wall and the heat transfer coefficient between the wall and the ambient are somewhat changed when an extended surface is attached to the wall, the efficiency defined by Eq. (3–174) is quite approximate. The error involved in this approximation depends on the length of the extended surfaces and the space between them. Since the changes in the wall temperature and the heat transfer coefficient affect equally the numerator and the denominator of Eq. (3–175), the efficiency defined by this equation is more realistic, and is often preferred in practice. However, rather than to demonstrate the increased heat transfer from a wall by the use of extended surfaces, this efficiency may better be employed to compare different extended surfaces. The particular values of these efficiencies for Example 3–9 are

$$\eta_b = \frac{\theta_0(hPkA)^{1/2}}{\theta_0 hA} = \left(\frac{kP}{hA}\right)^{1/2},$$

$$\eta = \lim_{L\to\infty} \frac{\theta_0(hPkA)^{1/2}}{\theta_0 hPL} = \lim_{L\to\infty}\left(\frac{kA}{hP}\right)\frac{1}{L} = \lim_{L\to\infty}\left(\frac{1}{mL}\right) \to 0,$$

and those for Example 3–10 are

$$\eta_b = \frac{\theta_0(hPkA)^{1/2}\tanh mL}{\theta_0 hA} = \left(\frac{kP}{hA}\right)^{1/2}\tanh mL,$$

$$\eta = \frac{\theta_0(hPkA)^{1/2}\tanh mL}{\theta_0 hPL} = \frac{\tanh mL}{mL}.$$

The efficiencies of extended surfaces have been extensively investigated in the literature. In practice, however, the technology involved may be a more important consideration than finding a 5–10% more efficient profile which is expensive to manufacture. For this reason, the efficiency of extended surfaces is not emphasized in this text. Detailed treatments may be found in References 8, 9, and 11.

Extended surfaces with variable cross sections. The general formulation of problems of extended surfaces with variable cross sections has already been given by Eq. (3–162). Since A and P are no longer constant, this equation now becomes a differential equation with variable coefficients whose general solution can be determined only when $A(x)$ and $P(x)$ are specified. In most cases, as mentioned in Section 3–6, Eq. (3–162) is reduced to a form of Bessel's equation; a special case is that leading to the equidimensional equation. Cases which do not lead to either of these equations may be treated individually by employing the power series solutions of differential equations. The next two examples illustrate extended surfaces with variable cross sections.

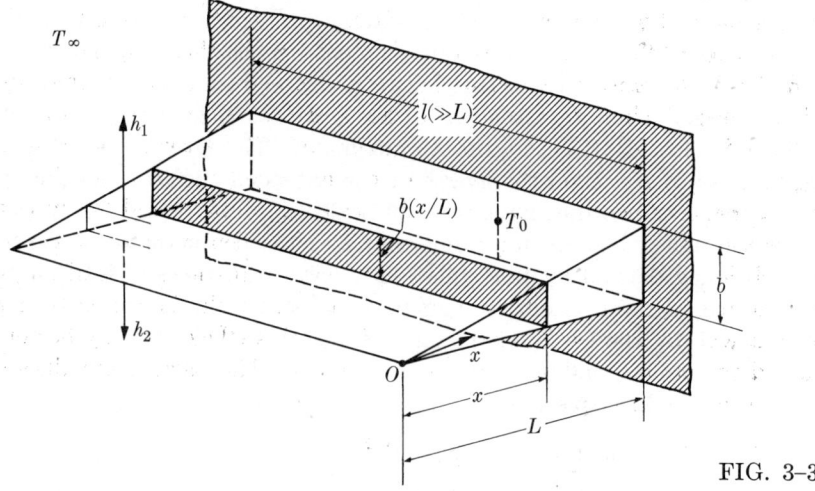

FIG. 3–30

Example 3–11. The geometry of a straight fin of triangular profile is described in Fig. 3–30. The base temperature T_0 of the fin is specified. We wish to find the temperature distribution in and the heat transfer from the fin.

We make the usual assumption for extended surfaces that $b/L \ll 1$, and furthermore, to make the temperature distribution of the present problem one-dimensional we assume either that $L/l \ll 1$ or that the ends in the l-direction are insulated.

Noting from Fig. 3–30 that $A = b(x/L)l$ and $hP = (h_1 + h_2)l$, inserting these values into Eq. (3–162), and rearranging the result, we get

$$\frac{d}{dx}\left(x \frac{d\theta}{dx}\right) - m^2 \theta = 0, \tag{3–176}$$

where $m^2 = (h_1 + h_2)L/kb$. Comparison of Eq. (3–176) with Table 3–1 gives

$$\alpha = 1, \quad \beta = 0, \quad \gamma = \pm im,$$
$$\nu = 0, \quad \mu = 2, \quad \nu/\mu = 0.$$

The general solution of Eq. (3–176) is then

$$\theta(x) = C_1 I_0(2mx^{1/2}) + C_2 K_0(2mx^{1/2}). \tag{3-177}$$

Since we have from Fig. 3–23(d)

$$\lim_{x \to 0} K_0(x) \to \infty,$$

the finiteness of tip temperature implies $C_2 = 0$. Next, the use of the base temperature yields $C_1 = \theta_0/I_0(2mL^{1/2})$, where, as before, $\theta_0 = T_0 - T_\infty$. Inserting the values of C_1 and C_2 into Eq. (3–177), we find that the temperature distribution in the fin is

$$\frac{\theta(x)}{\theta_0} = \frac{I_0(2mx^{1/2})}{I_0(2mL^{1/2})}. \tag{3-178}$$

Again, the heat transfer from the fin may conveniently be obtained by considering the conduction through its base. Thus

$$q = -[-kA(d\theta/dx)_{x=L}],$$

which may be evaluated from Eq. (3–178) by means of Eq. (3–137). It follows, in terms of $\xi = 2mx^{1/2}$, that

$$\frac{d}{dx}[I_0(\xi)] = \frac{d}{d\xi}[I_0(\xi)]\frac{d\xi}{dx} = \frac{m}{x^{1/2}} I_1(2mx^{1/2}),$$

and the heat transfer from the fin is

$$\frac{q}{kA\theta_0/L} = \frac{(mL^{1/2})I_1(2mL^{1/2})}{I_0(2mL^{1/2})}. \tag{3-179}$$

Example 3–12. An empty skillet is forgotten on a hot plate (Fig. 3–31). It may be assumed that the bottom of the skillet is subjected to a uniform heat flux q''. The ambient temperature is T_∞, and the heat transfer coefficients are h_1 and h_2. The thermal conductivity, thickness, radius, and height of the skillet are k, δ, R, and L, respectively. We wish to find the temperature distribution in the skillet.

Neglecting the temperature change across the thickness, we may assume that the temperature distribution in the skillet is one-dimensional, radial at the bottom, and axial in the side walls.

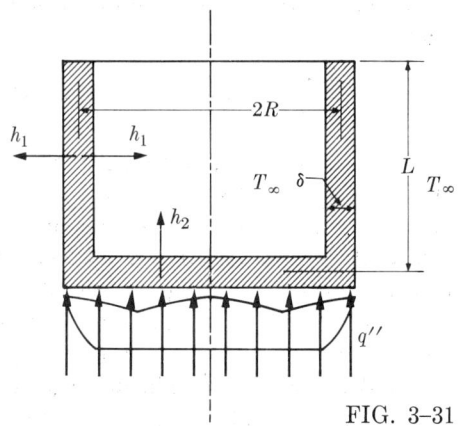

FIG. 3–31

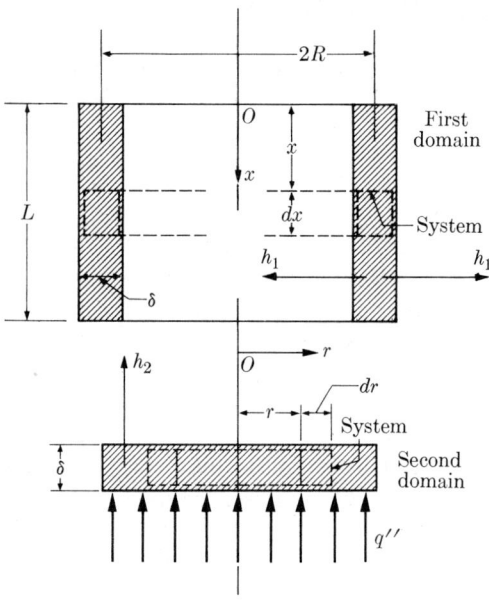

FIG. 3-32

This suggests that the problem be investigated in terms of two domains for mathematical convenience (Fig. 3-32).

The first two steps of the formulation, applied to the one-dimensional system considered for the side walls (Fig. 3-33), give (the general law)

$$0 = +q_x A - \left(q_x + \frac{dq_x}{dx} dx\right) A - q_{c_1} P_1 \, dx - q_{c_2} P_2 \, dx. \quad (3\text{-}180)$$

The particular law of the third step inserted into Eq. (3-180), then the governing equation resulting from the fourth step rearranged by the part of the fifth step related to the definition of heat transfer coefficient, give for the side walls

$$\frac{d^2 \theta_1}{dx^2} - m_1^2 \theta_1 = 0, \quad (3\text{-}181)$$

assuming that $\delta \ll R$ so that $P_1 \cong P_2$. Here $m_1^2 = 2h_1/k\delta$, $\theta_1 = T_1 - T_\infty$, and T_1 is the local temperature of the side walls. As expected, Eq. (3-181) is identical to the formulation of problems of extended surfaces with constant cross sections given by Eq. (3-163), except for the definition of m_1^2. The finite geometry of the side walls suggests a general

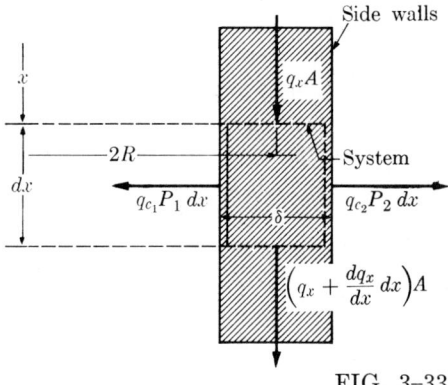

FIG. 3-33

solution for Eq. (3–181) in the form of Eq. (3–165). Thus we have

$$\theta_1(x) = C_3 \cosh m_1 x + C_4 \sinh m_1 x. \qquad (3\text{--}182)$$

Let us now consider the formulation related to the bottom of the skillet. The first two steps of the formulation, interpreted in terms of the one-dimensional system shown in Fig. 3–34, give

$$0 = +q_r A - \left[q_r A + \frac{d}{dr}(q_r A) \, dr \right] - q_c P \, dr + q'' P \, dr. \qquad (3\text{--}183)$$

Again, introducing the particular law from the third step of the formulation into Eq. (3–183) and rearranging the governing equation resulting from the fourth step by the part of the fifth step associated with the heat transfer coefficient, we get the formulation of the bottom as follows:

$$\frac{d}{dr}\left(r \frac{d\theta_2}{dr} \right) - m_2^2 r \left(\theta_2 - \frac{n}{m_2^2} \right) = 0, \qquad (3\text{--}184)$$

where $m_2^2 = h_2/k\delta$, $n = q''/k\delta$, $\theta_2 = T_2 - T_\infty$, and T_2 is the local temperature of the bottom. Note that Eq. (3–184) is homogeneous in terms of $\theta_2 - n/m_2^2$. Comparing Eq. (3–184) with Table 3–1 yields

$$\alpha = 1, \quad \beta = 1, \quad \gamma = \pm im,$$
$$\nu = 0, \quad \mu = 1, \quad \nu/\mu = 0.$$

Hence the solution for the bottom of the skillet is found to be

$$\theta_2(r) = n/m_2^2 + D_1 I_0(m_2 r) + D_2 K_0(m_2 r). \qquad (3\text{--}185)$$

Neglecting the heat transfer from the top of the side walls, and referring to conveniently selected origins for the axis of the bottom and for that of the side

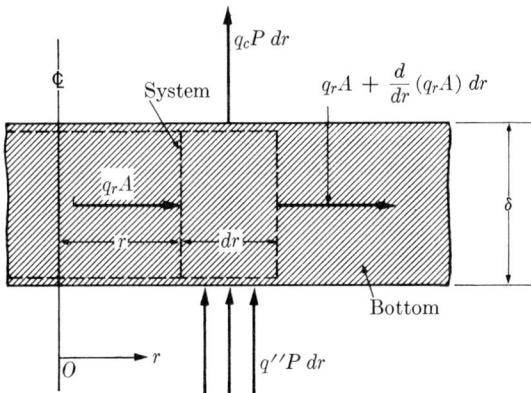

FIG. 3–34

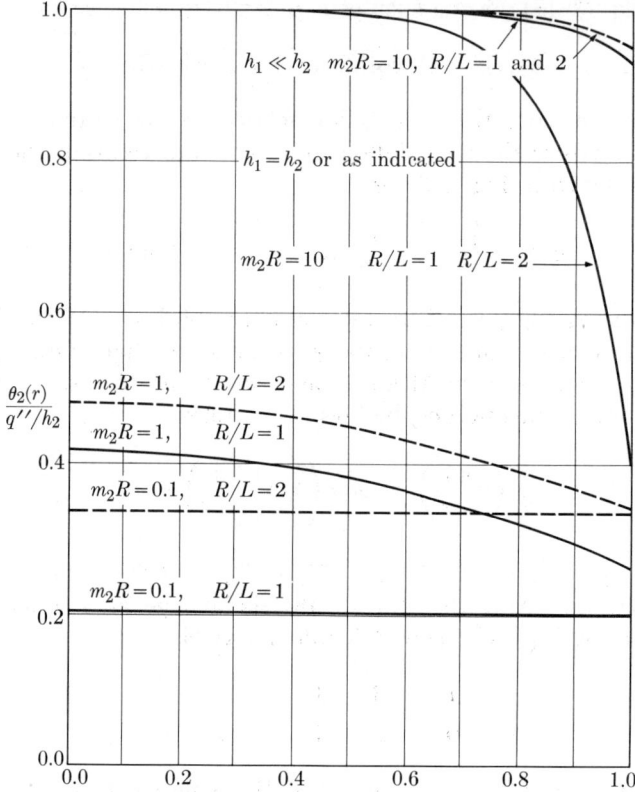

FIG. 3-35

walls (Fig. 3-32), we may now write the boundary conditions of the problem in the form

$$\frac{d\theta_1(0)}{dx} \cong 0, \quad (3\text{-}186)$$

$$\theta_1(L) = \theta_2(R), \quad (3\text{-}187)$$

$$\left[-k\frac{d\theta_1(L)}{dx}\right] + \left[-k\frac{d\theta_2(R)}{dr}\right] = 0, \quad (3\text{-}188)$$

$$\theta_2(0) = \text{finite} \quad \left(\text{or } \frac{d\theta_2(0)}{dr} = 0\right). \quad (3\text{-}189)$$

Equations (3-182) and (3-185) considered with Eqs. (3-186) and (3-189), respectively, give $C_4 = 0$ and $D_2 = 0$. Then the results introduced into Eqs. (3-187) and (3-188) yield

$$C_3 \cosh m_1 L = n/m_2^2 + D_1 I_0(m_2 R),$$

$$C_3 m_1 \sinh m_1 L + D_1 m_2 I_1(m_2 R) = 0.$$

Solving these algebraic equations for the constants C_3 and D_1, and inserting the resulting values together with C_4 and D_2 into Eqs. (3–182) and (3–185), we are led to the temperature distribution in the skillet as follows:

$$\frac{\theta_1(x)}{q''/h_2} = \frac{\cosh m_1 x/\cosh m_1 L}{1 + (m_1/m_2)[I_0(m_2 R)/I_1(m_2 R)]\tanh m_1 L}, \quad (3\text{–}190)$$

$$\frac{\theta_2(r)}{q''/h_2} = 1 - \frac{I_0(m_2 r)/I_0(m_2 R)}{1 + (m_2/m_1)[I_1(m_2 R)/I_0(m_2 R)]\coth m_1 L}. \quad (3\text{–}191)$$

[What are the limiting forms of Eqs. (3–190) and (3–191) as $m_2 \to 0$ and $m_2 \to \infty$? Explain the physics of the corresponding cases.]

In conjunction with the foregoing limiting cases, let us plot the temperature of the bottom as given by Eq. (3–191). Because of the many parameters involved, a complete parametric study is somewhat lengthy, and is unnecessary for our purpose. Here we consider only the practical case in which $h_1/h_2 \sim 1$ and R/L varies between 1 and 2. In Fig. 3–35 the values of $\theta_2(r)/(q''/h_2)$ are plotted against r/R for the values 0.1, 1, and 10 of $m_2 R$, and for the values 1 and 2 of R/L corresponding to $h_1/h_2 = 1$. Inspection of this figure reveals the practical values which may conveniently be used in place of the mathematical limits, $m_2 \to 0$ and $m_2 \to \infty$. Thus when $h_1/h_2 = 1$, we obtain $m_2 R \le 0.1$ for the lower limit. In this case $k \gg h$ and the temperature of the entire skillet can be lumped. It follows readily from the lumped formulation of the problem that

$$\frac{\theta}{q''/h} = \frac{1}{1 + 4(L/R)}, \quad \text{when} \quad h_1/h_2 = 1 \quad \text{and} \quad m_2 R \le 0.1.$$

For the upper limit we find $m_2 R \ge 10$. However, this case yields no simplification in the formulation.

Now, if we assume a thin layer of water at the bottom of the skillet, $h_1/h_2 \sim 1/200$, and when $m_2 R \ge 10$, the bottom temperature of the skillet may be lumped within 5% error or better as shown in the upper right-hand corner of Fig. 3–35. Then the lumped temperature of the bottom is found to be

$$\theta_2 = q''/h_2, \quad \text{when} \quad h_1 \ll h_2 \quad \text{and} \quad m_2 R \ge 10.$$

Employing this value as the base temperature, we may reduce the formulation of the side walls to that of Example 3–10. Thus the temperature distribution in the side walls becomes

$$\frac{\theta_1(x)}{q''/h_2} = \frac{\cosh m_1 x}{\cosh m_1 L}, \quad \text{when} \quad h_1 \ll h_2 \quad \text{and} \quad m_2 R \ge 10.$$

Here we find another opportunity to stress the importance of physical reasoning in the simplification of complex problems.

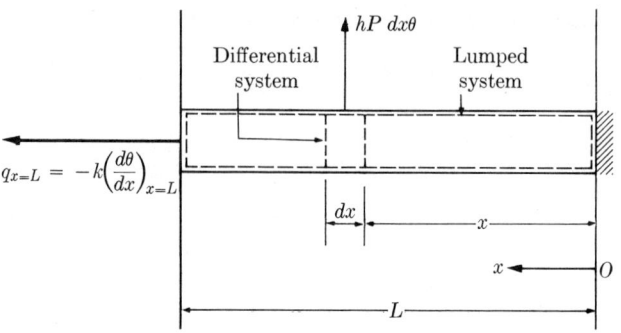

FIG. 3-36

3-9. Approximate Solutions for Extended Surfaces

In this section we obtain approximate solutions of some examples which have already been considered for extended surfaces. These solutions are based on the selection of temperature profiles satisfying the boundary conditions of the problem in terms of simple functions. The parameters of the selected profiles are determined from the integral formulation of the problem.

Example 3-13. We wish to find a first-order approximate solution for the problem of Example 3-10.

The integral formulation of the problem is readily obtained in terms of the lumped and the differential systems shown in Fig. 3-36. Thus we have

$$0 = -\left[-kA\left(\frac{d\theta}{dx}\right)_{x=L}\right] - hP\int_0^L \theta\, dx. \tag{3-192}$$

Equation (3-192) may be rearranged in the form

$$0 = \left(\frac{d\theta}{dx}\right)_{x=L} - m^2 \int_0^L \theta\, dx, \tag{3-193}$$

which is identical to the integral of Eq. (3-163) over the interval $(0, L)$, as expected. Equation (3-193) may further be rearranged in terms of $\xi = x/L$ and $\mu = mL$ for convenience in the evaluation of the temperature profiles. Thus

$$0 = \left(\frac{d\theta}{d\xi}\right)_{\xi=1} - \mu^2 \int_0^1 \theta\, d\xi. \tag{3-194}$$

A first-order Ritz profile which satisfies the boundary conditions of the problem may be written in terms of the following parabola* (Fig. 3-37):

$$\theta(\xi)/\theta_0 = 1 - (1 - \xi^2)a_0, \tag{3-195}$$

---

* It is also possible but less convenient to define this parabola in terms of the tip temperature, $b_0\theta_0$, in the form $\theta(\xi)/\theta_0 = \xi^2 + (1 - \xi^2)b_0$.

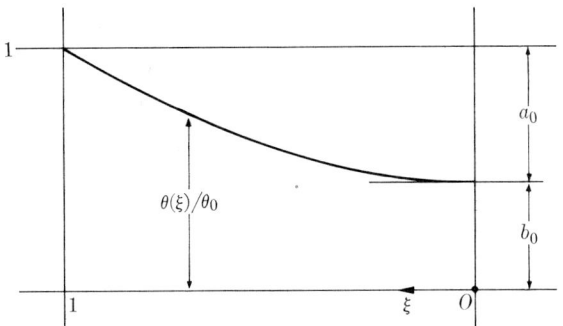

FIG. 3-37

where a_0 is the unknown parameter to be determined. Insertion of Eq. (3-195) into Eq. (3-194) and subsequent integration gives

$$a_0 = \frac{\mu^2/2}{1 + \mu^2/3}.$$

Thus we are led to a first-order Ritz approximation of the fin temperature in the form

$$\frac{\theta(\xi)}{\theta_0} = 1 - \frac{\mu^2/2}{1 + \mu^2/3}(1 - \xi^2). \tag{3-196}$$

Let us now compare the exact and the approximate solutions of the problem given by Eqs. (3-172) and (3-196). When we consider a particular location on the fin, the effect of ξ may be eliminated from this comparison. Since it is the temperature gradient, rather than the temperature itself, that is satisfied at the tip by the approximate profile, the maximum discrepancy between the exact and approximate temperatures is anticipated at the tip of the fin. Inserting $\xi = 0$ into the dimensionless form of Eq. (3-172) and into Eq. (3-196), we find that the exact and approximate tip temperatures are

$$\frac{\theta(0)}{\theta_0} = \frac{1}{\cosh \mu}, \qquad \frac{\theta(0)}{\theta_0} = \frac{1 - \mu^2/6}{1 + \mu^2/3}.$$

These temperatures are compared in Table 3-2 for some values of μ. Inspection of Table 3-2 reveals the close agreement between the exact and approximate solutions for small values of μ. To understand the poor agreement for large values of μ we note that the exact temperature distribution of the present problem behaves like $\theta(\xi)/\theta_0 = e^{-\mu\xi}$ as $\mu \to \infty$* which, for example, may be interpreted as $L \to \infty$. Now if we recall the fact that a function, say $e^{-\mu\xi}$, cannot well be approximated over a large interval by another function, say Eq. (3-196), the increased discrepancy between the two solutions as $\mu \to \infty$ becomes evident.

---

* The exact solution of the problem in terms of ξ measured from the base of the fin is $\theta(\xi)/\theta_0 = \cosh \mu(1 - \xi)/\cosh \mu$. As $\mu \to \infty$ this solution approaches $e^{\mu(1-\xi)}/e^{\mu} = e^{-\mu\xi}$, which is the dimensionless form of Eq. (3-168), as expected.

TABLE 3–2

μ	$\frac{1}{2}$	1	2	4
Exact	0.8868	0.6481	0.2658	0.0366
Approx.	0.8846	0.6250	0.1429	−0.2632
Error, %	0.248	3.56	46.2	819

In terms of heat transfer from extended surfaces, it is more appropriate to compare the exact and the approximate heat losses than the exact and approximate temperatures, to determine the limitation of the approximate solution given above. These losses are

$$\frac{q}{kA\theta_0/L} = \mu \tanh \mu, \qquad \frac{q}{kA\theta_0/L} = \frac{\mu^2}{1 + \mu^2/3}.$$

The latter is evaluated by use of $q = hP\int_0^L \theta\, dx$ rather than

$$q = -[-kA(d\theta/dx)_{x=L}].$$

(Why? Would not the two methods lead to the same answer?) The exact and approximate heat losses are compared in Table 3–3 for some values of μ. (Why is the error between the exact and approximate heat losses smaller than that between the exact and approximate temperatures for a given μ?) Thus when $\mu < 2$, the error in the approximate heat loss is less than 11.1%. Employing the limiting value $\mu = 2$ we may obtain the permissible length for a solid rod of radius R as $L = (2kR/h)^{1/2}$, for which the existing approximate solution holds.

TABLE 3–3

μ	$\frac{1}{2}$	1	2	4
Exact	0.2311	0.7616	1.9281	3.9973
Approx.	0.2308	0.7500	1.7143	2.5263
Error, %	0.130	1.52	11.1	36.8

As an example we may consider a steel rod ($k \cong 10$ Btu/ft·hr·°F) of $R = \frac{1}{2}$ in. transferring heat by free convection to a gaseous medium with $h = 1$ Btu/ft²·hr·°F. The allowable length for this case is $L \cong 0.91$ ft. If the rod is made of copper ($k \cong 200$ Btu/ft·hr·°F), the length becomes $L \cong 4.1$ ft. When the heat transfer is by free convection to liquids or is by forced convection, h increases many times, and the present approximate solution no longer leads to an appreciable length.

On the basis of the algebra involved in the exact and approximate methods of solution considered above, we may question the practicability of approximate solutions. However, the foregoing approximate solution has been considered primarily for further experience in the use of the integral method. It should be kept in mind that the algebra of the approximate solutions remains practically unchanged for complicated problems, whereas the complexity of the methods for exact solutions increases rapidly. This makes the integral method convenient, and often indispensable, for complicated problems.

Although we must defer until Chapter 5 and 7 the solution of unsteady one-dimensional problems by the differential method, we may easily demonstrate at this point the selection of approximate profiles for such problems by the integral method.

Example 3–14. Reconsider the problem of Examples 3–10 and 3–13, and assume that the initial temperature of the fin is uniform and equal to that of the ambient, T_∞. Assume next that the base temperature is suddenly changed to T_0 and held constant thereafter. We wish to find a first-order solution giving the approximate temperature variation in the fin.

Since the problem involves the penetration depth,* its formulation may conveniently be given in two successive time domains. In the first time domain the penetration depth is less than or equal (as a limit) to the length of the fin; in the second time domain the tip temperature rises from zero to its steady value.

Considering the appropriate lumped control volume and the differential system shown in Fig. 3–38, we find that the integral formulation of the problem for the first time domain is

$$\frac{1}{a}\frac{d}{dt}\int_0^{\tau_0} \theta\, dx = \left(\frac{\partial \theta}{\partial x}\right)_{x=\tau_0} - m_2 \int_0^{\tau_0} \theta\, dx, \quad (3\text{–}197)$$

where $\theta = T - T_\infty$, and x has its origin at the penetration depth. Noting that the length of the fin does not affect the formulation of the first time domain, we retain Eq. (3–197) rather than make it dimensionless.

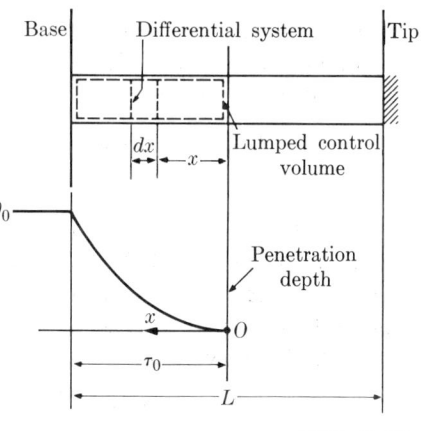

FIG. 3–38

We find it convenient to select a spacewise parabolic, timewise unspecified first-order Kantorovich profile which satisfies the boundary conditions of the problem and which is expressed in terms of the penetration depth τ_0. Thus we have

$$\frac{\theta(x,t)}{\theta_0} = \left(\frac{x}{\tau_0}\right)^2. \quad (3\text{–}198)$$

---

* Recall the integral formulation of Example 2–3.

Insertion of Eq. (3–198) into Eq. (3–197) and subsequent integration results in the nonlinear differential equation

$$\frac{d\tau_0}{dt} + am^2\tau_0 = \frac{6a}{\tau_0},$$

which can be made linear in terms of $\tau_0^2/2$ as follows:

$$\frac{d(\tau_0^2/2)}{dt} + 2am^2(\tau_0^2/2) = 6a. \tag{3-199}$$

Equation (3–199) is to be satisfied by

$$\tau_0(0) = 0. \tag{3-200}$$

The solution of Eq. (3–199), subject to Eq. (3–200), is

$$\tau_0^2 = \frac{6}{m^2}(1 - e^{-2am^2 t}). \tag{3-201}$$

Equation (3–201) may now be employed to evaluate the penetration time, t_0, at which $\tau_0(t_0) = L$. It follows then that

$$t_0 = \frac{1}{2am^2} \ln\left(\frac{1}{1 - \mu^2/6}\right), \tag{3-202}$$

where, as before, $\mu = mL$.

Finally, introducing Eq. (3–201) into Eq. (3–198) and rearranging gives the temperature of the fin in the first time domain,

$$\frac{\theta(x,t)}{\theta_0} = \frac{m^2 x^2}{6[1 - \exp(-2am^2 t)]}. \tag{3-203}$$

The differential and lumped systems suitable to the second time domain are indicated in Fig. 3–39. However, since the tip of the fin is insulated, rather than referring to Fig. 3–39 we may readily obtain the integral formulation of the problem for the second time domain from Eq. (3–199) by simply replacing τ_0 by L. Thus we have

$$\frac{1}{a}\frac{d}{dt}\int_0^L \theta\,dx = \left(\frac{\partial\theta}{\partial x}\right)_{x=L} - m^2\int_0^L \theta\,dx. \tag{3-204}$$

It is convenient, for this domain, to introduce the dimensionless variable $\xi = x/L$ and the parameter $\mu = mL$. Equation (3–204) may then be rearranged in the form

$$\frac{L^2}{a}\frac{d}{dt}\int_0^1 \theta\,d\xi = \left(\frac{\partial\theta}{\partial\xi}\right)_{\xi=1} - \mu^2\int_0^1 \theta\,d\xi. \tag{3-205}$$

The first-order Ritz profile employed for the steady problem of Example 3–13, Eq. (3–195), may again be considered, provided the constant parameter

a_0 is now a time-dependent parameter function, $a_0(t)$, to be determined. It follows that

$$\theta(\xi, t)/\theta_0 = 1 - (1 - \xi^2)a_0(t). \tag{3-206}$$

Equation (3–206) is a first-order Kantorovich profile.

Insertion of Eq. (3–206) into Eq. (3–205) and subsequent integration results in

$$\frac{da_0}{dt} + \frac{3a}{L^2}\left(1 + \frac{\mu^2}{3}\right)a_0 = \frac{3a}{L^2}\left(\frac{\mu^2}{2}\right). \tag{3-207}$$

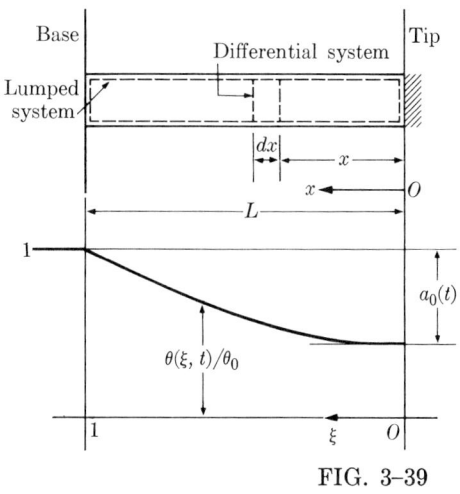

FIG. 3-39

The initial condition that is to be imposed on Eq. (3–207) is

$$a_0(t_0) = 1. \tag{3-208}$$

The solution of Eq. (3–207) which satisfies Eq. (3–208) is

$$a_0(t) = \frac{\mu^2/2}{1 + \mu^2/3} + \frac{1 - \mu^2/6}{1 + \mu^2/3} \exp\left[-3\left(1 + \frac{\mu^2}{3}\right)\frac{a(t - t_0)}{L^2}\right]. \tag{3-209}$$

Finally, introducing Eq. (3–209) into Eq. (3–206), we obtain the temperature of the fin in the second time domain as follows:

$$\frac{\theta(\xi, t)}{\theta_0} = 1 - (1 - \xi^2)\left\{\frac{\mu^2/2}{1 + \mu^2/3} + \frac{1 - \mu^2/6}{1 + \mu^2/3} \exp\left[-3\left(1 + \frac{\mu^2}{3}\right)\frac{a(t - t_0)}{L^2}\right]\right\}. \tag{3-210}$$

As $t \to t_0$ Eq. (3–210) approaches the upper limit of the first time domain solution, Eq. (3–198) for $\tau_0 = L$, and as $t \to \infty$ it tends to the steady solution given by Eq. (3–196). The error in Eq. (3–210) is expected to be on the order of that involved in Eq. (3–196).

Having extended the use and thus learned the importance of first-order steady profiles for unsteady problems, we next consider the ways of improving these profiles.

3-10. Higher-Order Approximations

Sometimes we wish to improve the result of first-order profiles because of the accuracy needed in the solution of our problems. Since the present chapter deals in the differential sense with steady one-dimensional problems only, let us carry on our discussion of higher-order profiles in terms of unsteady multidimensional problems.

Let the $(n+1)$th approximation of a three-dimensional Kantorovich profile be

$$\theta(\mathcal{V}, t) = \theta[\mathcal{V}, \tau_0(t), \tau_1(t), \ldots, \tau_n(t)], \tag{3-211}$$

where $\tau_0(t), \tau_1(t), \ldots, \tau_n(t)$ are the unknown parameter functions to be determined and $\mathcal{V}$ denotes the volume. For the sake of illustration only, assume that the general problem under consideration applies to a homogeneous isotropic stationary solid. The governing differential formulation is then

$$\frac{\partial \theta}{\partial t} = a \nabla^2 \theta.$$

The use of the integral formulation of the problem written, for example, in the form

$$\int_{\mathcal{V}} \left(\nabla^2 \theta - \frac{1}{a} \frac{\partial \theta}{\partial t} \right) \partial \mathcal{V} = 0 \tag{3-212}$$

gives one relation among the $(n+1)$ unknown parameter functions. There are two general methods for the evaluation of the remaining n relations among these parameter functions. The first method is to consider n relations to be obtained by multiplying the integrand of the integral formulation by the continuous but otherwise largely arbitrary functions $F_i(\mathcal{V})$. Thus we have

$$\int_{\mathcal{V}} \left(\nabla^2 \theta - \frac{1}{a} \frac{\partial \theta}{\partial t} \right) F_i(\mathcal{V}) \, d\mathcal{V} = 0, \quad i = 1, 2, 3, \ldots, n, \tag{3-213}$$

where $F_i(\mathcal{V})$ may most conveniently be assumed from the successive approximations of profiles already selected in the form of Eq. (3–211). The concept behind the establishment of Eq. (3–213) has its roots in the calculus of variations. Therefore, the discussion on this point is necessarily deferred to Chapter 8. Let us now note that the profiles selected do not yet satisfy the differential formulation. The second method is based on satisfying the differential formulation and, if necessary, its convenient space derivatives at n suitable points, P_j ($j = 1, 2, 3, \ldots, n$). It follows then that

$$\frac{\partial^m}{\partial x_k^m} \left(\nabla^2 \theta - \frac{1}{a} \frac{\partial \theta}{\partial t} \right)_{P_j} = 0; \quad j = 1, 2, 3, \ldots, n; \quad m = 0, 1, 2, \ldots, \tag{3-214}$$

where x_k denotes an appropriate direction in space. Because of the well-known smoothing nature of the process of integration, the use of Eq. (3–213) rather than Eq. (3–214) would, in general, yield more accurate profiles. However, when a second- or third-order approximation is desired, if Eq. (3–214) is satisfied at one or two points at which a maximum discrepancy is anticipated between the selected and exact profiles, it may give almost as accurate a result as that obtained by employing Eq. (3–213).

Let us now illustrate the use of these two methods in the evaluation of second-order profiles of a simple steady problem.

Example 3-15. We wish to find an approximate solution for the problem of Example 3-10 in terms of a second-order Ritz profile.

The suggested profile which satisfies the boundary conditions of the problem may readily be written by adding one term to the first-order approximation given by Eq. (3-195). Thus we have

$$\theta(\xi)/\theta_0 = 1 - (1 - \xi^2)(a_0 + a_1\xi^2). \tag{3-215}$$

Insertion of Eq. (3-215) into the integral formulation of the problem, Eq. (3-194) or the integral of dimensionless Eq. (3-163), and subsequent integration gives the first relation between the constant parameters as follows:

$$(1 + \mu^2/3)a_0 + (1 + \mu^2/15)a_1 = \mu^2/2. \tag{3-216}$$

For a second relation we first consider the appropriate form of Eq. (3-213) for $i = 1$. Assuming that $F_1(\xi), F_2(\xi), \ldots$ may be written from Eq. (3-215), we have $F_1(\xi) = (1 - \xi^2), F_2(\xi) = (1 - \xi^2)\xi^2, \ldots$ The needed second relation becomes then

$$\int_0^1 \left(\frac{d^2\theta}{d\xi^2} - \mu^2\theta\right)(1 - \xi^2)\, d\xi = 0. \tag{3-217}$$

Introducing Eq. (3-215) into Eq. (3-217) and integrating the result, we find a second relation to be

$$(1 + 2\mu^2/5)a_0 + (1/5 + 2\mu^2/35)a_1 = \mu^2/2. \tag{3-218}$$

Before evaluating a_0 and a_1, let us establish an alternative second relation to be obtained from Eq. (3-214). Since the approximate profile is constructed to satisfy, according to given boundary conditions, the gradient of temperature at the tip rather than the temperature itself, the maximum deviation between the exact and approximate profiles is expected at this location. Thus we proceed to find a second condition which improves the approximate profile especially in the neighborhood of the tip. This may be accomplished by satisfying the dimensionless differential formulation of the problem, $d^2\theta/d\xi^2 - \mu^2\theta = 0$, by Eq. (3-215) at $\xi = 0$. The result is

$$(1 + \mu^2/2)a_0 - a_1 = \mu^2/2. \tag{3-219}$$

It follows from the simultaneous consideration of Eqs. (3-216) and (3-218) that

$$a_0 = \frac{(\mu^2/2)(1 + \mu^2/84)}{1 + 9\mu^2/21 + \mu^4/105}, \quad a_1 = \frac{\mu^4/24}{1 + 9\mu^2/21 + \mu^4/105},$$

and of Eqs. (3-216) and (3-219) that

$$a_0 = \frac{(\mu^2/2)(1 + \mu^2/30)}{1 + 9\mu^2/20 + \mu^4/60}, \quad a_1 = \frac{\mu^4/24}{1 + 9\mu^2/20 + \mu^4/60}.$$

TABLE 3-4

μ	$\frac{1}{2}$	1	2	4
Exact	0.2311	0.7616	1.9281	3.9973
First Approx.	0.2308	0.7500	1.7143	2.5263
Error, %	0.130	1.52	11.1	36.8
Second Approx. I	0.2311	0.7616	1.9269	3.9223
Error, %	0.000	0.000	0.062	1.88
Second Approx. II	0.2311	0.7614	1.9130	3.6791
Error, %	0.000	0.026	0.783	7.96

Inserting these values into Eq. (3–215) and rearranging the result, we obtain two second-order Ritz profiles for our problem in the form

$$\frac{\theta(\xi)}{\theta_0} = 1 - \left(\frac{\mu^2/2}{1 + 9\mu^2/21 + \mu^4/105}\right)(1 - \xi^2)\left[\left(1 + \frac{\mu^2}{84}\right) + \frac{\mu^2\xi^2}{12}\right] \quad (3\text{–}220)$$

and

$$\frac{\theta(\xi)}{\theta_0} = 1 - \left(\frac{\mu^2/2}{1 + 9\mu^2/20 + \mu^4/60}\right)(1 - \xi^2)\left[\left(1 + \frac{\mu^2}{30}\right) + \frac{\mu^2\xi^2}{12}\right]. \quad (3\text{–}221)$$

For reasons explained in Example 3–13 we now compare the heat losses rather than the temperatures. Also, as indicated in the same example, we evaluate these losses from $q = hP\int_0^L \theta\, dx$. Thus we have

$$\frac{q}{kA\theta_0/L} = \frac{\mu^2(1 + 2\mu^2/21)}{1 + 9\mu^2/21 + \mu^4/105}, \quad \frac{q}{kA\theta_0/L} = \frac{\mu^2(1 + 7\mu^2/60)}{1 + 9\mu^2/20 + \mu^4/60}.$$

These heat losses, together with the corresponding first-order approximation, are compared with the exact heat losses in Table 3–4, where the suffixes I and II denote the second-order approximations based on Eq. (3–213) and on Eq. (3–214), respectively. Inspection of Table 3–4 reveals that the errors involved are smaller for the approximation calculated through Eq. (3–213). However, we obtain this higher accuracy only at the cost of some reasonably involved algebra. For a particular problem, one of these approximations often becomes more convenient than the other, the choice depending on the accuracy required in the solution and on the inherent complexity of the problem. For example, Eq. (3–213), which readily gives as many relations as desired for the unknown parameters, is more convenient for higher-order approximations, whereas for

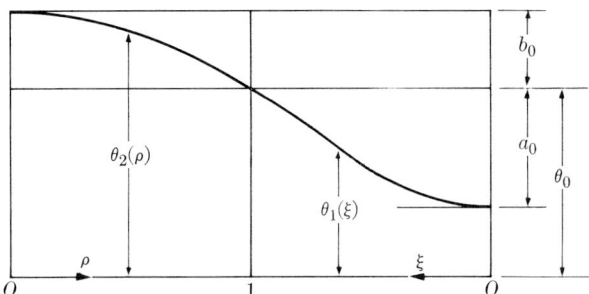

FIG. 3-40

the use of Eq. (3-214) it may be difficult to find enough locations at which a large discrepancy is anticipated between the exact and approximate profiles.

For the same reason as was given in Example 3-13, the discrepancy between the exact and second-order approximate solutions increases as $\mu \to \infty$. However, the second-order profiles may be employed for a larger range of the parameter μ than the first-order profile. This range is $\mu \leq 4$ within an error of less than 2% and 8% in the second-order approximations I and II, respectively.

The unsteady problem corresponding to the present example (i.e., including a sudden change in the base temperature) is left to the reader as an exercise.

Having learned how to select and evaluate approximate profiles we conclude this chapter with an example of the order of profiles appropriate for a given problem.

Example 3-16. We wish to select profiles for an approximate solution to the problem of Example 3-12.

By means of polynomials, for example, the temperature distribution in the bottom and the side walls of the skillet may be approximated in terms of parabolas (Fig. 3-40). Thus we have first-order Ritz profiles in the form

$$\theta_1(\xi) = \theta_0 - (1 - \xi^2)a_0, \tag{3-222}$$

$$\theta_2(\rho) = \theta_0 + (1 - \rho^2)b_0, \tag{3-223}$$

where θ_0, a_0, and b_0 are the unknown corner, bottom center, and side-wall tip temperatures, respectively, and $\xi = x/L$, $\rho = r/R$. Equations (3-222) and (3-223) satisfy the boundary conditions of the problem, Eqs. (3-186) through (3-189), except the condition given by Eq. (3-188). Employing now the dimensionless form of Eq. (3-188), we obtain $a_0 = b_0$, which states the equality of temperature drops in the bottom and side walls. This certainly is a serious restriction which may not be allowed according to the physics of the problem.*
Therefore, the first-order approximations given by Eqs. (3-222) and (3-223) may not be suitable to our problem. In this case the simplest pair of physically meaningful profiles must necessarily be based on the second-order approximations, to be considered for one or both of these profiles.

---

* See the discussion related to Fig. 3-35.

References

1. Th. von Kármán and M. A. Biot, *Mathematical Methods in Engineering*. New York: McGraw-Hill, 1940.
2. N. W. McLachlan, *Bessel Functions for Engineers*. Oxford: Clarendon Press, 1955.
3. F. B. Hildebrand, *Advanced Calculus for Engineers*. New York: Prentice-Hall, 1948.
4. W. T. Martin and E. Reissner, *Elementary Differential Equations*. Reading, Mass.: Addison-Wesley, 1956.
5. W. Kaplan, *Ordinary Differential Equations*. Reading, Mass.: Addison-Wesley, 1958.
6. W. M. Rohsenow, Class Notes on "Advanced Heat Transfer." MIT, 1956.
7. H. S. Carslaw and J. C. Jaeger, *Conduction of Heat in Solids*. Oxford: Clarendon Press, 1959.
8. E. Schmidt, "Die Wärmeübertragung durch Rippen." *Zeitschr. VDI*, **70**, 885, 947 (1926).
9. W. B. Harper and D. R. Brown, "Mathematical Equations for Heat Conduction in the Fins of Air-Cooled Engines." *NACA Report 158*, 679 (1922).
10. K. A. Gardner, "Efficiency of Extended Surfaces." *Trans. ASME*, **67**, 621 (1945).
11. M. Avrami and J. B. Little, "Diffusion of Heat Through a Rectangular Bar and the Cooling and Insulating Effect of Fins." *Journ. Appl. Phys.*, **13**, 255 (1942).
12. M. Jakob, *Heat Transfer I*. New York: Wiley, 1949.

Problems

3–1. Reconsider the general problem of Section 3–1 and the heat transfer

$$q = \frac{T_1 - T_2}{(1/k)\int_{s_1}^{s_2} ds/A(s)}. \tag{3-4}$$

(a) Show that when Eq. (3–4) is written in the form

$$q = k\bar{A}\,\frac{T_1 - T_2}{s_2 - s_1},$$

$\bar{A}$ corresponds to the actual heat transfer area of the cartesian geometry,

$$\bar{A} = A;$$

to the *logarithmic mean* value of the inner and outer heat transfer areas of the cylindrical geometry,

$$\bar{A} = \frac{A_2 - A_1}{\ln(A_2/A_1)};$$

and to the *geometric mean* value of the inner and outer heat transfer areas of the spherical geometry,

$$\bar{A} = (A_2 A_1)^{1/2}.$$

(b) Show that as $A_2 \to A_1$ the logarithmic and geometric mean values approach the *arithmetic mean* value

$$\bar{A} = (A_2 + A_1)/2.$$

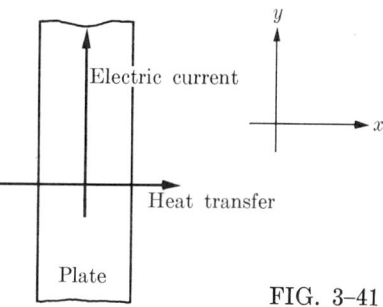

FIG. 3-41

3-2. Consider a flat plate in which internal energy is generated by means of an external electric circuit (Fig. 3-41). (a) Show that the one-dimensional cartesian form of the conduction equation applied to this plate can be written as

$$\frac{d}{dx}\left(k\frac{dT}{dx}\right) + k_e\left(\frac{dE}{dy}\right)^2 = 0,$$

where k_e is the electrical conductivity per unit length and per unit area in y, and E the electrical potential. (b) Show that the three-dimensional form of the foregoing equation is

$$\nabla \cdot (k\,\nabla T) + k_e(\nabla E)^2 = 0.$$

3-3. The outside wall of a house is composed of two parallel plates (Fig. 3-42). The thermal conductivity and the thickness of the plates are k_1, k_2 and L_1, L_2, respectively. The inside and outside ambient temperatures and the heat transfer coefficients are the same, say T_∞ and h. The net radiation between the sun and the outer surface of the wall is q''. (a) Calculate the heat transfer to the house. (b) Draw the analogous electric circuit for part (a).

3-4. Neglecting the effect of curvature, assume that the brake of a vehicle may be simulated by a flat plate (brake drum) moving on a composite plate (brake shoe) with a constant velocity V (Fig. 3-43). The constant and uniform interface pressure is p, the coefficient of dry friction is μ, the ambient temperature is T_∞, and the heat transfer coefficients are h_1, h_2. The thermal conductivities and the thicknesses of the plates are

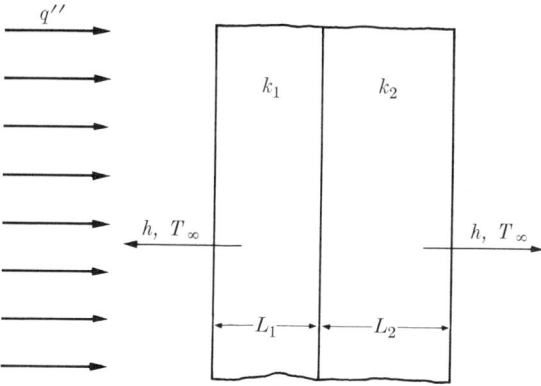

FIG. 3-42

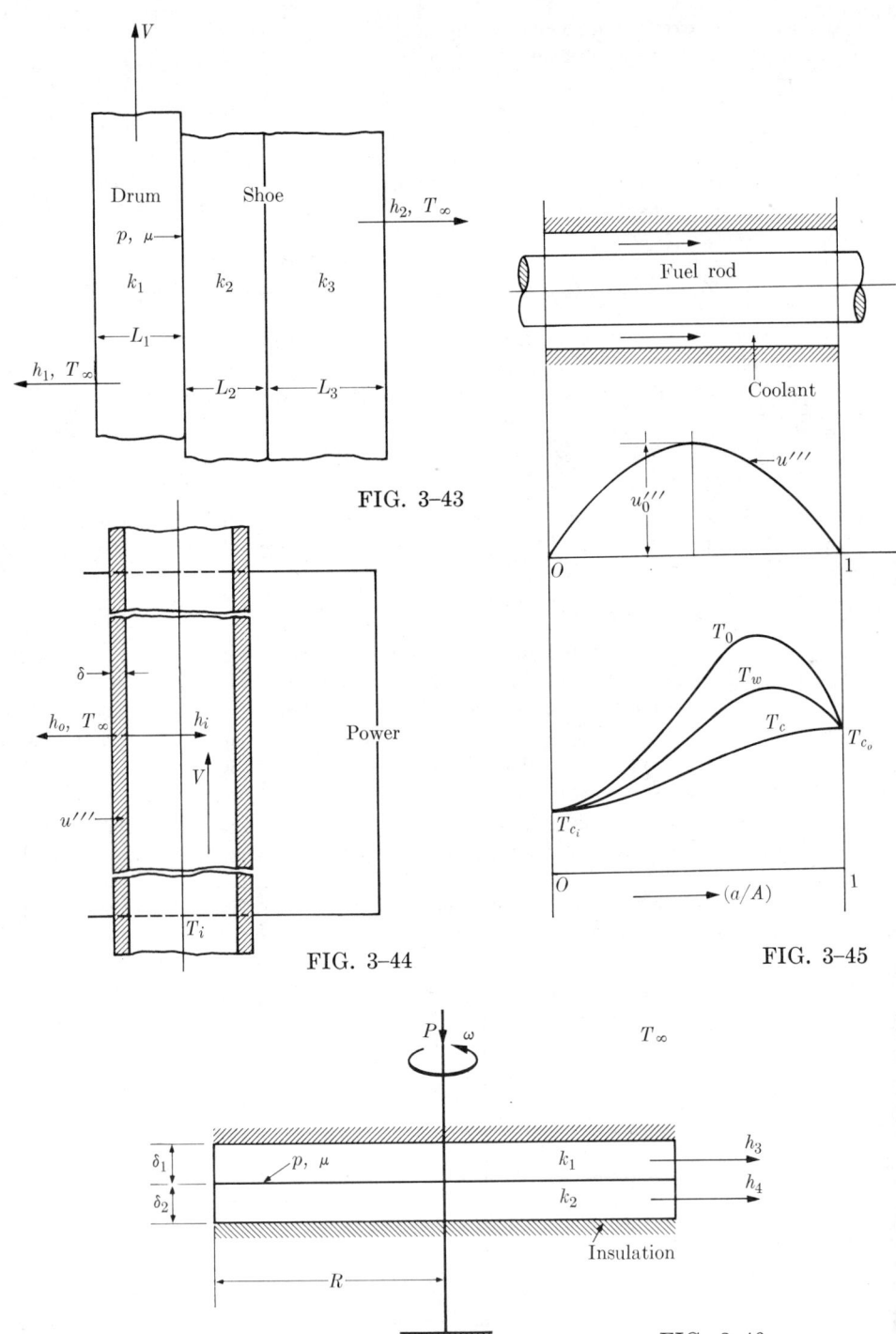

FIG. 3-43

FIG. 3-44

FIG. 3-45

FIG. 3-46

k_1, k_2, k_3, and L_1, L_2, L_3, respectively. (a) Calculate the heat transfer to the drum and to the shoe. (b) Find the maximum temperature of the brake. (c) Draw the analogous electric circuit for part (a).

3–5. Pressurized water flows with velocity V through a bare tube (Fig. 3–44). The inlet water temperature is T_i, and the surrounding air temperature is T_∞. The inside and outside heat transfer coefficients are h_i and h_o, respectively. The constant internal energy u''' is uniformly generated in the tube walls. The axial conduction is negligible. Find the radially lumped, axially distributed temperature of the system (a) according to the false assumption that $h_i \sim h_o$, and (b) noting that $h_i \gg h_o$.

3–6. A unit cell of a liquid-cooled reactor core may be simulated by a coolant which flows coaxially over a fuel rod (Fig. 3–45). The internal energy generation in the rod is $u''' = u_0''' \sin \pi(a/A)$, where a and A are the local and total heat transfer areas, respectively. Neglecting the axial conduction, and assuming a constant heat transfer coefficient h between the fuel rod and the coolant, show that

$$\frac{T_c - T_{c_i}}{T_{c_o} - T_{c_i}} = \frac{1}{2}\left(1 - \cos \pi \frac{a}{A}\right),$$

$$\frac{T_w - T_{c_i}}{T_{c_o} - T_{c_i}} = \frac{1}{2}\left(1 - \cos \pi \frac{a}{A}\right) + \frac{\pi}{2}\left(\frac{wc}{Ah}\right) \sin \pi \frac{a}{A},$$

$$\frac{T_0 - T_{c_i}}{T_{c_o} - T_{c_i}} = \frac{1}{2}\left(1 - \cos \pi \frac{a}{A}\right) + \frac{\pi}{2}\left(\frac{wc}{Ah}\right)\left(1 + \frac{hR}{2k}\right) \sin \pi \frac{a}{A}.$$

where w is the flow rate of the coolant, T_c the local temperature of the coolant, T_{c_i} and T_{c_o} are the inlet and outlet temperatures of the coolant, c is the specific heat of the coolant, k the thermal conductivity of the rod, R the outer radius of the rod, T_w the coolant-rod interface temperature, and T_0 the centerline temperature of the rod.

3–7. A thin-walled disk rotates with an angular velocity ω on another thin-walled but stationary disk (Fig. 3–46). The upper and lower surfaces of the system are insulated. The interface pressure is p, the coefficient of dry friction is μ, the peripheral heat transfer coefficients are h_3 and h_4, and the ambient temperature is T_∞. The radius, thickness, and thermal conductivity of the disks are R, δ_1, δ_2 and k_1, k_2, respectively. Find the temperature distribution in the disks.

3–8. The local heat transfer coefficient inside a tube is to be experimentally determined. For this purpose let us reconsider Problem 3–5, but let us now assume that the tube is insulated and that its temperature is radially as well as axially distributed. According to the definition of heat transfer coefficients in tubes, we have locally

$$h = \left(\frac{R_o^2 - R_i^2}{2R_i}\right) \frac{u'''}{(T_{w_i} - T_b)},$$

where R_i and R_o are the inner and outer radii of the tube, respectively, T_{w_i} is the inside wall temperature, and T_b the bulk temperature of the fluid (Fig. 3–47). The evaluation of the local bulk temperature T_b in terms of its measured inlet value T_i constitutes part (b) of Problem 3–4. Since the inside wall temperature T_{w_i} is difficult to measure (why?), assume that the outside wall temperature T_{w_o} is measured instead. Find an expression which relates the local values of T_{w_i} to those of the measured T_{w_o}.

3-9. The inside surface of a bomb calorimeter is subjected to the heat flux q'' resulting from an exothermic chemical process (Fig. 3-48). The inner and outer radii of the calorimeter are R_i and R_o, respectively. The outside heat transfer coefficient is h, the ambient temperature T_∞. (a) Evaluate the inside surface temperature of the calorimeter. (b) Is it possible to lower this temperature without changing R_i, R_o, q'' and T_∞?

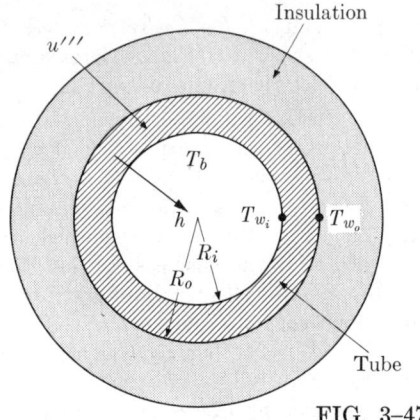

FIG. 3-47

3-10. The fuel element of a reactor is a solid sphere of radius R. The uniform internal energy u''' is generated in the sphere. The heat transfer coefficient is h, and the ambient temperature is T_∞. The centerline temperature T_0 is higher than an interphase temperature T_i at which the fuel changes its structure from the α- to the β-phase (Fig. 3-49). The thermal conductivities of the two phases are different, say k_1 and k_2. Find the interphase radius R_i which separates the α-phase from the β-phase of the sphere.

3-11. The internal energy u''' is uniformly generated in a plate. The left-hand surface of the plate is subjected to the uniform heat flux q_1'' which acts at a distance; the uniform heat flux q_2'' is extracted from the right-hand surface of the plate (Fig. 3-50). The heat transfer coefficient and the ambient temperature on the left side of the plate are h and T_∞, respectively. Using the principle of superposition, separate the problem into an appropriate number of simpler problems. (Do not solve these problems.)

3-12. (a) Start from the statement of Fourier's conduction law and show that in the absence of any internal energy generation, when thermal conductivity linearly

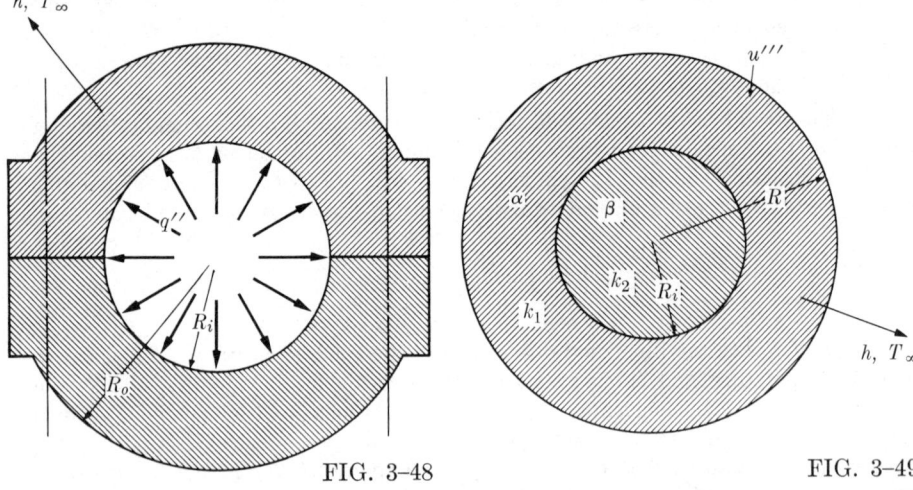

FIG. 3-48   FIG. 3-49

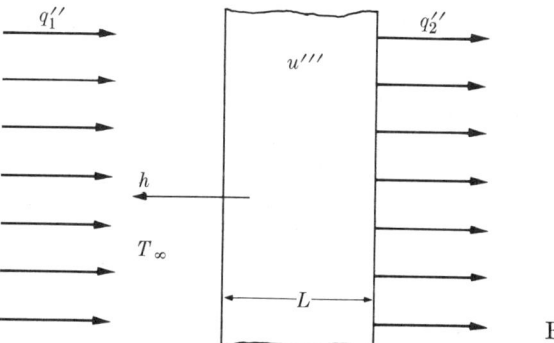

FIG. 3-50

depends on the temperature as $k = k_0(1 + \beta T)$, the one-dimensional conduction in a flat plate can be given in the form

$$q = k_m A \frac{T_1 - T_2}{x_2 - x_1},$$

where $k_m = (k_1 + k_2)/2$, x_1 and x_2 are the coordinates of the plate surfaces, and T_1 and T_2 the corresponding temperatures. Find the temperature distribution of the plate (b) by direct integration, (c) by Kirchhoff's method.

3-13. When the thermal conductivity depends on the temperature, in addition to Kirchhoff's method there exists another well-known procedure applicable to steady one-dimensional problems with an insulated boundary, or with a symmetry plane or axis. The procedure is based on the expansion of the unknown temperature into a Taylor series around the insulated boundary, or around the symmetry plane or axis. The terms of the series are then evaluated by means of the governing equation of the problem and its successive derivatives. To illustrate the method reconsider Example 3-8. Expand the temperature of the plate around $x = 0$, for which the Taylor series is reduced to the Maclaurin series. Obtain the terms of this series from the governing equation of the problem,

$$\frac{dk}{dT}\left(\frac{dT}{dx}\right)^2 + k(T)\frac{d^2T}{dx^2} + u''' = 0,$$

and its successive derivatives with the consideration that $(dT/dx)_{x=0} = 0$. Show that the result is

$$T(x) = T_0 - \frac{u'''}{k(T_0)}\frac{x^2}{2!} - \frac{3u'''^2}{k^3(T_0)}\left(\frac{dk}{dT}\right)_{T=T_0}\frac{x^4}{4!} + \cdots,$$

where T_0 is the temperature of the plate at $x = 0$.

The maximum truncation error ϵ_{max} of the foregoing series is

$$\epsilon_{max} < \left(\frac{d^6T}{dx^6}\right)_{max}\frac{L^6}{6!}.$$

Note that $(d^6T/dx^6)_{max}$ is difficult to estimate. However, it can be closely approximated by $(d^6T/dx^6)_{x=0}$ because of the smooth variation of the temperature. In general the surface temperature T_∞ of the plate (or another boundary condition) is given rather than T_0. A trial-and-error procedure is therefore necessary to relate the unknown T_0 to the given T_∞. The case of constant thermal conductivity may be used as a first approximation. When the value of T_0 thus obtained is introduced into the Maclaurin series it gives a surface temperature different than T_∞. This procedure is repeated by changing T_0 until the difference between the given and found surface temperatures becomes smaller than a desired value.

3-14. Using the method of solution of Problem 3-13, re-solve Problem 3-8 with the assumption that the thermal and electrical conductivities of the tube material depend linearly on the temperature.

3-15. Reconsider the spherical fuel element of Example 3-7. Assume that the nuclear internal energy generation in the fissionable material is uniform. The thermal conductivity of this material depends linearly on the temperature, while that of the cladding is constant. Using the method of Problem 3-13 find the temperature distribution in the fuel element.

3-16. Starting from the general form of the conduction equation for homogeneous isotropic continua,
$$\frac{dT}{dt} = a\nabla^2 T + \frac{u'''}{\rho c},$$
obtain the governing equation of extended surfaces with constant cross section
$$\frac{d^2T}{dx^2} - \left(\frac{hP}{kA}\right)(T - T_\infty) = 0.$$

3-17. Consider an infinite fin composed of two semi-infinite fins (Fig. 3-51). Uniform heat flux q'' is generated at the joint. The geometry of the fins is identical. The thermal conductivities are k_1 and k_2. Find the temperature distribution in the system.

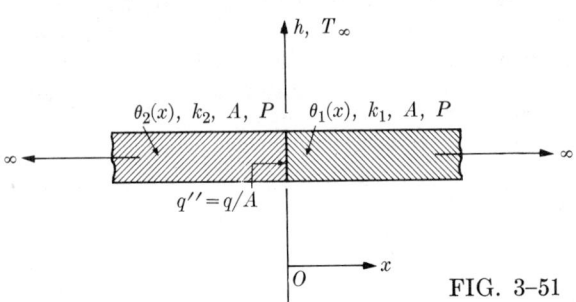

FIG. 3-51

3-18. A flat plate in an ambient at temperature T_∞ receives from a source the net radiant heat flux q'' (Fig. 3-52). The thickness of the plate δ is much less than its length $2L$; the third dimension of the plate extends to infinity. The upward and downward heat transfer coefficients are h_1, h_2, respectively. The heat transfer coefficient from the ends of the plate, if needed, may be taken to be $h_3 \neq h_1 \neq h_2$. Find the temperature distribution in the system.

3-19. Consider an infinite fin. Internal energy u''' is generated uniformly in part of the fin as shown in Fig. 3-53. The thermal conductivity, cross-sectional area, and periphery of the fin are k, A and P, respectively. The heat transfer coefficient is h, the ambient temperature T_∞. Find the steady temperature of the fin.

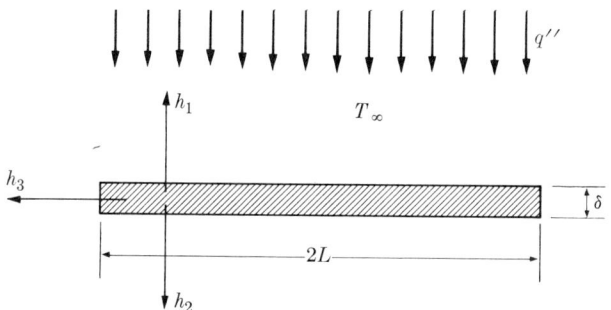

FIG. 3–52

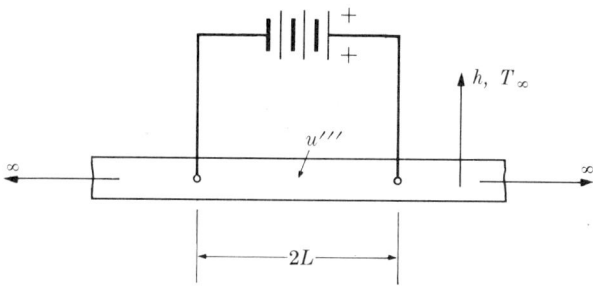

FIG. 3–53

3–20. A spoon in a cup of tea may be approximated as a rod of constant cross section (Fig. 3–54). The thermal conductivity, length, periphery, and cross-sectional area of the spoon are k, $2L$, P, and A, respectively. The heat transfer coefficients are h and h_0. One-half of the spoon is in the tea. Assuming that the temperature of the tea remains constant and that the ends of the spoon are insulated, find the steady temperature of the spoon.

3–21. Find the temperature distribution in the square rod shown in Fig. 3–55. The thermal conductivity of the upper part of the rod and that of the lower part, say k_1 and k_2, are different. The total electrical resistivities of the upper and lower parts are R_1 and R_2, respectively. The power drawn from a battery is $\mathcal{P}$.

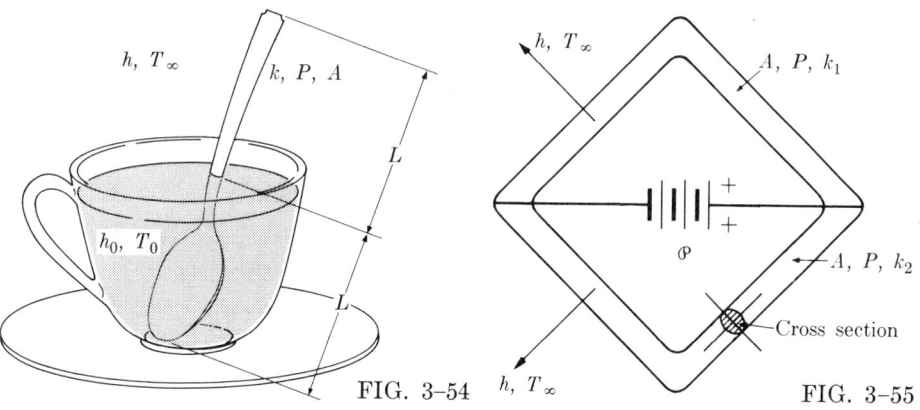

FIG. 3–54   FIG. 3–55

174 STEADY ONE-DIMENSIONAL PROBLEMS

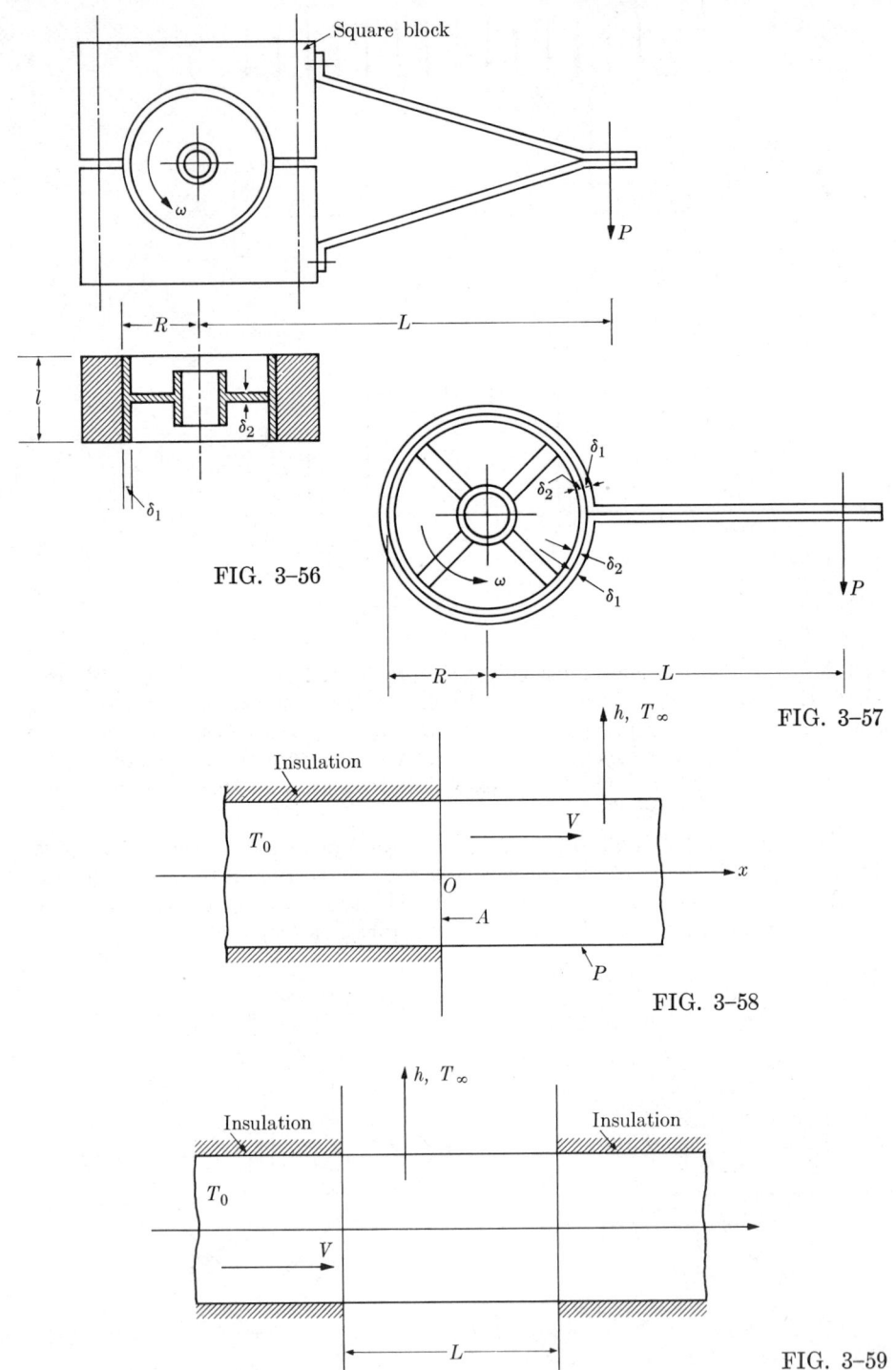

FIG. 3-56

FIG. 3-57

FIG. 3-58

FIG. 3-59

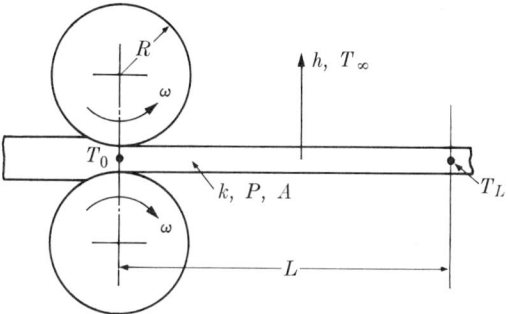

FIG. 3-60

3-22. Under steady conditions, using a Prony brake, the torque of an engine is balanced by force P at angular velocity ω (Fig. 3-56). The coefficient of dry friction is μ. The thermal conductivity of the square block is low compared to that of the flywheel. Thus all the heat generated may be assumed to flow to the flywheel. Calculate the temperature distribution in the flywheel.

3-23. Re-solve Problem 3-22, substituting the Prony brake shown in Fig. 3-57. Assume that all thermal conductivities are now of the same order of magnitude.

3-24. A constant-property inviscous liquid having velocity V and far upstream temperature T_0 flows steadily through an infinitely long tube of cross-sectional area A and periphery P (Fig. 3-58). The wall thickness of the tube is negligible. The downstream half of the tube transfers heat to an ambient at temperature T_∞ while the upstream half is insulated. The heat transfer coefficient is h. Find the axial temperature distribution in the liquid.

3-25. Re-solve Problem 3-24 considering that a length L rather than one-half of the tube transfers heat to the ambient at temperature T_∞ (Fig. 3-59).

3-26. A hot-drawing machine is shown in Fig. 3-60. The heat is transferred from the drawn wire with heat transfer coefficient h to an ambient at temperature T_∞. At location L, we need to know the temperature T_L for a chemical treatment. The inlet temperature of the wire T_0 is specified. Calculate the angular velocity of the rotating dies. Analyze the problem by using a system as well as a control volume.

3-27. Consider a straight fin of parabolic profile (Fig. 3-61). The thermal conductivity, base thickness, and length of the fin are k, $2b$, and L, respectively. The heat transfer coefficient is h and the ambient temperature T_∞. Find the steady temperature of and the total heat transfer from the fin, assuming that the parabola is given by (a) $y = Cx^{1/2}$, (b) $y = Cx^2$, where C is a constant.

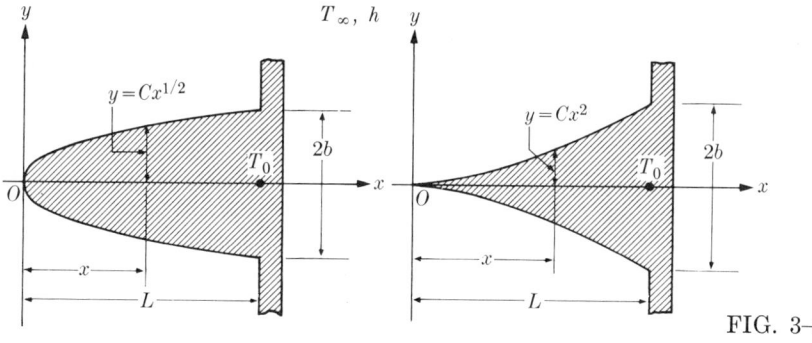

FIG. 3-61

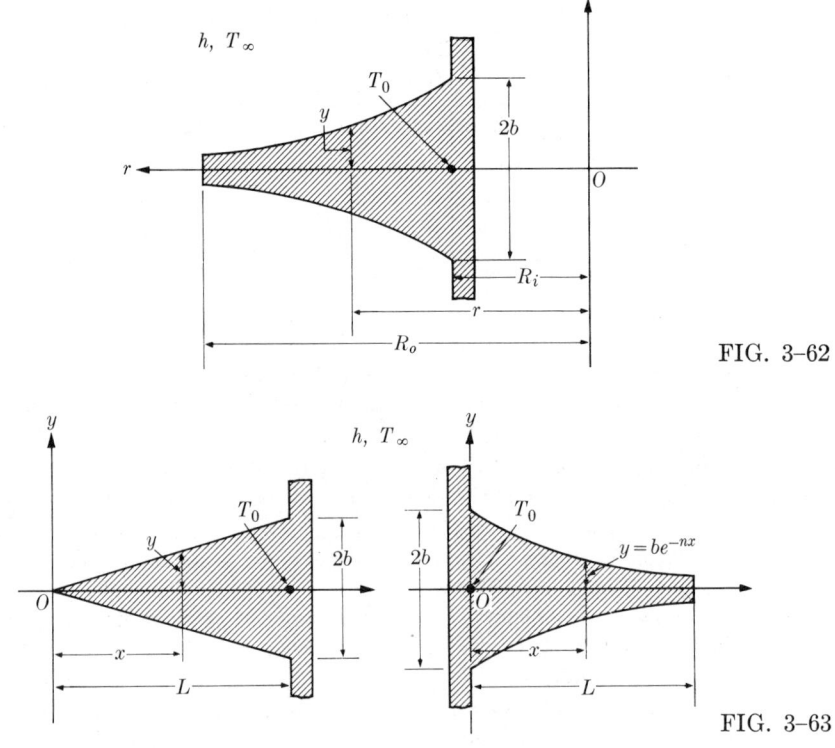

FIG. 3-62

FIG. 3-63

3-28. Consider an annular fin of hyperbolic profile (Fig. 3-62). The thermal conductivity, base thickness, and inner and outer radii of the fin are k, $2b$, R_i, and R_o, respectively. The heat transfer coefficient is h and the ambient temperature T_∞. Find the steady temperature of and the total heat transfer from the fin, assuming that the hyperbola is given by (a) $yr^{1/2} = C$, (b) $yr^2 = C$, where C is a constant.

3-29. Consider a spine (pin fin) of variable cross section (Fig. 3-63). The thermal conductivity, base thickness, and length of the spine are k, $2b$, and L, respectively. The heat transfer coefficient is h and the ambient temperature T_∞. Find the steady temperature of and the total heat loss from the spine, assuming that the cross section varies (a) linearly, (b) exponentially.

3-30. A thin-walled disk rotates with angular velocity ω on a stationary disk (Fig. 3-64). The interface pressure is p, the dry friction coefficient μ, and the ambient temperature T_∞. The upward, downward, and peripheral heat transfer coefficients are h_1, h_2 and h_3, h_4, respectively. The radius, thickness, and thermal conductivity of the plates are R, δ_1, δ_2 and k_1, k_2. Find the steady temperature of the system.

3-31. Condensing steam at temperature T_0 is flowing through a finned tube (Fig. 3-65). The outer radius of the tube and that of the fins are R_0 and R, respectively. The thickness of the fins is δ, the heat transfer coefficient between the fins and the ambient h, and the ambient temperature T_∞. Find the temperature distribution in the fins.

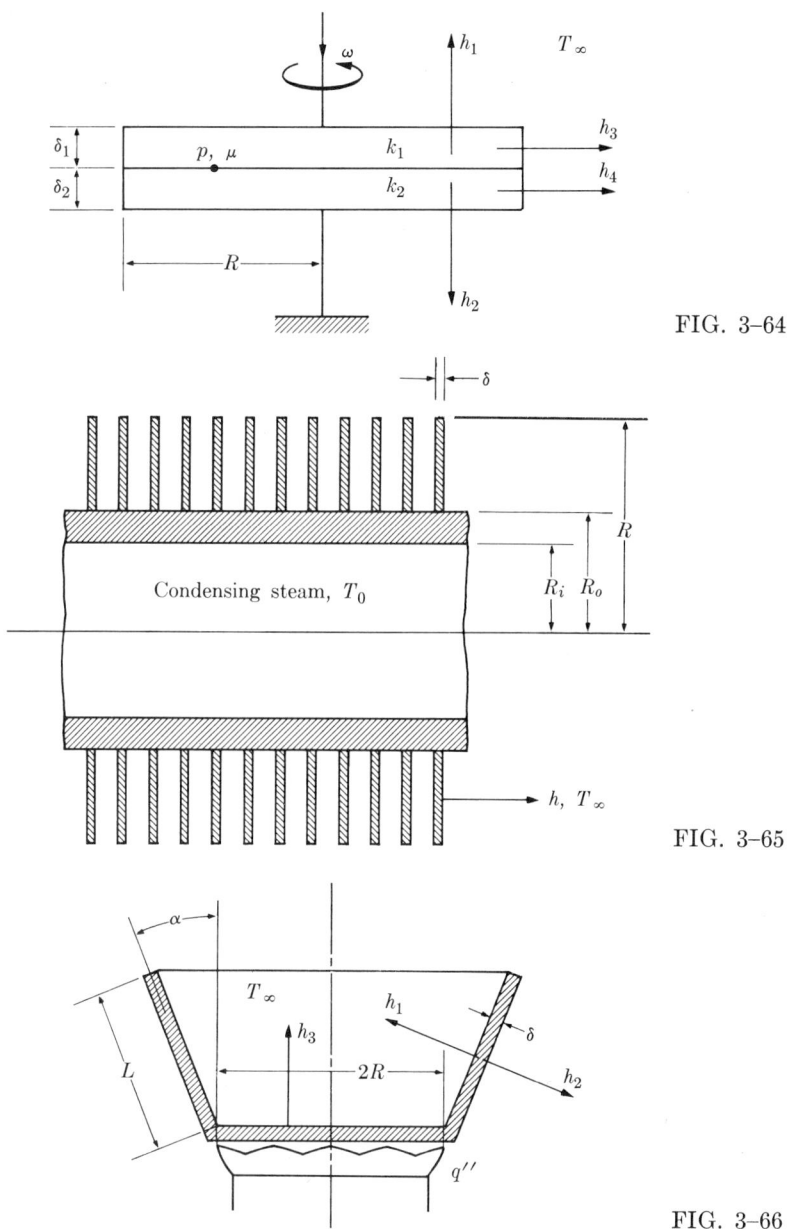

FIG. 3-64

FIG. 3-65

FIG. 3-66

3-32. An empty skillet is heated from the bottom by uniform heat flux q'' (Fig. 3-66). The ambient temperature is T_∞, and the heat transfer coefficients are h_1, h_2, h_3. The thermal conductivity, length, bottom radius, and thickness of the skillet are k, L, R, and δ, respectively. Using the simplest possible formulation of the problem, find the steady temperature of the skillet.

178 STEADY ONE-DIMENSIONAL PROBLEMS

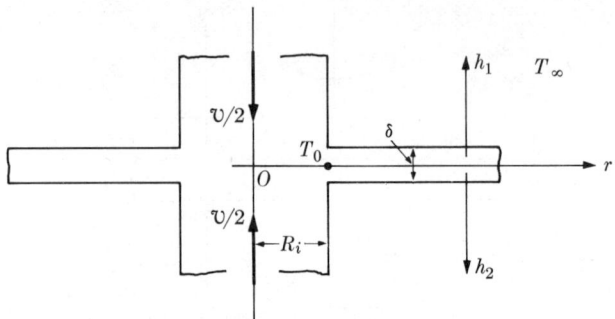

FIG. 3-67

3-33. An incompressible fluid flows steadily and radially between two concentric parallel disks of negligible thickness (Fig. 3-67). The distance δ between the disks is small. The inlet temperature of the fluid T_0 is specified. The volumetric flow rate is $\mathcal{V}$, the thermal conductivity of the fluid is k, the ambient temperature is T_∞, the inner radius of disks is R_i, and the upward and downward heat transfer coefficients are h_1, h_2. Find the radial temperature distribution in the fluid.

3-34. Industrial fins (straight, annular, or pin) are often made of thin metal plates of uniform thickness welded to bare tubes. Because of the decreasing temperature difference between the fin and ambient along the length of the fin, heat transfer to the ambient does not remain constant, and poor use of the material results. We wish to correct this situation by changing the cross-sectional area of the fin in such a way that the heat flux, q/A, remains constant along the length of the fin. Find the profile of straight, annular, and pin fins with zero tip thickness which satisfies the foregoing condition.

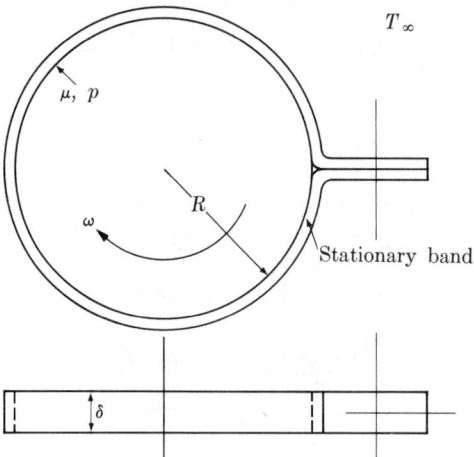

FIG. 3-68

3-35. Let us simulate a Prony brake by a solid disk of radius R and thickness δ peripherally surrounded by a band of negligible thickness (Fig. 3-68). The interface pressure is p and the coefficient of dry friction μ. The brake is at the ambient temperature T_∞ initially; then it suddenly assumes the constant angular velocity ω. Find the unsteady temperature of the brake.

3-36. Consider a lighted candle. The initial and instantaneous lengths of the candle are L and $X(t)$, respectively, the periphery is P, the cross-sectional area A (Fig. 3-69). The melting temperature of the candle wax is T_m and the ambient temperature T_∞. Find the melting rate of the candle.

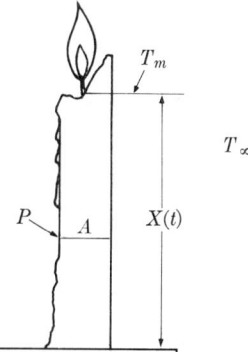

FIG. 3-69

3-37. Reconsider Example 3-10. Find the temperature distribution of the extended surface by assuming that its thermal conductivity depends on (a) the temperature, say $k = k(T)$, (b) the space, say $k = k_0 x^n$; distinguish the following three cases: $0 < n < 1$, $n = 2$, and $n \neq 2$ but an integer.

CHAPTER 4

STEADY TWO- AND THREE-DIMENSIONAL PROBLEMS. SEPARATION OF VARIABLES. ORTHOGONAL FUNCTIONS.

Thus far our discussion has been confined to steady one-dimensional problems. Since these yield ordinary differential equations of second order, they are, in principle, always solvable. On the other hand, steady two- and three-dimensional and unsteady problems result in partial differential equations for which no general method of solution is available. In this chapter and in Chapters 5, 6, and 7, some methods for exact solution of these problems will be developed.

When the boundaries of a multidimensional conduction problem correspond to the coordinate surfaces in a system of orthogonal coordinates, such as cartesian, cylindrical, or spherical coordinates, an exact solution by analytical methods becomes possible. One common method is based on the *separation of variables*, another on the *Laplace transforms* of the operational calculus. There are convincing arguments as to which of these two most-used methods might be better suited for a particular problem; however, the Laplace transforms, though convenient for the solution of complicated problems, require the knowledge of more advanced mathematics. Thus, delaying the Laplace transforms to Chapter 7, we devote this chapter to the method of separation of variables. In the next four sections we review the mathematics necessary for this method.

4–1. Boundary-Value Problems. Characteristic-Value Problems

Consider first an ordinary differential equation of second order which may result from the differential formulation of a steady one-dimensional conduction problem. The solution of this equation involves two arbitrary constants which are determined by two conditions, each specified at one boundary of the problem. Problems of this type are called *boundary-value problems* to distinguish them from *initial-value problems*, in which all conditions are specified at one location. An example of an initial-value problem is the free fall of a body, mentioned in Chapter 1.†

Next consider the homogeneous‡ linear equation of second order

$$\frac{d^2y}{dx^2} + f_1(x)\frac{dy}{dx} + f_2(x)y = 0, \tag{4-1}$$

which vanishes at the boundaries $x = a$ and $x = b$,

$$y(a) = 0, \quad y(b) = 0. \tag{4-2}$$

---

† An unsteady problem becomes an initial-value problem when it is lumped, and an initial- and boundary-value problem when it is distributed.
‡ See Section 3–4 for the definition of homogeneous and nonhomogeneous differential equations and boundary conditions.

The general solution of this equation may be written in the form

$$y = C_1 y_1(x) + C_2 y_2(x), \qquad (4\text{--}3)$$

where $y_1(x)$ and $y_2(x)$ are linearly independent solutions, and C_1 and C_2 are arbitrary constants. This solution, combined with boundary conditions, gives

$$\begin{aligned} C_1 y_1(a) + C_2 y_2(a) &= 0, \\ C_1 y_1(b) + C_2 y_2(b) &= 0. \end{aligned} \qquad (4\text{--}4)$$

One possible solution of these homogeneous equations is $C_1 = C_2 = 0$, leading to the *trivial solution* $y \equiv 0$. If the determinant of the coefficients of C_1 and C_2 does not vanish, then this is the *only* solution. If the determinant of the coefficients does vanish, then

$$\begin{vmatrix} y_1(a) & y_2(a) \\ y_1(b) & y_2(b) \end{vmatrix} = 0. \qquad (4\text{--}5)$$

Now the two equations of (4–4) become identical and one constant can be expressed as a multiple of the other by use of either equation, the second constant being arbitrary. Thus *if Eq. (4–5) is satisfied* and if we discard, for example, the second equation of (4–4), we obtain from the first equation $C_2 y_2(a) = -C_1 y_1(a)$. Defining a new constant C by $C_1 = C y_2(a)$ yields $C_2 = -C y_1(a)$, and Eq. (4–3) becomes

$$y = C[y_2(a) y_1(x) - y_1(a) y_2(x)]. \qquad (4\text{--}6)$$

It may readily be seen that Eq. (4–6) satisfies the boundary conditions. One condition, $y(a) = 0$, is obtained directly from Eq. (4–6). For the other, $y(b) = 0$, Eq. (4–6) gives $y_2(a) y_1(b) - y_1(a) y_2(b) = 0$, which is the negative of Eq. (4–5). It should be remarked that Eq. (4–6) is a nontrivial solution only if $y_1(a)$ and $y_2(a)$ are not both zero. If $y_1(a) = y_2(a) = 0$, the first equation of (4–4) is a trivial identity, in which case only the second equation may be used to relate C_1 and C_2. This gives a nontrivial solution in the form

$$y = C[y_2(b) y_1(x) - y_1(b) y_2(x)],$$

provided $y_1(b)$ and $y_2(b)$ are not both zero. If $y_1(x)$ and $y_2(x)$ both vanish at $x = a$ and $x = b$, then Eq. (4–3) satisfies Eq. (4–2) for arbitrary values of both C_1 and C_2. If Eq. (4–5) is not satisfied, the only solution is the trivial one $y \equiv 0$.

The solution of two- and three-dimensional steady conduction problems, and that of one- and multidimensional unsteady problems may be reduced to the solution of Eq. (4–1), whose coefficients $f_1(x)$ and/or $f_2(x)$ depend on a parameter λ. In such problems the determinant of Eq. (4–5) may vanish only for certain values of λ, say $\lambda_1, \lambda_2, \lambda_3, \ldots$; these values are called the *characteristic values*. For each such value of λ a solution similar to Eq. (4–6) is obtained; these particular solutions are the *characteristic functions* of the problem, and problems of this kind are known as *characteristic-value problems*. The terms *eigenvalues*, *eigenfunctions*, and *eigenvalue problems* are also used frequently.

The foregoing generalized procedure can best be explained by an illustrative example. Reconsider the differential equation

$$\frac{d^2y}{dx^2} + y = 0$$

which was used in Section 3–6 to show the method of power series solution of differential equations. Furthermore, assume that this homogeneous equation involves a parameter λ as

$$\frac{d^2y}{dx^2} + \lambda^2 y = 0, \qquad (4\text{--}7)$$

and is subject to homogeneous boundary conditions

$$y(0) = 0, \qquad (4\text{--}8)$$

$$y(L) = 0. \qquad (4\text{--}9)$$

The general solution of Eq. (4–7) is

$$y = C_1 \sin \lambda x + C_2 \cos \lambda x. \qquad (4\text{--}10)$$

The use of Eq. (4–8) results in $C_2 = 0$ and

$$y = C_1 \sin \lambda x, \qquad (4\text{--}11)$$

and Eq. (4–9), combined with Eq. (4–11), gives $0 = C_1 \sin \lambda L$. The problem has nontrivial solutions only if λ satisfies the equation $0 = \sin \lambda L$. Therefore,

$$\lambda_n = n\pi/L, \qquad n = 1, 2, 3, \ldots, \qquad (4\text{--}12)$$

and the corresponding solutions of Eq. (4–11) are

$$y = C_1 \varphi_n(x), \qquad \varphi_n(x) = \sin(n\pi/L)x. \qquad (4\text{--}13)$$

Note that no new solutions are obtained when n assumes negative integer values.

Thus the foregoing boundary-value problem, Eqs. (4–7), (4–8), and (4–9), has no solution other than the trivial solution $y \equiv 0$, unless λ assumes one of the characteristic values given by Eq. (4–12). Corresponding to each characteristic value of λ_n there exists a characteristic function $\varphi_n(x)$ given by Eq. (4–13), such that any constant multiple of this function is a solution of the problem. It is important to note that the boundary-value problem given by

$$\frac{d^2y}{dx^2} - \lambda^2 y = 0; \qquad y(0) = 0, \quad y(L) = 0$$

has no solution other than the trivial solution $y \equiv 0$ corresponding to $\lambda = 0$. Hence there does not exist any set of characteristic values and characteristic functions for this problem. This illustrates the fact that *a boundary-value problem may or may not be a characteristic-value problem. A boundary-value problem is a characteristic-value problem when it has particular solutions that are periodic in nature; the period and amplitude of these solutions may or may not be constant.* Examples are the circular functions and Bessel functions of the first and second kinds, of any order. Since the starting point of a characteristic-value problem is a boundary-value problem, *a characteristic-value problem is always a boundary-value problem.*

In the next three sections the general properties of characteristic functions are investigated.

4–2. Orthogonality of Characteristic Functions

By definition, two functions $\varphi_n(x)$ and $\varphi_m(x)$ are said to be *orthogonal with respect to a weighting function* $w(x)$, over a finite interval (a, b), if the integral of the product $w\varphi_n\varphi_m$ over that interval vanishes as

$$\int_a^b w(x)\varphi_n(x)\varphi_m(x)\, dx = 0, \qquad m \neq n. \tag{4-14}$$

Furthermore, a set of functions is said to be *orthogonal* in (a, b) if all pairs of distinct functions in the set are orthogonal in (a, b). The word *orthogonality* comes from vector analysis. Let $\varphi_m(x_i)$ denote a vector in three-dimensional space whose rectangular components are $\varphi_m(x_1)$, $\varphi_m(x_2)$, and $\varphi_m(x_3)$. Two vectors, $\varphi_m(x_i)$ and $\varphi_n(x_i)$, are said to be orthogonal, or perpendicular to each other, if

$$\varphi_m(x_i) \cdot \varphi_n(x_i) = \sum_{i=1}^{3} \varphi_m(x_i)\varphi_n(x_i) = 0.$$

When the units of length on the coordinate axes vary from one axis to another, the foregoing scalar product assumes the form

$$\varphi_m(x_i) \cdot \varphi_n(x_i) = \sum_{i=1}^{3} w(x_i)\varphi_m(x_i)\varphi_n(x_i),$$

where the *weighting numbers* $w(x_1)$, $w(x_2)$, and $w(x_3)$ depend upon the units of length used along the three axes. Similarly, the vectors in an N-dimensional space having components $\varphi_m(x_i)$, $\varphi_n(x_i)$, $i = 1, 2, 3, \ldots, N$ are said to be orthogonal with respect to the weighting numbers $w(x_i)$ when

$$\varphi_m(x_i) \cdot \varphi_n(x_i) = \sum_{i=1}^{N} w(x_i)\varphi_m(x_i)\varphi_n(x_i) = 0. \tag{4-15}$$

If the vector space has an infinite number of dimensions, the components $\varphi_m(x_i)$ and $\varphi_n(x_i)$ become continuously distributed and x_i is no longer a discrete number but a continuous variable, say x; in this case Eq. (4–15) becomes identical to Eq. (4–14).

It will now be shown that *the characteristic functions of a characteristic-value problem are orthogonal over a finite interval with respect to a weighting function.* To establish this fact, consider the characteristic-value problem composed of the linear homogeneous second-order differential equation of the general form

$$\frac{d^2y}{dx^2} + f_1(x)\frac{dy}{dx} + [f_2(x) + \lambda^2 f_3(x)]y = 0$$

and two homogeneous boundary conditions prescribed at the ends of the finite interval (a, b). This equation, multiplied through by the factor $\exp[\int f_1(x)\, dx] = p(x)$ and with the functions defined as $f_2(x)p(x) = q(x)$ and $f_3(x)p(x) = w(x)$, may be rearranged in the form

$$\frac{d}{dx}\left[p(x)\frac{dy}{dx}\right] + [q(x) + \lambda^2 w(x)]y = 0, \tag{4–16}$$

which is more convenient for the following discussion.

Let λ_m, λ_n be any two distinct characteristic numbers, that is, $m \neq n$, and let $\varphi_m(x)$, $\varphi_n(x)$ be the corresponding characteristic functions. Since $y = \varphi_m(x)$ and $y = \varphi_n(x)$ are solutions of Eq. (4–16),

$$\frac{d}{dx}\left(p\frac{d\varphi_m}{dx}\right) + (q + \lambda_m^2 w)\varphi_m = 0,$$

$$\frac{d}{dx}\left(p\frac{d\varphi_n}{dx}\right) + (q + \lambda_n^2 w)\varphi_n = 0.$$

Multiplying the first equation by φ_n and the second by φ_m, then subtracting the second of the resulting equations from the first one gives

$$\varphi_n \frac{d}{dx}\left(p\frac{d\varphi_m}{dx}\right) - \varphi_m \frac{d}{dx}\left(p\frac{d\varphi_n}{dx}\right) + (\lambda_m^2 - \lambda_n^2)w\varphi_m\varphi_n = 0.$$

Integrating this equation over the finite interval (a, b) yields

$$(\lambda_n^2 - \lambda_m^2)\int_a^b w\varphi_m\varphi_n\, dx = \int_a^b \left[\varphi_n \frac{d}{dx}\left(p\frac{d\varphi_m}{dx}\right) - \varphi_m \frac{d}{dx}\left(p\frac{d\varphi_n}{dx}\right)\right]dx,$$

and integration by parts for the right-hand member results in

$$(\lambda_n^2 - \lambda_m^2)\int_a^b w\varphi_m\varphi_n\, dx = \left\{p(x)\left[\varphi_n(x)\frac{d\varphi_m(x)}{dx} - \varphi_m(x)\frac{d\varphi_n(x)}{dx}\right]\right\}\bigg|_a^b. \tag{4–17}$$

Since both $y = \varphi_m(x)$ and $y = \varphi_n(x)$ are particular solutions of Eq. (4–16), the right-hand side of Eq. (4–17) vanishes when one of the following conditions is prescribed at each end of the interval (a, b):

$$y = 0, \qquad (4\text{–}18)$$

$$dy/dx = 0, \qquad (4\text{–}19)$$

$$dy/dx + By = 0, \qquad (4\text{–}20)$$

where B is an arbitrary parameter. The fact that Eq. (4–17) vanishes when Eq. (4–20) is satisfied may be clarified by rearranging the right-hand member of Eq. (4–17) in the form

$$\varphi_n \varphi'_m - \varphi_m \varphi'_n = \varphi_n \varphi'_m - \varphi_m \varphi'_n \pm B\varphi_m \varphi_n = \varphi_n(\varphi'_m + B\varphi_m) - \varphi_m(\varphi'_n + B\varphi_n).$$

Particularly, if $p(x) = 0$ when $x = a$ or $x = b$, the right-hand side of Eq. (4–17) vanishes, and the condition given by Eq. (4–18), (4–19), or (4–20) satisfied at $x = a$ or $x = b$ can be dropped from the problem provided y and (dy/dx) are finite at that point. If $p(b) = p(a)$, the orthogonality continues to exist when the boundary conditions are replaced by the conditions $y(b) = y(a)$ and $y'(b) = y'(a)$, which are called the *periodic boundary conditions*.

As an example, reconsider the characteristic-value problem given by Eqs. (4–7), (4–8), and (4–9). Comparison of Eqs. (4–7) and (4–16) gives $w(x) = 1$, and the condition of orthogonality for this problem is

$$\int_0^L \varphi_m(x)\varphi_n(x)\,dx = \int_0^L \sin(m\pi x/L)\sin(n\pi x/L)\,dx = 0, \qquad m \neq n,$$

which can also be verified independently by direct integration.

4–3. Expansion of Arbitrary Functions in Series of Orthogonal Functions

Let $\varphi_n(x)$ be a set of functions orthogonal with respect to a weighting function $w(x)$ over a finite interval (a, b). We wish to expand an arbitrary function $f(x)$ into a series of this set as

$$f(x) = b_0\varphi_0(x) + b_1\varphi_1(x) + b_2\varphi_2(x) + \cdots = \sum_{n=0}^{\infty} b_n\varphi_n(x). \qquad (4\text{–}21)$$

Assuming that such an expansion exists, we may evaluate the coefficients b_n. Multiplying both sides of Eq. (4–21) by $w(x)\varphi_m(x)$ and integrating the result over the interval (a, b) with the *assumption* that the integral of the infinite sum is equivalent to the sum of the integrals, we have

$$\int_a^b w(x)f(x)\varphi_m(x)\,dx = \sum_{n=0}^{\infty} b_n \int_a^b w(x)\varphi_m(x)\varphi_n(x)\,dx, \qquad (4\text{–}22)$$

where $\varphi_m(x)$ is the mth term in the set. Using the orthogonality of the set, we find that all terms in the sum on the right of Eq. (4–22) are zero except the term corresponding to $n = m$. Hence Eq. (4–22) gives

$$b_n = \frac{\int_a^b w(x)f(x)\varphi_n(x)\,dx}{\int_a^b w(x)\varphi_n^2(x)\,dx}. \tag{4–23}$$

The general problem of determining whether the expansion given by Eq. (4–21) represents the function $f(x)$ is beyond the scope of this text. Let us note, however, the generality of such expansions. Unlike the Taylor and Maclaurin series expansions of a function $f(x)$, which require that the function and all its derivatives be continuous, if $f(x)$ is piecewise differentiable† in a finite interval (a, b), the series given by Eq. (4–21) converges to $f(x)$ at all points of continuity, and to the mean value $\frac{1}{2}[f(x+) + f(x-)]$ of $f(x)$ at any discontinuity.

In the next section we show that the well-known Fourier series actually are special cases of the expansion in terms of a set of orthogonal functions involving circular functions.

4–4. Fourier Series

We have learned in Section 4–1 that the boundary-value problem given by

$$\frac{d^2y}{dx^2} + \lambda^2 y = 0; \quad y(0) = 0, \quad y(L) = 0$$

leads to a characteristic-value problem which has the following characteristic values and functions

$$\lambda_n = n\pi/L, \quad n = 1, 2, 3, \ldots, \tag{4–12}$$

$$\varphi_n = \sin(n\pi/L)x. \tag{4–13}$$

We now wish to expand an arbitrary function $f(x)$ into a series of these characteristic functions as

$$f(x) = \sum_{n=1}^{\infty} b_n \sin(n\pi/L)x, \quad 0 < x < L, \tag{4–24}$$

where the coefficients b_n are calculated from Eq. (4–23):

$$b_n = \frac{\int_0^L f(x)\sin(n\pi/L)x\,dx}{\int_0^L \sin^2(n\pi/L)x\,dx}. \tag{4–25}$$

It can readily be shown by direct integration that the denominator of Eq. (4–25)

---

† A function $f(x)$ is said to be piecewise differentiable over a finite range if it is possible to divide that range into a finite number of intervals such that $f(x)$ is differentiable inside each interval and $f'(x)$ approaches finite values from either side of a discontinuity.

is equal to $L/2$ regardless of the value of n. Hence

$$b_n = \frac{2}{L} \int_0^L f(x) \sin\left(\frac{n\pi}{L}\right) x \, dx. \tag{4-26}$$

The series given by Eq. (4–24) is the definition of the *Fourier sine series* of $f(x)$ over the interval $(0, L)$. All terms of this series are periodic with the period $2L$, which is twice the length of the interval. If x is replaced by $-x$, the sign of each term is reversed. Hence in the interval $(-L, 0)$ the series represents the function $-f(-x)$. The behavior of the series in the interval $(-L, L)$ is repeated periodically for all values of x. If $f(x)$ is an odd function of x, the series given by Eq. (4–24) represents $f(x)$ not only in the interval $(0, L)$ but also in the interval $(-L, L)$. If, in addition, $f(x)$ is periodic, of period $2L$, the series represents $f(x)$ everywhere.

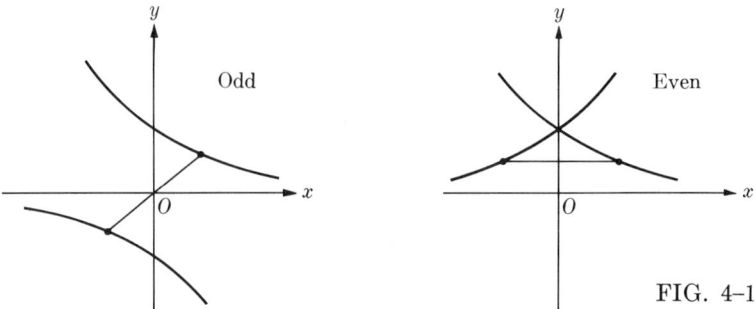

FIG. 4–1

A function $f(x)$ is said to be an *odd function* if $f(-x) = -f(x)$ and an *even function* if $f(-x) = f(x)$. An odd function is symmetrical with respect to the origin, whereas an even function is symmetrical with respect to the y-axis (Fig. 4–1).

Consider, as an example, the Fourier sine series of the function

$$f(x) = \begin{cases} 0, & -\infty < x < 0 \text{ and } L/2 < x < \infty \\ 1, & 0 < x < L/2 \end{cases} \tag{4-27}$$

over the interval $(0, L)$ (Fig. 4–2). The coefficients of the series are

$$b_n = \frac{2}{L} \int_0^L f(x) \sin\left(\frac{n\pi}{L}\right) x \, dx = \frac{2}{L} \int_0^{L/2} 1 \cdot \sin\left(\frac{n\pi}{L}\right) x \, dx$$

$$= \frac{2}{n\pi}\left(1 - \cos\frac{n\pi}{2}\right);$$

hence the series is

$$f(x) = \frac{2}{\pi} \sum_{n=1}^{\infty} \frac{1}{n}\left(1 - \cos\frac{n\pi}{2}\right) \sin\left(\frac{n\pi}{L}\right) x. \tag{4-28}$$

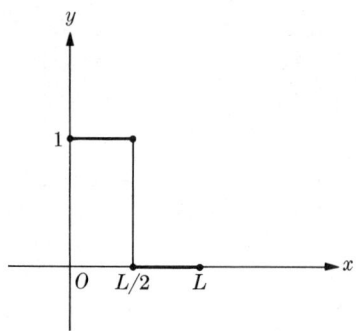

FIG. 4-2

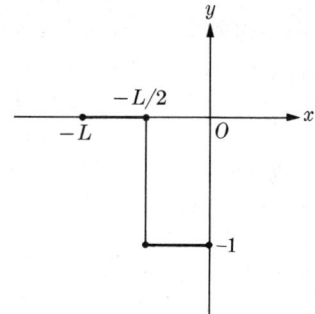

FIG. 4-3

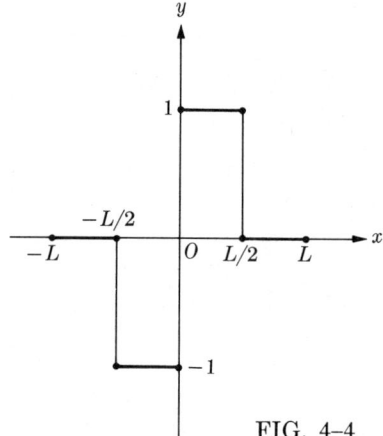

FIG. 4-4

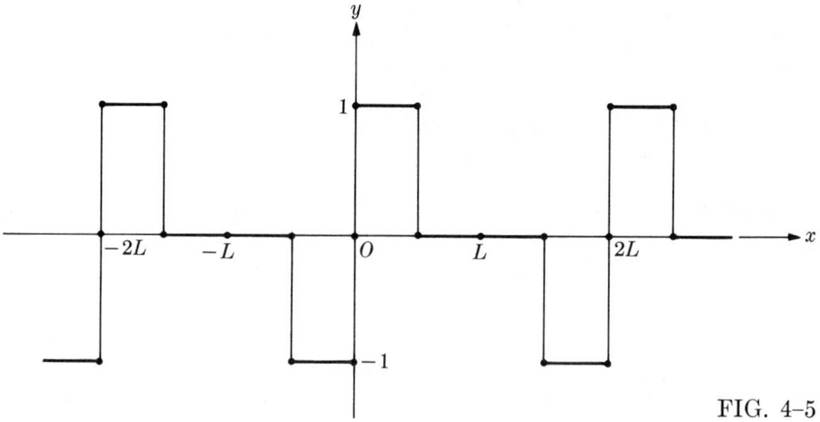

FIG. 4-5

The same series also represents the function

$$-f(x) = f(-x) = \begin{cases} 0, & -\infty < x < -L/2 \text{ and } 0 < x < \infty \\ -1, & -L/2 < x < 0 \end{cases}$$

over the interval $(-L, 0)$ (Fig. 4–3). Thus the Fourier sine series of the odd function defined by

$$f(x) = \begin{cases} -1, & -L/2 < x < 0 \\ 0, & -\infty < x < L/2 \text{ and } L/2 < x < \infty \\ 1, & 0 < x < L/2 \end{cases}$$

represents $f(x)$ in the interval $(-L, L)$ (Fig. 4–4) except at points of discontinuity, and repeats this behavior periodically for all values of x (Fig. 4–5).

A series expansion involving cosine terms rather than sine terms may be obtained by considering instead the boundary-value problem

$$\frac{d^2y}{dx^2} + \lambda^2 y = 0; \quad y'(0) = 0, \quad y'(L) = 0. \tag{4-29}$$

Following the procedure of the previous characteristic-value problem, we may readily obtain the characteristic values and functions in the form

$$\lambda_n = n\pi/L, \quad n = 0, 1, 2, \ldots; \quad \varphi_n = \cos(n\pi/L)x. \tag{4-30}$$

Note that $\varphi_0(x) = 1$ is a member of the set of characteristic functions given in Eq. (4-30) corresponding to $\lambda_0 = 0$. Now, expanding an arbitrary function $f(x)$ into a series of the foregoing set, we have

$$f(x) = a_0 + \sum_{n=1}^{\infty} a_n \cos(n\pi/L)x, \quad 0 < x < L, \tag{4-31}$$

where the coefficients a_n are calculated from Eq. (4-23) in the form

$$a_0 = \frac{\int_0^L f(x)\,dx}{\int_0^L dx}, \quad a_n = \frac{\int_0^L f(x)\cos(n\pi/L)x\,dx}{\int_0^L \cos^2(n\pi/L)x\,dx}. \tag{4-32}$$

Furthermore, noting that

$$\int_0^L \cos^2\left(\frac{n\pi}{L}\right)x\,dx = \begin{cases} L, & n = 0 \\ L/2, & n = 1, 2, 3, \ldots \end{cases}$$

we may rearrange Eq. (4-32) as

$$a_0 = \frac{1}{L}\int_0^L f(x)\,dx, \quad a_n = \frac{2}{L}\int_0^L f(x)\cos\left(\frac{n\pi}{L}\right)x\,dx. \tag{4-33}$$

The series given by Eq. (4–31) is the definition of the *Fourier cosine series* of $f(x)$ over the interval $(0, L)$. All terms of Eq. (4–31) are even functions of x, and they are periodic with the period $2L$. This series represents the function $f(-x)$ in the interval $(-L, 0)$. If $f(x)$ is an even function of x, the series converges to $f(x)$ not only in the interval $(0, L)$ but also in the interval $(-L, 0)$. If, in addition, $f(x)$ is periodic, of period $2L$, the series represents $f(x)$ everywhere.

Consider now the Fourier cosine series of the previous example, Eq. (4–27). The coefficients of the series are

$$a_0 = \frac{1}{L} \int_0^L f(x)\, dx = \frac{1}{L} \int_0^{L/2} 1 \cdot dx = \tfrac{1}{2},$$

$$a_n = \frac{2}{L} \int_0^L f(x) \cos\left(\frac{n\pi}{L}\right) x\, dx = \frac{2}{L} \int_0^{L/2} 1 \cdot \cos\left(\frac{n\pi}{L}\right) x\, dx$$

$$= \frac{2}{n\pi} \sin\left(\frac{n\pi}{2}\right).$$

Therefore,

$$f(x) = \frac{1}{2} + \frac{2}{\pi} \sum_{n=1}^{\infty} \frac{1}{n} \sin\left(\frac{n\pi}{2}\right) \cos\left(\frac{n\pi}{L}\right) x. \tag{4–34}$$

The same series also represents the function

$$f(x) = f(-x) = \begin{cases} 0, & -\infty < x < -L/2 \text{ and } 0 < x < \infty \\ -1, & -L/2 < x < 0 \end{cases}$$

over the interval $(-L, 0)$ (Fig. 4–6). Thus the Fourier cosine series of the even function defined by

$$f(x) = \begin{cases} 0, & -\infty < x < -L/2 \text{ and } L/2 < x < \infty \\ 1, & -L/2 < x < L/2 \end{cases}$$

represents $f(x)$ in the interval $(-L, L)$ (Fig. 4–7) except at points of discontinuity, and repeats this behavior periodically for all values of x (Fig. 4–8).

So far, we have seen that any piecewise continuous function can be expressed in the interval $(0, L)$ by a series consisting of sines or cosines with the common period $2L$. When the function is odd, the sine series representation is valid in the interval $(-L, L)$, whereas for an even function the cosine series representation holds in the same interval.

We now wish to express a function $f(x)$, which is piecewise continuous in the interval $(-L, L)$, in terms of both sines and cosines having the common

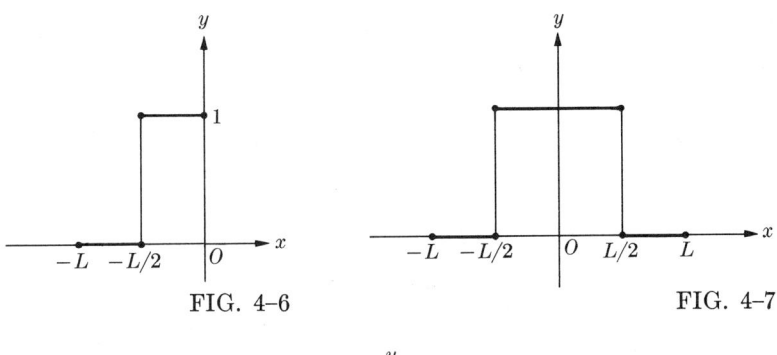

FIG. 4-6   FIG. 4-7

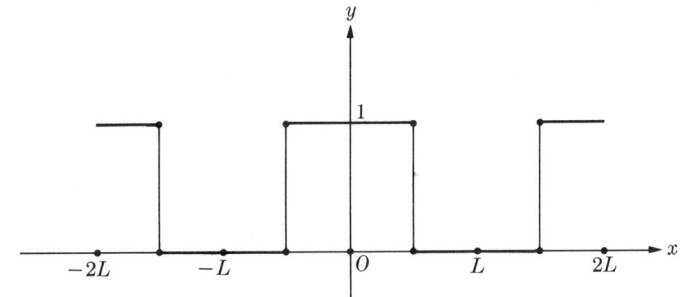

FIG. 4-8

period 2L. Noting that

$$f(x) = \tfrac{1}{2}[f(x) + f(-x)] + \tfrac{1}{2}[f(x) - f(-x)],$$

where the function in the first brackets is even and that in the second is odd, we arrive at the fact that an arbitrary function can be expressed as the sum of an even function and an odd function. Hence

$$f(x) = f_e(x) + f_o(x). \tag{4-35}$$

Expressing $f_e(x)$ in terms of cosines and $f_o(x)$ in terms of sines in the interval $(-L, L)$, we have

$$f_e(x) = a_0 + \sum_{n=1}^{\infty} a_n \cos (n\pi/L)x, \tag{4-36}$$

$$f_o(x) = \sum_{n=1}^{\infty} b_n \sin (n\pi/L)x, \tag{4-37}$$

where

$$a_0 = \frac{1}{L} \int_0^L f_e(x)\, dx, \qquad a_n = \frac{2}{L} \int_0^L f_e(x) \cos\left(\frac{n\pi}{L}\right) x\, dx,$$

$$b_n = \frac{2}{L} \int_0^L f_o(x) \sin\left(\frac{n\pi}{L}\right) x\, dx.$$

Since the integrands of these equations are even functions of x, replacing $\int_0^L$ by $\frac{1}{2}\int_{-L}^{L}$ gives

$$a_0 = \frac{1}{2L}\int_{-L}^{L} f(x)\,dx, \qquad a_n = \frac{1}{L}\int_{-L}^{L} f(x)\cos\left(\frac{n\pi}{L}\right)x\,dx,$$

$$b_n = \frac{1}{L}\int_{-L}^{L} f(x)\sin\left(\frac{n\pi}{L}\right)x\,dx. \tag{4-38}$$

It is clear that the odd part of the first two integrals and the even part of the third integral vanish identically. Hence introducing Eqs. (4–36) and (4–37) into Eq. (4–35) yields

$$f(x) = a_0 + \sum_{n=1}^{\infty}[a_n\cos(n\pi/L)x + b_n\sin(n\pi/L)x], \qquad -L < x < L. \tag{4-39}$$

This series is the definition of the *complete Fourier series* of $f(x)$ over the interval $(-L, L)$. When $f(x)$ is even, the resulting series involves cosine terms only; a series involving only sines results when $f(x)$ is odd. Otherwise Eq. (4–39) is in terms of both sines and cosines of period $2L$ in the interval $(-L, L)$, and repeats this behavior periodically for all values of x.

Finally, let us find the complete Fourier series of the function given by Eq. (4–27). The coefficients of this series are

$$a_0 = \frac{1}{2L}\int_{-L}^{L} f(x)\,dx = \frac{1}{2L}\int_{0}^{L/2} 1\cdot dx = \tfrac{1}{4},$$

$$a_n = \frac{1}{L}\int_{-L}^{L} f(x)\cos\left(\frac{n\pi}{L}\right)x\,dx = \frac{1}{L}\int_{0}^{L/2} 1\cdot\cos\left(\frac{n\pi}{L}\right)x\,dx = \frac{1}{n\pi}\sin\frac{n\pi}{2},$$

$$b_n = \frac{1}{L}\int_{-L}^{L} f(x)\sin\left(\frac{n\pi}{L}\right)x\,dx$$

$$= \frac{1}{L}\int_{0}^{L/2} 1\cdot\sin\left(\frac{n\pi}{L}\right)x\,dx = \frac{1}{n\pi}\left(1 - \cos\frac{n\pi}{2}\right).$$

Therefore,

$$f(x) = \frac{1}{4} + \frac{1}{\pi}\sum_{n=1}^{\infty}\frac{1}{n}\left[\sin\left(\frac{n\pi}{2}\right)\cos\left(\frac{n\pi}{L}\right)x + \left(1 - \cos\frac{n\pi}{2}\right)\sin\left(\frac{n\pi}{L}\right)x\right], \tag{4-40}$$

which converges for all values of x, in the sense described, to the periodic function shown in Fig. 4–9. [Compare Eq. (4–40) with the sum of the sine and cosine series expansions of the same function obtained by adding Figs. (4–5) and (4–8) geometrically, or Eqs. (4–28) and (4–34) algebraically.]

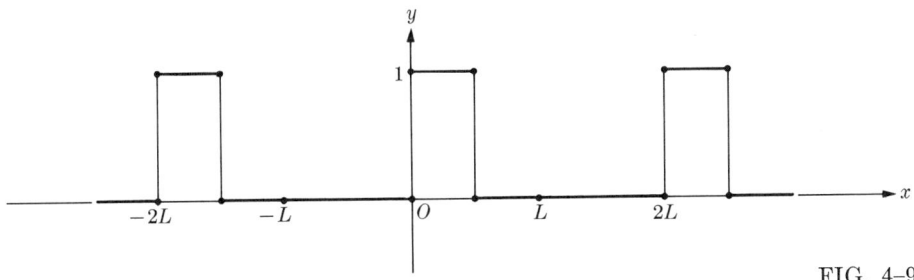

FIG. 4-9

Having gained the foregoing mathematical background we now proceed to the solution of problems by the method of separation of variables.

4-5. Separation of Variables. Steady Two-Dimensional Cartesian Geometry

When the boundary conditions of a problem are in terms of specified T, $\partial T/\partial n$, or $\partial T/\partial n + BT$, where n is the normal to the boundary and B a constant, the solution may be expressed as a product of functions of each coordinate separately. This allows the boundary conditions to be expressed in terms of a single variable, and reduces the partial differential equation to a set of ordinary differential equations.

The essential features of the method will now be illustrated by means of a steady two-dimensional example. Consider the second-order partial differential equation

$$a_1(x)\frac{\partial^2 T}{\partial x^2} + a_2(x)\frac{\partial T}{\partial x} + a_3(x)T + b_1(y)\frac{\partial^2 T}{\partial y^2} + b_2(y)\frac{\partial T}{\partial y} + b_3(y)T = 0. \tag{4-41}$$

A more generalized form of this equation which involves coefficients as functions of both independent variables is not suitable for the separation of variables.

Assume the existence of a product solution

$$T(x, y) = X(x)Y(y), \tag{4-42}$$

where X is a function of x alone and Y is a function of y. This assumption becomes meaningful when the two functions X and Y actually satisfy separate differential equations.

Introducing Eq. (4-42) into Eq. (4-41) and dividing the result by XY yields

$$\left[a_1(x)\frac{d^2X}{dx^2} + a_2(x)\frac{dX}{dx} + a_3 X\right]\frac{1}{X} = -\left[b_1(y)\frac{d^2Y}{dy^2} + b_2(y)\frac{dY}{dy} + b_3(y)Y\right]\frac{1}{Y}. \tag{4-43}$$

The left-hand side of this equation is independent of y and the right-hand side is independent of x. Since x and y can vary independently, both sides of Eq. (4–43) must be independent of either variable; that is, they must be equal to a constant, say $+\lambda^2$ or $-\lambda^2$. This constant is called the *separation parameter*. Hence the partial differential equation of Eq. (4–41) is reduced to the following two ordinary differential equations:

$$a_1(x) \frac{d^2 X}{dx^2} + a_2(x) \frac{dX}{dx} + [a_3(x) \pm \lambda^2]X = 0,$$

$$b_1(y) \frac{d^2 Y}{dy^2} + b_2(y) \frac{dY}{dy} + [b_3(y) \mp \lambda^2]Y = 0.$$

(4–44)

The method of separation of variables is applicable to steady two-dimensional problems if and when (i) *one of the directions of the problem is expressed by a homogeneous differential equation subject to homogeneous boundary conditions* (the homogeneous direction), *while the other direction is expressed by a homogeneous differential equation subject to one homogeneous and one nonhomogeneous boundary condition* (the nonhomogeneous direction), and (ii) *the sign of λ^2 is chosen such that the boundary-value problem of the homogeneous direction leads to a characteristic-value problem*.

The solutions obtained by the separation of variables are in the form of a sum or integral, depending on whether the homogeneous direction is finite or extends to infinity, respectively. This chapter and the next one are devoted to series solutions which are applicable to problems homogeneous in finite directions. Problems requiring homogeneity in infinite domains are suitable to integral solutions. These, being easier to solve by the Laplace transforms, are delayed to Chapter 7. (see Section 7–5).

The result of the present section may readily be extended to steady three-dimensional problems and to unsteady problems (see Section 4–11 and Chapter 5, respectively). We shall now illustrate the method of separation of variables by a number of examples.

Example 4–1. Consider an infinitely long two-dimensional fin of thickness l (Fig. 4–10). The base temperature of the fin is $F(y)$, the ambient temperature T_∞. The heat transfer coefficient is large. We wish to find the steady temperature of the fin.

The differential formulation of the problem, according to the selected reference frame, is

$$\frac{\partial^2 T}{\partial x^2} + \frac{\partial^2 T}{\partial y^2} = 0,$$

$$T(0, y) = F(y), \quad T(\infty, y) = T_\infty,$$

$$T(x, 0) = T_\infty, \quad T(x, l) = T_\infty.$$

We now seek a solution by the method of separation of variables, which requires that the differential equation and three of the boundary conditions be homogeneous. Although the problem expressed in T does not satisfy these conditions, the simple transformation

$$\theta = T - T_\infty$$

reduces three of the nonhomogeneous boundary conditions to homogeneous conditions without affecting the homogeneity of the differential equation. Thus the formulation of the problem in θ becomes

$$\frac{\partial^2 \theta}{\partial x^2} + \frac{\partial^2 \theta}{\partial y^2} = 0, \quad (4\text{-}45)$$

$$\theta(0, y) = F(y) - T_\infty = f(y), \quad (4\text{-}46)$$

$$\theta(\infty, y) = 0, \quad (4\text{-}47)$$

$$\theta(x, 0) = 0, \quad (4\text{-}48)$$

$$\theta(x, l) = 0. \quad (4\text{-}49)$$

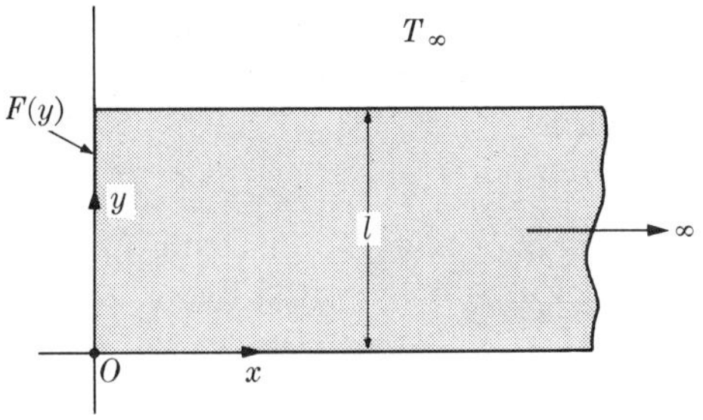

FIG. 4–10

Assuming the existence of a product solution of the form

$$\theta(x, y) = X(x)Y(y), \quad (4\text{-}50)$$

then introducing Eq. (4–50) into Eq. (4–45) and dividing each term by XY, we obtain

$$\frac{1}{X}\frac{d^2 X}{dx^2} = -\frac{1}{Y}\frac{d^2 Y}{dy^2} = \pm \lambda^2. \quad (4\text{-}51)$$

Here the sign of λ^2 must be chosen such that the homogeneous y-direction results in a characteristic-value problem. The selection of $-\lambda^2$ yields particular solutions in y expressible by hyperbolic functions which, as indicated in Section 4–1, cannot be made orthogonal; hence $+\lambda^2$ is suitable to our problem.

Further use of Eq. (4–50) reduces the two-dimensional homogeneous boundary conditions of the problem to one-dimensional conditions. This may readily be illustrated using one of these conditions, say Eq. (4–47). Thus, since $Y(y)$ is arbitrary, $\theta(\infty, y) = X(\infty)Y(y) = 0$ implies $X(\infty) = 0$.

Finally, we have the problem separately expressed in the x- and y-directions as follows:

$$\frac{d^2 Y}{dy^2} + \lambda^2 Y = 0; \quad Y(0) = 0, \quad Y(l) = 0, \quad (4\text{-}52)$$

$$\frac{d^2 X}{dx^2} - \lambda^2 X = 0; \quad X(\infty) = 0. \quad (4\text{-}53)$$

The nonhomogeneous boundary condition in the x-direction, being nonseparable, is left to the final stage of the solution.

The characteristic-value problem of Eq. (4–52) has already been solved in Section 4–1. [See Eqs. (4–7) through (4–13).] The resulting characteristic functions are $\varphi_n(y) = \sin \lambda_n y$, and the characteristic values are $\lambda_n = n\pi/l$, $n = 1, 2, 3, \ldots$. Thus the solution of Eq. (4–52) is $Y_n(y) = C_n \varphi_n(y)$, where C_n is an arbitrary constant. On the other hand, the general solution of Eq. (4–53) expressed conveniently in terms of exponential functions is

$$X_n(x) = B_n e^{-\lambda_n x}, \tag{4-54}$$

where B_n is an arbitrary constant.

Thus, using Eq. (4–50), we obtain the product solution of the problem in the form

$$\theta(x, y) = a_n e^{-\lambda_n x} \sin \lambda_n y, \tag{4-55}$$

where $B_n C_n$, always appearing as a product and depending on the particular value of n, is represented by a_n.

Now we must determine which values of n should be used in Eq. (4–55) for the solution of the problem. Clearly, Eq. (4–55) satisfies the differential equation, Eq. (4–45), and the homogeneous boundary conditions, Eqs. (4–47), (4–48), (4–49), for all positive integer values of n. On the other hand, the nonhomogeneous boundary condition, Eq. (4–46), combined with Eq. (4–55), requires that

$$f(y) = a_n \sin \lambda_n y,$$

which, in general, may not be satisfied by a single value of n. This difficulty may, however, be circumvented by considering the linearity of the problem and utilizing the principle of superposition. Hence, in terms of all positive values of n, we seek a solution of the form

$$\theta(x, y) = \sum_{n=1}^{\infty} a_n e^{-\lambda_n x} \sin \lambda_n y \tag{4-56}$$

rather than that given by Eq. (4–55).

Thus the nonhomogeneous boundary condition employed with Eq. (4–56) yields the expansion

$$f(y) = \sum_{n=1}^{\infty} a_n \sin \lambda_n y, \tag{4-57}$$

which is possible according to Section 4–3. Here, noting that the weighting function is unity, we may calculate the value of a_n from Eq. (4–23). The result is

$$a_n = \frac{\int_0^l f(y) \sin \lambda_n y \, dy}{\int_0^l \sin^2 \lambda_n y \, dy}. \tag{4-58}$$

The denominator of Eq. (4–58) is equal to $l/2$ for positive integer values of n. Hence Eq. (4–58) becomes

$$a_n = \frac{2}{l} \int_0^l f(y) \sin \lambda_n y \, dy. \qquad (4\text{--}59)$$

Finally, inserting Eq. (4–59) into Eq. (4–56), we find the solution of the problem to be

$$\theta(x, y) = \frac{2}{l} \sum_{n=1}^{\infty} \left[\int_0^l f(\eta) \sin \lambda_n \eta \, d\eta \right] e^{-\lambda_n x} \sin \lambda_n y, \qquad (4\text{--}60)$$

where the dummy variable y of Eq. (4–59) is replaced by η for clarity.

In particular, if the initial temperature is uniform, that is, if

$$\theta(0, y) = f(y) = \theta_0,$$

the result of Eq. (4–59) is

$$a_n = \frac{2\theta_0}{\lambda_n l} [1 - (-1)^n]. \qquad (4\text{--}61)$$

Thus, if we introduce Eq. (4–61) into Eq. (4–56), the solution of the problem corresponding to a constant base temperature becomes

$$\frac{\theta(x, y)}{\theta_0} = \frac{2}{l} \sum_{n=1}^{\infty} \frac{[1 - (-1)^n]}{\lambda_n} e^{-\lambda_n x} \sin \lambda_n y. \qquad (4\text{--}62)$$

Since only the odd integer values of n contribute to the temperature, by defining the integer k as $n = 2k + 1$ and employing $\lambda_n = n\pi/l$, we may write Eq. (4–62) more conveniently in the form

$$\frac{\theta(x, y)}{\theta_0} = \frac{4}{\pi} \sum_{k=0}^{\infty} \frac{e^{-(2k+1)(\pi/l)x}}{(2k + 1)} \sin (2k + 1) \frac{\pi}{l} y, \qquad (4\text{--}63)$$

where k assumes all positive integer values including zero.

Example 4–2. To show the effect of the choice of a coordinate system on the problem, let us find the solution of Example 4–1 in a new coordinate system (Fig. 4–11) for the case of uniform base temperature θ_0.

The thermal symmetry of the problem with respect to the new abscissa suggests the use of $2l$ for the thickness of the fin. The formulation of the

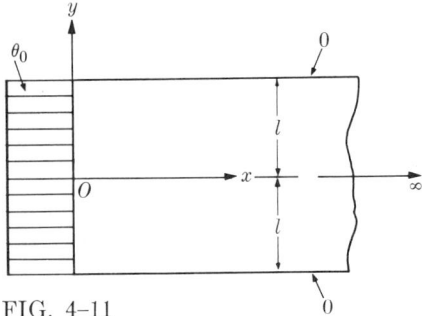

FIG. 4–11

problem in terms of θ is

$$\frac{\partial^2 \theta}{\partial x^2} + \frac{\partial^2 \theta}{\partial y^2} = 0,$$

$$\theta(0, y) = \theta_0, \qquad \theta(\infty, y) = 0,$$

$$\frac{\partial \theta(x, 0)}{\partial y} = 0 \quad \text{or} \quad \theta(x, -l) = 0,$$

$$\theta(x, l) = 0.$$

Employing the product solution of Eq. (4–50) and $+\lambda^2$, we reduce the differential equation and homogeneous boundary conditions to the following system:

$$\frac{d^2 Y}{dy^2} + \lambda^2 Y = 0; \qquad \frac{dY(0)}{dy} = 0, \quad Y(l) = 0, \qquad (4\text{–}64)$$

$$\frac{d^2 X}{dx^2} - \lambda^2 X = 0; \qquad X(\infty) = 0. \qquad (4\text{–}53)$$

Here the condition of middle-plane symmetry in the y-direction, being more convenient, is preferred.

The solution of the characteristic-value problem of the y-direction given by Eq. (4–64) is

$$Y_n(y) = C_n \varphi_n(y) \qquad (4\text{–}65)$$

together with

$$\varphi_n(y) = \cos \lambda_n y, \qquad \text{characteristic functions}, \qquad (4\text{–}66)$$

$$\lambda_n = (2n + 1)\pi/2l, \quad n = 0, 1, 2, \ldots, \qquad \text{characteristic values}, \qquad (4\text{–}67)$$

where C_n is an arbitrary constant.

The solution of the boundary-value problem in x, being identical to that of the previous example, Eq. (4–53), is

$$X_n(x) = B_n e^{-\lambda_n x}. \qquad (4\text{–}54)$$

Thus the product solution of Eq. (4–50), combined with Eqs. (4–54), (4–65), (4–66), and (4–67), yields

$$\theta(x, y) = \sum_{n=0}^{\infty} a_n e^{-\lambda_n x} \cos \lambda_n y, \qquad (4\text{–}68)$$

where again $a_n = B_n C_n$.

Finally, using the nonhomogeneous boundary condition given by $\theta(0, y) = \theta_0$, we may reduce Eq. (4–68) to

$$\theta_0 = \sum_{n=0}^{\infty} a_n \cos \lambda_n y. \qquad (4\text{–}69)$$

Then, as before, using Eq. (4–23), we find that

$$a_n = (-1)^n \frac{2\theta_0}{\lambda_n l}. \tag{4-70}$$

Thus, introducing Eq. (4–70) into Eq. (4–68), we find that the solution of the problem is

$$\frac{\theta(x, y)}{\theta_0} = \frac{2}{l} \sum_{n=0}^{\infty} \frac{(-1)^n e^{-\lambda_n x}}{\lambda_n} \cos \lambda_n y. \tag{4-71}$$

Equations (4–63) and (4–71) represent the same temperature in two different coordinate systems. Except in trivial problems, the selection of coordinates is very important and is the first step toward a solution. This point will be explored in Section 4–6.

Example 4–3. Find the temperature distribution of the extended surface of Example 4–2 for a finite heat transfer coefficient h (Fig. 4–12).

This problem replaces the one-dimensional problem of Example 3–9 when the temperature distribution across the thickness of the fin cannot be neglected compared to that in the longitudinal direction. In this case the formulation of the problem in terms of the coordinate system of Example 4–2 becomes

$$\frac{\partial^2 \theta}{\partial x^2} + \frac{\partial^2 \theta}{\partial y^2} = 0,$$

$$\theta(0, y) = \theta_0, \qquad \theta(\infty, y) = 0,$$

$$\frac{\partial \theta(x, 0)}{\partial y} = 0, \qquad -k \frac{\partial \theta(x, l)}{\partial y} = h\theta(x, l).$$

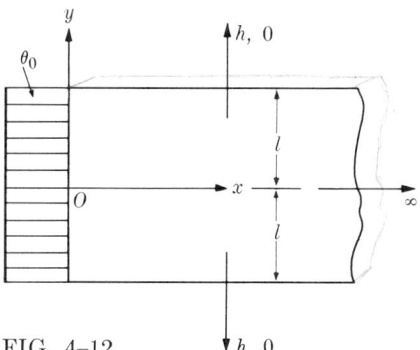

FIG. 4–12

The product solution given by Eq. (4–50) converts the formulation to the set of equations

$$\frac{d^2 Y}{dy^2} + \lambda^2 Y = 0; \qquad \frac{dY(0)}{dy} = 0, \qquad -k \frac{dY(l)}{dy} = hY(l), \tag{4-72}$$

$$\frac{d^2 X}{dx^2} - \lambda^2 X = 0; \qquad X(\infty) = 0. \tag{4-53}$$

The characteristic-value problem of the y-direction, Eq. (4–72), results in the solution

$$Y_n(y) = C_n \varphi_n(y), \tag{4-73}$$

with

$$\varphi_n(y) = \cos \lambda_n y, \qquad \text{characteristic functions,} \tag{4-74}$$

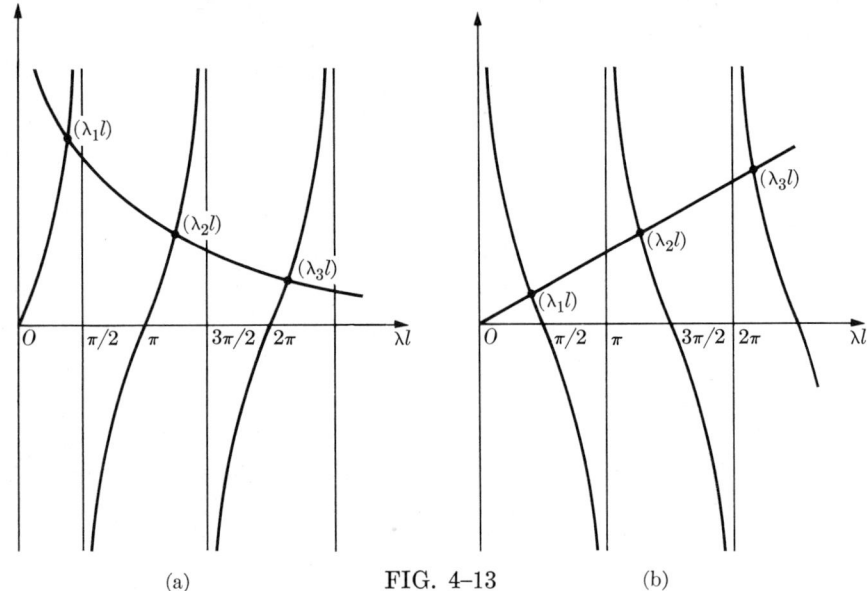

FIG. 4–13

where the characteristic values are the roots of the transcendental equation

$$k\lambda_n \sin \lambda_n l = h \cos \lambda_n l. \tag{4-75}$$

Equation (4–75) may be rearranged in either one of the forms

$$\tan \lambda_n l = \text{Bi}/\lambda_n l, \qquad \cot \lambda_n l = \lambda_n l/\text{Bi}; \qquad \text{Bi} = hl/k,$$

which are shown graphically in Fig. 4–13.†

The boundary-value problem of the x-direction is the same as that of Examples 4–1 and 4–2. Hence

$$X_n(x) = B_n e^{-\lambda_n x}, \tag{4-54}$$

and the product solution of the problem is

$$\theta(x, y) = \sum_{n=1}^{\infty} a_n e^{-\lambda_n x} \cos \lambda_n y. \tag{4-76}$$

Finally, using the remaining (nonhomogeneous) boundary condition, we have

$$\theta_0 = \sum_{n=1}^{\infty} a_n \cos \lambda_n y, \tag{4-77}$$

where the value of a_n obtained from Eq. (4–23) is

$$a_n = \frac{2\theta_0 \sin \lambda_n l}{\lambda_n l + \sin \lambda_n l \cos \lambda_n l}. \tag{4-78}$$

---

† See Appendix IV of Reference 3 for the numerical values of λ_n.

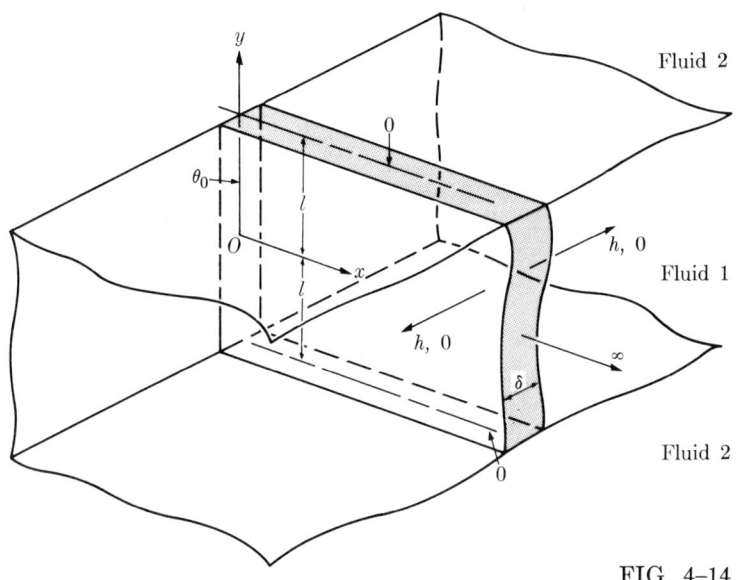

FIG. 4–14

Introducing Eq. (4–78) into Eq. (4–76), we find the temperature distribution of the problem to be

$$\frac{\theta(x,y)}{\theta_0} = 2 \sum_{n=1}^{\infty} \left(\frac{\sin \lambda_n l}{\lambda_n l + \sin \lambda_n l \cos \lambda_n l} \right) e^{-\lambda_n x} \cos \lambda_n y. \quad (4\text{--}79)$$

[What happens if we let $h \to \infty$ in Eq. (4–79)?]

Example 4–4. The extended surface of Example 4–2 is assumed to be a thin plate of thickness δ and width $2l$ (Fig. 4–14). The vertical heat transfer coefficient is large, and the horizontal heat transfer coefficient has the moderate value h.† The temperature distribution across the thickness δ is negligible. We wish to find the steady temperature of the plate.

The formulation of the problem in terms of θ is

$$\frac{\partial^2 \theta}{\partial x^2} + \frac{\partial^2 \theta}{\partial y^2} - m^2 \theta = 0,$$

$$\theta(0, y) = \theta_0, \qquad \theta(\infty, y) = 0,$$

$$\frac{\partial \theta(x, 0)}{\partial y} = 0, \qquad \theta(x, l) = 0,$$

where $m^2 = 2h/k\delta$.

---

† This is accomplished by using two fluids which yield different heat transfer coefficients, or by the same fluid appearing in two phases in the vertical direction only.

Proceeding as before, we obtain by means of a product solution†

$$\frac{d^2 Y}{dy^2} + \lambda^2 Y = 0; \qquad \frac{dY(0)}{dy} = 0, \quad Y(l) = 0, \qquad (4\text{-}64)$$

[handwritten: $Y'(l) = -\frac{h_3}{k} Y(l)$]

$$\frac{d^2 X}{dx^2} - (\lambda^2 + m^2) X = 0; \qquad X(\infty) = 0. \qquad (4\text{-}80)$$

The y-direction of the problem, Eq. (4-64), is the same as that of Example 4-2. Hence

$$Y_n(y) = C_n \cos \lambda_n y; \qquad \lambda_n = (2n+1)\pi/2l, \quad n = 0, 1, 2, \ldots$$

The x-direction, on the other hand, yields

$$X_n(x) = B_n e^{-(\lambda_n^2 + m^2)^{1/2} x}, \qquad (4\text{-}81)$$

and thus differs from Example 4-2. However, since the x-direction does not play any role in the evaluation of a_n, the value of a_n may be taken from Eq. (4-70).

Hence the solution of the problem may readily be written in the form

$$\frac{\theta(x, y)}{\theta_0} = \frac{2}{l} \sum_{n=0}^{\infty} \frac{(-1)^n}{\lambda_n} e^{-(\lambda_n^2 + m^2)^{1/2} x} \cos \lambda_n y. \qquad (4\text{-}82)$$

When $m \to 0$, which may be interpreted as $h \to 0$ or $\delta \to \infty$, Eq. (4-82) approaches Eq. (4-71), as expected.

Example 4-5. We wish to know what is the effect of motion on the solution of Example 4-2. Assume that the extended surface moves with the constant velocity V along the x-axis, in which conduction is negligible (Fig. 4-15). Find the steady temperature distribution with respect to the stationary coordinates used in Example 4-2.

The formulation is

$$2s \frac{\partial \theta}{\partial x} = \frac{\partial^2 \theta}{\partial y^2},$$

$$\theta(0, y) = \theta_0, \qquad \frac{\partial \theta(x, 0)}{\partial y} = 0, \qquad \theta(x, l) = 0,$$

---

† It is also possible to obtain the differential equations

$$\frac{d^2 Y}{dy^2} + (\lambda^2 + m^2) Y = 0, \qquad \frac{d^2 X}{dx^2} - \lambda^2 X = 0$$

by proper separation of the variables. However, the solution of this set, being more involved algebraically, is not preferred.

where $2s = \rho c V/k$. The product solution results in

$$\frac{d^2 Y}{dy^2} + \lambda^2 Y = 0; \qquad \frac{dY(0)}{dy} = 0, \quad Y(l) = 0, \qquad (4\text{–}64)$$

$$2s\frac{dX}{dx} + \lambda^2 X = 0. \qquad (4\text{–}83)$$

Again the y-direction has the same formulation as that of Examples 4–2 and 4–4.

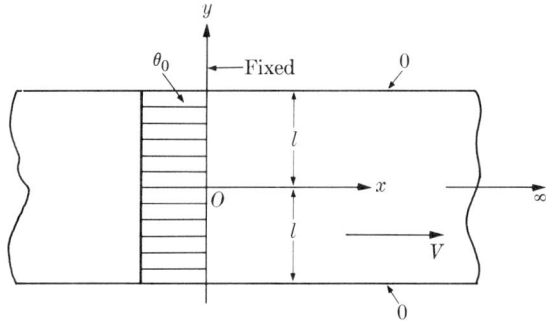

FIG. 4–15

It is now apparent that *when the physics and geometry of a problem in one direction are not altered, the separation of variables yields the same formulation for this direction regardless of the physics and geometry of the other direction.*

The solution in the x-direction is

$$X_n(x) = B_n e^{-\lambda_n^2 x/2s}.$$

Thus the product solution becomes

$$\theta(x, y) = \sum_{n=0}^{\infty} a_n e^{-\lambda_n^2 x/2s} \cos \lambda_n y. \qquad (4\text{–}84)$$

Finally, the use of the nonhomogeneous boundary condition reduces Eq. (4–84) to Eq. (4–69), from which follows the value of a_n, Eq. (4–70). Inserting this value into Eq. (4–84), we obtain the temperature of the fin in the form

$$\frac{\theta(x, y)}{\theta_0} = \frac{2}{l} \sum_{n=0}^{\infty} \frac{(-1)^n}{\lambda_n} e^{-\lambda_n^2 x/2s} \cos \lambda_n y. \qquad (4\text{–}85)$$

The foregoing example is related to one of the most celebrated problems of linear diffusion, the *Graetz problem*, which we now present.

Example 4–6. The solid of Example 4–5 is replaced by the fully developed laminar flow of a viscous fluid between two parallel plates (Fig. 4–16). The thickness of the plates is negligible. The upstream halves of the plates are in-

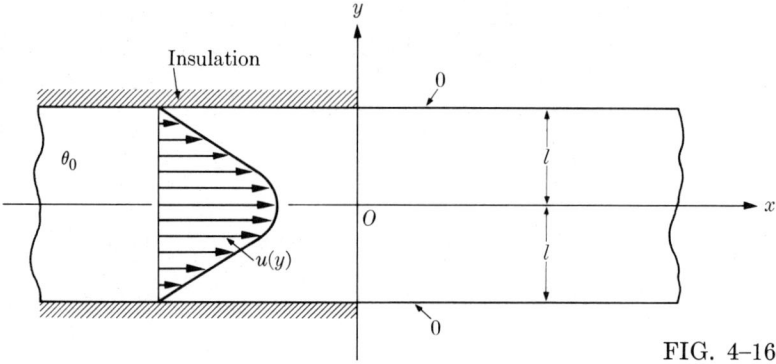

FIG. 4-16

sulated while the downstream halves are kept at zero temperature. Axial conduction is negligible. The upstream fluid temperature is uniform, say θ_0. We wish to find the temperature distribution in the fluid.

The formulation of the problem is identical to that of Example 4–5 provided the uniform velocity of that example is replaced by the velocity of the fully developed laminar flow between two parallel plates†

$$u(y) = \frac{(-dp/dx)l^2}{2\mu}\left[1 - \left(\frac{y}{l}\right)^2\right]. \quad (1\text{-}13)$$

This velocity may conveniently be expressed in terms of its mean value u_m and the dimensionless transversal variable $\eta = y/l$ as follows:

$$\frac{u(\eta)}{u_m} = \frac{3}{2}(1 - \eta^2) \quad (4\text{-}86)$$

where $u_m = (-dp/dx)l^2/3\mu$.

Because of the intended parametric study we consider the dimensionless formulation of the problem. Replacing V by Eq. (4–86) in the formulation of Example 4–5, and introducing the appropriate dimensionless numbers, we have

$$\tfrac{3}{2}(1 - \eta^2)\frac{\partial \psi}{\partial \xi} = \frac{\partial^2 \psi}{\partial \eta^2}, \quad (4\text{-}87)$$

where $\xi = (x/l)/\text{Pé}$, $\text{Pé} = u_m l/a$, and $\psi = \theta/\theta_0$. The inverse of ξ is called the *Graetz modulus* and is denoted by $\text{Gz} = 1/\xi = \text{Pé}/(x/l)$. The boundary conditions to be satisfied by Eq. (4–87) are

$$\psi(0, \eta) = 1, \quad \frac{\partial \psi(\xi, 0)}{\partial \eta} = 0, \quad \psi(\xi, 1) = 0.$$

---

† See Example 1–3.

The existence of a product solution, $\psi(\xi, \eta) = X(\xi)Y(\eta)$, readily separates the problem to give

$$\frac{d^2 Y}{d\eta^2} + \tfrac{3}{2}\lambda^2(1 - \eta^2)Y = 0; \qquad \frac{dY(0)}{d\eta} = 0, \quad Y(1) = 0, \qquad (4\text{--}88)$$

$$\frac{dX}{d\xi} + \lambda^2 X = 0. \qquad (4\text{--}89)$$

Unlike the corresponding equations of the previous examples, the solution of Eq. (4–88) cannot be expressed in terms of simple functions. Thus we are forced to employ the method of power series solutions developed in Section 3–6. Leaving the explicit forms of these solutions to the reader as an exercise, we may implicitly express the general solution as

$$Y(\eta) = A G_0^{(1)}(\lambda \eta) + B G_0^{(2)}(\lambda \eta), \qquad (4\text{--}90)$$

where $G_0^{(1)}(\lambda \eta)$ and $G_0^{(2)}(\lambda \eta)$ denote *cartesian Graetz functions of the first and second kind, of order zero.*

We may now recall from mathematics that a general solution of fluctuating nature such as Eq. (4–90) is usually composed of an even function and an odd function. Let $G_0^{(1)}(\lambda \eta)$ and $G_0^{(2)}(\lambda \eta)$ denote the even and odd functions, respectively. The first boundary condition in Eq. (4–88) implies then that $B \equiv 0$, and the characteristic functions are $\varphi_n(\eta) = G_0^{(1)}(\lambda_n \eta)$. The second boundary condition gives the characteristic values, the roots of $G_0^{(1)}(\lambda_n) = 0$. By means of the general solution of Eq. (4–89), $e^{-\lambda^2 \xi}$, we may now write the product solution of the problem in the form

$$\psi(\xi, \eta) = \sum_{n=1}^{\infty} a_n e^{-\lambda_n^2 \xi} G_0^{(1)}(\lambda_n \eta). \qquad (4\text{--}91)$$

Finally, the use of the nonhomogeneous boundary condition of the problem, $\psi(0, \eta) = 1$, yields

$$1 = \sum_{n=1}^{\infty} a_n G_0^{(1)}(\lambda_n \eta),$$

where, according to Eq. (4–23)†,

$$a_n = \frac{\int_0^1 (1 - \eta^2) G_0^{(1)}(\lambda_n \eta)\, d\eta}{\int_0^1 (1 - \eta^2)[G_0^{(1)}(\lambda_n \eta)]^2\, d\eta}. \qquad (4\text{--}92)$$

It is possible, by mathematical manipulations, to express a_n in a more convenient form. Noting that $G_0^{(1)}(\lambda_n \eta)$ is a particular solution of Eq. (4–88), re-

---

† Note that the weighting function is $w(x) = \tfrac{3}{2}(1 - \eta^2)$.

placing $Y(\eta)$ by $G_0^{(1)}(\lambda_n\eta)$ in this equation, and integrating the result over the interval $(0, 1)$ gives a simpler expression for the numerator of Eq. (4–92) in the form

$$\int_0^1 (1 - \eta^2) G_0^{(1)}(\lambda_n\eta) \, d\eta = -\frac{2}{3\lambda_n^2} \left.\frac{dG_0^{(1)}(\lambda_n\eta)}{d\eta}\right|_{\eta=1}. \tag{4–93}$$

The rearrangement of the denominator of Eq. (4–92) is somewhat more involved. First, we multiply Eq. (4–88) expressed in terms of $G_0^{(1)}(\lambda_n\eta)$ by $dG_0^{(1)}(\lambda_n\eta)/d\lambda_n$, and integrate the result over the interval $(0, 1)$ to obtain

$$\left.\left(\frac{dG_0^{(1)}}{d\eta}\right)\left(\frac{dG_0^{(1)}}{d\lambda_n}\right)\right|_{\eta=1} - \int_0^1 \frac{dG_0^{(1)}}{d\eta} \frac{d}{d\eta}\left(\frac{dG_0^{(1)}}{d\lambda_n}\right) d\eta$$

$$- \tfrac{3}{4}\lambda_n^2 \frac{d}{d\lambda_n} \int_0^1 (1 - \eta^2)[G_0^{(1)}(\lambda_n\eta)]^2 \, d\eta = 0. \tag{4–94}$$

Second, we multiply Eq. (4–88) expressed in terms of $G_0^{(1)}(\lambda_n\eta)$ by $G_0^{(1)}(\lambda_n\eta)$, and integrate the result over the same interval to get

$$\left.G_0^{(1)} \frac{dG_0^{(1)}}{d\eta}\right|_{\eta=1} - \int_0^1 \left(\frac{dG_0^{(1)}}{d\eta}\right)^2 d\eta + \tfrac{3}{2}\lambda_n^2 \int_0^1 (1 - \eta^2)[G_0^{(1)}(\lambda_n\eta)]^2 \, d\eta = 0. \tag{4–95}$$

Differentiation of Eq. (4–95) with respect to λ_n gives

$$\left.\left(\frac{dG_0^{(1)}}{d\lambda_n}\right)\left(\frac{dG_0^{(1)}}{d\eta}\right)\right|_{\eta=1} + \left.G_0^{(1)} \frac{d}{d\lambda_n}\left(\frac{dG_0^{(1)}}{d\eta}\right)\right|_{\eta=1} - 2\int_0^1 \frac{dG_0^{(1)}}{d\eta} \frac{d}{d\lambda_n}\left(\frac{dG_0^{(1)}}{d\eta}\right) d\eta$$

$$+ 3\lambda_n \int_0^1 (1 - \eta^2)[G_0^{(1)}(\lambda_n\eta)]^2 \, d\eta$$

$$+ \tfrac{3}{2}\lambda_n^2 \frac{d}{d\lambda_n}\int_0^1 (1 - \eta^2)[G_0^{(1)}(\lambda_n\eta)]^2 \, d\eta = 0. \tag{4–96}$$

Noting that the order of differentiation with respect to η and λ_n may be interchanged, dividing Eq. (4–96) by 2, and subtracting the result from Eq. (4–94) yields

$$\int_0^1 (1 - \eta^2)[G_0^{(1)}(\lambda_n\eta)]^2 \, d\eta = \frac{1}{3\lambda_n^2} \left.\frac{dG_0^{(1)}(\lambda_n\eta)}{d\lambda_n} \frac{dG_0^{(1)}(\lambda_n\eta)}{d\eta}\right|_{\eta=1}. \tag{4–97}$$

Inserting Eqs. (4–93) and (4–97) into Eq. (4–92), we obtain

$$a_n = -\frac{2}{\lambda_n \, dG_0^{(1)}(\lambda_n\eta)/d\lambda_n|_{\eta=1}}.$$

Thus the temperature distribution in the fluid is found to be

$$\psi(\xi, \eta) = -2 \sum_{n=1}^{\infty} \frac{e^{-\lambda_n^2 \xi} G_0^{(1)}(\lambda_n \eta)}{\lambda_n \, dG_0^{(1)}(\lambda_n \eta)/d\lambda_n|_{\eta=1}} . \qquad (4\text{--}98)$$

However, from the standpoint of heat transfer calculations, the heat loss from the fluid is more important to know than the temperature of the fluid. In this connection we may first discuss the dimensionless numbers associated with heat transfer from a surface.

As we saw in Chapter 2, the heat transfer q_n from a surface σ may be evaluated by means of a heat transfer coefficient as follows:

$$q_n = h(T_\sigma - T_\infty),$$

where T_σ and T_∞ denote the surface and the ambient temperatures, respectively. In Chapter 2 we also expressed q_n in terms of conduction from or to the solid. Thus, according to Fig. 4–17(a), we have

$$-k_s \left(\frac{\partial T_s}{\partial n}\right)_\sigma = h(T_\sigma - T_\infty),$$

where T_s and k_s are the temperature and the thermal conductivity of the solid, respectively. Rearranging the foregoing equation we obtain the *Biot modulus*,

$$\text{Bi} = \frac{hL}{k_s} = -\frac{[\partial T_s/\partial(n/L)]_\sigma}{T_\sigma - T_\infty}, \qquad (4\text{--}99)$$

where L is a characteristic length. Clearly, q_n may also be expressed in terms of conduction to or from the fluid as shown in Fig. 4–17(b). Thus we have

$$-k_f \left(\frac{\partial T_f}{\partial n}\right)_\sigma = h(T_\sigma - T_\infty),$$

which may be rearranged in the form

$$\text{Nu} = \frac{hL}{k_f} = -\frac{[\partial T_f/\partial(n/L)]_\sigma}{T_\sigma - T_\infty}, \qquad (4\text{--}100)$$

where Nu is defined to be the *Nusselt modulus*. When the geometry of the fluid is finite, T_∞ of Eqs. (4–99) and (4–100) is conveniently replaced by a mean fluid temperature, say the bulk temperature,

$$T_b = \frac{1}{AU} \int_A u(\mathbf{r}, t) T(\mathbf{r}, t) \, dA, \qquad (4\text{--}101)$$

where A is the transversal area of the fluid, and U the transversally averaged value of the fluid velocity.

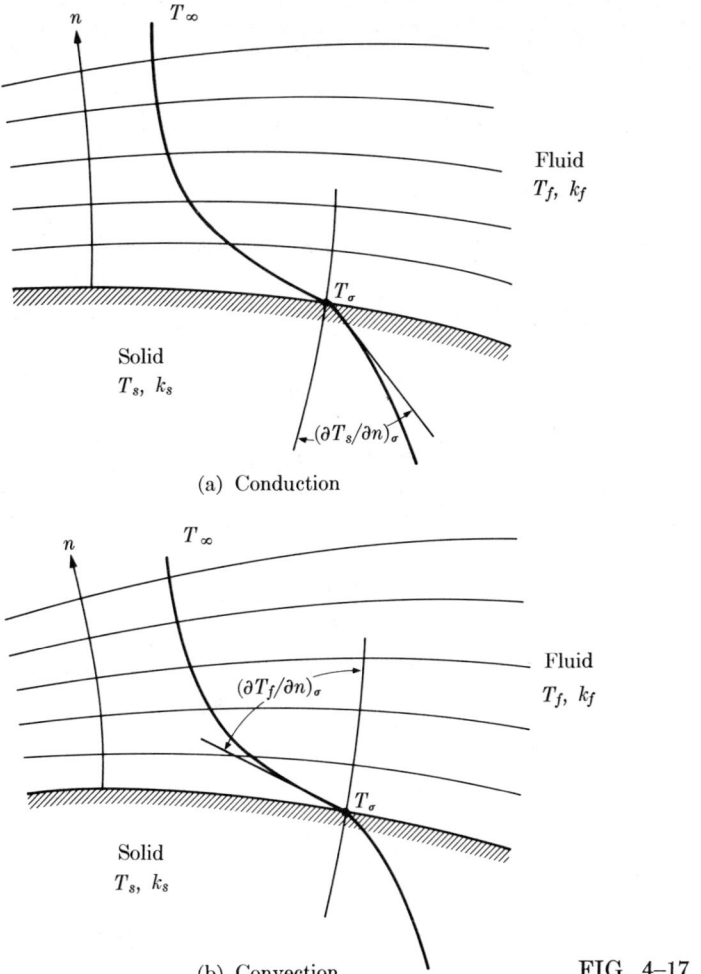

(a) Conduction

(b) Convection

FIG. 4-17

Note the difference in the use of Eqs. (4-99) and (4-100). In conduction problems h and T_∞ (or T_b) are given, and Eq. (4-99) is employed as a boundary condition; in convection problems T_∞ is given but h is not. Convection problems are solved in terms of boundary conditions unrelated to h; then the resulting solutions are introduced into Eq. (4-100) to get the value of h.

We may now return to our problem. The appropriate forms of Eqs. (4-100) and (4-101) are

$$\mathrm{Nu} = \frac{hl}{k} = \frac{-(\partial \psi/\partial \eta)_{\eta=1}}{\psi_b}$$

and

$$\psi_b = \tfrac{3}{2} \int_0^1 (1 - \eta^2)\psi(\xi, \eta)\, d\eta.$$

From Eq. (4–98) we have immediately

$$\left(\frac{\partial \psi}{\partial \eta}\right)_{\eta=1} = -2 \sum_{n=1}^{\infty} \frac{dG_0^{(1)}/d\eta}{\lambda_n \, dG_0^{(1)}/d\lambda_n}\bigg|_{\eta=1} e^{-\lambda_n^2 \xi}.$$

Next, introducing Eq. (4–98) into ψ_b, and noting Eq. (4–93), we obtain

$$\psi_b = 2 \sum_{n=1}^{\infty} \frac{dG_0^{(1)}/d\eta}{\lambda_n^3 \, dG_0^{(1)}/d\lambda_n}\bigg|_{\eta=1} e^{-\lambda_n^2 \xi}.$$

Thus the Nusselt modulus is found to be

$$\text{Nu} = \frac{\displaystyle\sum_{n=1}^{\infty} \frac{dG_0^{(1)}/d\eta}{\lambda_n \, dG_0^{(1)}/d\lambda_n}\bigg|_{\eta=1} e^{-\lambda_n^2 \xi}}{\displaystyle\sum_{n=1}^{\infty} \frac{dG_0^{(1)}/d\eta}{\lambda_n^3 \, dG_0^{(1)}/d\lambda_n}\bigg|_{\eta=1} e^{-\lambda_n^2 \xi}}. \tag{4–102}$$

Nu is plotted against ξ in Fig. 4–18†. Thus we learn that Nu $\to \infty$ as $\xi \to 0$ and Nu $\to 1.885$ as $\xi \to \infty$. [Obtain the foregoing limits of Nu by physical reasoning rather than referring to the results of the foregoing analysis.] The

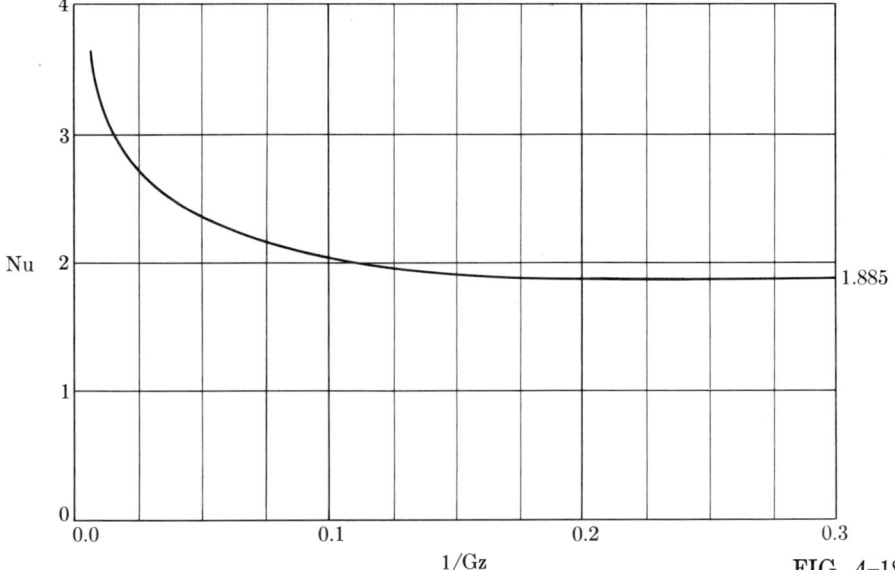

FIG. 4–18

† From J. A. Prins, J. Mulder, and J. Schenk, "Heat Transfer in Laminary Flow between Parallel Plates," *Appl. Sci. Res.*, **A2**, 431 (1951). Used by permission.

Graetz problem has been extensively studied in the literature. For a brief review of the subject see References 6 through 17.

In the next example we include the effect of axial conduction in our problem. However, because of the increased complexity, let us reconsider the case of an ideal fluid (or a moving extended surface).

Example 4–7. Resolve Example 4–5 including the effects of conduction in the direction of motion.

As explained in Example 3–3, there are two effects of axial conduction. The first, often considered as the only effect of axial conduction, is the addition of $\partial^2 \theta / \partial x^2$ into the differential equation; the second and equally important effect, hidden in the physics of the problem, is the penetration of conduction out of the former integration domain. Thus we have a two-domain rather than a single-domain problem (Fig. 4–19).† Since the location $x = 0$ now becomes the interface of two domains, the temperature at $x = 0$ can no longer be specified. We must, rather, specify the natural boundary conditions, in other words, the equality of temperatures and of heat fluxes. At large negative distances from $x = 0$, as well as at the surface of the fin for $x < 0$, the temperature may be specified as θ_0. Using $\theta(x, y)$ for the temperature distribution of the region $x > 0$ and $\psi(x, y)$ for that of $x < 0$, we obtain the formulation of the problem in the form

$$\frac{\partial^2 \psi}{\partial x^2} - 2s \frac{\partial \psi}{\partial x} + \frac{\partial^2 \psi}{\partial y^2} = 0, \quad x < 0$$

$$\frac{\partial^2 \theta}{\partial x^2} - 2s \frac{\partial \theta}{\partial x} + \frac{\partial^2 \theta}{\partial y^2} = 0, \quad x > 0$$

$$\psi(-\infty, y) = \theta_0, \qquad \theta(\infty, y) = 0,$$

$$\psi(0, y) = \theta(0, y), \qquad \frac{\partial \psi(0, y)}{\partial x} = \frac{\partial \theta(0, y)}{\partial x},$$

$$\frac{\partial \psi(x, 0)}{\partial y} = 0, \qquad \frac{\partial \theta(x, 0)}{\partial y} = 0,$$

$$\psi(x, l) = \theta_0, \qquad \theta(x, l) = 0.$$

This formulation comprises two homogeneous differential equations, and four homogeneous plus four nonhomogeneous boundary conditions. One of the nonhomogeneous conditions, $\psi(x, l) = \theta_0$, introduces a nonhomogeneity into the only possible finite orthogonal direction and thus precludes the use of separation of variables for the region $x < 0$. However, the transformation $\psi(x, y) = \theta_0 + \phi(x, y)$ for $x < 0$ eliminates this difficulty. Hence in terms of

---

† If axial conduction is negligible, the temperature of the new domain $-\infty < x \leq 0$ is identical to θ_0, and the problems stated by Figs. 4–15 and 4–19 yield the same temperature distribution. This is not the case when axial conduction is appreciable.

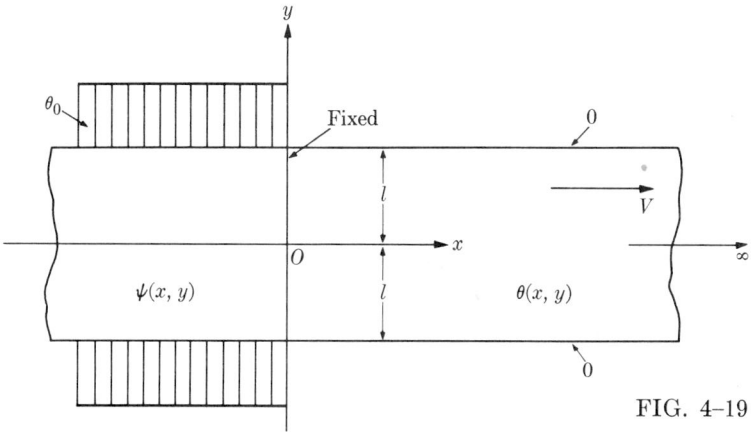

FIG. 4–19

$\phi(x, y)$, the region $x < 0$ is satisfied by the equations

$$\frac{\partial^2 \phi}{\partial x^2} - 2s \frac{\partial \phi}{\partial x} + \frac{\partial^2 \phi}{\partial y^2} = 0, \quad x < 0$$

$$\phi(-\infty, y) = 0,$$

$$\theta_0 + \phi(0, y) = \theta(0, y), \quad \frac{\partial \phi(0, y)}{\partial x} = \frac{\partial \theta(0, y)}{\partial x},$$

$$\frac{\partial \phi(x, 0)}{\partial y} = 0, \quad \phi(x, l) = 0.$$

while the remaining equations of the previous formulation for the domain $x > 0$, being unchanged, are not repeated.

The solution of the homogeneous y-direction of $\theta(x, y)$ and that of $\phi(x, y)$ are the same as in Examples 4–2, 4–4, and 4–5 regardless of the x-direction of these. The general solution in the x-direction,

$$X_n(x) = A_n e^{[s+(s^2+\lambda_n^2)^{1/2}]x} + B_n e^{[s-(s^2+\lambda_n^2)^{1/2}]x}, \quad (4\text{--}103)$$

applies equally to both regions; however, the first term of Eq. (4–103) satisfies the physics of the region $x < 0$, whereas the second term is suitable to the region $x > 0$.

Hence the solution of $\psi(x, y)$ and that of $\theta(x, y)$ may be written as

$$\psi(x, y) = \theta_0 + \sum_{n=0}^{\infty} a_n e^{[s+(s^2+\lambda_n^2)^{1/2}]x} \cos \lambda_n y, \quad (4\text{--}104)$$

$$\theta(x, y) = \sum_{n=0}^{\infty} b_n e^{[s-(s^2+\lambda_n^2)^{1/2}]x} \cos \lambda_n y. \quad (4\text{--}105)$$

Here a_n and b_n are calculated from the interface conditions at $x = 0$; the equality of interface temperatures gives

$$\theta_0 + \sum_{n=0}^{\infty} a_n \cos \lambda_n y = \sum_{n=0}^{\infty} b_n \cos \lambda_n y. \tag{4-106}$$

Expanding θ_0 into a series in terms of the same orthogonal set, Eq. (4–66), as

$$\theta_0 = \frac{2\theta_0}{l} \sum_{n=0}^{\infty} \frac{(-1)^n}{\lambda_n} \cos \lambda_n y, \tag{4-107}$$

and introducing Eq. (4–107) into Eq. (4–106) results in

$$(-1)^n \frac{2\theta_0}{\lambda_n l} + a_n = b_n. \tag{4-108}$$

The equality of heat fluxes, on the other hand, yields

$$[s + (s^2 + \lambda_n^2)^{1/2}]a_n = [s - (s^2 + \lambda_n^2)^{1/2}]b_n. \tag{4-109}$$

Finally, solving Eqs. (4–108) and (4–109) for a_n and b_n, inserting the result into Eqs. (4–104) and (4–105), and introducing the dimensionless parameters

$$\text{Pé (or } P) = Vl/a, \quad \text{Gz} = 1/\xi = \text{Pé}/(x/l),$$

$$\eta = y/l, \quad \mu_n = (2n+1)\pi/2, \quad n = 0, 1, 2, \ldots,$$

we find that the dimensionless temperatures of the two regions are

$$\frac{\psi(\xi, \eta, P)}{\theta_0} = 1 - \sum_{n=0}^{\infty} \frac{(-1)^n}{\mu_n} \left[1 - \frac{1}{[1 + (2\mu_n/P)^2]^{1/2}}\right]$$

$$\times \cos \mu_n \eta \, \exp\left(\frac{1}{2}\left\{1 + \left[1 + \left(\frac{2\mu_n}{P}\right)^2\right]^{1/2}\right\} P^2 \xi\right), \quad \xi < 0, \tag{4-110}$$

$$\frac{\theta(\xi, \eta; P)}{\theta_0} = \sum_{n=0}^{\infty} \frac{(-1)^n}{\mu_n} \left\{1 + \frac{1}{[1 + (2\mu_n/P)^2]^{1/2}}\right\}$$

$$\times \cos \mu_n \eta \, \exp\left(\frac{1}{2}\left\{1 - \left[1 + \left(\frac{2\mu_n}{P}\right)^2\right]^{1/2}\right\} P^2 \xi\right), \quad \xi > 0. \tag{4-111}$$

Since, by definition (see Example 3–3),

$$\frac{\text{Axial enthalpy flow}}{\text{Axial conduction}} = \text{Pé},$$

as Pé → ∞ the effect of axial conduction diminishes, and Eqs. (4–110) and (4–111) take the form

$$\frac{\psi(\xi, \eta)}{\theta_0} = 0, \quad \xi < 0, \tag{4-112}$$

$$\frac{\theta(\xi, \eta)}{\theta_0} = 2 \sum_{n=0}^{\infty} \frac{(-1)^n}{\mu_n} e^{-\mu_n^2 \xi} \cos \mu_n \eta, \quad \xi > 0. \tag{4-113}$$

Equation (4–113) is the dimensionless form of Eq. (4–85), as expected.

The dimensionless form of the temperatures $\psi(x, y)$ and $\theta(x, y)$ given by Eqs. (4–110) and (4–111) now allows the following numerical study.

In Fig. 4–20 the interface temperature, that is, the temperature at the location $\xi = 0$ (Gz = ∞), is plotted against η for the values 0.05, 0.3, 1, 2, 5, and 50 of Pé. The resulting curves show the progressive effect of axial conduction as Pé decreases.

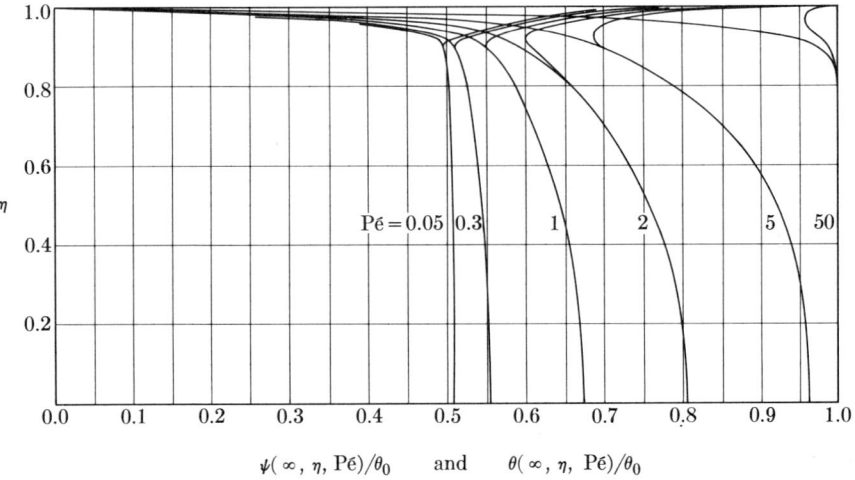

FIG. 4–20

In Fig. 4–21 the middle-plane temperature is plotted against 1/Gz for the values 0.05, 0.1, 0.2, 0.3, 1, and 2 of Pé. As expected, the penetration of the temperature distribution into the domain $x < 0$ increases as Pé decreases.

Finally, the heat transfer coefficients based on l may be written in terms of

$$\theta_b = \int_0^1 \theta \, d\eta \quad \text{and} \quad \psi_b = \int_0^1 \psi \, d\eta$$

as follows:

$$\mathrm{Nu} = \frac{(\partial \psi / \partial \eta)_{\eta=1}}{\psi_w - \psi_b}, \quad \xi < 0,$$

$$\mathrm{Nu} = \frac{-(\partial \theta / \partial \eta)_{\eta=1}}{\theta_b - \theta_w}, \quad \xi > 0,$$

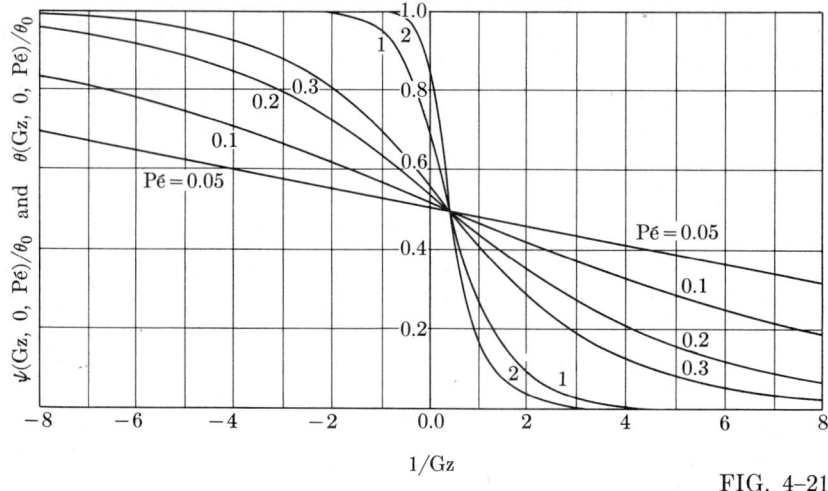

FIG. 4–21

where $\psi_w = \psi(\xi, 1; P) = \theta_0$ and $\theta_w = \theta(\xi, 1; P) = 0$. Introducing Eqs. (4–110) and (4–111) into the foregoing equations, we obtain

$$\mathrm{Nu} = \frac{\sum_{n=0}^{\infty} \left\{1 - \frac{1}{[1 + (2\mu_n/P)^2]^{1/2}}\right\} \exp\left(\frac{1}{2}\left\{1 + \left[1 + \left(\frac{2\mu_n}{P}\right)^2\right]^{1/2}\right\} P^2 \xi\right)}{\sum_{n=0}^{\infty} \frac{1}{\mu_n^2}\left\{1 - \frac{1}{[1 + (2\mu_n/P)^2]^{1/2}}\right\} \exp\left(\frac{1}{2}\left\{1 + \left[1 + \left(\frac{2\mu_n}{P}\right)^2\right]^{1/2}\right\} P^2 \xi\right)},$$

$$\xi < 0, \quad (4\text{–}114)$$

$$\mathrm{Nu} = \frac{\sum_{n=0}^{\infty} \left\{1 + \frac{1}{[1 + (2\mu_n/P)^2]^{1/2}}\right\} \exp\left(\frac{1}{2}\left\{1 - \left[1 + \left(\frac{2\mu_n}{P}\right)^2\right]^{1/2}\right\} P^2 \xi\right)}{\sum_{n=0}^{\infty} \frac{1}{\mu_n^2}\left\{1 + \frac{1}{[1 + (2\mu_n/P)^2]^{1/2}}\right\} \exp\left(\frac{1}{2}\left\{1 - \left[1 + \left(\frac{2\mu_n}{P}\right)^2\right]^{1/2}\right\} P^2 \xi\right)},$$

$$\xi > 0. \quad (4\text{–}115)$$

Equations (4–114) and (4–115) are plotted in Fig. 4–22 against $1/\mathrm{Gz}$ for the values 0.05, 0.1, 0.2, 0.3, 1, and 2 of Pé. Note that the development of the heat transfer coefficients slows down as the importance of axial conduction increases.

4–6. Selection of Coordinate Axes

So far, a number of steady two-dimensional cartesian problems have been solved without any discussion of the coordinate system selected. Although the solutions of Example 4–1 (for constant base temperature) and Example 4–2,

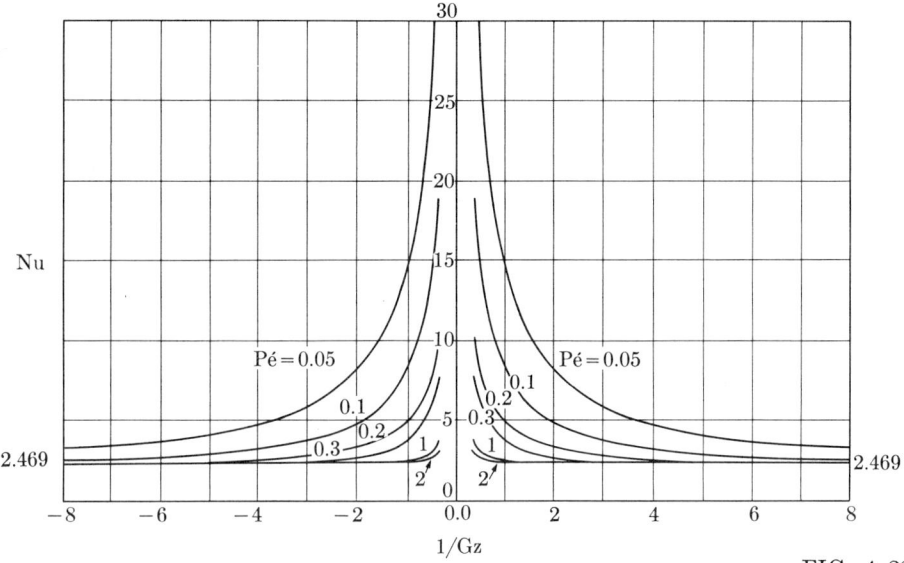

FIG. 4-22

which both describe the same physical problem, were given in terms of two different coordinate systems, the solutions were comparable in algebraic complexity and hence did not prompt an investigation of the selection of coordinates. The choice of coordinate axes may, however, be important in other cases, as illustrated by the following example.

Example 4–8. Re-solve Example 4–2 assuming that the fin has, instead, the finite length L (Fig. 4–23).

The problem is readily separated in a manner similar to that of the previous problems. The solution in the y-direction (the orthogonal direction) is identical to that of Examples 4–2, 4–4, 4–5, and 4–7. The solution in the x-direction (the nonorthogonal direction) will now be treated explicitly. The formulation of this direction together with the separable boundary condition is

$$\frac{d^2 X}{dx^2} - \lambda^2 X = 0, \quad (4\text{-}116)$$

$$X(L) = 0. \quad (4\text{-}117)$$

Since hyperbolic functions are conveniently used for finite intervals, the general solution of Eq. (4–116) may be written in the form

$$X(x) = A \cosh \lambda x + B \sinh \lambda x. \quad (4\text{-}118)$$

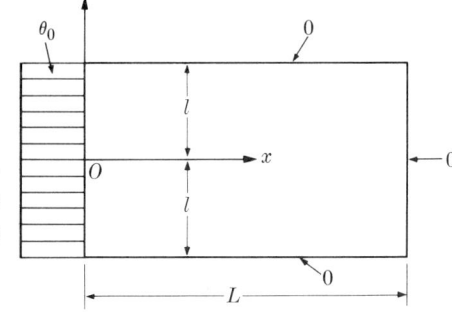

FIG. 4-23

Employing Eq. (4–117), we have from Eq. (4–118)

$$0 = A \cosh \lambda L + B \sinh \lambda L. \tag{4–119}$$

Solving Eq. (4–119) for A or B, say for B, and inserting the result into Eq. (4–118) gives

$$X(x) = A\left(\cosh \lambda x - \frac{\cosh \lambda L}{\sinh \lambda L} \sinh \lambda x\right),$$

which may be rearranged as

$$X(x) = A^* \sinh \lambda(L - x), \qquad A^* = A/\sinh \lambda L.$$

Then the product solution of the problem is

$$\theta(x, y) = \sum_{n=0}^{\infty} a_n^* \sinh \lambda_n(L - x) \cos \lambda_n y.$$

The evaluation of a_n^* is similar to that in previous examples. Thus we have the temperature distribution in the fin as follows:

$$\frac{\theta(x, y)}{\theta_0} = 2 \sum_{n=0}^{\infty} \frac{(-1)^n}{(\lambda_n l)} \left[\frac{\sinh \lambda_n(L - x)}{\sinh \lambda_n L}\right] \cos \lambda_n y. \tag{4–120}$$

Let us now re-solve the same problem in terms of the coordinate system shown in Fig. 4–24. The y-direction remains the same. However, the formulation in the x-direction now yields, in terms of the new variable ξ,

$$\frac{d^2 X}{d\xi^2} - \lambda^2 X = 0, \tag{4–121}$$

$$X(0) = 0. \tag{4–122}$$

The general solution of Eq. (4–121) is identical to that of Eq. (4–118), provided x is replaced by ξ in Eq. (4–118). The use of Eq. (4–122) gives $A = 0$, and the new product solution becomes

$$\theta(x, y) = \sum_{n=0}^{\infty} b_n \sinh \lambda_n \xi \cos \lambda_n y.$$

Evaluating b_n, we find that the temperature of the fin is

$$\frac{\theta(\xi, y)}{\theta_0} = 2 \sum_{n=0}^{\infty} \frac{(-1)^n}{(\lambda_n l)} \left(\frac{\sinh \lambda_n \xi}{\sinh \lambda_n L}\right) \cos \lambda_n y.\dagger \tag{4–123}$$

---

† Equation (4–123) may be converted to Eq. (4–120), or vice versa, by the transformation $x + \xi = L$, as expected. Furthermore, these equations show the effect of finite length on Example 4–2.

Hence, because of the simplicity of the solution in the x-direction, the coordinate system of Fig. 4–24 is more suitable to the present problem than that of Fig. 4–23.

The foregoing result for the nonhomogeneous direction will now be generalized to include a similar argument for the choice of the origin in the homogeneous direction.

◀ In the formulation of a problem, the selection of the most convenient axes is an important task. We learned in the discussion of the separation of variables and the theory of orthogonal functions that the first boundary condition considered in the homogeneous direction of any problem determines the characteristic functions depending on the selected coordinate system. The second boundary condition combined with the result obtained from the first boundary condition gives the characteristic values which are invariant to coordinate axes (see Problem 4–4). Therefore, the simplest possible characteristic functions correspond to the coordinate axis (in the homogeneous direction) measured from the simplest homogeneous boundary condition.

We have also learned that the homogeneous boundary condition of the nonhomogeneous direction gives a relation between the particular solutions in this direction. Therefore, the simplest solution in this direction corresponds to the coordinate axis (in the same direction) measured from the separable boundary condition.

The procedure just outlined is the reverse of that used for the choice of coordinate axes in one-dimensional problems. This is not surprising, since one-dimensional problems are not characteristic-value problems, and their boundary conditions are used for the evaluation of the unknown coefficients of the particular solutions rather than for characteristic functions and characteristic values. Hence in one-dimensional problems, by taking the origin at the most complicated boundary we may simplify considerably one or two conditions involving these coefficients. ▶

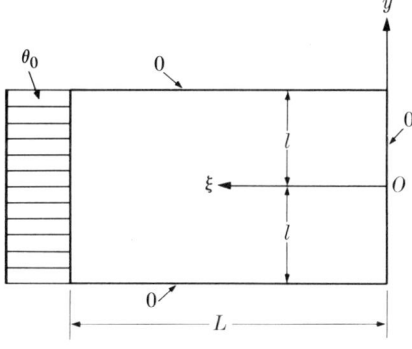

FIG. 4–24

4–7. Nonhomogeneity

So far, the steady two-dimensional cartesian problems that we have solved by the method of separation of variables were those involving a homogeneous differential equation subject to two homogeneous boundary conditions in the finite direction, and one homogeneous plus one nonhomogeneous boundary condition in the remaining (finite or infinite) direction. Most two-dimensional problems,

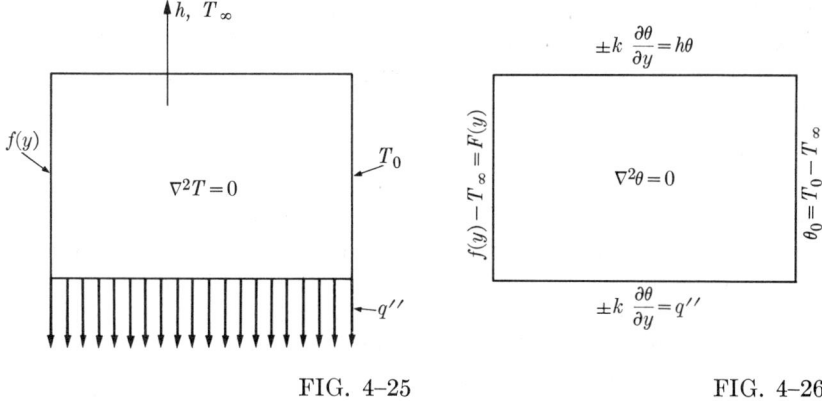

FIG. 4-25   FIG. 4-26

however, do not *a priori* satisfy these conditions. In such cases the problem must be transformed, if possible, to one that does satisfy them. One way of doing this is to shift the temperature level, a procedure which has been employed with previous examples in this chapter. If this is impossible, then using the principle of superposition, we may divide the problem into a number of simpler problems each satisfying the required homogeneity conditions and all adding up to the posed problem.

Nonhomogeneities may result from nonhomogeneous boundaries and/or nonhomogeneous differential equations. Since nonhomogeneous boundaries are easier, we shall treat these first.

Example 4-9. We wish to find the steady temperature of the problem given by Fig. 4-25.

The vertical axis is arbitrarily selected as the y-direction without altering the generality of the problem. Figure 4-25 may be divided into four problems, each one having one nonhomogeneous and three homogeneous boundary conditions. However, the simple transformation $\theta = T - T_\infty$ (or $\psi = T - T_0$) readily converts the nonhomogeneous condition h, T_∞ (or T_0) to a homogeneous condition. Using, for example, $\theta = T - T_\infty$, we may transform the problem of Fig. 4-25 to the problem of Fig. 4-26. This problem is expressible in terms of three suitable problems rather than four. Note that *the transformation of a nonhomogeneous boundary condition to a homogeneous boundary condition should not violate the physics of the boundary condition.* In other words, the only homogeneous form of $\pm k(\partial T/\partial n) = h(T - T_\infty)$ is $\pm k(\partial T/\partial n) = hT$, that of T_∞ is 0, and that of $\pm k(\partial T/\partial n) = q''$ is $(\partial T/\partial n) = 0$.

Hence the problem of Fig. 4-26 can be expressed in terms of the three problems shown in Fig. 4-27, such that the sum $\theta_1 + \theta_2 + \theta_3$ satisfies the differential equation and boundary conditions of Fig. 4-26. The solution of θ_1, θ_2, and θ_3 may readily be obtained following the procedure of the previous examples, and will not be given here. The axes suitable to each problem are shown in

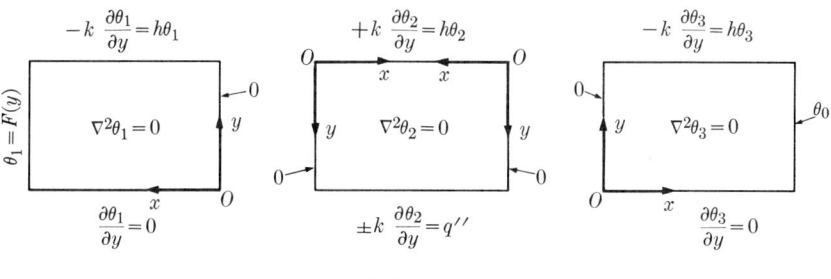

FIG. 4-27

Fig. 4-27. *The solutions of these problems can be added to each other only when they are expressed in a common coordinate system.*

The second type of nonhomogeneity, the nonhomogeneous differential equation, is related to internal energy generation.† The following example demonstrates the method of solution of these problems.

Example 4-10. Find the steady temperature of the electric heater of Example 2-1, given that the heat transfer coefficient is large.

The formulation of the problem in terms of Fig. 4-28 is

$$\frac{\partial^2 \theta}{\partial x^2} + \frac{\partial^2 \theta}{\partial y^2} + \frac{u'''}{k} = 0, \qquad (4\text{-}124)$$

$$\frac{\partial \theta(0, y)}{\partial x} = 0, \qquad \theta(L, y) = 0, \qquad (4\text{-}125)$$

$$\frac{\partial \theta(x, 0)}{\partial y} = 0, \qquad \theta(x, l) = 0. \qquad (4\text{-}126)$$

The differential equation, being nonhomogeneous, is not separable.

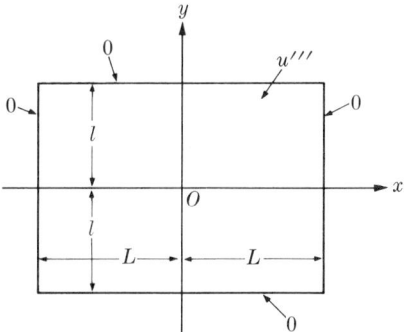

FIG. 4-28

The solution of the problem is now assumed to be

$$\theta(x, y) = \psi(x, y) + \phi(x) \tag{4-127}$$

or

$$\theta(x, y) = \psi(x, y) + \phi(y). \tag{4-128}$$

The use of either one of these forms is arbitrary in this case.

With the inclusion of the internal energy generation u''' in the formulation of the one-dimensional problem, $\phi(x)$ or $\phi(y)$, the differential equation to be satisfied by the two-dimensional problem, $\psi(x, y)$, can be made homogeneous.

---

† The surface heat flux imposed in a direction in which the temperature is lumped also yields a nonhomogeneous differential equation. However, this case, since it has little practical application, will not be considered here.

Thus $\psi(x, y)$ is suitable for separation of variables. However, the complete formulation of ϕ, say $\phi(x)$, and of $\psi(x, y)$ requires that the boundary conditions of these be specified. Here, $\phi(x)$ is assumed to satisfy the one-dimensional form of Eq. (4–125). Hence

$$\frac{d^2\phi}{dx^2} + \frac{u'''}{k} = 0; \qquad \frac{d\phi(0)}{dx} = 0, \quad \phi(L) = 0. \qquad (4\text{–}129)$$

Then, combining Eqs. (4–124), (4–125), (4–126), (4–127), and (4–129), we find that $\psi(x, y)$ is satisfied by

$$\frac{\partial^2 \psi}{\partial x^2} + \frac{\partial^2 \psi}{\partial y^2} = 0, \qquad (4\text{–}130)$$

$$\frac{\partial \psi(0, y)}{\partial x} = 0, \qquad \psi(L, y) = 0, \qquad (4\text{–}131)$$

$$\frac{\partial \psi(x, 0)}{\partial y} = 0, \qquad \psi(x, l) = -\phi(x). \qquad (4\text{–}132)$$

Thus the solution of the nonseparable problem $\theta(x, y)$ is reduced to that of the separable problem $\psi(x, y)$. The details of this solution are left to the reader. The result, including $\phi(x)$, is†

$$\frac{\theta(x, y)}{u'''L^2/k} = \frac{1}{2}\left[1 - \left(\frac{x}{L}\right)^2\right] - 2\sum_{n=0}^{\infty} \frac{(-1)^n}{(\lambda_n L)^3}\left(\frac{\cosh \lambda_n y}{\cosh \lambda_n l}\right)\cos \lambda_n x, \qquad (4\text{–}133)$$

where $\lambda_n L = (2n + 1)\pi/2$, $n = 0, 1, 2, \ldots$. Note that the foregoing procedure, being applicable to cylindrical and spherical geometries and to unsteady problems, is general.

The study of steady two-dimensional cartesian problems will now be terminated by an example in which one of the foregoing problems is solved approximately by the integral method. Since the first-order approximations of Example 4–10 have already been given by Eqs. (2–136) and (2–143), we now consider the second-order approximations.

Example 4–11. Re-solve Example 4–10 by the integral method, using the second-order approximations of the Ritz and Kantorovich profiles.

The thermal symmetry of the problem with respect to the coordinate axes of Fig. 4–28 suggests the generalized Ritz profile

$$\theta(x, y) = (L^2 - x^2)(l^2 - y^2)(a_0 + a_1 x^2 + b_1 y^2 + a_2 x^4 + b_2 y^4 + c_2 x^2 y^2 + \cdots). \qquad (4\text{–}134)$$

---

† If Eq. (4–128) were used for the reduction of $\theta(x, y)$, the solution of $\theta(x, y)$ would have a different form. Write this solution from inspection of Eq. (4–133).

In Eq. (4–134), either $(a_0 + a_1x^2)$ or $(a_0 + b_1y^2)$ may be used as the last term of the second-order approximation. The choice depends on the ratio of L/l, and will be discussed in Chapter 8 (see Example 8–5). Here

$$\theta(x, y) = (L^2 - x^2)(l^2 - y^2)(a_0 + a_1x^2) \qquad (4\text{–}135)$$

is used arbitrarily. Inspection of the result, however, readily yields the solution of the alternative choice.

Equation (4–135), having two unknowns a_0 and a_1 to be determined, requires that two conditions be specified. As before, the integral form of the governing differential equation, Eq. (2–132) or Eq. (2–133), may be used as one of these conditions. As we learned in Section 3–10, there are two ways of getting the second condition, one related to the integral formulation, the other to the differential formulation of the problem. [See Eqs. (3–213) and (3–214).] Leaving the former to the reader, we here consider the latter. Since Eq. (4–135) satisfies only the boundary conditions of the problem, the maximum discrepancy between Eq. (4–135) and the exact solution occurs at the point farthest from the boundaries, namely, at the origin of the coordinate axes. Hence the differential formulation of the problem, satisfied at the origin by Eq. (4–135), becomes a convenient second condition.

The first condition,

$$4\int_0^L \int_0^l \{-2(l^2 - y^2)[(a_0 - a_1L^2) + 6a_1x^2] \\ - 2(a_0L^2 + a_1L^2x^2 - a_0x^2 - a_1x^4) + (u'''/k)\}\,dx\,dy = 0,$$

results in

$$(l^2 + L^2)a_0 + L^2(l^2 + L^2/5)a_1 = 3u'''/4k, \qquad (4\text{–}136)$$

and the second gives

$$(l^2 + L^2)a_0 - L^2a_1 = u'''/2k. \qquad (4\text{–}137)$$

Solving Eqs. (4–136) and (4–137) for a_0 and a_1 yields

$$a_0 = \left(\frac{u'''}{4kl^2}\right) \frac{5 + (\tfrac{2}{5})(L/l)^2}{[1 + (L/l)^2][2 + (\tfrac{1}{5})(L/l)^2]},$$

$$a_1 = \left(\frac{u'''}{4kl^2}\right) \frac{(1/L)^2}{2 + (\tfrac{1}{5})(L/l)^2}.$$

Hence the second-order Ritz profile is found to be

$$\frac{\theta(x, y)}{u'''L^2/k} = \frac{1}{4} \frac{[1 - (x/L)^2][1 - (y/l)^2]}{[2 + (\tfrac{1}{5})(L/l)^2]} \left[\frac{5 + (\tfrac{2}{5})(L/l)^2}{1 + (L/l)^2} + \left(\frac{x}{L}\right)^2\right]. \qquad (4\text{–}138)$$

Higher-order approximations of the Kantorovich profile, again depending on the ratio of (L/l) (see Example 8–6), may be selected in the form

$$\theta(x, y) = (l^2 - y^2)[X_1(x) + y^2 X_2(x) + y^4 X_3(x) + \cdots] \qquad (4\text{--}139)$$

or

$$\theta(x, y) = (L^2 - x^2)[Y_1(y) + x^2 Y_2(y) + x^4 Y_3(y) + \cdots]. \qquad (4\text{--}140)$$

Here

$$\theta(x, y) = (l^2 - y^2)[X_1(x) + y^2 X_2(x)] \qquad (4\text{--}141)$$

will arbitrarily be considered as the second-order approximation.

Insertion of Eq. (4–141) into the integral form of the differential equation given by Eq. (2–133),

$$4\int_0^L \int_0^l [l^2 X_1'' - y^2 X_1'' + l^2 y^2 X_2'' - y^4 X_2'' - 2X_1 + 2l^2 X_2 - 12 y^2 X_2 + (u'''/k)]\, dx\, dy = 0,$$

and subsequent integration in the y-direction yields

$$4\int_0^L \left[\tfrac{2}{3} l^3 X_1'' + \tfrac{2}{15} l^5 X_2'' - 2l X_1 - 2l^3 X_2 + \frac{u''' l}{k}\right] dx = 0,$$

which implies that the integrand is identically zero:

$$\tfrac{1}{3} l^2 X_1'' + \tfrac{1}{15} l^4 X_2'' - X_1 - l^2 X_2 = -\frac{u'''}{2k}. \qquad (4\text{--}142)$$

Next, the differential equation satisfied at the origin by Eq. (4–141) gives

$$l^2 X_1'' - 2X_1 + 2l^2 X_2 = -\frac{u'''}{k}. \qquad (4\text{--}143)$$

Introducing the operator $D \equiv d/dx$, we may rearrange Eqs. (4–142) and (4–143) in the equivalent form

$$\left(\frac{l^2}{3} D^2 - 1\right) X_1 + l^2 \left(\frac{l^2}{15} D^2 - 1\right) X_2 = -\frac{u'''}{2k}, \qquad (4\text{--}144)$$

$$(l^2 D^2 - 1) X_1 + 2l^2 X_2 = -\frac{u'''}{k}. \qquad (4\text{--}145)$$

Solving Eqs. (4–144) and (4–145) for X_1 and X_2 by the theory of determinants or by elimination results in

$$\left(\frac{l^4}{15} D^4 - \frac{9 l^2}{5} D^2 + 4\right) X_1 = 2\frac{u'''}{k}, \qquad (4\text{--}146)$$

$$\left(\frac{l^4}{15} D^4 - \frac{9 l^2}{5} D^2 + 4\right) X_2 = 0. \qquad (4\text{--}147)$$

The particular solution of Eq. (4–146) is $u'''/2k$. The general solution of Eq. (4–147) and that of the homogeneous part of Eq. (4–146) may conveniently be written in terms of four hyperbolic functions $\cosh \alpha(x/l)$, $\sinh \alpha(x/l)$, $\cosh \beta(x/l)$, and $\sinh \beta(x/l)$. Here

$$\alpha = [27/2 - (489/4)^{1/2}]^{1/2}, \qquad \beta = [27/2 + (489/4)^{1/2}]^{1/2}$$

and $\pm \alpha/l$, $\pm \beta/l$ are the roots of

$$(1/15)r^4 - (9/5)r^2 + 4 = 0.$$

Hence the general solutions of X_1 and X_2, excluding the thermally nonsymmetric solutions $\sinh \alpha(x/l)$ and $\sinh \beta(x/l)$, are

$$X_1 = \frac{u'''}{2k} + A \cosh \alpha\left(\frac{x}{l}\right) + B \cosh \beta\left(\frac{x}{l}\right), \qquad (4\text{–}148)$$

$$X_2 = C \cosh \alpha\left(\frac{x}{l}\right) + D \cosh \beta\left(\frac{x}{l}\right). \qquad (4\text{–}149)$$

Here the four unknown constants A, B, C, D, being the result of the elimination process considered between Eqs. (4–144) and (4–145), are not independent. To determine the relationship among these constants Eqs. (4–148) and (4–149) are introduced into Eq. (4–143) and the result is required to be identically zero. Thus we obtain

$$(\alpha^2 - 2)A + 2l^2 C = 0,$$
$$(\beta^2 - 2)B + 2l^2 D = 0.$$

The same conditions would also result from the use of Eq. (4–142). Thus only two of the four constants are truly arbitrary. Expressing, say, C in A and D in B, and inserting the results into Eqs. (4–148) and (4–149) gives

$$X_1 = \frac{u'''}{2k} + A \cosh \alpha\left(\frac{x}{l}\right) + B \cosh \beta\left(\frac{x}{l}\right), \qquad (4\text{–}150)$$

$$X_2 = -\left(\frac{\alpha^2 - 2}{2l^2}\right) A \cosh \alpha\left(\frac{x}{l}\right) - \left(\frac{\beta^2 - 2}{2l^2}\right) B \cosh \beta\left(\frac{x}{l}\right). \qquad (4\text{–}151)$$

Finally, satisfying the boundaries in the x-direction by letting

$$X_1(L) = 0, \qquad X_2(L) = 0,$$

we get

$$A = -\left(\frac{\beta^2 - 2}{\beta^2 - \alpha^2}\right) \frac{u'''/2k}{\cosh \alpha(x/l)}, \qquad B = \left(\frac{\alpha^2 - 2}{\beta^2 - 2}\right) \frac{u'''/2k}{\cosh \beta(x/l)}. \qquad (4\text{–}152)$$

Hence by combining Eqs. (4–141), (4–150), (4–151), and (4–152), we find the second-order Kantorovich profile to be

$$\frac{\theta(x,y)}{u'''l^2/k} = \frac{1}{2}\left[1 - \left(\frac{y}{l}\right)^2\right]\left\{1 - \left(\frac{\beta^2 - 2}{\beta^2 - \alpha^2}\right)\left[1 + \left(1 - \frac{\alpha^2}{2}\right)\left(\frac{y}{l}\right)^2\right]\frac{\cosh \alpha(x/l)}{\cosh \alpha(L/l)} \right.$$
$$\left. + \left(\frac{\alpha^2 - 2}{\beta^2 - \alpha^2}\right)\left[1 + \left(1 - \frac{\beta^2}{2}\right)\left(\frac{y}{l}\right)^2\right]\frac{\cosh \beta(x/l)}{\cosh \beta(L/l)}\right\}.$$
(4–153)

Further discussion of the approximate solutions is delayed to Chapter 8. Having finished the study of cartesian problems we now proceed to steady two-dimensional cylindrical problems.

4–8. Steady Two-Dimensional Cylindrical Geometry. Solutions by Fourier Series

The inherent nature of cylindrical coordinates implies three types of two-dimensional problems in the form $T(r, \varphi)$, $T(r, z)$, and $T(\varphi, z)$ (Fig. 4–29). Since $T(\varphi, z)$ has no physical significance (except in thin-walled tubes, which can be investigated in terms of cartesian coordinates), it is not considered here. $T(r, z)$ may depend on the expansion of an arbitrary function into a series in terms of cylindrical (Bessel) functions, and will be discussed in the next section. Problems of the type $T(r, \varphi)$, on the other hand, require no further mathematical background than that needed for cartesian geometry. Such problems are illustrated by the following examples.

Example 4–12. The surface temperature $f(\varphi)$ of an infinitely long solid rod of radius R is specified (Fig. 4–30). We wish to find the steady temperature of the rod.

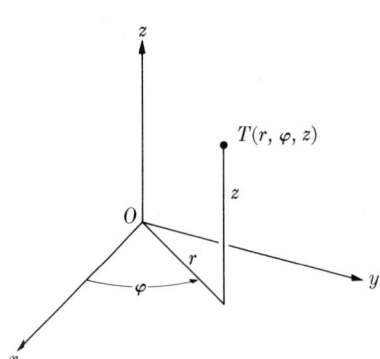

FIG. 4–29

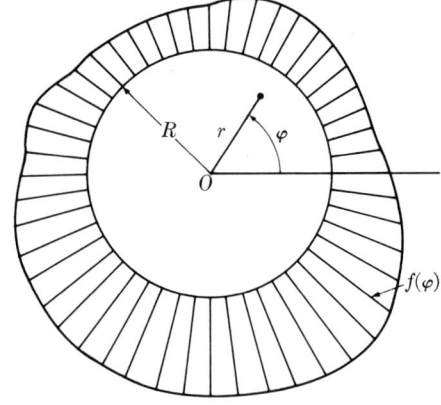

FIG. 4–30

The formulation of the problem is†

$$\frac{\partial^2 T}{\partial r^2} + \frac{1}{r}\frac{\partial T}{\partial r} + \frac{1}{r^2}\frac{\partial^2 T}{\partial \varphi^2} = 0, \qquad (4\text{--}154)$$

$$T(0, \varphi) = \text{finite}, \qquad (4\text{--}155)$$

$$T(R, \varphi) = f(\varphi), \qquad (4\text{--}156)$$

$$T(r, \varphi) = T(r, \varphi + 2\pi), \qquad (4\text{--}157)$$

$$\frac{\partial T(r, \varphi)}{r\, \partial \varphi} = \frac{\partial T(r, \varphi + 2\pi)}{r\, \partial \varphi}. \qquad (4\text{--}158)$$

The r-direction cannot be made orthogonal by any transformation. This leaves φ as the only possible orthogonal direction. Hence the product solution $T(r, \varphi) = \mathcal{R}(r)\phi(\varphi)$ with the proper choice of the separation constant yields

$$\frac{d^2\phi}{d\varphi^2} + \lambda^2 \phi = 0, \qquad (4\text{--}159)$$

$$\phi(\varphi) = \phi(\varphi + 2\pi), \qquad (4\text{--}160)$$

$$\frac{d\phi(\varphi)}{d\varphi} = \frac{d\phi(\varphi + 2\pi)}{d\varphi}, \qquad (4\text{--}161)$$

and

$$r^2 \frac{d^2 \mathcal{R}}{dr^2} + r \frac{d\mathcal{R}}{dr} - \lambda^2 \mathcal{R} = 0, \qquad (4\text{--}162)$$

$$\mathcal{R}(0) = \text{finite}. \qquad (4\text{--}163)$$

The general solution of Eq. (4–159) is

$$\phi = A \cos \lambda \varphi + B \sin \lambda \varphi. \qquad (4\text{--}164)$$

The physics of the problem requires that ϕ be single-valued in the rod. This condition, which is expressed by Eq. (4–160), can be satisfied when the circular functions of Eq. (4–164) have a common period 2π. The same requirement also serves to determine the permissible values (characteristic values) of the separation constant,

$$\lambda = n, \qquad n = 0, 1, 2, \ldots. \qquad (4\text{--}165)$$

Thus Eq. (4–164) becomes

$$\phi = A_0 + A_n \cos n\varphi + B_n \sin n\varphi. \qquad (4\text{--}166)$$

It is clear that ϕ, besides being single-valued, is continuous and thus automatically satisfies the second boundary condition given by Eq. (4–161).

---

† See Problem 2–6.

Equation (4–162) is an equidimensional equation [see Eq. (3–130)] with the general solution

$$\mathcal{R} = Cr^\lambda + Dr^{-\lambda}. \qquad (4\text{--}167)$$

Employing Eqs. (4–163) and (4–165), we reduce Eq. (4–167) to

$$\mathcal{R} = \begin{cases} C_0, & \text{when } n = 0, \\ C_n r^n, & \text{when } n = 1, 2, 3, \ldots \end{cases} \qquad (4\text{--}168)$$

Hence the product solution written in the form

$$T(r, \varphi) = \mathcal{R}_0 \phi_0 + \sum_{n=1}^{\infty} \mathcal{R}_n \phi_n$$

gives

$$T(r, \varphi) = a_0 + \sum_{n=1}^{\infty} r^n (a_n \cos n\varphi + b_n \sin n\varphi), \qquad (4\text{--}169)$$

where $a_0 = A_0 C_0$, $a_n = A_n C_n$, and $b_n = B_n C_n$.

Finally, using the nonhomogeneous surface temperature, we obtain from Eq. (4–169)

$$f(\varphi) = a_0 + \sum_{n=1}^{\infty} R^n (a_n \cos n\varphi + b_n \sin n\varphi), \qquad (4\text{--}170)$$

which is the complete Fourier series representation of $f(\varphi)$ (see Section 4–4). Thus we have

$$a_0 = \frac{1}{2\pi} \int_0^{2\pi} f(\varphi) \, d\varphi, \qquad a_n R^n = \frac{1}{\pi} \int_0^{2\pi} f(\varphi) \cos n\varphi \, d\varphi,$$

$$b_n R^n = \frac{1}{\pi} \int_0^{2\pi} f(\varphi) \sin n\varphi \, d\varphi. \qquad (4\text{--}171)$$

Specifically, if the surface temperature has the particular form given by Fig. 4–31, Eq. (4–171) yields

$$a_0 = \tfrac{1}{2} T_0, \qquad a_n = 0, \qquad b_n R^n = 2T_0/n\pi; \qquad n = 1, 3, 5, \ldots,$$

and the solution becomes

$$\frac{T(r, \varphi)}{T_0} = \frac{1}{2} + 2 \sum_{n=1}^{\infty} \frac{1}{n\pi} \left(\frac{r}{R}\right)^n \sin n\varphi. \qquad (4\text{--}172)$$

The results of the foregoing problem may readily be extended to the problem consisting of an infinite domain with a cylindrical hole (Fig. 4–32) whose surface temperature, $f(\varphi)$, is specified. The solution of this problem is left to the reader.

[4–8] SOLUTIONS BY FOURIER SERIES 227

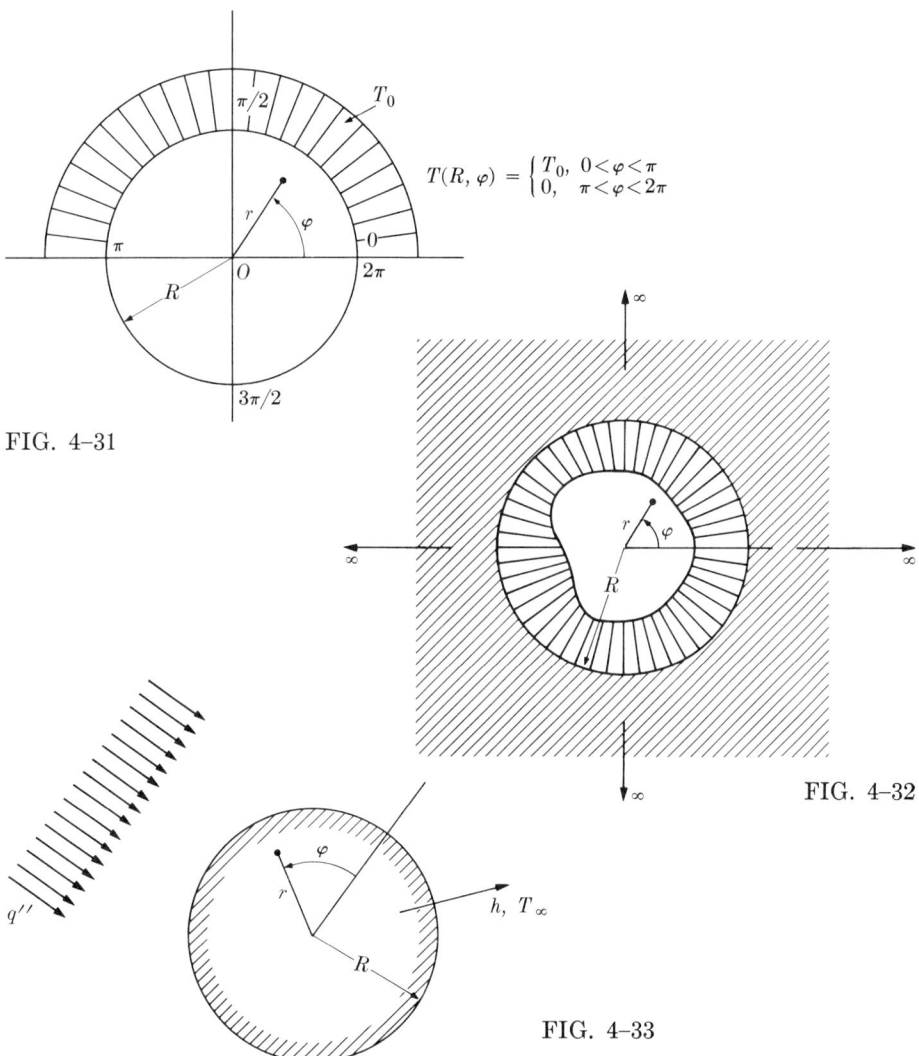

$$T(R, \varphi) = \begin{cases} T_0, & 0 < \varphi < \pi \\ 0, & \pi < \varphi < 2\pi \end{cases}$$

FIG. 4–31

FIG. 4–32

FIG. 4–33

Example 4–13. The pole of a television antenna is made up by a long rod of radius R (Fig. 4–33), receiving on one side the net radiant heat flux $q''(\varphi)$ from the sun while losing heat from the entire periphery by convection to the ambient at temperature T_∞. The heat transfer coefficient is h. Assuming that the problem is quasi-steady, that is, neglecting the rotation of the earth, (a) find the steady temperature of the pole, and (b) apply the results of part (a) to the special case

$$q''(\varphi) = \begin{cases} q_0'' \sin \varphi, & 0 < \varphi < \pi, \\ 0, & \pi < \varphi < 2\pi. \end{cases} \qquad (4\text{–}173)$$

The formulation of the problem is identical, except for the surface boundary condition, to the previous problem. Hence the temperature of the rod $\theta(r, \varphi)$ measured above the ambient is

$$\theta(r, \varphi) = a_0 + \sum_{n=1}^{\infty} r^n (a_n \cos n\varphi + b_n \sin n\varphi). \qquad (4\text{–}174)$$

Introducing Eq. (4–174) into the boundary condition

$$+ k \frac{\partial \theta(R, \varphi)}{\partial r} = q''(\varphi) - h\theta(R, \varphi) \qquad (4\text{–}175)$$

yields

$$\frac{q''(\varphi)}{h} = a_0 + \sum_{n=1}^{\infty} R^n \left(1 + \frac{nk}{hR}\right) (a_n \cos n\varphi + b_n \sin n\varphi), \qquad (4\text{–}176)$$

which is the complete Fourier series of $q''(\varphi)/h$. Hence the unknowns a_0, a_n, and b_n are evaluated from

$$a_0 = \frac{1}{2\pi h} \int_0^{2\pi} q''(\varphi)\, d\varphi,$$

$$\left(1 + \frac{nk}{hR}\right) R^n a_n = \frac{1}{\pi h} \int_0^{2\pi} q''(\varphi) \cos n\varphi\, d\varphi, \qquad (4\text{–}177)$$

$$\left(1 + \frac{nk}{hR}\right) R^n b_n = \frac{1}{\pi h} \int_0^{2\pi} q''(\varphi) \sin n\varphi\, d\varphi.$$

In particular, if the heat flux is specified by Eq. (4–173), the foregoing coefficients become

$$a_0 = \frac{q_0''}{\pi h}, \quad a_1 = 0, \quad a_n = -\left(\frac{q_0''}{\pi h}\right) \frac{1 + (-1)^n}{(n^2 - 1)(1 + nk/hR) R^n};$$

$$n = 2, 3, 4, 5, \ldots,$$

$$b_1 = \left(\frac{q_0''}{2h}\right) \frac{1}{(1 + k/hR) R}, \quad b_n = 0; \quad n = 2, 3, 4, 5, \ldots,$$

and the solution is found to be

$$\frac{\theta(r, \varphi)}{q_0''/h} = \frac{1}{\pi} + \frac{(r/R)}{2(1 + k/hR)} \sin \varphi - \frac{2}{\pi} \sum_{m=1}^{\infty} \frac{(r/R)^{2m}}{(4m^2 - 1)(1 + 2mk/hR)} \cos 2m\varphi,$$

$$(4\text{–}178)$$

where $m = n/2$.

Example 4–14. Consider an infinitely long cylindrical shell of angular section φ_0 (Fig. 4–34). The inner and outer radii of the shell are R_i, R_o, respectively.

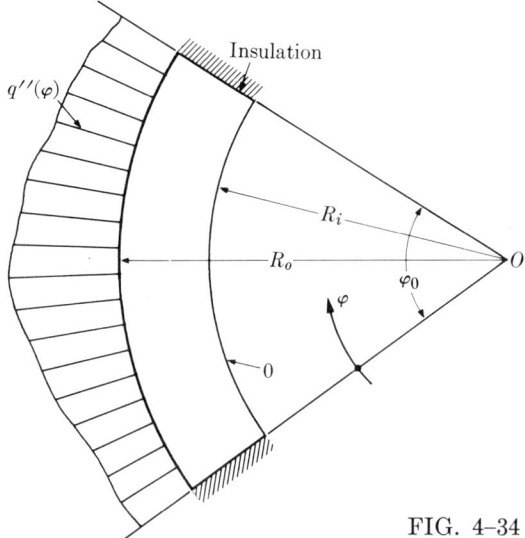

FIG. 4–34

The outer surface receives the net radiant heat flux $q''(\varphi)$ while the inner surface is maintained at a uniform temperature, say zero. The ends of the shell at $\varphi = 0$ and $\varphi = \varphi_0$ are insulated. We wish to find the steady temperature of the shell.

The formulation of the problem is

$$\frac{\partial^2 \theta}{\partial r^2} + \frac{1}{r} \frac{\partial \theta}{\partial r} + \frac{1}{r^2} \frac{\partial^2 \theta}{\partial \varphi^2} = 0,$$

$$\theta(R_i, \varphi) = 0, \qquad + k \frac{\partial \theta(R_o, \varphi)}{\partial r} = q''(\varphi),$$

$$\frac{\partial \theta(r, 0)}{r\, \partial \varphi} = 0, \qquad \frac{\partial \theta(r, \varphi_0)}{r\, \partial \varphi} = 0.$$

Noting that only the φ-direction can be made orthogonal, and choosing the sign of the separation parameter accordingly, we can separate the problem as follows:

$$\frac{d^2 \phi}{d\varphi^2} + \lambda^2 \phi = 0; \qquad \frac{d\phi(0)}{d\varphi} = 0, \qquad \frac{d\phi(\varphi_0)}{d\varphi} = 0, \tag{4-179}$$

and

$$r^2 \frac{d^2 \mathcal{R}}{dr^2} + r \frac{d\mathcal{R}}{dr} - \lambda^2 \mathcal{R} = 0; \qquad \mathcal{R}(R_i) = 0. \tag{4-180}$$

The solution of Eq. (4–179) is $\phi(\varphi) = C_n \psi_n(\varphi)$, with

$\psi_n(\varphi) = \cos \lambda_n \varphi,$ characteristic functions,

$\lambda_n = n\pi/\varphi_0, \quad n = 0, 1, 2, \ldots,$ characteristic values.

The solution of Eq. (4–180) is

$$\mathcal{R}_n(r) = \begin{cases} A_0 \ln(r/R_i), & \lambda_0 = 0, \\ A_n[(r/R_i)^{\lambda_n} - (r/R_i)^{-\lambda_n}], & \lambda_n = n\pi/\varphi_0, \quad n = 1, 2, 3, \ldots \end{cases}$$

We have then the product solution

$$\theta(r, \varphi) = a_0 \ln(r/R_i) + \sum_{n=1}^{\infty} a_n[(r/R_i)^{\lambda_n} - (r/R_i)^{-\lambda_n}] \cos \lambda_n \varphi, \quad (4\text{–}181)$$

where $a_0 = A_0 C_0$ and $a_n = A_n C_n$.

Finally, the use of the nonseparable boundary condition in the r-direction gives

$$q''(\varphi) R_o/k = a_0 + \sum_{n=1}^{\infty} a_n \lambda_n [(R_o/R_i)^{\lambda_n} + (R_o/R_i)^{-\lambda_n}] \cos \lambda_n \varphi, \quad (4\text{–}182)$$

from which we obtain, according to the theory of Fourier series,

$$a_0 = (R_o/k\varphi_0) \int_0^{\varphi_0} q''(\varphi)\, d\varphi,$$
$$\quad (4\text{–}183)$$
$$a_n \lambda_n [(R_o/R_i)^{\lambda_n} + (R_o/R_i)^{-\lambda_n}] = (2R_o/k\varphi_0) \int_0^{\varphi_0} q''(\varphi) \cos \lambda_n \varphi\, d\varphi.$$

We now terminate the study of problems in the form $T(r, \varphi)$ and proceed to those in the form $T(r, z)$. If the z-direction is orthogonal, these problems do not require additional mathematics, and can be solved by using circular functions in z and the modified Bessel functions in r [see Problem 4–19(a)]. On the other hand, when the r-direction is orthogonal, the expansion of an arbitrary function $f(r)$ into a series in terms of Bessel functions becomes necessary. This is done in the next section, and the results are applied to a number of problems.

4–9. Steady Two-Dimensional Cylindrical Geometry. Fourier-Bessel Series

When the steady two-dimensional cylindrical problem of type $T(r, z)$ is orthogonal in the r-direction, it can be solved by the proper choice of the separation constant leading to a second-order differential equation in z satisfied by hyperbolic functions, and to Bessel's equation in r,

$$r\frac{d}{dr}\left(r\frac{d\mathcal{R}}{dr}\right) + (\lambda^2 r^2 - \nu^2)\mathcal{R} = 0, \quad (4\text{–}184)$$

whose solution is expressible in terms of Bessel's functions as

$$\mathcal{R}(r) = A J_\nu(\lambda r) + B Y_\nu(\lambda r). \quad (4\text{–}185)$$

[See Eqs. (3–108) and (3–109).] Here the discussion will be restricted to problems of the solid cylinder. The procedure for hollow-cylinder problems, being analogous but algebraically complex, will not be considered here (see Problem 4–31).

The condition of finiteness or centerline symmetry at $r = 0$ requires that $B = 0$ in Eq. (4–185). Furthermore, dividing Eq. (4–184) by r and comparing the result with the general form of the second-order differential equation given by Eq. (4–16) yields

$$p(x) = r, \qquad q(x) = -\nu^2/r, \qquad w(x) = r.$$

Hence the characteristic functions $J_\nu(\lambda_n r)$ of the solid cylinder of radius R are orthogonal with respect to the weighting function $w(x) = r$ over the interval $(0, R)$. This implies that

$$\int_0^R r J_\nu(\lambda_m r) J_\nu(\lambda_n r) \, dr = 0, \qquad \lambda_m \neq \lambda_n. \tag{4-186}$$

The characteristic values, λ_n, are the roots of (i) $J_\nu(\lambda_n R) = 0$ when $\Re(R) = 0$ (zero surface temperature), (ii) $J'_\nu(\lambda_n R) = 0$ when $\Re'(R) = 0$ (zero surface heat flux), (iii) $J'_\nu(\lambda_n R) + B J_\nu(\lambda_n R) = 0$ when $\Re'(R) + B\Re(R) = 0$ (heat transfer to the ambient at zero temperature). In the last case B is a parameter.

Note that the roots of any one of these three equations exist in pairs, symmetrically located with respect to the center, $r = 0$. However, since the replacement of λ_n by $-\lambda_n$ in $J_\nu(\lambda_n r)$ either does not change $J_\nu(\lambda_n r)$ or multiplies it by -1, the negative values of λ_n need not be considered. If $\lambda_0 = 0$ is a characteristic number, $J_\nu(\lambda_0 r)$ becomes identically zero, and is no longer a characteristic function except in the case when $\nu = 0$ and $J'_0(\lambda_n R) = 0$. With that exception, the set of characteristic functions $J_\nu(\lambda_n r)$ should therefore be considered for *positive characteristic numbers* λ_n ($n = 1, 2, 3, \ldots$) *only*. For the exceptional case, the characteristic function $J_0(\lambda_0 r) = 1$ corresponding to $\lambda_0 = 0$ should also be included into the set.

We now return to the basic concern of this section, expansion of an arbitrary function $f(r)$ into a series of Bessel functions. This expansion, except in the case where $\nu = 0$ and $J'_0(\lambda_n R) = 0$, may be written in the form

$$f(r) = \sum_{n=1}^{\infty} a_n J_\nu(\lambda_n r), \qquad 0 < r < R, \tag{4-187}$$

which is the *Fourier-Bessel series* of $f(r)$ over the interval $(0, R)$. Here, noting that the weighting function is $w(r) = r$, we may readily obtain the coefficient a_n from Eq. (4–23) as follows:

$$a_n = \frac{\int_0^R r f(r) J_\nu(\lambda_n r) \, dr}{\int_0^R r J_\nu^2(\lambda_n r) \, dr}. \tag{4-188}$$

When $\nu = 0$ and $J_0'(\lambda_n R) = 0$, the Fourier-Bessel series of $f(r)$ becomes

$$f(r) = a_0 + \sum_{n=1}^{\infty} a_n J_0(\lambda_n r), \qquad (4\text{-}189)$$

where a_0 and a_n are the special cases of Eq. (4-188) corresponding to $\nu = 0$. Thus we have

$$a_n = \frac{\int_0^R rf(r)J_0(\lambda_n r)\,dr}{\int_0^R rJ_0^2(\lambda_n r)\,dr}, \qquad (4\text{-}190)$$

and since $\lambda_0 = 0$ and $J_0(0) = 1$, from Eq. (4-190)

$$a_0 = \frac{2}{R^2}\int_0^R rf(r)\,dr. \qquad (4\text{-}191)$$

Furthermore, in terms of the outer boundary condition, the integral of the denominator of Eq. (4-188) can be determined once and for all. Noting that the characteristic function $\varphi_n(r) = J_\nu(\lambda_n r)$ is a solution of Eq. (4-184), we have

$$r\frac{d}{dr}\left(r\frac{dJ_\nu}{dr}\right) + (\lambda_n^2 r^2 - \nu^2)J_\nu = 0. \qquad (4\text{-}192)$$

Multiplying Eq. (4-192) by $2(dJ_\nu/dr)$ and rearranging gives

$$\frac{d}{dr}\left(r\frac{dJ_\nu}{dr}\right)^2 = -(\lambda_n^2 r^2 - \nu^2)\frac{dJ_\nu^2}{dr}. \qquad (4\text{-}193)$$

It follows from the integration of Eq. (4-193) over $(0, R)$ and the rearrangement of the right-hand side by integration by parts that

$$\int_0^R rJ_\nu^2(\lambda_n r)\,dr = \frac{1}{2\lambda_n^2}\left\{(\lambda_n^2 r^2 - \nu^2)J_\nu^2(\lambda_n r) + \left[r\frac{dJ_\nu(\lambda_n r)}{dr}\right]^2\right\}_{r=R}. \qquad (4\text{-}194)$$

The left-hand side of Eq. (4-194) is identical to the denominator of Eq. (4-188). Special values of the denominator may now be summarized as follows.

(i) When λ_n is a root of $J_\nu(\lambda_n R) = 0$, noting from Eq. (3-139) that

$$\frac{dJ_\nu(\lambda_n r)}{dr} = -\lambda_n J_{\nu+1}(\lambda_n r) + \frac{\nu}{r}J_\nu(\lambda_n r),$$

we see that the right-hand side of Eq. (4-194), or the denominator of Eq. (4-188), yields

$$\frac{R^2}{2}J_{\nu+1}^2(\lambda_n R). \qquad (4\text{-}195)$$

(ii) When λ_n is a root of $J'_\nu(\lambda_n R) = 0$, the denominator of Eq. (4–188) becomes

$$\frac{\lambda_n^2 R^2 - \nu^2}{2\lambda_n^2} J_\nu^2(\lambda_n R). \tag{4–196}$$

(iii) When λ_n is a root of $J'_\nu(\lambda_n R) + BJ_\nu(\lambda_n R) = 0$, the denominator is

$$\frac{(\lambda_n^2 + B^2) R^2 - \nu^2}{2\lambda_n^2} J_\nu^2(\lambda_n R). \tag{4–197}$$

The denominator of Eq. (4–190) is the special case of Eqs. (4–195), (4–196), and (4–197) corresponding to $\nu = 0$.

The numerator of Eq. (4–188) requires that $f(r)$ be specified. For most practical cases this numerator can be evaluated with the help of the identity

$$\int_0^R r^{\nu+1} J_\nu(\lambda_n r) \, dr = \frac{R^{\nu+1}}{\lambda_n} J_{\nu+1}(\lambda_n R). \dagger \tag{4–198}$$

Next we illustrate the use of the Fourier-Bessel series by a number of steady two-dimensional cylindrical problems homogeneous in the r-direction.

Example 4–15. Consider a semi-infinite solid cylinder of radius R (Fig. 4–35). The heat transfer coefficient is large. The temperature of the ambient is T_∞, and that of the base is specified as $f(r)$. We need to know the steady two-dimensional temperature of the cylinder.

The formulation of the problem in terms of $\theta = T - T_\infty$ may readily be found to be

$$\frac{\partial}{\partial r}\left(r \frac{\partial \theta}{\partial r}\right) + r \frac{\partial^2 \theta}{\partial z^2} = 0,$$

$$\frac{\partial \theta(0, z)}{\partial r} = 0 \quad \text{or} \quad \theta(0, z) = \text{finite},$$

$$\theta(R, z) = 0,$$

$$\theta(r, 0) = f(r) - T_\infty = F(r),$$

$$\theta(r, \infty) = 0 \text{ (finite)}.$$

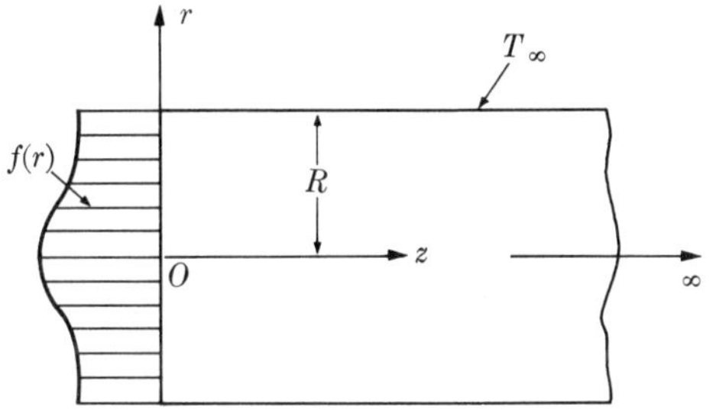

FIG. 4–35

We now seek a product solution in the form $\theta(r, z) = \mathcal{R}(r) Z(z)$. The separation constant is selected such that Bessel functions of the first kind are obtained in r, and exponential (or hyperbolic) functions are obtained in z. Hence the

---

† Equation (4–198) is the integral of Eq. (3–135).

problem reduces to

$$\frac{d}{dr}\left(r\frac{d\mathcal{R}}{dr}\right) + \lambda^2 r\mathcal{R} = 0; \quad \frac{d\mathcal{R}(0)}{dr} = 0 \text{ or } \mathcal{R}(0) = \text{finite}, \quad \mathcal{R}(R) = 0,$$
(4–199)

and

$$\frac{d^2 Z}{dz^2} - \lambda^2 Z = 0, \quad Z(\infty) = 0.$$
(4–200)

As before, the use of the nonhomogeneous (therefore nonseparable) boundary condition in z is delayed to the end of the solution.

The solution of Eq. (4–199) is

$$\mathcal{R}_n = A_n \varphi_n(r), \quad \varphi_n(r) = J_0(\lambda_n r), \quad \text{characteristic functions},$$

and the characteristic values, λ_n, are the zeros of $J_0(\lambda_n R) = 0$. [See Fig. 3–24(a).] On the other hand, Eq. (4–200) yields

$$Z_n = C_n e^{-\lambda_n z}.$$

Hence the product solution of the problem becomes

$$\theta(r, z) = \sum_{n=1}^{\infty} a_n e^{-\lambda_n z} J_0(\lambda_n r),$$
(4–201)

where $a_n = A_n C_n$. This satisfies the original differential equation and the three homogeneous boundary conditions of the problem. Now we must determine the coefficients a_n in such a way that the nonhomogeneous boundary condition, the base temperature $\theta(r, 0) = F(r)$, is satisfied. This gives

$$F(r) = \sum_{n=1}^{\infty} a_n J_0(\lambda_n r).$$
(4–202)

Thus the problem is reduced to the expansion of $F(r)$ into a Fourier-Bessel series. Since the characteristic values are the zeros of $J_0(\lambda_n R) = 0$, using Eqs. (4–188) and (4–195) for $\nu = 0$, we have

$$a_n = 2 \frac{\int_0^R r F(r) J_0(\lambda_n r)\, dr}{R^2 J_1^2(\lambda_n R)}.$$
(4–203)

Finally, introducing Eq. (4–203) into Eq. (4–201), we find that the solution of the problem is

$$\theta(r, z) = 2 \sum_{n=1}^{\infty} \left[\int_0^R \rho F(\rho) J_0(\lambda_n \rho)\, d\rho\right] \frac{e^{-\lambda_n z} J_0(\lambda_n r)}{R^2 J_1^2(\lambda_n R)}.$$
(4–204)

To distinguish the actual variable r of Eq. (4–204) from the dummy variable r of Eq. (4–203) the latter is replaced by ρ in Eq. (4–204).

In particular, if the base temperature is uniform, say θ_0, the integral of Eq. (4–204) may readily be evaluated from Eq. (4–198) for $\nu = 0$. Thus,

$$\int_0^R \rho J_0(\lambda_n \rho)\, d\rho = \frac{R}{\lambda_n} J_1(\lambda_n R), \qquad (4\text{–}205)$$

and the temperature distribution becomes

$$\frac{\theta(r, z)}{\theta_0} = 2 \sum_{n=1}^{\infty} \frac{e^{-\lambda_n z} J_0(\lambda_n r)}{(\lambda_n R) J_1(\lambda_n R)}. \qquad (4\text{–}206)$$

The physics of Example 4–2 and that of the foregoing problem are the same. The only difference is in the geometry, the latter involving the effect of curvature.

Example 4–16. Re-solve Example 4–15 for the uniform base temperature θ_0, assuming a finite length L in the z-direction.

The most convenient coordinate axes (see Example 4–8) are shown in Fig. 4–36. Since the physics in the r-direction does not change, Eq. (4–199) of the preceding example applies equally to the present case. The z-direction, on the other hand, now becomes

$$\frac{d^2 Z}{dz^2} - \lambda^2 Z = 0, \qquad Z(0) = 0. \qquad (4\text{–}207)$$

The solution of Eq. (4–207) may conveniently be expressed in terms of hyperbolic functions as

$$Z_n = C_n \sinh \lambda_n z.$$

Hence the product solution of the problem is

$$\theta(r, z) = \sum_{n=1}^{\infty} a_n \sinh \lambda_n z \, J_0(\lambda_n r) \qquad (4\text{–}208)$$

which, when satisfied by the base temperature θ_0, gives

$$\theta_0 = \sum_{n=1}^{\infty} a_n \sinh \lambda_n L \, J_0(\lambda_n r). \qquad (4\text{–}209)$$

This again is a Fourier-Bessel expansion of θ_0.

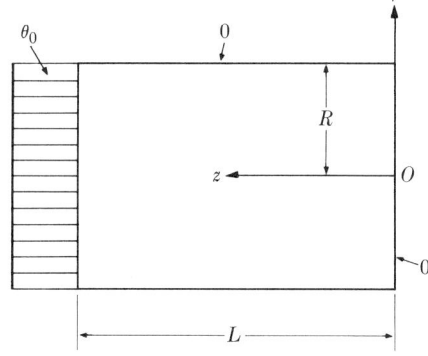

FIG. 4–36

Evaluating a_n from Eq. (4–209) and introducing the result into Eq. (4–208) yields the temperature of the cylinder in the form

$$\frac{\theta(r, z)}{\theta_0} = 2 \sum_{n=1}^{\infty} \left(\frac{\sinh \lambda_n z}{\sinh \lambda_n L}\right) \frac{J_0(\lambda_n r)}{(\lambda_n R) J_1(\lambda_n R)}. \qquad (4\text{–}210)$$

Examples 4–8 and 4–16 represent the same physics in cartesian and cylindrical geometries, respectively.

Example 4–17. Re-solve the special case of Example 4–15 corresponding to the uniform base temperature θ_0 for a finite heat transfer coefficient h.

The formulation of the problem is identical to that of Example 4–15 except for the peripheral boundary condition, which now becomes

$$-k \frac{\partial \theta(R, z)}{\partial r} = h\theta(R, z).$$

Hence, for the r-direction, the product solution yields

$$\frac{d}{dr}\left(r \frac{d\Re}{dr}\right) + \lambda^2 r \Re = 0, \tag{4-211}$$

$$\frac{d\Re(0)}{dr} = 0 \quad \text{or} \quad \Re(0) = 0, \quad \Re(R) + B\Re(R) = 0,$$

where $B = h/k$, while the z-direction remains unchanged as compared with that of Example 4–15. Therefore, the z-direction again satisfies Eq. (4–200).

The solution of Eq. (4–211) is

$$\Re_n = A_n \varphi_n(r), \quad \varphi_n(r) = J_0(\lambda_n r), \quad \text{characteristic functions,}$$

and the characteristic values, λ_n, are the zeros of $J_0'(\lambda_n R) + BJ_0(\lambda_n R) = 0$.

The product solution then becomes

$$\theta(r, z) = \sum_{n=1}^{\infty} a_n e^{-\lambda_n z} J_0(\lambda_n r), \tag{4-212}$$

which is identical to that of Example 4–15 except for the value of the coefficient a_n. Employing Eqs. (4–188) and (4–197) for $\nu = 0$, and Eq. (4–205), we have

$$a_n = \frac{2\theta_0 (\lambda_n R) J_1(\lambda_n R)}{[\lambda_n^2 R^2 + (\text{Bi})^2] J_0^2(\lambda_n R)},$$

where $\text{Bi} = hR/k$. This equation may further be rearranged by considering the relation $J_0'(\lambda_n R) + BJ_0(\lambda_n R) = 0$ which, according to Eq. (3–137), is equivalent to

$$(\lambda_n R) J_1(\lambda_n R) = \text{Bi } J_0(\lambda_n R). \tag{4-213}$$

The result is

$$a_n = \frac{2\theta_0 \text{ Bi}}{[\lambda_n^2 R^2 + (\text{Bi})^2] J_0(\lambda_n R)}. \tag{4-214}$$

Finally, introducing Eq. (4–214) into Eq. (4–212), we find that the temperature of the cylinder is

$$\frac{\theta(r, z)}{\theta_0} = 2 \sum_{n=1}^{\infty} \frac{\text{Bi } e^{-\lambda_n z} J_0(\lambda_n r)}{[\lambda_n^2 R^2 + (\text{Bi})^2] J_0(\lambda_n R)}. \tag{4-215}$$

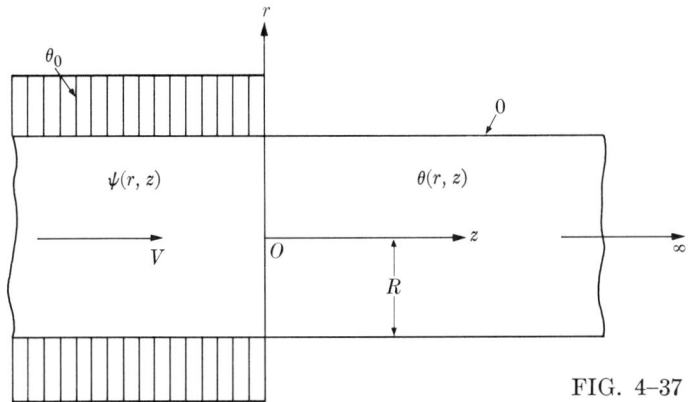

FIG. 4-37

Again the only difference between the solutions given by Eqs. (4–79) and (4–215) is the curvature effect included in the latter.

Example 4–18. We wish to know the effect of the uniform velocity V on Example 4–15 with respect to the fixed coordinate system shown in Fig. 4–37. The motion is in the z-direction, axial conduction is not negligible, and the surface temperature of the incoming rod is uniform, θ_0.

This problem is the cylindrical case of Example 4–7. As explained in Example 4–7, because of the penetration of axial conduction in the negative z-direction, the problem becomes one of two domains.

The formulation of the problem and the solution procedure, provided the effect of curvature is added, are identical to those of Example 4–7, and will not be elaborated here. The result, before the use of the interface boundary conditions, is

$$\psi(r, z) = \theta_0 + \sum_{n=1}^{\infty} a_n e^{[s+(s^2+\lambda_n^2)^{1/2}]z} J_0(\lambda_n r), \qquad z < 0, \qquad (4\text{–}216)$$

$$\theta(r, z) = \sum_{n=1}^{\infty} b_n e^{[s-(s^2+\lambda_n^2)^{1/2}]z} J_0(\lambda_n r), \qquad z > 0, \qquad (4\text{–}217)$$

where the characteristic values are the zeros of $J_0(\lambda_n R) = 0$, and again $2s = \rho c V/k$.

Now let us consider the interface conditions. The equality of interface temperatures,

$$\theta_0 + \sum_{n=1}^{\infty} a_n J_0(\lambda_n r) = \sum_{n=1}^{\infty} b_n J_0(\lambda_n r),$$

if we expand θ_0 into the appropriate Fourier-Bessel series as

$$\theta_0 = 2\theta_0 \sum_{n=1}^{\infty} \frac{J_0(\lambda_n r)}{(\lambda_n R) J_1(\lambda_n R)},$$

gives
$$2\theta_0/(\lambda_n R)J_1(\lambda_n R) + a_n = b_n, \qquad (4\text{--}218)$$

and the equality of heat fluxes yields
$$[s + (s^2 + \lambda_n^2)^{1/2}]a_n = [s - (s^2 + \lambda_n^2)^{1/2}]b_n. \qquad (4\text{--}219)$$

Finally, solving Eqs. (4–218) and (4–219) for a_n and b_n, introducing the result into Eqs. (4–216) and (4–217), and using the dimensionless numbers

$$\mu_n = \lambda_n R, \quad \text{Pé (or } P) = VR/a, \quad \text{Gz} = 1/\zeta = \text{Pé}/(x/R), \quad \text{and} \quad \rho = r/R,$$

we find that the dimensionless temperatures of the two regions are

$$\frac{\psi(\rho, \zeta; P)}{\theta_0} = 1 - \sum_{n=1}^{\infty}\left\{1 - \frac{1}{[1 + (\mu_n/P)^2]^{1/2}}\right\}\frac{J_0(\mu_n\rho)}{\mu_n J_1(\mu_n)}$$
$$\times \exp\left(\left\{1 + \left[1 + \left(\frac{\mu_n}{P}\right)^2\right]^{1/2}\right\}P^2\zeta\right), \quad \zeta < 0, \qquad (4\text{--}220)$$

$$\frac{\theta(\rho, \zeta; P)}{\theta_0} = \sum_{n=1}^{\infty}\left\{1 + \frac{1}{[1 + (\mu_n/P)^2]^{1/2}}\right\}\frac{J_0(\mu_n\rho)}{\mu_n J_1(\mu_n)}$$
$$\times \exp\left(\left\{1 - \left[1 + \left(\frac{\mu_n}{P}\right)^2\right]^{1/2}\right\}P^2\zeta\right), \quad \zeta > 0. \qquad (4\text{--}221)$$

Here the dimensionless forms of the temperatures are used for a direct comparison with Eqs. (4–110) and (4–111). The interested reader, by evaluating the Nusselt number and the interface and centerline temperatures from Eqs. (4–220) and (4–221), and comparing them with Figs. 4–20, 4–21, and 4–22, may investigate the effect of curvature on the problem.

Example 4–19. Both ends of a solid rod of radius R and length $2L$ are insulated. Halfway between the ends of this rod a metal sleeve of negligible thickness and length $2l$ rotates with constant angular velocity ω (Fig. 4–38). The

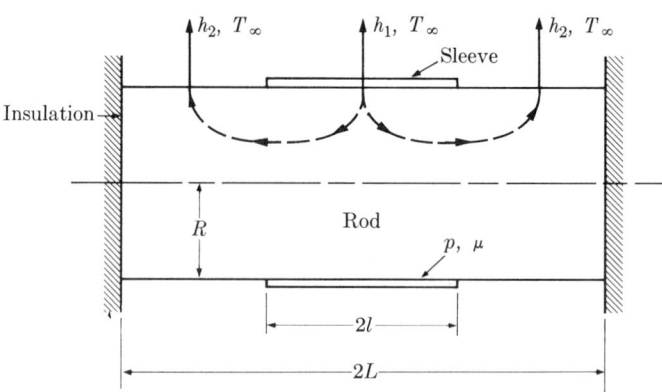

FIG. 4–38

pressure between the sleeve and the rod is p, and the coefficient of dry friction is μ. Let us consider two cases: (i) a single heat transfer coefficient for the entire system, $h_1 = h_2 = h$; (ii) two heat transfer coefficients, one for the rotating sleeve, h_1, and the other for the stationary rod, h_2. We wish to find the steady temperature of the system.

Although the ends of the rod are insulated, it follows from the heat flux paths shown in Fig. 4–38 that the problem is two-dimensional. Let us first consider the case of a single heat transfer coefficient. The formulation of the problem is then

$$\frac{1}{r}\frac{\partial}{\partial r}\left(r\frac{\partial \theta}{\partial r}\right) + \frac{\partial^2 \theta}{\partial z^2} = 0,$$

$$\frac{\partial \theta(0, z)}{\partial r} = 0, \quad -k\frac{\partial \theta(R, z)}{\partial r} + \begin{Bmatrix} \mu p \omega R, & 0 \le z < l \\ 0, & l < z \le L \end{Bmatrix} = h\theta(R, z),$$

$$\frac{\partial \theta(r, 0)}{\partial z} = 0, \quad \frac{\partial \theta(r, L)}{\partial z} = 0.$$

Assuming the r-axis as the homogeneous direction, we may solve this problem as a two-domain problem in a manner similar to that of Example 4–18. Since the peripheral boundary conditions of the two domains differ only in homogeneity, the characteristic values of both domains are identical. Leaving the steps of this solution to the reader as an exercise, we here proceed to a one-domain solution by taking the z-axis as the homogeneous direction. For this case, the separation of variables gives

$$\frac{d^2Z}{dz^2} + \lambda^2 Z = 0; \quad \frac{dZ(0)}{dz} = 0, \quad \frac{dZ(L)}{dz} = 0,$$

$$\frac{d}{dr}\left(r\frac{d\mathcal{R}}{dr}\right) - \lambda^2 r \mathcal{R} = 0; \quad \frac{d\mathcal{R}(0)}{dr} = 0.$$

The characteristic functions in z are $\varphi_n(z) = \cos \lambda_n z$, and the characteristic values are $\lambda_n L = n\pi$, $n = 0, 1, 2, \ldots$ The appropriate particular solution in r which satisfies the centerline symmetry is $\mathcal{R}_n(r) = I_0(\lambda_n r)$. The product solution may then be written in the form

$$\theta(r, z) = a_0 + \sum_{n=1}^{\infty} a_n I_0(\lambda_n r) \cos \lambda_n z. \tag{4–222}$$

Finally, the use of the nonhomogeneous boundary condition in r yields

$$-k \sum_{n=1}^{\infty} a_n \lambda_n I_1(\lambda_n R) \cos \lambda_n z + \begin{Bmatrix} \mu p \omega R, & 0 \le z < l \\ 0, & l < z \le L \end{Bmatrix}$$

$$= h a_0 + h \sum_{n=1}^{\infty} a_n I_0(\lambda_n R) \cos \lambda_n z,$$

from which there follows

$$a_0 = \frac{\mu p\omega Rl}{hL}, \qquad a_n = \frac{2(\mu p\omega R^2/k)\sin\lambda_n l}{\lambda_n L[\text{Bi } I_0(\lambda_n R) + (\lambda_n R)I_1(\lambda_n R)]},$$

where $\text{Bi} = hR/k$. Introducing these values into Eq. (4–222) we find that the temperature distribution in the system is

$$\frac{\theta(r,z)}{\mu p\omega R^2/k} = \frac{l/L}{\text{Bi}} + 2\sum_{n=1}^{\infty}\frac{\sin\lambda_n l\; I_0(\lambda_n r)}{\lambda_n L[\text{Bi } I_0(\lambda_n R) + (\lambda_n R)I_1(\lambda_n R)]}\cos\lambda_n z. \qquad (4\text{–}223)$$

When $h_1 \neq h_2$, we obtain two sets of characteristic values, and the problem can no longer be solved as a one-domain problem. This case requires a new concept, the expansion of an arbitrary function into a set of nonorthogonal functions, which is beyond the scope of the text.†

With the foregoing example, we terminate our study of cylindrical geometry, and proceed to spherical geometry.

4–10. Steady Two-Dimensional Spherical Geometry. Legendre Polynomials. Fourier-Legendre Series

When a spherical problem depends on the cone angle θ (Fig. 4–39), its solution can be reduced to the expansion of an arbitrary function into a series of *Legendre polynomials*. We shall first develop the necessary mathematics, and then apply the results to a problem.

The linear second-order differential equation with variable coefficients

$$(1-x^2)\frac{d^2y}{dx^2} - 2x\frac{dy}{dx} + n(n+1)y = 0 \qquad (4\text{–}224)$$

is known as *Legendre's equation*, and its solutions are known as *Legendre functions*. In particular, when n is zero or a positive integer, the solutions of Eq. (4–224) are called *Legendre polynomials*. Here we shall discuss briefly only the latter case, which is encountered when the separation of variables is applied to spherical conduction problems.

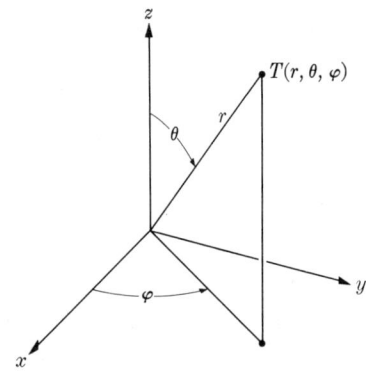

FIG. 4–39

The solution of Eq. (4–224) may be obtained by the method of power series (see Section 3–6). The result is

$$y(x) = a_0 P_n(x) + a_1 Q_n(x), \qquad (4\text{–}225)$$

---

† See, for example, Reference 4, Chapter 1, p. 44, Section 1–3.

where $P_n(x)$ is the so-called *Legendre polynomial of degree n, of the first kind*, and $Q_n(x)$ the *Legendre polynomial of degree n, of the second kind*. The expansion of $P_n(x)$ has the form

$$P_n(x) = (-1)^{n/2} \frac{1 \cdot 3 \cdot 5 \cdots (n-1)}{2 \cdot 4 \cdot 6 \cdots n}$$
$$\times \left[1 - \frac{n(n+1)}{2!} x^2 + \frac{n(n-2)(n+1)(n+3)}{4!} x^4 + \cdots \right], \quad (4\text{-}226)$$

where n is odd and equal to or greater than 2, and

$$P_n(x) = (-1)^{(n-1)/2} \frac{1 \cdot 3 \cdot 5 \cdots n}{2 \cdot 4 \cdot 6 \cdots (n-1)}$$
$$\times \left[x - \frac{(n-1)(n+2)}{3!} x^3 + \frac{(n-1)(n-3)(n+2)(n+4)}{5!} x^5 + \cdots \right],$$
$$(4\text{-}227)$$

when n is even and equal to or greater than 3. The first six of these polynomials are

$$P_0(x) = 1, \qquad P_1(x) = x,$$
$$P_2(x) = \tfrac{1}{2}(3x^2 - 1), \qquad P_3(x) = \tfrac{1}{2}(5x^3 - 3x), \qquad (4\text{-}228)$$
$$P_4(x) = \tfrac{1}{8}(35x^4 - 30x^2 + 3), \qquad P_5(x) = \tfrac{1}{8}(63x^5 - 70x^3 + 15x).$$

Equations (4–227) and (4–228) converge only when $|x| < 1$.† Similarly, the expansion of $Q_n(x)$ has the form

$$Q_n(x) = (-1)^{(n+1)/2} \frac{2 \cdot 4 \cdot 6 \cdots (n-1)}{1 \cdot 3 \cdot 5 \cdots n}$$
$$\times \left[1 - \frac{n(n+1)}{2!} x^2 + \frac{n(n-2)(n+1)(n+3)}{4!} x^4 + \cdots \right], \quad (4\text{-}229)$$

when n is odd and equal to or greater than 3, and

$$Q_n(x) = (-1)^{n/2} \frac{2 \cdot 4 \cdot 6 \cdots n}{1 \cdot 3 \cdot 5 \cdots (n-1)}$$
$$\times \left[x - \frac{(n-1)(n+2)}{3!} x^3 + \frac{(n-1)(n-3)(n+2)(n+4)}{5!} x^5 + \cdots \right],$$
$$(4\text{-}230)$$

when n is even and equal to or greater than 2.

---

† See Eq. (4–240) for the case of $|x| > 1$.

When $|x| < 1$ and m is a positive integer, the functions

$$P_n^m(x) = (1 - x^2)^{m/2} \frac{d^m P_n(x)}{dx^m},$$

$$Q_n^m(x) = (1 - x^2)^{m/2} \frac{d^m Q_n(x)}{dx}$$

(4–231)

are called the *associated Legendre polynomials of degree n and order m, of the first and second kinds*, respectively. When $|x| > 1$, the definition of Eq. (4–231) is modified by replacing $(1 - x^2)$ by $(x^2 - 1)$.

The special case $m = 0$ yields

$$P_n^0(x) = P_n(x),$$
$$Q_n^0(x) = Q_n(x).$$

(4–232)

In this case the Legendre polynomials are the *associated Legendre polynomials of order 0*.

The associated Legendre polynomials satisfy the differential equation

$$(1 - x^2)\frac{d^2 y}{dx^2} - 2x\frac{dy}{dx} + \left[n(n+1) - \frac{m^2}{1 - x^2}\right] y = 0. \quad (4\text{–}233)$$

When $|x| < 1$, by introducing $x = \cos\theta$, we may write Legendre's equation, Eq. (4–224), in the form

$$\frac{1}{\sin\theta}\frac{d}{d\theta}\left(\sin\theta \frac{dy}{d\theta}\right) + n(n+1)y = 0, \quad (4\text{–}234)$$

which appears in two-dimensional spherical problems. The general solution of Eq. (4–234) is

$$y = a_0 P_n(\cos\theta) + a_1 Q_n(\cos\theta). \quad (4\text{–}235)$$

The following properties of Legendre polynomials are frequently used:

$$P_n(-x) = (-1)^n P_n(x), \quad (4\text{–}236)$$

$$(n+1)P_{n+1}(x) + nP_{n-1}(x) = (2n+1)xP_n(x), \quad n = 1, 2, 3, \ldots, \quad (4\text{–}237)$$

$$P_n(x) = \frac{1}{2^n n!}\frac{d^n}{dx^n}(x^2 - 1)^n, \quad \text{Rodrigues' formula}, \quad (4\text{–}238)$$

$$P'_{n+1}(x) - P'_{n-1}(x) = (2n+1)P_n(x), \quad n = 1, 2, 3, \ldots, \quad (4\text{–}239)$$

$$Q_0(x) = \begin{cases} \frac{1}{2} \ln\left(\frac{1+x}{1-x}\right), & \text{when} \quad |x| < 1, \\ \frac{1}{2} \ln\left(\frac{x+1}{x-1}\right), & \text{when} \quad |x| > 1, \end{cases}$$

$$Q_1(x) = Q_0(x) P_1(x) - 1,$$
$$Q_2(x) = Q_0(x) P_2(x) - \tfrac{3}{2}x,$$
$$Q_3(x) = Q_0(x) P_3(x) - \tfrac{5}{2}x^2 + \tfrac{2}{3},$$
$$Q_4(x) = Q_0(x) P_4(x) - \tfrac{35}{8}x^3 + \tfrac{55}{24}x^2,$$
$$\vdots$$
$$Q_n(x) = Q_0(x) P_n(x) - \frac{(2n-1)}{1 \cdot n} P_{n-1}(x) - \frac{(2n-5)}{3(n-1)} P_{n-3}(x) - \cdots,$$

(4–240)

$$P_0(\cos\theta) = 1,$$
$$P_1(\cos\theta) = \cos\theta,$$
$$P_2(\cos\theta) = \tfrac{1}{4}(3\cos 2\theta + 1),$$
$$P_3(\cos\theta) = \tfrac{1}{8}(5\cos 3\theta + 3\cos\theta),$$
$$\vdots$$
$$P_n(\cos\theta) = \frac{1 \cdot 3 \cdots (2n-1)}{n! \, 2^{n-1}} \left[\cos n\theta + \frac{1 \cdot n}{1 \cdot (2n-1)} \cos(n-2)\theta \right. $$
$$\left. + \frac{1 \cdot 3 \cdot n(n-1)}{1 \cdot 2 \cdot (2n-1)(2n-3)} \cos(n-4)\theta + \cdots \right].$$

(4–241)

Next we consider the orthogonality of Legendre polynomials. However, noting from Eq. (4–240) that $Q_n(x) \to \infty$ as $x \to \pm 1$, we exclude the physically insignificant $Q_n(x)$ from our discussion.†

Legendre's equation, Eq. (4–224), rearranged in the form

$$\frac{d}{dx}\left[(1-x^2) \frac{dy}{dx} \right] + n(n-1) y = 0 \qquad (4\text{–}242)$$

and compared with Eq. (4–16) gives

$$p(x) = 1 - x^2, \quad q(x) = 0, \quad \lambda^2 = n(n+1), \quad w(x) = 1.$$

---

† Expansion in terms of the associated Legendre function $P_n^m(x)$ which arises in connection with three-dimensional problems will not be discussed here. See, for example, Reference 5.

Since $p(x) = 0$ when $x = \pm 1$, according to Eq. (4–17) of Section 4–2, no boundary conditions are needed for Eq. (4–242) to form a characteristic-value problem over the interval $(-1 \leq x \leq 1)$ provided the characteristic functions and their first derivatives are finite at the ends of the interval. As we learned in our review of the properties of Legendre functions and polynomials, Eq. (4–242) has finite solutions at $x = \pm 1$ when these solutions are in terms of $P_n(x)$, and only if n is a positive integer or zero,

$$n = 0, 1, 2, 3, \ldots \qquad (4\text{–}243)$$

Hence the Legendre polynomials $P_n(x)$ are the characteristic functions of the characteristic-value problem stated by Eq. (4–242). These polynomials form an orthogonal set with respect to the weighting function $w(x) = 1$ over the interval $(-1, 1)$; that is,

$$\int_{-1}^{1} P_m(x) P_n(x)\, dx = 0 \quad \text{if} \quad m \neq n. \qquad (4\text{–}244)$$

Now the expansion of an arbitrary function $f(x)$ in terms of appropriate Legendre polynomials, the *Fourier-Legendre series*, may be written in the form

$$f(x) = \sum_{n=0}^{\infty} a_n P_n(x), \quad -1 < x < 1. \qquad (4\text{–}245)$$

Here the coefficient a_n again follows from Eq. (4–23) as

$$a_n = \frac{\int_{-1}^{1} f(x) P_n(x)\, dx}{\int_{-1}^{1} P_n^2(x)\, dx}. \qquad (4\text{–}246)$$

The denominator of Eq. (4–246) is unlike the Fourier or Fourier-Bessel series, since it does not depend on the type of boundary condition. Therefore, it may be evaluated once and for all. Also, the numerator may be rearranged in an alternative form for convenience. Let us start with the latter, and using Eq. (4–238), write the numerator in the form

$$\int_{-1}^{1} f(x) P_n(x)\, dx = \frac{1}{2^n n!} \int_{-1}^{1} f(x) \frac{d^n}{dx^n} (x^2 - 1)^n\, dx. \qquad (4\text{–}247)$$

If $f(x)$ and its first n derivatives are continuous in the interval, integrating the right-hand side of Eq. (4–247) n times by parts gives

$$\int_{-1}^{1} f(x) P_n(x)\, dx = \frac{(-1)^n}{2^n n!} \int_{-1}^{1} (x^2 - 1)^n \frac{d^n f(x)}{dx^n}\, dx. \qquad (4\text{–}248)$$

Now, replacing $f(x)$ by $P_n(x)$ in Eq. (4–248), and employing the nth derivative of Eq. (4–238),

$$\frac{d^n P_n(x)}{dx^n} = \frac{(2n)!}{2^n n!},$$

we find the denominator to be

$$\int_{-1}^{1} P_n^2(x)\, dx = \frac{(2n)!}{2^{2n}(n!)^2} \int_{-1}^{1} (1 - x^2)^n\, dx. \qquad (4\text{--}249)$$

The right-hand side of Eq. (4–249) integrated n times by parts yields

$$\int_{-1}^{1} (1 - x^2)^n\, dx = \frac{2^{2n+1}(n!)^2}{(2n + 1)!}. \qquad (4\text{--}250)$$

Introducing Eq. (4–250) into Eq. (4–249), we obtain

$$\int_{-1}^{1} P_n^2(x)\, dx = \frac{2}{2n + 1}. \qquad (4\text{--}251)$$

Hence the coefficient a_n becomes

$$a_n = \begin{cases} \dfrac{2n + 1}{2} \displaystyle\int_{-1}^{1} f(x) P_n(x)\, dx, \\[2ex] \dfrac{2n + 1}{2^{n+1} n!} \displaystyle\int_{-1}^{1} (1 - x^2)^n \dfrac{d^n f(x)}{dx^n}\, dx. \end{cases} \qquad (4\text{--}252)$$

The second form of Eq. (4–252) can be used only if $f(x)$ and its first n derivatives are continuous in $(-1, 1)$.

Furthermore, noting that $P_n(x)$ is an even function of x when n is even, and an odd function when n is odd, we have

$$a_n = \begin{cases} (2n + 1) \displaystyle\int_0^1 f(x) P_n(x)\, dx, & n \text{ even}, \\ 0, & n \text{ odd}, \end{cases} \qquad (4\text{--}253)$$

for an *even function* $f(x)$, and

$$a_n = \begin{cases} (2n + 1) \displaystyle\int_0^1 f(x) P_n(x)\, dx, & n \text{ odd}, \\ 0, & n \text{ even}, \end{cases} \qquad (4\text{--}254)$$

for an *odd function* $f(x)$.

The series expansion given by Eq. (4–245), considered with Eq. (4–253) or Eq. (4–254), represents $f(x)$ inside the interval $(0, 1)$. Also, Eq. (4–245), together with Eq. (4–253), describes $f(-x)$, and together with Eq. (4–254), describes $-f(-x)$ in the interval $(-1, 0)$. The use of Fourier-Legendre series is illustrated by the following example.

Example 4–20. The surface temperature of a sphere of radius R is specified in the form $f(\theta)$. We wish to find the steady temperature of the sphere.

The formulation of the problem (see Problem 2-6) is

$$\frac{\partial}{\partial r}\left(r^2 \frac{\partial T}{\partial r}\right) + \frac{1}{\sin\theta}\frac{\partial}{\partial \theta}\left(\sin\theta \frac{\partial T}{\partial \theta}\right) = 0, \qquad (4\text{-}255)$$

$$T(0, \theta) = \text{finite}, \qquad (4\text{-}256)$$

$$T(R, \theta) = f(\theta). \qquad (4\text{-}257)$$

The missing boundaries in the θ-direction will be discussed later.

Since θ is the only possible orthogonal direction, with the appropriate choice of separation constant the product solution $T(r, \theta) = \Re(r)\vartheta(\theta)$ yields

$$\frac{1}{\sin\theta}\frac{d}{d\theta}\left(\sin\theta \frac{d\vartheta}{d\theta}\right) + \lambda\vartheta = 0, \qquad (4\text{-}258)$$

and

$$r^2 \frac{d^2\Re}{dr^2} + 2r\frac{d\Re}{dr} - \lambda\Re = 0, \qquad (4\text{-}259)$$

$$\Re(0) = \text{finite}. \qquad (4\text{-}260)$$

Now Eq. (4-258), first rearranged in the form

$$\frac{1}{\sin\theta}\frac{d}{d\theta}\left(\frac{1-\cos^2\theta}{\sin\theta}\frac{d\vartheta}{d\theta}\right) + \lambda\vartheta = 0, \qquad (4\text{-}261)$$

then transformed with $x = \cos\theta$, may be written as

$$\frac{d}{dx}\left[(1-x^2)\frac{d\vartheta}{dx}\right] + n(n+1)\vartheta = 0, \qquad (4\text{-}262)$$

where $n(n+1) = \lambda$. This is a Legendre equation. Its particular solution, finite at $x = \pm 1(\theta = 0, \pi)$, is

$$\vartheta_n = A_n \psi_n(\theta), \quad \psi_n(\theta) = P_n(\cos\theta), \quad \text{characteristic functions},$$
$$\qquad (4\text{-}263)$$

$$n = 0, 1, 2, 3, \ldots, \quad \text{characteristic values}.$$

Here the condition of finiteness specifying the characteristic functions and characteristic values takes care of the two missing boundary conditions in the θ-direction.

The general solution of the equidimensional equation given by Eq. (4-259) is

$$\Re_n(r) = C_n r^n + D_n r^{-(n+1)}, \qquad (4\text{-}264)$$

where $n = -\frac{1}{2} + (\lambda + \frac{1}{4})^{1/2}$.

The particular form of Eq. (4-264) which satisfies Eq. (4-260) is

$$\Re_n(r) = C_n r^n. \qquad (4\text{-}265)$$

Thus the product solution of the problem is

$$T(r, \theta) = \sum_{n=0}^{\infty} a_n r^n P_n(\cos \theta), \qquad (4\text{--}266)$$

where $a_n = A_n C_n$. The use of Eq. (4–257) reduces Eq. (4–266) to

$$f(\theta) = \sum_{n=0}^{\infty} a_n R^n P_n(\cos \theta), \qquad (4\text{--}267)$$

which is the expansion of $f(\theta)$ into a Fourier-Legendre series. Here the coefficient a_n is readily obtained from Eq. (4–252) in the form

$$a_n R^n = \frac{2n+1}{2} \int_0^\pi f(\theta) P_n(\cos \theta) \sin \theta \, d\theta. \qquad (4\text{--}268)$$

In particular, if the surface temperature is specified† as

$$T(R, \theta) = f(\theta) = \begin{cases} T_0, & 0 < \theta < \pi/2, \\ 0, & \pi/2 < \theta < \pi, \end{cases}$$

Eq. (4–268) in terms of x becomes

$$a_n R^n = T_0 \left(\frac{2n+1}{2} \right) \int_0^1 P_n(x) \, dx. \qquad (4\text{--}269)$$

Integrating Eq. (4–269) by means of Eq. (4–228) gives

$$a_0 = \tfrac{1}{2} T_0 \int_0^1 dx = \tfrac{1}{2} T_0,$$

$$a_1 R = \tfrac{3}{2} T_0 \int_0^1 x \, dx = \tfrac{3}{4} T_0,$$

$$a_2 R^2 = 0,$$

$$a_3 R^3 = \tfrac{7}{2} \cdot \tfrac{1}{2} T_0 \int_0^1 (5x^3 - 3x) \, dx = -\tfrac{7}{16} T_0,$$

$$a_4 R^4 = 0,$$

$$a_5 R^5 = \tfrac{11}{2} \cdot \tfrac{1}{8} T_0 \int_0^1 (63x^5 - 70x^3 + 15x) \, dx = \tfrac{11}{32} T_0,$$

$$\vdots$$

---

† What happens if $f(\theta)$ is given as

$$f(\theta) = \begin{cases} T_0, & 0 < \theta < \pi/2, \\ 0, & \pi/2 < \theta < 2\pi? \end{cases}$$

What is your conclusion?

Hence the solution of the problem is

$$\frac{T(r,\theta)}{T_0} = \frac{1}{2} + \frac{3}{4}\left(\frac{r}{R}\right)P_1(\cos\theta) - \frac{7}{16}\left(\frac{r}{R}\right)^3 P_3(\cos\theta)$$

$$+ \frac{11}{32}\left(\frac{r}{R}\right)^5 P_5(\cos\theta) + \cdots \quad (4\text{--}270)$$

Equations (4–270) and (4–172) denote the solution of the same problem applied to spherical and cylindrical geometries, respectively. The present problem of a solid sphere can easily be extended to that of the outside of a spherical hole of radius R. This is left to the reader.

4–11. Steady Three-Dimensional Geometry

These problems, except the expansion of a function in terms of associated Legendre polynomials for a three-dimensional spherical geometry, require no additional mathematics. Therefore, rather than giving an extensive treatment, we shall demonstrate only the solution procedure.

Example 4–21. Consider a semi-infinite rod of rectangular cross section ($2L \times 2l$) (Fig. 4–40). The base temperature of the rod is θ_0, the ambient temperature 0. The heat transfer coefficient is large. We wish to find the steady temperature of the rod.†

The formulation of the problem is

$$\frac{\partial^2 \theta}{\partial x^2} + \frac{\partial^2 \theta}{\partial y^2} + \frac{\partial^2 \theta}{\partial z^2} = 0,$$

$$\frac{\partial \theta(0, y, z)}{\partial x} = 0, \qquad \theta(L, y, z) = 0,$$

$$\frac{\partial \theta(x, 0, z)}{\partial y} = 0, \qquad \theta(x, l, z) = 0,$$

$$\theta(x, y, 0) = \theta_0, \qquad \theta(x, y, \infty) = 0 \text{ (finite)}.$$

The problem is homogeneous in both the x- and y-directions. Using the product solution $\theta(x, y, z) = X(x)Y(y)Z(z)$, we may separate the differential equation in the form

$$-\frac{1}{X}\frac{d^2 X}{dx^2} = \frac{1}{Y}\frac{d^2 Y}{dy^2} + \frac{1}{Z}\frac{d^2 Z}{dz^2} = \lambda^2, \quad (4\text{--}271)$$

which leads in the x-direction to the first characteristic-value problem:

$$\frac{d^2 X}{dx^2} + \lambda^2 X = 0; \qquad \frac{dX(0)}{dx} = 0, \quad X(L) = 0. \quad (4\text{--}272)$$

---

† This problem is the three-dimensional form of Example 4–2.

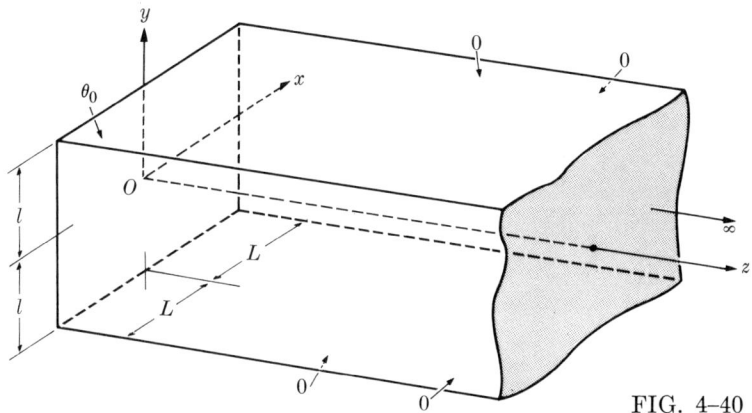

FIG. 4-40

Furthermore, rearranging the second equality of Eq. (4-271) in the form

$$-\frac{1}{Y}\frac{d^2Y}{dy^2} = \frac{1}{Z}\frac{d^2Z}{dz^2} - \lambda^2 = \mu^2 \tag{4-273}$$

yields in the y-direction the second characteristic-value problem:

$$\frac{d^2Y}{dy^2} + \mu^2 Y = 0; \quad \frac{dY(0)}{dy} = 0, \quad Y(l) = 0. \tag{4-274}$$

Thus the z-direction, from the second equality of Eq. (4-273), is found to satisfy

$$\frac{d^2Z}{dz^2} - (\lambda^2 + \mu^2)Z = 0; \quad Z(\infty) = 0 \text{ (finite)}, \tag{4-275}$$

where, as before, the second and nonseparable boundary condition, $\theta(x, y, 0) = \theta_0$, is left to the end of the solution.

The solution of Eq. (4-272) is

$$X_n(x) = A_n \varphi_n(x), \quad \varphi_n(x) = \cos \lambda_n x, \quad \text{characteristic functions,}$$

$$\lambda_n L = (2n+1)\pi/2, \quad n = 0, 1, 2, 3, \ldots, \quad \text{characteristic values.}$$

Similarly, the solution of Eq. (4-274) is

$$Y_m(y) = B_m \psi_m(y), \quad \psi_m(y) = \cos \mu_m y, \quad \text{characteristic functions,}$$

$$\mu_m l = (2m+1)\pi/2, \quad m = 0, 1, 2, 3, \ldots, \quad \text{characteristic values.}$$

Finally, the solution of Eq. (4-275) is

$$Z_{mn}(z) = C_{mn} e^{-(\lambda_n^2 + \mu_m^2)^{1/2} z}.$$

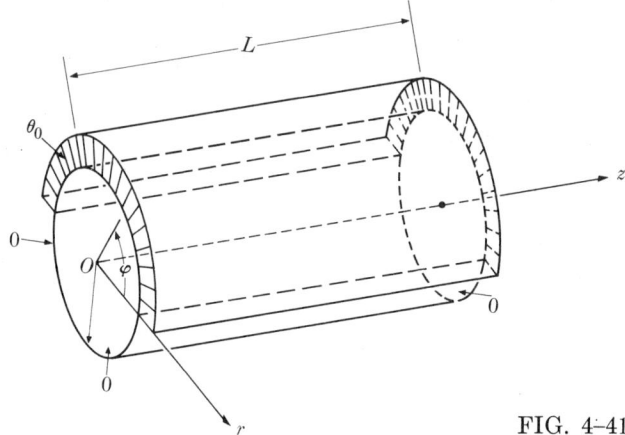

FIG. 4-41

Thus the product solution leads to

$$\theta(x, y, z) = \sum_{n=0}^{\infty} \sum_{m=0}^{\infty} a_{mn} e^{-(\lambda_n^2 + \mu_m^2)^{1/2} z} \cos \lambda_n x \cos \mu_m y, \qquad (4\text{-}276)$$

where $a_{mn} = A_n B_m C_{mn}$. Equation (4-276), by means of the nonseparable boundary condition, becomes

$$\theta_0 = \sum_{n=0}^{\infty} \sum_{m=0}^{\infty} a_{mn} \cos \lambda_n x \cos \mu_m y. \qquad (4\text{-}277)$$

Equation (4-277) is the *double Fourier cosine-series expansion* of θ_0 over the cross section of the rod. Here the coefficient a_{mn} may readily be obtained by simply extending the methods used earlier in Fourier series (see Sections 4-3, 4-4). Thus multiplying both sides of Eq. (4-277) by $\cos \lambda_p x \cos \mu_q y$, integrating the result over the cross section, and using the orthogonality in both the x- and y-directions gives for $p = n$, $q = m$

$$a_{mn} = \theta_0 \frac{\int_0^L \int_0^l \cos \lambda_n x \cos \mu_m y \, dx \, dy}{\int_0^L \int_0^l \cos^2 \lambda_n x \cos^2 \mu_m y \, dx \, dy},$$

which results in

$$a_{mn} = \frac{4\theta_0 (-1)^{n+m}}{(\lambda_n L)(\mu_m l)}. \qquad (4\text{-}278)$$

Finally, introducing Eq. (4-278) into Eq. (4-276), we find that the solution is

$$\frac{\theta(x, y, z)}{\theta_0} = 4 \sum_{n=0}^{\infty} \sum_{m=0}^{\infty} \frac{(-1)^{n+m} e^{-(\lambda_n^2 + \mu_m^2)^{1/2} z}}{(\lambda_n L)(\mu_m l)} \cos \lambda_n x \cos \mu_m y. \qquad (4\text{-}279)$$

The foregoing example is the three-dimensional form of Example 4-2.

Example 4–22. Consider a finite solid rod of radius R and length L. The temperature of one half of the peripheral surface is kept at the uniform temperature θ_0, while the other half and the ends are at 0 (Fig. 4–41). We need to know the steady temperature of the rod.

The formulation of the problem yields

$$\frac{1}{r}\frac{\partial}{\partial r}\left(r\frac{\partial\theta}{\partial r}\right) + \frac{1}{r^2}\frac{\partial^2\theta}{\partial\varphi^2} + \frac{\partial^2\theta}{\partial z^2} = 0,$$

$$\theta(0,\varphi,z) = \text{finite}, \qquad \theta(R,\varphi,z) = \begin{cases} \theta_0, & 0 < \varphi < \pi, \\ 0, & \pi < \varphi < 2\pi, \end{cases}$$

$$\theta(r,\varphi,z) = \theta(r,\varphi+2\pi,z)\dagger, \qquad \frac{\partial\theta(r,\varphi,z)}{r\,\partial\varphi} = \frac{\partial\theta(r,\varphi+2\pi,z)}{r\,\partial\varphi},$$

$$\theta(r,\varphi,0) = 0, \qquad \theta(r,\varphi,L) = 0.$$

The proper use of the product solution $\theta(r,\varphi,z) = \mathcal{R}(r)\phi(\varphi)Z(z)$ separates the problem as follows:

$$\frac{d^2Z}{dz^2} + \lambda^2 Z = 0; \qquad Z(0) = 0, \quad Z(L) = 0, \tag{4-280}$$

$$\frac{d^2\phi}{d\varphi^2} + \mu^2\phi = 0; \qquad \phi(\varphi) = \phi(\varphi+2\pi), \tag{4-281}$$

$$r\frac{d}{dr}\left(r\frac{d\mathcal{R}}{dr}\right) - (\lambda^2 r^2 + \mu^2)\mathcal{R} = 0; \qquad \mathcal{R}(0) = \text{finite}. \tag{4-282}$$

The solution of Eq. (4–280) is

$$Z_n(z) = A_n\varphi_n(z), \qquad \varphi_n(z) = \sin\lambda_n z, \quad \text{characteristic functions,}$$
$$\lambda_n L = n\pi, \qquad n = 1, 2, 3, \ldots, \quad \text{characteristic values.}$$

The solution of Eq. (4–281) gives

$$\phi_\mu(\varphi) = B_\mu\cos\mu\varphi + C_\mu\sin\mu\varphi,$$
$$\cos\mu\varphi, \quad \sin\mu\varphi, \quad \text{characteristic functions,}$$
$$\mu = 0, 1, 2, 3, \ldots, \quad \text{characteristic values.}$$

Finally, Eq. (4–282) has the solution

$$\mathcal{R}_{n\mu}(r) = D_{n\mu}I_\mu(\lambda_n r)$$

---

† This boundary condition, as seen in Example 4–12, determines both the characteristic functions and the characteristic values in the φ-direction. Thus the use of the next boundary condition in φ becomes unnecessary.

[see Eqs. (3–115) and (3–122)]. Thus the product solution becomes

$$\theta(r, \varphi, z) = \sum_{n=1}^{\infty} a_{n0} I_0(\lambda_n r) \sin \lambda_n z$$

$$+ \sum_{n=1}^{\infty} \sum_{\mu=1}^{\infty} (a_{n\mu} \cos \mu\varphi + b_{n\mu} \sin \mu\varphi) I_\mu(\lambda_n r) \sin \lambda_n z, \qquad (4\text{–}283)$$

where $a_{n0} = A_n B_0 D_{n0}$, $a_{n\mu} = A_n B_\mu D_{n\mu}$, and $b_{n\mu} = A_n C_\mu D_{n\mu}$. Then the use of the nonseparable (peripheral) boundary condition gives

$$\theta_0 = \sum_{n=1}^{\infty} a_{n0} I_0(\lambda_n R) \sin \lambda_n z$$

$$+ \sum_{n=1}^{\infty} \sum_{\mu=1}^{\infty} (a_{n\mu} \cos \mu\varphi + b_{n\mu} \sin \mu\varphi) I_\mu(\lambda_n R) \sin \lambda_n z. \qquad (4\text{–}284)$$

Equation (4–284) is the double Fourier series representation of θ_0 (complete in φ but only in terms of sine in z).

The coefficients of this series may be calculated as follows. Assuming Eq. (4–284) as a simple Fourier series in φ and noting $\theta_0 = 0$ when $\pi < \varphi < 2\pi$, we have from Eqs. (4–169) and (4–171)

$$\sum_{n=1}^{\infty} a_{n0} I_0(\lambda_n R) \sin \lambda_n z = \frac{\theta_0}{2\pi} \int_0^\pi d\varphi,$$

$$\sum_{n=1}^{\infty} a_{n\mu} I_\mu(\lambda_n R) \sin \lambda_n z = \frac{\theta_0}{\pi} \int_0^\pi \cos \mu\varphi \, d\varphi,$$

$$\sum_{n=1}^{\infty} b_{n\mu} I_\mu(\lambda_n R) \sin \lambda_n z = \frac{\theta_0}{\pi} \int_0^\pi \sin \mu\varphi \, d\varphi,$$

or

$$\sum_{n=1}^{\infty} a_{n0} I_0(\lambda_n R) \sin \lambda_n z = \frac{\theta_0}{2},$$

$$\sum_{n=1}^{\infty} a_{n\mu} I_\mu(\lambda_n R) \sin \lambda_n z = 0,$$

$$\sum_{n=1}^{\infty} b_{n\mu} I_\mu(\lambda_n R) \sin \lambda_n z = \frac{2\theta_0}{\mu\pi}, \qquad \mu = 1, 3, 5, \ldots$$

Thus the problem is reduced to three Fourier sine-series expansions in z. The three coefficients a_{n0}, $a_{n\mu}$, and $b_{n\mu}$ may now be evaluated in the usual manner.

The result is
$$a_{n0} I_0(\lambda_n R) = \frac{2\theta_0}{n\pi}, \qquad n = 1, 3, 5, \ldots,$$
$$a_{n\mu} = 0, \tag{4-285}$$
$$b_{n\mu} I_\mu(\lambda_n R) = \frac{8\theta_0}{(\mu\pi)(n\pi)}.$$

Introducing Eq. (4-285) into Eq. (4-283), we find that the solution of the problem is

$$\frac{\theta(r, \varphi, z)}{\theta_0} = \sum_{n=1}^{\infty} \left(\frac{2}{n\pi}\right) \frac{I_0(\lambda_n r)}{I_0(\lambda_n R)} \sin \lambda_n z$$
$$+ 2 \sum_{n=1}^{\infty} \sum_{\mu=1}^{\infty} \left(\frac{2}{n\pi}\right)\left(\frac{2}{\mu\pi}\right) \frac{I_\mu(\lambda_n r)}{I_\mu(\lambda_n R)} \sin \lambda_n z \sin \mu\varphi. \tag{4-286}$$

The foregoing problem is the three-dimensional form of the special case of Example 4-12 (Fig. 4-31).

References

1. R. V. CHURCHILL, *Fourier Series and Boundary Value problems*. New York: McGraw-Hill, 1963.

2. F. B. HILDEBRAND, *Advanced Calculus for Engineers*. Englewood Cliffs: Prentice-Hall, 1956.

3. H. S. CARSLAW and J. C. JAEGER, *Conduction of Heat in Solids*. Oxford: Clarendon Press, 1959.

4. L. V. KANTOROVICH and V. I. KRYLOV, *Approximate Methods of Higher Analysis*. New York-Groningen, Holland: Interscience-Noordhoff, 1958.

5. E. J. WHITTAKER and G. N. WATSON, *Modern Analysis*. Cambridge, England: Cambridge University Press, 1927.

6. L. GRAETZ, "Über die Wärmeleitfähigkeit von Flüssigkeiten." *Ann. Phys.*, **18,** 79 (1883), **25,** 337 (1885).

7. W. NUSSELT, "Die Abhängigkeit der Wärmeübergangszahl von der Rohrlänge." *Zeitschr. VDI*, **54,** 1154 (1910).

8. M. A. LEVÊQUE, "Les Lois de la Transmission de la Chaleur par Convection." *Ann. Mines*, **13,** 201, 305, 381 (1928).

9. J. A. PRINS, J. MULDER, and J. SCHENK, "Heat Transfer in Laminary Flow between Parallel Plates." *Appl. Sci. Res.*, **A2,** 431 (1951).

10. J. A. W. VAN DER DOES DE BYE and J. SCHENK, "Heat Transfer in Laminary Flow between Parallel Plates." *Appl. Sci. Res.*, **A3,** 308 (1952).

11. J. R. SELLARS, M. TRIBUS, and J. S. KLEIN, "Heat Transfer to Laminar Flow in a Round Tube or Flat Conduit—The Graetz Problem Extended." *Trans. ASME*, **78,** 441 (1956).

12. P. J. Schneider, "Effect of Axial Fluid Conduction on Heat Transfer in the Entrance Regions of Parallel Plates and Tubes." *Trans. ASME*, **79**, 765 (1957).

13. S. N. Singh, "The Determination of Eigen-Functions of a Certain Sturm-Liouville Equation and its Application to Problems of Heat Transfer." *Appl. Sci. Res.*, **A7**, 237 (1958).

14. S. N. Singh, "Heat Transfer by Laminar Flow in a Cylindrical Tube." *Appl. Sci. Res.*, **A7**, 325 (1958).

15. R. Siegel, E. M. Sparrow, and T. M. Hallman, "Steady Laminar Heat Transfer in a Circular Tube with Prescribed Wall Heat Flux." *Appl. Sci. Res.*, **A7**, 386 (1958).

16. R. D. Cess and E. C. Shaffer, "Heat Transfer to Laminar Flow between Parallel Plates with a Prescribed Wall Heat Flux." *Appl. Sci. Res.*, **A8**, 339 (1959).

17. H. C. Agrawal, "Heat Transfer in Laminar Flow between Parallel Plates at Small Péclèt Numbers." *Appl. Sci. Res.*, **A9**, 177 (1960).

Problems

4–1. Consider a semi-infinite plate of thickness δ and width L (Fig. 4–42). The thickness δ is small. The base temperature is T_0, the ambient temperature T_∞. The heat transfer coefficients are h_1, h_2, and h_3. Write the steady temperature of the plate from inspection of the problems solved in the text and assuming that (a) the sides of the plate are insulated, (b) the heat is transferred from the sides of the plate.

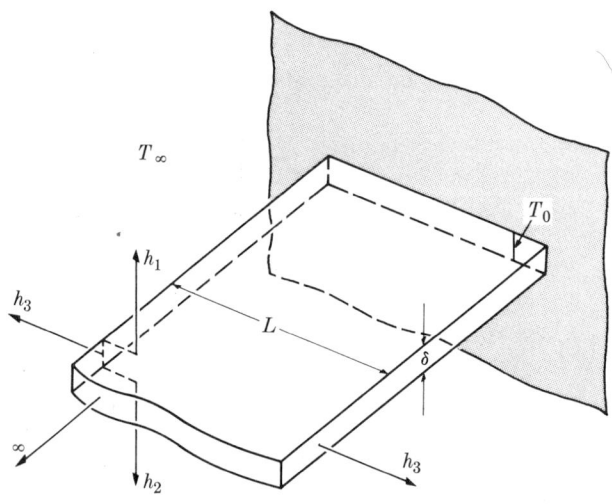

FIG. 4–42

4–2. The uniform internal energy u''' is generated in a $2L \times 2l \times \delta$ rectangular flat plate (Fig. 4–43). The upward and downward heat transfer coefficients are h_1 and h_2, respectively. The peripheral heat transfer coefficient is large. The ambient temperature is T_∞. Find the steady temperature of the plate.

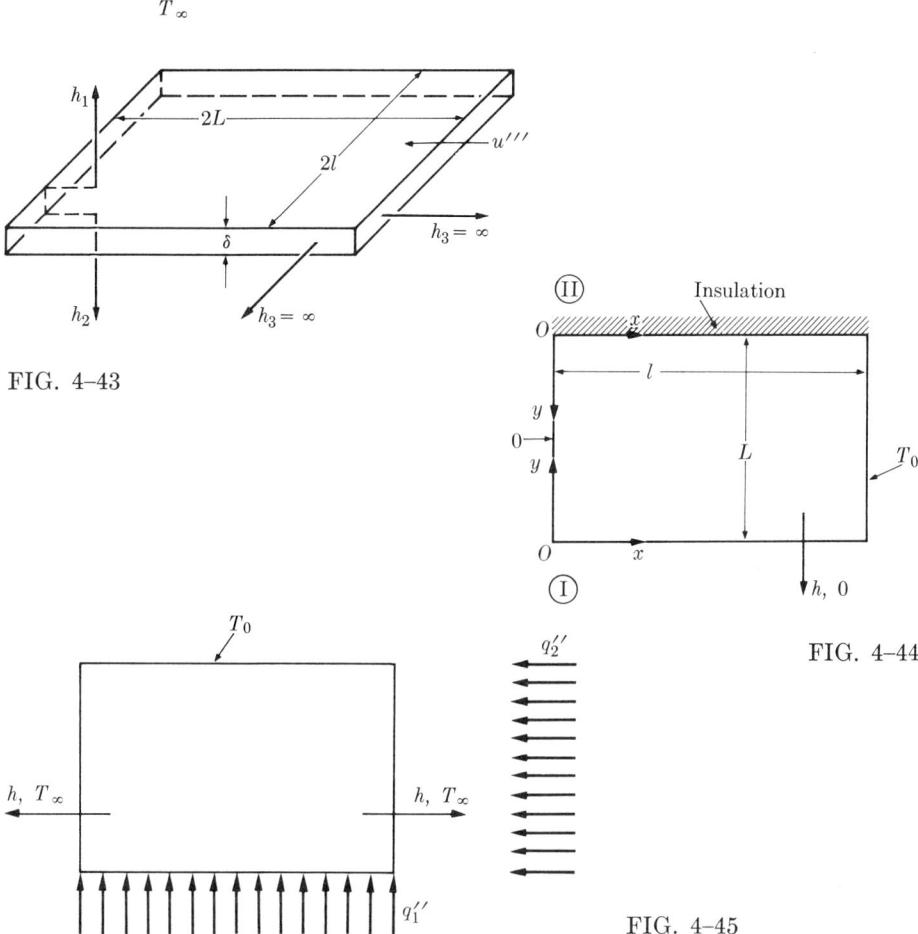

FIG. 4-43

FIG. 4-44

FIG. 4-45

4-3. The temperature is maintained constant, say zero, along three sides of an infinitely long rod of square cross section $(L \times L)$, while the fourth side has the uniform temperature T_0. Calculate the steady temperature at the intersection of the diagonals.

4-4. Consider an infinitely long rod of rectangular cross section $(L \times l)$. The boundary conditions are specified as shown in Fig. 4-44. Find the steady temperature of the rod (a) in terms of coordinate system I and the characteristic functions based on the insulated boundary, (b) in terms of coordinate system II and the characteristic functions based on the heat transfer boundary. (c) Compare the algebraic complexity of the solutions.

4-5. An infinitely long rod of rectangular cross section $(L \times l)$ is subjected to the boundary conditions given in Fig. 4-45. Determine the most convenient coordinate axes, and transform the problem to one suitable to the separation of variables.

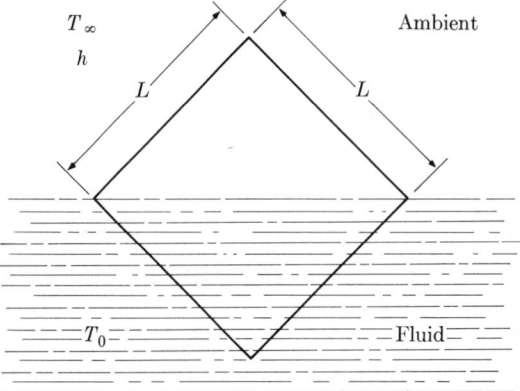

FIG. 4-46

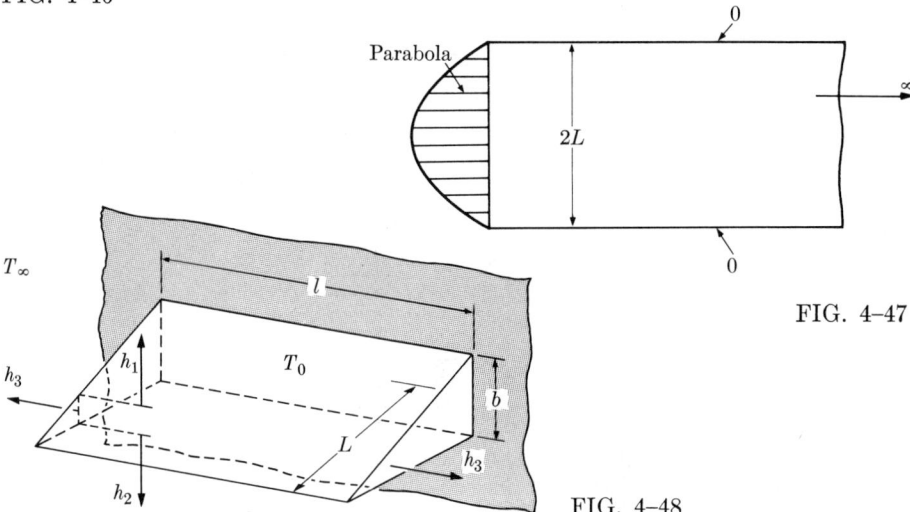

FIG. 4-47

FIG. 4-48

4-6. An infinitely long rod of square cross section ($L \times L$) floats in a fluid (Fig. 4-46). The heat transfer coefficient between the rod and the fluid is large compared with that between the rod and the ambient. The fluid and the ambient temperatures are T_0 and T_∞, respectively. Find the steady temperature of the rod.

4-7. Consider a semi-infinite plate of thickness $2L$. The end of the plate is subjected to a parabolic temperature distribution while the surfaces are maintained at constant temperature, say zero (Fig. 4-47). Using the integral technique, obtain a first and second approximation to the temperature of the plate in terms of the Ritz and Kantorovich profiles.

4-8. Consider a wedge-shaped rectangular plate (Fig. 4-48). The base thickness of the plate is b, the base temperature T_0, and the ambient temperature T_∞. The upward, downward, and side heat transfer coefficients are h_1, h_2, and h_3. Find the steady temperature of the plate. (*Note:* This problem is the two-dimensional form of Example 3-11.)

PROBLEMS

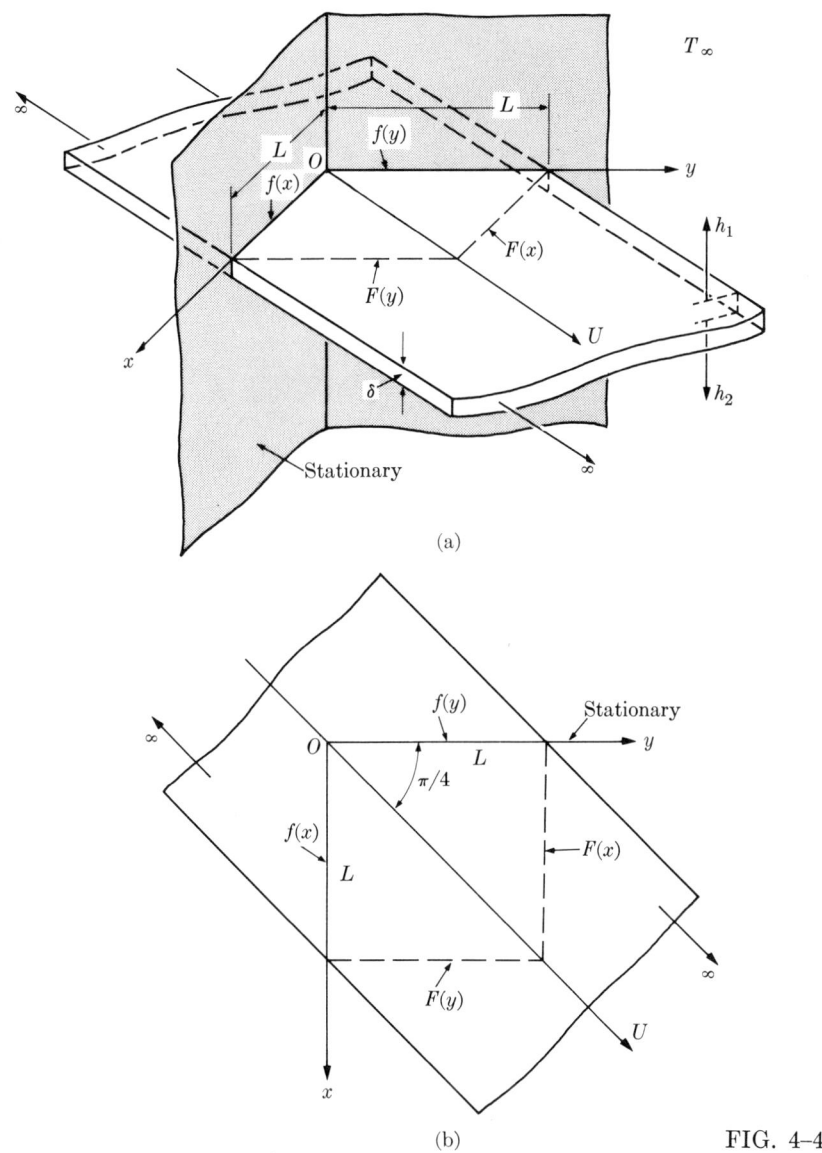

FIG. 4–49

4–9. A flat plate of thickness δ moves with the constant velocity U relative to the coordinate system shown in Fig. 4–49. Approaching the coordinate system, the plate has the uniform temperature T_0. The boundary temperatures $f(x)$, $f(y)$, $F(x)$, $F(y)$ are *measured*, not imposed. The ambient temperature is T_∞. The upward and downward heat transfer coefficients are h_1 and h_2, respectively. Find the steady temperature of that part of the plate which instantaneously occupies the square area $(L \times L)$ measured from the stationary coordinates.

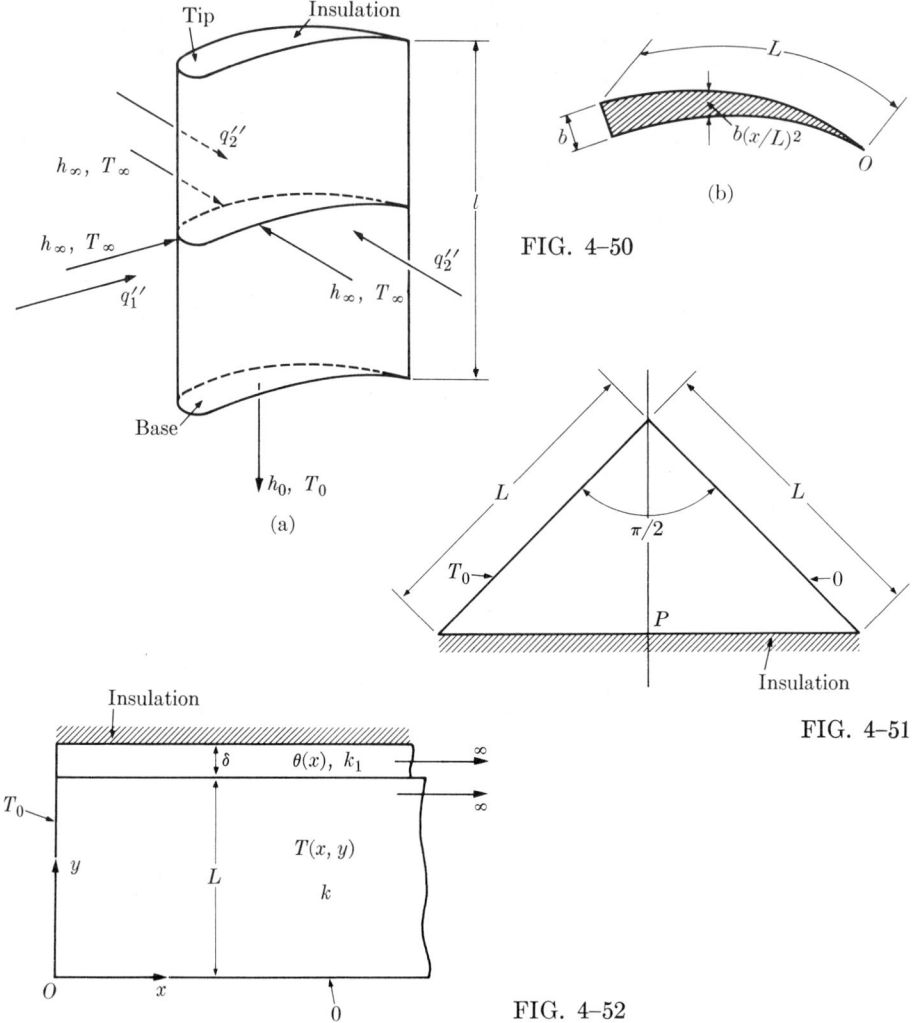

FIG. 4-50

FIG. 4-51

FIG. 4-52

4–10. A blade of a gas turbine is being cooled from its base while receiving convective and radiant heats from combustion products (Fig. 4–50a). The tip of the blade is insulated. The cross section of the blade may be idealized as shown in Fig. 4–50(b). Show the basic steps in the solution of this problem.

4–11. Consider an infinitely long rod of triangular cross section. One surface of the rod has the uniform temperature T_0, the second zero, while the hypotenuse is insulated (Fig. 4–51). Calculate the steady temperature at point P of the hypotenuse.

4–12. A composite wall is made up of two parallel plates of thermal conductivities k_1, k and thicknesses $\delta, L (\delta \ll L)$. The boundary conditions are as shown in Fig. 4–52. We wish to find the steady temperature of the wall, and we proceed according to the following method of solution.

The temperature variation across the thickness of the thin plate is neglected. Then the temperatures $T(x, y)$, $\theta(x)$ of the plates satisfy

$$\frac{\partial^2 T}{\partial x^2} + \frac{\partial^2 T}{\partial y^2} = 0, \qquad (4\text{-}287)$$

$$\frac{d^2\theta}{dx^2} - m\left(\frac{\partial T}{\partial y}\right)_{y=L} = 0, \qquad (4\text{-}288)$$

where $m = (k/k_1\delta)$. *Assuming that the y-direction can be made orthogonal*, we seek a product solution for the thick plate in the form $T(x, y) = X(x)Y(y)$. This gives

$$\frac{d^2 Y}{dy^2} + \lambda^2 Y = 0; \qquad Y(0) = 0, \qquad (4\text{-}289)$$

$$\frac{d^2 X}{dx^2} - \lambda^2 X = 0; \qquad X(\infty) = 0. \qquad (4\text{-}290)$$

The remaining boundary condition of Eq. (4-289) and the nonseparable boundary condition in x will be considered later. The solution of Eq. (4-289) is

$$Y(y) = A\varphi(y), \qquad \varphi(y) = \sin \lambda y, \quad \text{characteristic functions.} \qquad (4\text{-}291)$$

The solution of Eq. (4-290) is

$$X(x) = Be^{-\lambda x}. \qquad (4\text{-}292)$$

Then the product solution of $T(x, y)$ may be written in the form

$$T(x, y) = be^{-\lambda x} \sin \lambda y, \qquad (4\text{-}293)$$

where $b = AB$. Furthermore, since $T(x, L) = \theta(x)$, Eq. (4-288) becomes

$$\lambda^2 T(x, L) - m\frac{\partial T(x, L)}{\partial y} = 0. \qquad (4\text{-}294)$$

Equation (4-294), by means of the product solution, may be rearranged as

$$\lambda^2 Y(L) - m\frac{dY(L)}{dy} = 0, \qquad (4\text{-}295)$$

which is the *second homogeneous boundary condition in the y-direction*. Thus the *y-direction*, involving a homogeneous differential equation and two homogeneous boundary conditions, *satisfies the previously made assumption*, that is, *the orthogonality of this direction*.

The characteristic values are the zeros of the equation obtained by inserting Eq. (4-291) into Eq. (4-295). The result is

$$(\lambda_n L) \sin \lambda_n L = (mL) \cos \lambda_n L. \qquad (4\text{-}296)$$

Hence Eq. (4-293), including all the characteristic values, becomes

$$T(x, y) = \sum_{n=1}^{\infty} b_n e^{-\lambda_n x} \sin \lambda_n y. \qquad (4\text{-}297)$$

Finally, employing the nonseparable base temperature T_0, from Eq. (4–297) we obtain

$$T_0 = \sum_{n=1}^{\infty} b_n \sin \lambda_n y. \qquad (4\text{–}298)$$

This is the expansion of T_0 into a Fourier sine series. The coefficients b_n can be evaluated in the usual manner. The result is

$$b_n = 2T_0 \left(\frac{1 - \cos \lambda_n L}{\lambda_n L - \sin \lambda_n L \cos \lambda_n L} \right). \qquad (4\text{–}299)$$

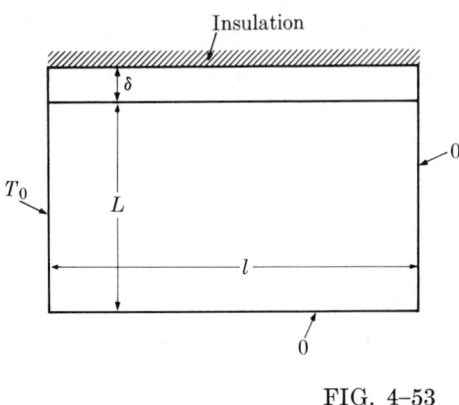

FIG. 4–53

Thus the solution of the thick plate is found to be

$$\frac{T(x,y)}{T_0} = 2 \sum_{n=1}^{\infty} \left(\frac{1 - \cos \lambda_n L}{\lambda_n L - \sin \lambda_n L \cos \lambda_n L} \right) e^{-\lambda_n x} \sin \lambda_n y. \qquad (4\text{–}300)$$

Introducing Eq. (4–300) into Eq. (4–288), integrating the result twice in x, and using the boundary conditions $\theta(0) = T_0$, $\theta(\infty) = 0$ yields the temperature of the thin plate.

The foregoing method of solution involves *one false argument* which results in *a wrong solution given by Eq. (4–300)*. Find this false argument.

4–13. Reconsider Problem 4–12, assuming instead that the length of the plates is finite, say l (Fig. 4–53). Find the steady temperature of the composite wall.

4–14. An infinitely long flat plate of thickness $2L$ moves with the constant velocity V relative to the stationary coordinate system of Fig. 4–54. One half of the x-axis, $x > 0$, is subjected to the uniform heat flux q''; the other half, $x < 0$, is insulated. Approaching the origin of the coordinate system, the plate has a uniform temperature, say zero. (a) Including the effect of axial conduction, find the steady temperature of the plate. (b) Calculate the Nusselt number and the temperature penetration depth in the region $x < 0$. (c) Plot the results of parts (a) and (b) in terms of the appropriate dimensionless numbers. (*Note:* This problem requires a conceptually different form of superposition which may be obtained by physics. If the reader has difficulty seeing the point, the problem should be delayed until Chapter 5 and reconsidered following Example 5–5.)

4–15. A fluid at velocity U flows steadily through a narrow rectangular channel of cross section $(2l \times \delta)$ and length L (Fig. 4–55). The inlet temperature of the fluid is T_0 and the ambient temperature T_∞. The upward and downward heat transfer coefficients are h_1 and h_2, respectively. The sides of the channel are insulated. The conduction in the direction of flow is negligible. The internal energy u''' is generated in the fractional volume $(2b \times \delta \times L)$ of the fluid. Find the steady fluid temperature at the exit of the channel.

4–16. Consider an infinitely long hollow cylinder (Fig. 4–56). The inner and outer radii and surface temperatures are R_i, R_o, $T_i(\varphi)$, $T_o(\varphi)$, respectively. (a) Find the

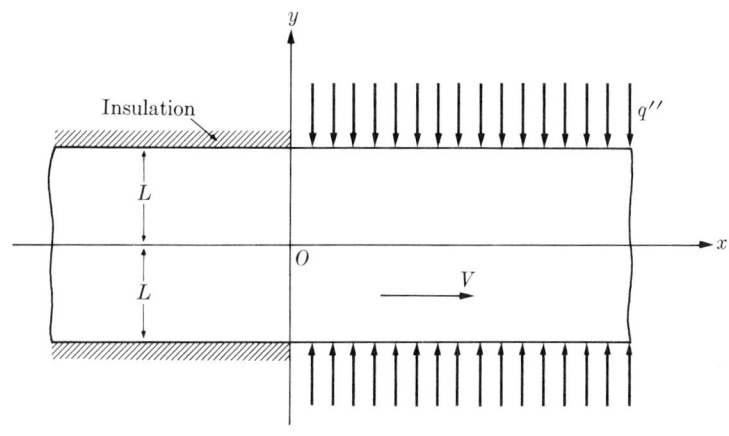

FIG. 4-54

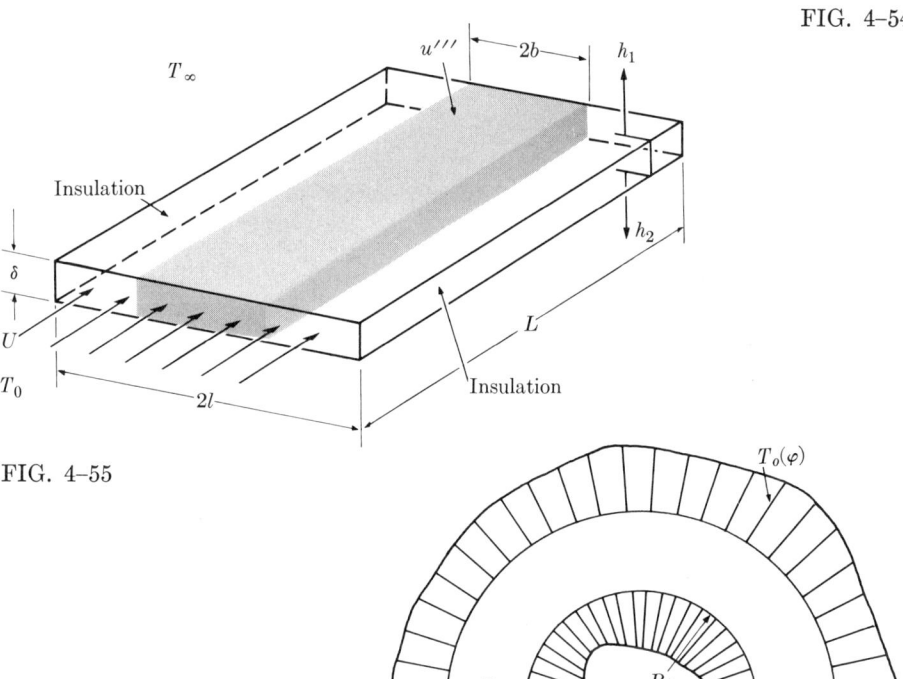

FIG. 4-55

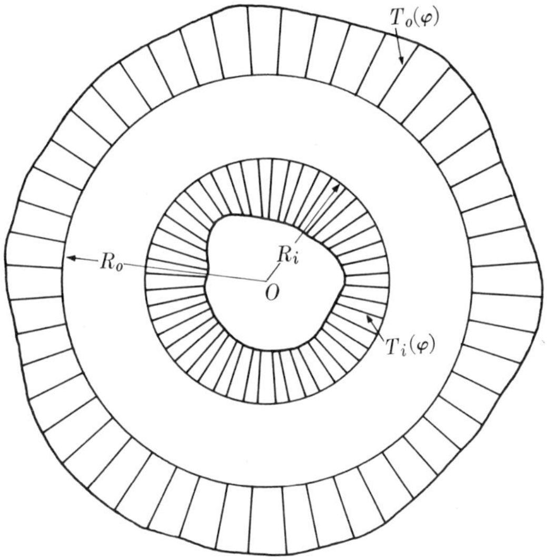

FIG. 4-56

262 STEADY TWO- AND THREE-DIMENSIONAL PROBLEMS

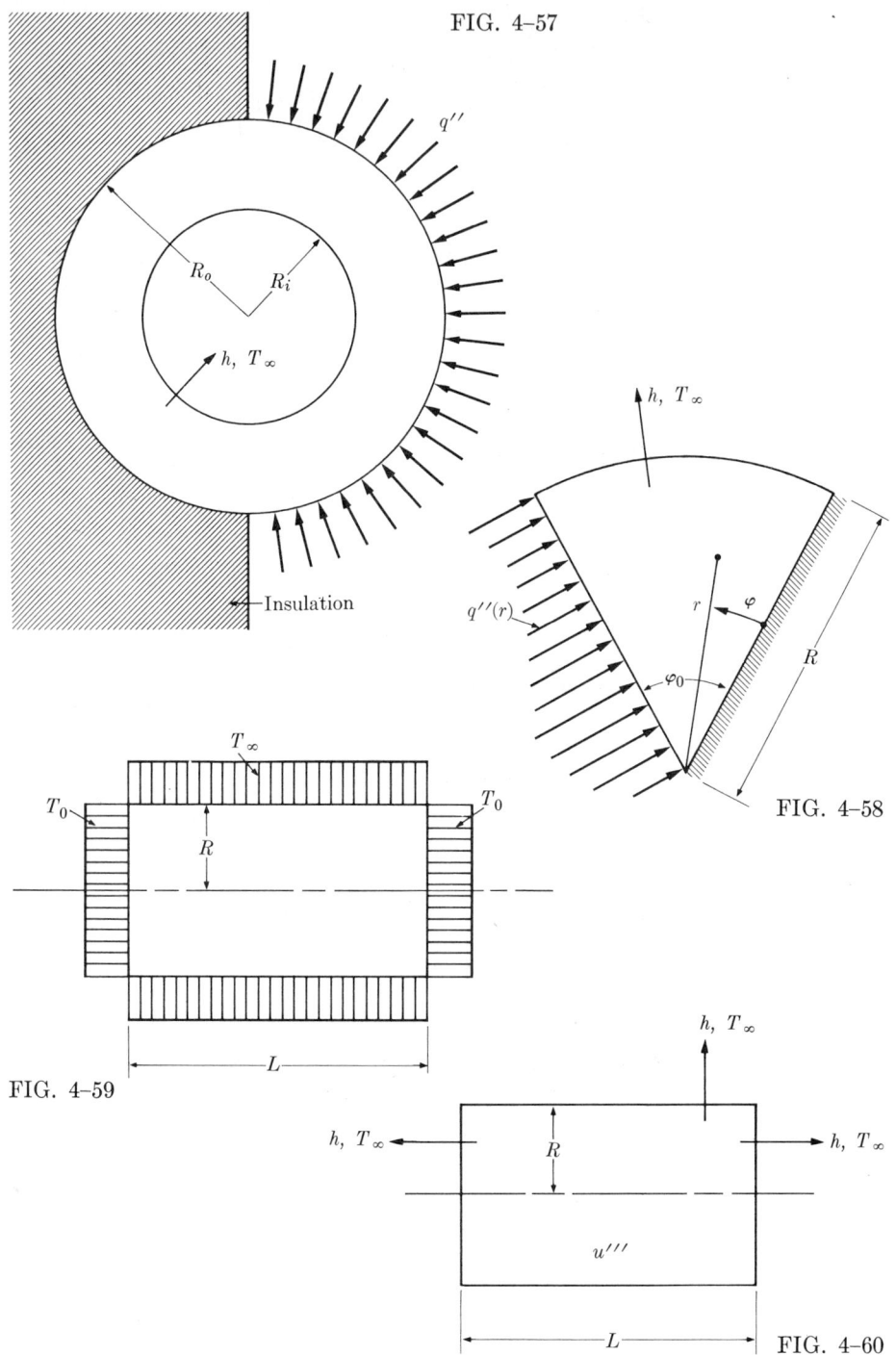

FIG. 4-57

FIG. 4-58

FIG. 4-59

FIG. 4-60

steady temperature of the cylinder. (b) Show that the problems of Example 4–12, the solid rod and the infinite domain with a hole, are special cases of part (a).

4–17. One half of a thick-walled boiler tube receives the uniform heat q'' while the other half is insulated (Fig. 4–57). The inner and outer radii are R_i, R_o, respectively. The temperature of the inside fluid is T_∞, and the inside heat transfer coefficient is large (boiling). Find the steady temperature of the tube.

4–18. Consider an infinitely long rod whose cross section is a sector of a circle of radius R (Fig. 4–58). One side surface of the rod is subjected to the heat flux $q''(r)$ while the other is insulated. The peripheral surface transfers heat to the ambient at temperature T_∞ with the heat transfer coefficient h. Find the particular form of $q''(r)$ for which the problem admits a product solution.

4–19. A finite solid rod of radius R and length L has the uniform temperature T_∞ on its peripheral surface and T_0 at the ends (Fig. 4–59). Find the steady temperature of the rod in terms of the temperatures (a) $\theta = T - T_0$, (b) $\psi = T - T_\infty$.

4–20. The internal energy u''' is generated uniformly in a finite solid rod of radius R and length L (Fig. 4–60). The ambient temperature is T_∞, and the heat transfer coefficient h. Find the steady temperature of the rod.

4–21. A solid rod of radius R and height H is peripherally insulated (Fig. 4–61). The top and the bottom temperatures of the rod are 0 and θ_0, respectively. Find the temperature distribution in the rod by (a) recognizing the physics of the problem, (b) following the formal mathematical procedure.

4–22. A thrust bearing may be idealized by a vertical rod of radius R rotating with a constant angular velocity ω on a base (Fig. 4–62). The thermal conductivity of the base is low compared with that of the rod. The unnoticed failure of the lubrication pump creates conditions of dry friction between the rod and the base. The interface pressure is p, the coefficient of friction μ, and the wear constant. *Note that the pressure exists only on the area* $\pi(R^2 - R_o^2)$. Find the ultimate temperature of the bearing.

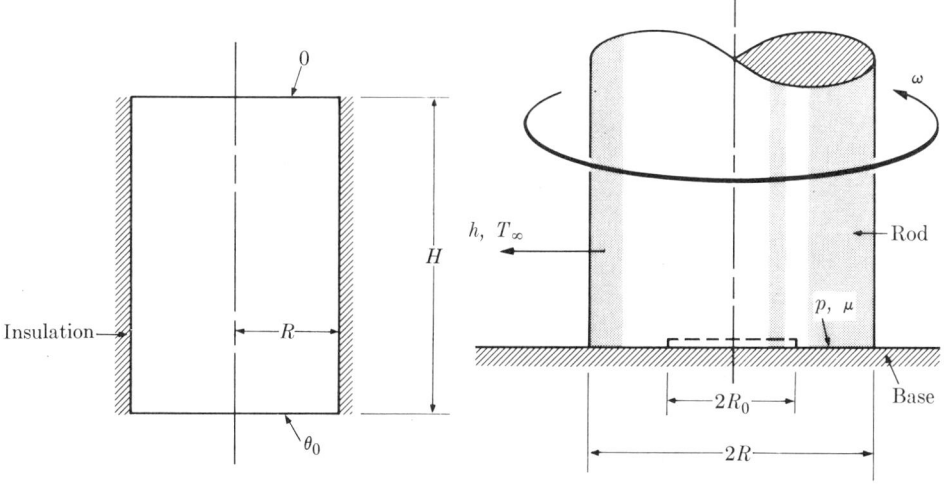

FIG. 4–61   FIG. 4–62

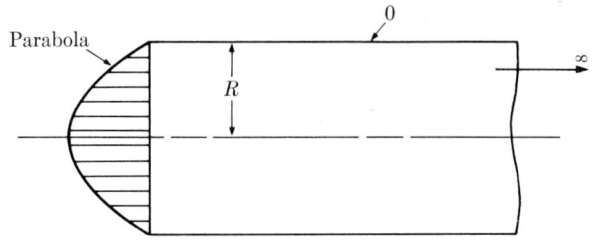

FIG. 4-63

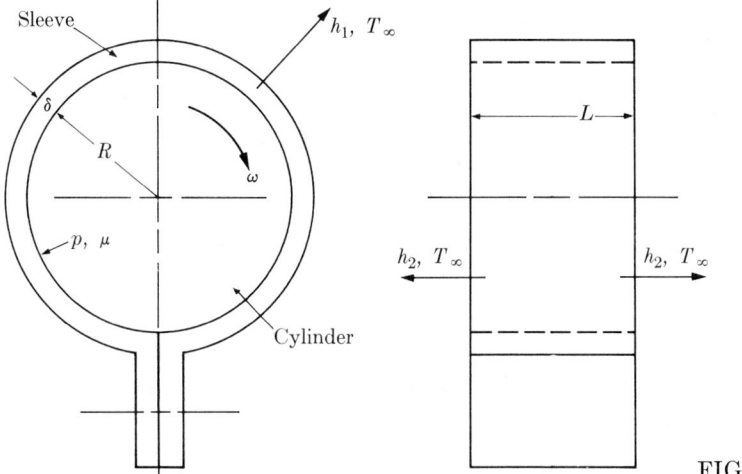

FIG. 4-64

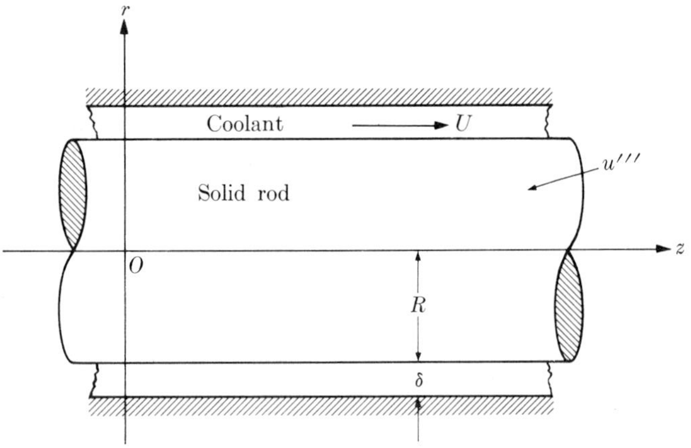

FIG. 4-65

4–23. A semi-infinite solid rod of radius R is subjected to a parabolic temperature distribution at one end while the peripheral surface is maintained at constant temperature, say zero (Fig. 4–63). (a) Using the integral method, find a first and second approximation to the steady temperature of the rod in terms of the Ritz and Kantorovich profiles. (b) Compare the results with those of Problem 4–7.

4–24. A solid cylinder of radius R and length L rotates with the constant angular velocity ω in a sleeve of thickness $\delta(\ll R)$ and of the same length as the cylinder (Fig. 4–64). The pressure p and the coefficient of dry friction μ between the cylinder and sleeve are assumed to be uniform. The peripheral and axial heat transfer coefficients are h_1 and h_2, respectively. The ambient temperature is T_∞. Find the steady temperature of the system.

4–25. A heterogeneous nuclear reactor is simulated by a solid rod along which the coolant flows coaxially (Fig. 4–65). The following assumptions are made: (a) The outward heat loss from the coolant may be neglected. (b) The coolant velocity is approximately uniform (slug flow). (This is particularly true for liquid metals, which have small Prandtl numbers). (c) The internal energy generation u''' in the rod is uniform. (In the actual case, the energy generation, being proportional to the neutron flux, is distributed cylindrically in the radial and hyperbolically in the axial directions.) (d) The coolant and rod have comparable thermal conductivities. (e) Axial conduction in both the coolant and the rod is negligible. (f) The radial thickness δ of the coolant is much less than the radius R of the rod. (g) The coolant inlet temperature is uniform, say T_0. Find the steady temperature of the system.

4–26. An infinitely long solid rod of radius R moves with constant velocity V relative to the stationary coordinate system of Fig. 4–66. One half of the z-axis, $z > 0$, is subjected to the uniform heat flux q'', while the other half, $z < 0$, is insulated. Approaching the origin of the coordinate system, the rod has a uniform temperature, say zero. (a) Including the effect of axial conduction, find the steady temperature of the rod. (b) Calculate the Nusselt number and the penetration depth of the temperature in the region $z < 0$. (c) Plot the results of parts (a) and (b) in terms of the appropriate dimensionless numbers. (d) Show the effect of curvature on the problem and compare the foregoing results to those of Problem 4–14. (*Note:* The solution of this problem depends on the reader's success on Problem 4–14.)

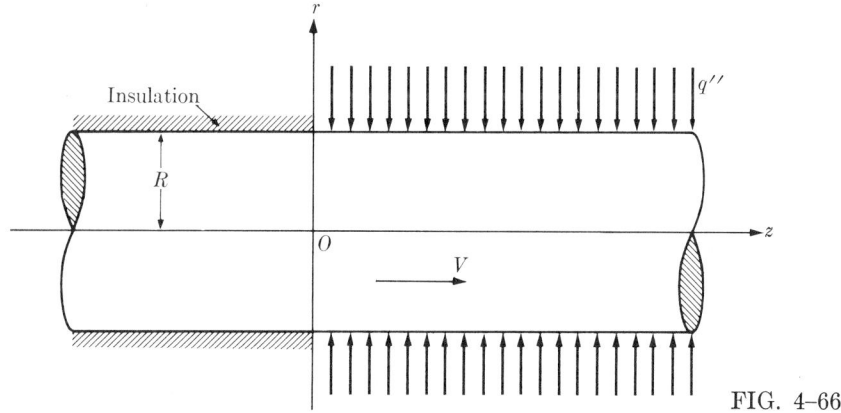

FIG. 4–66

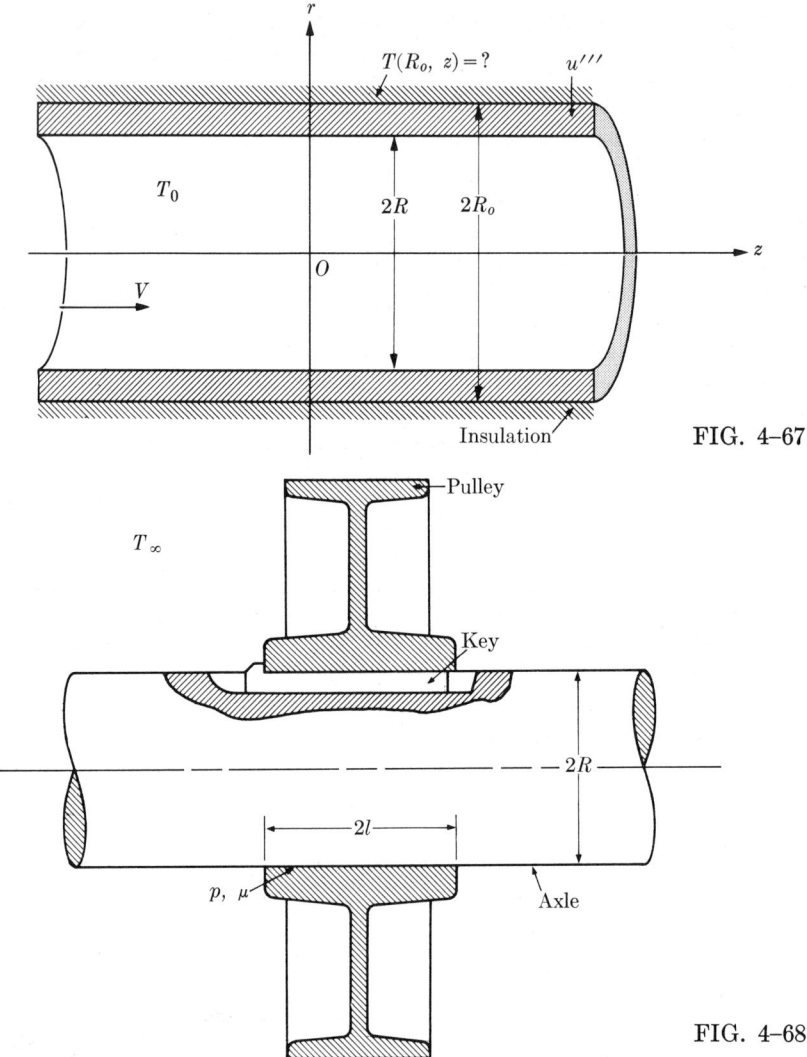

FIG. 4-67

FIG. 4-68

4-27. An incompressible fluid flows steadily through an insulated tube (Fig. 4-67). The constant internal energy u''' is uniformly generated in the downstream half of the tube. The inner and outer radii of the tube are R and R_o, respectively. The mean velocity and the upstream temperature of the fluid are V and T_0. *Axial conduction may be neglected both in the fluid and the tube walls.* Find the steady temperature of the outer surface of the tube. (*Note:* This problem is conceptually identical to Problem 4-14.)

4-28. Because of a sudden load increase in an industrial transmission, a pulley shears off its key and starts rotating over its axle with a uniform angular velocity ω (Fig. 4-68). Assume that the heat capacity of the pulley is negligible compared to that of the axle, the pressure between the pulley and the axle is uniform, say p, and the coefficient of dry friction is μ. Find the temperature distribution in the axle.

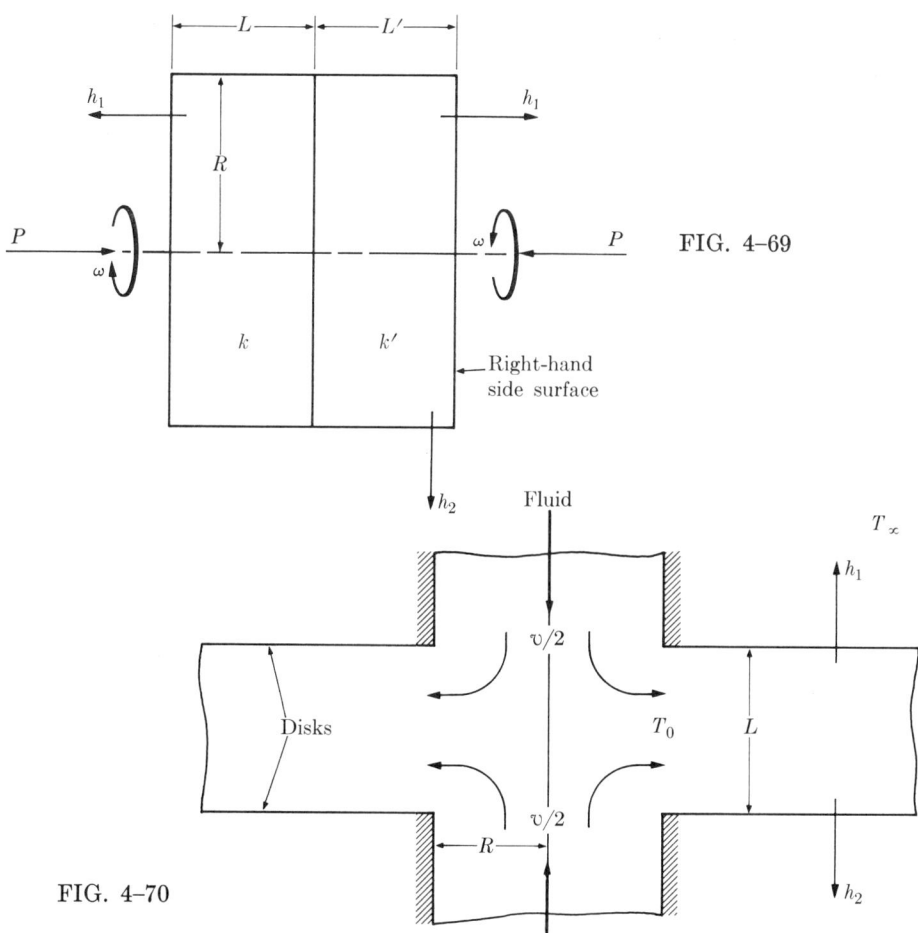

FIG. 4-69

FIG. 4-70

4-29. Two solid cyclinders are rotated reversely by a constant angular velocity ω and are pressed together by a force P (Fig. 4-69). The radius, thickness, and thermal conductivity of the cylinders are R, (L, L'), and (k, k'), respectively. The side and peripheral heat transfer coefficients are h_1 and h_2. The ambient temperature is T_∞. Which of the following cases can be solved by the techniques of Chapter 4? (a) $k = k'$ and $L = L'$; (b) $k = k'$ and $L \neq L'$; (c) $k \neq k'$ and $L = L'$; (d) $L' \ll L$ or $k' \gg k$; (e) the right-hand side surface is insulated; (f) one cylinder is stationary. (Note that the problem involves modified forms of Example 2-4.)

4-30. An incompressible fluid flows steadily and radially between two concentric parallel disks of negligible thickness (Fig. 4-70). The distance between the disks is L, the volumetric flow rate is $\mathcal{V}$, the thermal conductivity of the fluid k, the ambient temperature T_∞, the inner radius of the disks R. The upward and downward heat transfer coefficients are h_1 and h_2, respectively. The inlet temperature of the disks is assumed to be uniform, say T_0. Neglecting the axial distribution of the velocity but not of the temperature, find the steady temperature of the fluid. (Note that this problem is the two-dimensional form of Problem 3-33.)

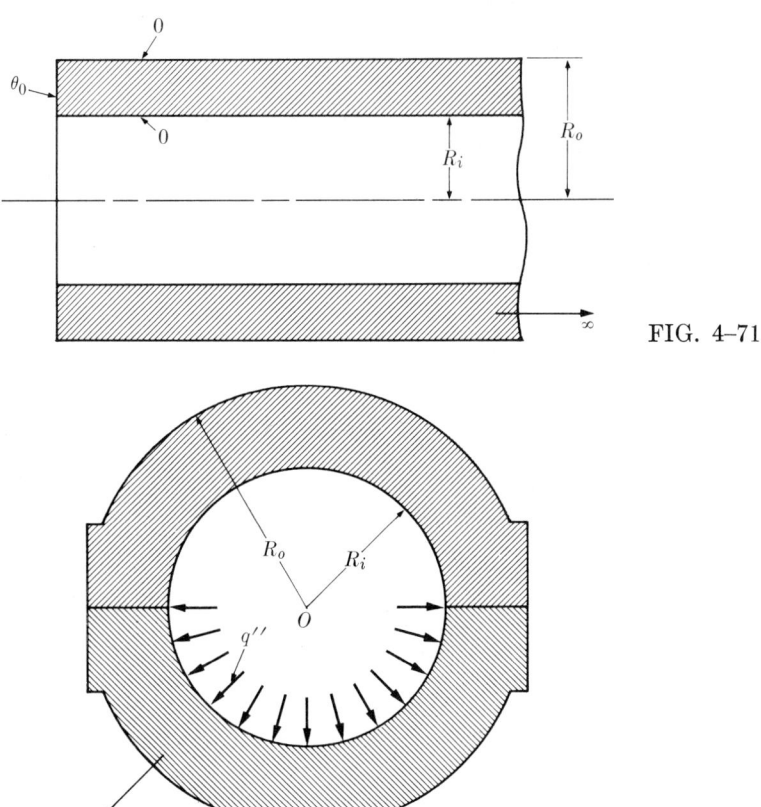

FIG. 4-71

FIG. 4-72

4-31. Extend the theory of Fourier-Bessel series, Section 4-9, to a hollow cylinder. Apply the results to the following problem. The inner and outer surfaces of a semi-infinite, hollow cylinder of radii R_i, R_o (Fig. 4-71) are kept at zero temperature. The base temperature is uniform, say θ_0. Find the steady temperature of the cylinder.

4-32. A ripening apple on its tree is assumed to be a sphere of radius R receiving the net radiant heat from the sun in the form

$$q''(\varphi) = \begin{cases} q_0'' \sin \varphi, & 0 < \varphi < \pi, \\ 0, & \pi < \varphi < 2\pi, \end{cases}$$

while losing heat by convection to the ambient at temperature T_∞ (see Fig. 4-33). The heat transfer coefficient is h. Assuming that the problem is quasi-steady, that is, neglecting the rotation of the earth, find the steady temperature of the apple. (Note that the physics of this problem is identical to that of Example 4-13.)

4-33. A high-pressure combustion vessel is made up of two thick-walled hemispherical shells (Fig. 4-72). The inner and outer radii of these are R_i and R_o, respectively. During a combustion process the inner surface of the shells is subjected to constant radiant heat flux q'', while the outer surface transfers heat to the ambient at

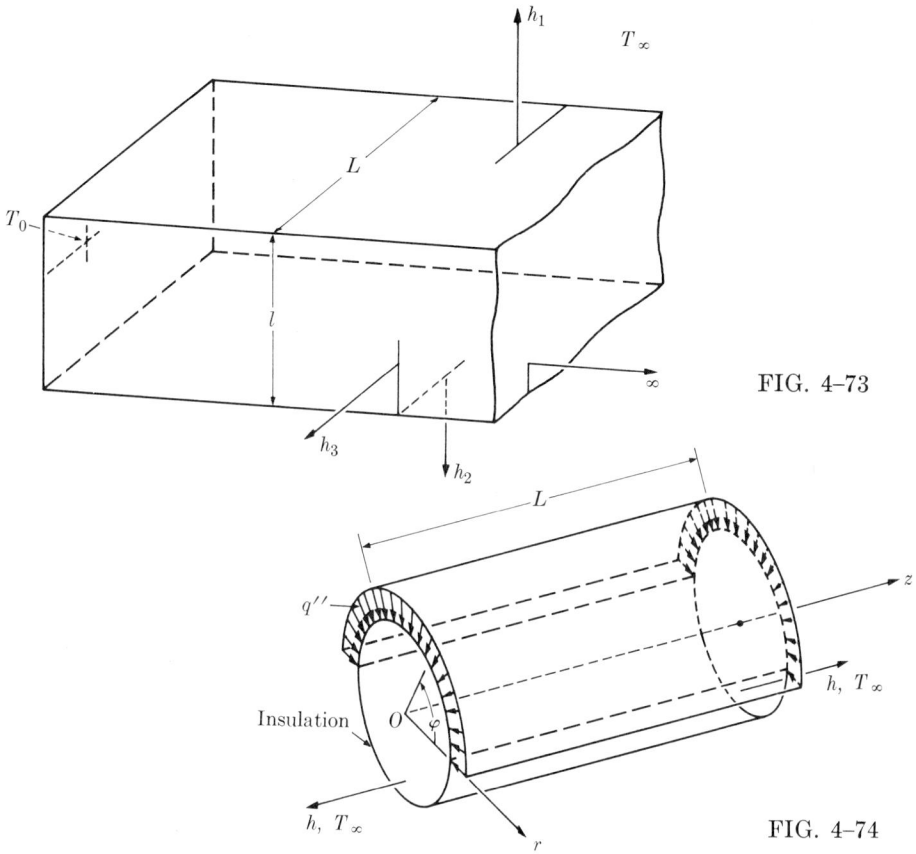

FIG. 4-73

FIG. 4-74

temperature T_∞. The heat transfer coefficient is h. The hemispherical shells have the same conductivity but different inner surface emissivity such that one shell receives negligible radiant heat compared with the other. Calculate the steady temperature of the vessel.

4-34. The base temperature T_0 of a semi-infinite rod of rectangular cross section ($L \times l$) is specified, while the sides transfer heat to the ambient at temperature T_∞ (Fig. 4-73). The upward, downward, and side heat transfer coefficients are h_1, h_2, and h_3, respectively. Find the steady temperature of the rod. (Note that the problem is the three-dimensional form of Examples 3-9 and 4-3.)

4-35. The uniform heat q'' is applied to one half of the peripheral surface of a finite rod of radius R and length L (Fig. 4-74). The other half of the periphery is insulated. Heat is transferred from the ends of the rod to the ambient with the heat transfer coefficient h. The ambient temperature is T_∞. Calculate the steady temperature of the rod.

CHAPTER 5

SEPARATION OF VARIABLES. UNSTEADY PROBLEMS.
ORTHOGONAL FUNCTIONS

In this chapter we essentially consider the unsteady problems of Chapter 2 as well as the steady problems of Chapters 3 and 4, now modified to include the effect of time. Recall that we have already obtained solutions to these problems in their lumped and integral formulations; hence we devote the chapter primarily to solutions for differential formulations. Since no further mathematics is needed, the content of the chapter is somewhat more applied in nature than that of the preceding chapters; except for a conceptual point or two (see Examples 5-5 and 5-6, and Section 5-3), it merely involves the solution of a number of illustrative examples.

In Section 2-2 we classified problems with respect to their dependence on space (as lumped or distributed), and since then we have formulated them accordingly. Now we may also classify these problems with respect to their dependence on time (as transient or periodic). Thus we have:

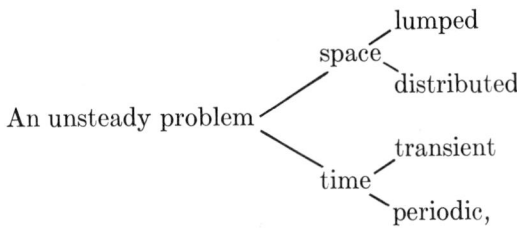

where each transient or periodic problem involves a *starting*, a *steady*, and an *ending* time interval as shown in Fig. 5-1.

A typical example of the transient case is the heating (or cooling) of ingots. From the variation of the temperature we can predict the time required for such bodies to attain predetermined temperature levels for purposes of melting, hot-working, heat treatment, and so on. The periodic case, on the other hand, is illustrated by the temperature fluctuations in the walls of engines. These fluctuations, although periodic, may not be harmonic; however, the linearity of conduction problems allows us to express any periodic disturbance in terms of its harmonics, and reduces the problem to one having harmonic disturbances.

As indicated in Chapter 2, convective heat transfer through boundaries is important in the formulation and solution of conduction problems. The dimensionless form of the boundary condition yields the Biot number, $\mathrm{Bi} = hL/k$, which is the ratio of external conductance to internal conductance. When the *internal resistance is negligible* (or when the internal conductance is large), $k/L \to \infty$ and $\mathrm{Bi} \to 0$. This case corresponds to a *small L* or *large k*, and permits the omission of the spatial temperature variation perpendicular to the boundary having this condition; thus it leads to a *lumped system* analysis. When the *external resistance is negligible* (or when the external conductance is large), which is the case in boiling, condensation, and highly turbulent flows, $h \to \infty$ and $\mathrm{Bi} \to \infty$. This implies that *boundary temperature* $\to$ *ambient temperature*. Hence the general boundary condition is simplified to that of a specified boundary temperature that is assumed to be approximately equal to the ambient temperature. When the *internal and external resistances* (or conductances) *are comparable*, the general boundary condition cannot be simplified and the problem must be solved in terms of this condition. This and the previous case require the *distributed system* approach.

Thus unsteady problems, like steady problems, are formulated according to a lumped or a distributed system analysis. The lumped model, which depends only on time as an independent variable, yields an *initial-value problem*.*

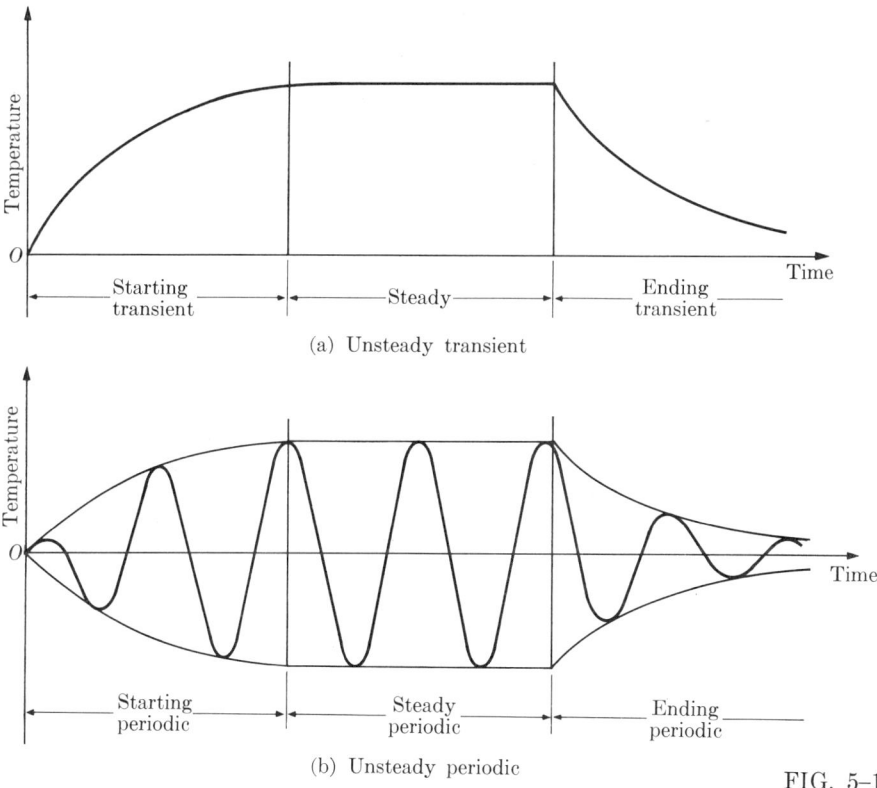

FIG. 5-1

---

* See Section 4-1 for the definition of initial-value and boundary-value problems.

This may readily be solved by employing the methods of ordinary first-order differential equations. The distributed model, on the other hand, depends on both time and space and hence results in an *initial- and boundary-value problem*. We can, by the method of separation of variables, reduce this to a characteristic-value problem in each space variable involved, to be solved by the techniques learned in Chapter 4.

Let us now consider a number of problems involving both lumped and distributed systems. We begin with a lumped problem.

Example 5–1. Reconsider the hot plate of Example 2–3. The thickness L of the plate is small enough that the temperature variation across the plate can be neglected. The starting transient case, resulting from the sudden application of the heat flux q'' to the bottom of the plate, was treated in Example 2–3. There we obtained

$$\frac{T - T_\infty}{q''/h} = 1 - e^{-mt}, \qquad (2\text{–}172)$$

where $m = h/\rho cL$. The ending transient corresponding to sudden removal of the heat flux q'', which will not be derived here, may readily be obtained in the form

$$\frac{T - T_\infty}{q''/h} = e^{-mt}.$$

The unsteady periodic case, however, deserves some attention. Let the heat flux q'', applied suddenly to the bottom of the plate, oscillate as $q'' \cos \omega t$. Here ω is the angular frequency. The formulation of the problem gives

$$\frac{d\theta}{dt} + m\theta = n \cos \omega t, \qquad (5\text{–}1)$$

$$\theta(0) = 0, \qquad (5\text{–}2)$$

where $\theta = T - T_\infty$, $m = h/\rho cL$, and $n = q''/\rho cL$. Equation (5–1) may be solved, for example, by the method of variation of parameters. The result that is satisfied by Eq. (5–2) is

$$\frac{T - T_\infty}{q''/h} = \frac{m}{(m^2 + \omega^2)^{1/2}} \cos(\omega t - \alpha) - \frac{m^2 e^{-mt}}{(m^2 + \omega^2)}, \qquad (5\text{–}3)$$

where $\alpha = \tan^{-1}(\omega/m)$. Equation (5–3) is the starting periodic solution of the problem. As $t \to \infty$ the second term of this equation vanishes, and the equation approaches asymptotically the steady periodic solution.

Let us now solve some distributed problems. The method of separation of variables, which may be employed conveniently with problems having stepwise disturbances, is not suitable to those having transient or periodic disturbances. Even the steady periodic solutions of the latter may be obtained for only a few simple cases and after a lengthy procedure (see Example 5–6). Other and more convenient techniques are, however, available. Thus the response of distributed systems to any transient or periodic disturbance may be obtained by *Duhamel's integral* (Section 5-3), steady periodic solutions by the method of *complex temperature* (Chapter 6), and starting periodic solutions by means of *Laplace transforms* (Chapter 7).

5-1. Distributed Systems Having Stepwise Disturbances

For distributed systems having stepwise disturbances, no mathematics beyond that introduced in Chapter 4 is needed. Difficulties arising from nonhomogeneous boundary conditions or differential equations may be eliminated, as with steady problems, by introducing a *change in temperature level* or by *superposition*. Some remarks on nonhomogeneities may, however, be helpful. In this connection, let us consider the following four problems.

Example 5-2. A plate of thickness $2L$ having the uniform initial temperature T_0 is plunged suddenly into a bath at the constant temperature T_∞. The heat transfer coefficient is large. We wish to find the unsteady temperature of the plate.

In terms of $\theta = T - T_\infty$ and the x-axis of Fig. 5-2, the formulation of the problem is

$$\frac{\partial \theta}{\partial t} = a \frac{\partial^2 \theta}{\partial x^2}, \qquad \theta(x, 0) = \theta_0 = T_0 - T_\infty,$$

$$\frac{\partial \theta(0, t)}{\partial x} = 0, \qquad \theta(L, t) = 0.$$

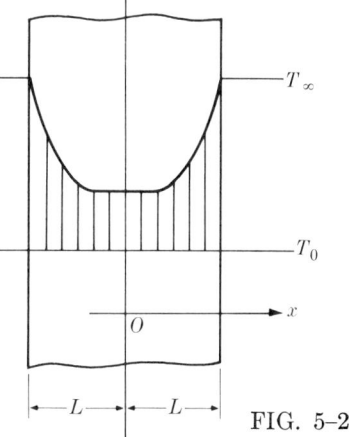

FIG. 5-2

Let us note that only the x-direction yields a characteristic-value problem; then, with the proper choice of separation constant, the product solution $\theta(x, t) = X(x)\tau(t)$ gives

$$\frac{d^2 X}{dx^2} + \lambda^2 X = 0; \qquad \frac{dX(0)}{dx} = 0, \quad X(L) = 0, \tag{5-4}$$

$$\frac{d\tau}{dt} + a\lambda^2 \tau = 0. \tag{5-5}$$

Here, as with steady problems, the nonhomogeneous (initial) condition is left to the end of the problem. This condition will be used when the coefficients of the resulting series expansion are to be determined.

The solution of Eq. (5-4) is

$$X_n(x) = A_n \varphi_n(x), \qquad \varphi_n(x) = \cos \lambda_n x, \qquad \text{characteristic functions,}$$
$$\lambda_n L = (2n + 1)\pi/2, \qquad n = 0, 1, 2, 3, \ldots, \qquad \text{characteristic values,}$$

and the solution of Eq. (5–5) is

$$\tau_n(t) = C_n e^{-a\lambda_n^2 t}. \tag{5-6}$$

Hence the product solution becomes

$$\theta(x, t) = \sum_{n=0}^{\infty} a_n e^{-a\lambda_n^2 t} \cos \lambda_n x, \tag{5-7}$$

where $a_n = A_n C_n$.

Finally, introducing the initial condition, $\theta(x, 0) = \theta_0$, into Eq. (5–7) gives

$$\theta_0 = \sum_{n=0}^{\infty} a_n \cos \lambda_n x. \tag{5-8}$$

Equation (5–8) is the Fourier cosine series expansion of θ_0 over the interval $(0, L)$. The coefficient a_n may be evaluated in the usual manner [see Eqs. (4–69) and (4–70)]. The result is

$$a_n = (-1)^n \frac{2\theta_0}{\lambda_n L}. \tag{5-9}$$

Introducing Eq. (5–9) into Eq. (5–7), we find the unsteady temperature of the plate to be

$$\frac{T(x, t) - T_\infty}{T_0 - T_\infty} = 2 \sum_{n=0}^{\infty} \frac{(-1)^n}{\lambda_n L} e^{-a\lambda_n^2 t} \cos \lambda_n x. \tag{5-10}$$

Example 5–3. We wish to re-solve Example 5–2, given a moderate heat transfer coefficient h.

Except for the surface boundary condition, the formulation of the problem remains the same. Hence the characteristic-value problem in the x-direction, Eq. (5–4), is modified by means of the new condition $-k\,[\partial\theta(L, t)/\partial x] = h\theta(L, t)$ to give

$$\frac{d^2 X}{dx^2} + \lambda^2 X = 0; \quad \frac{dX(0)}{dx} = 0, \quad \frac{dX(L)}{dx} + \frac{h}{k} X(L) = 0. \tag{5-11}$$

The solution of Eq. (5–11) is

$$X_n(x) = A_n \varphi_n(x), \quad \varphi_n(x) = \cos \lambda_n x, \quad \text{characteristic functions,}$$

and the zeros of $(\lambda_n L) \sin \lambda_n L = \text{Bi} \cos \lambda_n L$, where $\text{Bi} = hL/k$, are the characteristic values (see Fig. 4–13).

The form of the product solution remains the same except for n, which now starts from $n = 1$. Thus we have

$$\theta(x, t) = \sum_{n=1}^{\infty} a_n e^{-a\lambda_n^2 t} \cos \lambda_n x. \tag{5-12}$$

Here we have, for the coefficient a_n,*

$$a_n = \frac{2\theta_0 \sin \lambda_n L}{\lambda_n L + \sin \lambda_n L \cos \lambda_n L}.$$

Hence the unsteady temperature of the plate is

$$\frac{T(x, t) - T_\infty}{T_0 - T_\infty} = 2 \sum_{n=1}^{\infty} \left(\frac{\sin \lambda_n L}{\lambda_n L + \sin \lambda_n L \cos \lambda_n L} \right) e^{-a\lambda_n^2 t} \cos \lambda_n x. \quad (5\text{-}13)$$

The last two examples, Example 5–2 with a surface temperature $T(L, t) = T_\infty$ and Example 5–3 with heat transfer to the ambient $-k[\partial T(L, t)/\partial t] = h[T(L, t) - T_\infty]$, were subject to nonhomogeneous boundary conditions which, as we know, are not suitable to the use of separation of variables. However, the transformation $\theta = T - T_\infty$ converted these conditions into homogeneous conditions, and thus eliminated the arising difficulties. The nonhomogeneity of the next two examples, on the other hand, cannot be handled by such a transformation.

Example 5–4. We wish to solve the differential formulation of Example 2–2, which we left untreated in Chapter 2. The heat transfer coefficient is assumed to be large (Fig. 5–3).

The formulation of the problem in terms of $\theta = T - T_\infty$ is

$$\frac{\partial \theta}{\partial t} = a \frac{\partial^2 \theta}{\partial x^2} + \frac{u'''}{\rho c}, \quad \theta(x, 0) = 0,$$

$$\frac{\partial \theta(0, t)}{\partial x} = 0, \quad \theta(L, t) = 0.$$

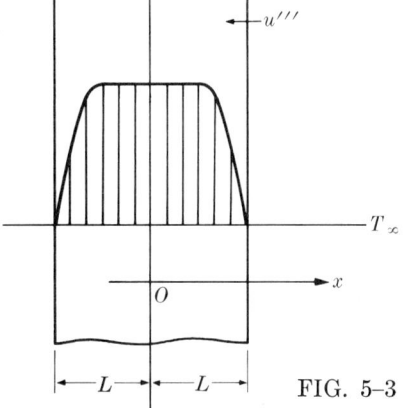

FIG. 5-3

Note that the shift in the temperature level has reduced the nonhomogeneous (surface) boundary condition to a homogeneous condition without affecting the nonhomogeneity of the differential equation. To eliminate the difficulty arising from the latter nonhomogeneity we now express the temperature of the plate as the sum of two temperatures in the form

$$\theta(x, t) = \psi(x, t) + \phi(x) \quad (5\text{-}14)$$

(see Section 4–7 and Example 4–10) so that the energy generation is included in the formulation of the one-dimensional (steady) problem. Hence $\phi(x)$ satisfies

$$\frac{d^2 \phi}{dx^2} + \frac{u'''}{k} = 0, \quad \frac{d\phi(0)}{dx} = 0, \quad \phi(L) = 0, \quad (5\text{-}15)$$

---

* See Eq. (4–78).

and in terms of $\phi(x)$, the formulation of the two-dimensional (unsteady) problem $\psi(x, t)$ becomes*

$$\frac{\partial \psi}{\partial t} = a \frac{\partial^2 \psi}{\partial x^2}, \qquad \psi(x, 0) = -\phi(x),$$

$$\frac{\partial \psi(0, t)}{\partial x} = 0, \qquad \psi(L, t) = 0. \tag{5-16}$$

The solution of Eq. (5-15) is

$$\frac{\phi(x)}{u'''L^2/k} = \frac{1}{2}\left[1 - \left(\frac{x}{L}\right)^2\right]. \tag{5-17}$$

The product solution $\psi(x, t) = X(x)\tau(t)$ applied to Eq. (5-16) results in Eqs. (5-4) and (5-5) of Example 5-2. However, the coefficient a_n of the product solution

$$\psi(x, t) = \sum_{n=0}^{\infty} a_n e^{-a\lambda_n^2 t} \cos \lambda_n x$$

must be determined from the Fourier cosine series,

$$-\phi(x) = \sum_{n=0}^{\infty} a_n \cos \lambda_n x, \tag{5-18}$$

where $\phi(x)$ is given by Eq. (5-17). The value of this coefficient is

$$a_n = -(-1)^n \left(\frac{u'''L^2}{k}\right) \frac{2}{(\lambda_n L)^3}. \tag{5-19}$$

Hence the unsteady temperature of the plate is

$$\frac{\theta(x, t)}{u'''L^2/k} = \frac{1}{2}\left[1 - \left(\frac{x}{L}\right)^2\right] - 2 \sum_{n=0}^{\infty} \frac{(-1)^n}{(\lambda_n L)^3} e^{-a\lambda_n^2 t} \cos \lambda_n x. \tag{5-20}$$

This problem may readily be extended to the case of a finite heat transfer coefficient h (see Problem 5-8).

Note that the one-dimensional mathematical function $\phi(x)$ is the actual (physical) steady temperature of the plate, while the two-dimensional mathematical function $\psi(x, t)$ is the instantaneous difference between the steady and

---

* Note that the transformation $\theta = T - T_\infty$ is not necessary for the solution of the problem. Eq. (5-14) would have to be replaced then by $T(x, t) = \psi(x, t) + \phi(x)$, and the formulation of $\phi(x)$ would include not only the nonhomogeneity of the differential equation but also the nonhomogeneity of the surface boundary condition as $\phi(L) = T_\infty$. The formulation of $\psi(x, t)$ would remain the same except for the initial condition, which would become $\psi(x, 0) = T_\infty - \phi(x)$.

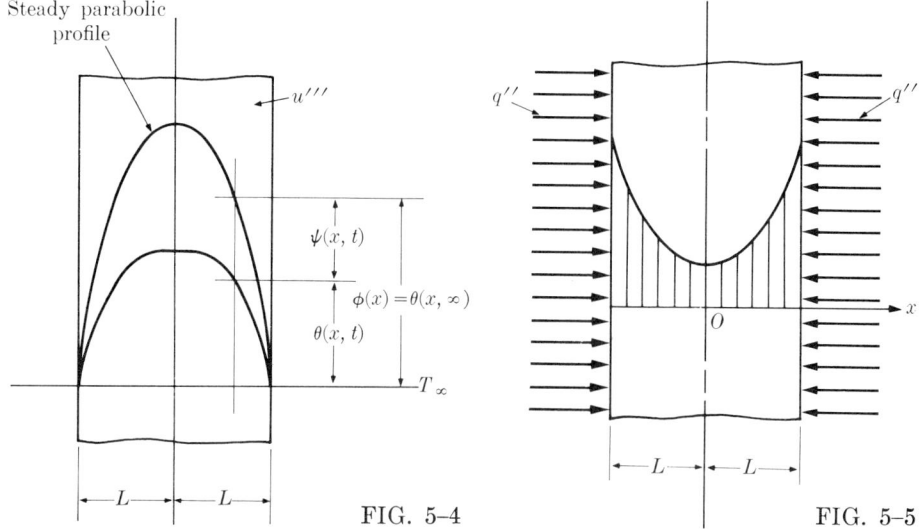

FIG. 5-4   FIG. 5-5

unsteady temperatures of the plate. Figure 5-4 illustrates the relation among the temperatures $\theta(x, t)$, $\psi(x, t)$, and $\phi(x)$, which are the terms of the superposition given by Eq. (5–14).

The next problem answers the question whether the superposition given by the relation $\theta(x, t) = \psi(x, t) + \phi(x)$ can always eliminate the difficulty arising from a nonhomogeneity.

Example 5–5. The constant heat flux q'' is applied to both surfaces of a flat plate of thickness $2L$ (Fig. 5–5). The initial temperature of the plate is T_∞. We wish to find the unsteady temperature of the plate.

The formulation of the problem in terms of $\theta = T - T_\infty$ is

$$\frac{\partial \theta}{\partial t} = a \frac{\partial^2 \theta}{\partial x^2}, \qquad \theta(x, 0) = 0,$$

$$\frac{\partial \theta(0, t)}{\partial x} = 0, \qquad +k \frac{\partial \theta(L, t)}{\partial x} = q''.$$

Let us try to solve the problem by the assumption of Eq. (5–14) that

$$\theta(x, t) = \psi(x, t) + \phi(x). \tag{5-14}$$

The separation constant that is selected to make the x-direction of $\psi(x, t)$ a characteristic-value problem forces $\psi(x, t)$ to be an exponentially decaying function in time; hence in the limit $\psi(x, t) \to 0$ and $\theta(x, t) \to \phi(x)$ as $t \to \infty$.*

---

* Furthermore, note that $\phi(x)$ with one boundary insulated and the other subjected to a prescribed heat flux cannot occur.

This result violates the physics of the problem; the temperature of the plate should increase without limit as $t \to \infty$. To satisfy this condition, we modify Eq. (5-14) by the addition of the term $\varphi(t)$ such that $\varphi(t) \to \infty$ as $t \to \infty$. Thus the proper assumption is

$$\theta(x, t) = \psi(x, t) + \phi(x) + \varphi(t). \dagger \qquad (5\text{-}21)$$

Now in terms of Eq. (5-21), the formulation of the problem becomes

$$\frac{\partial \psi}{\partial t} = a \frac{\partial^2 \psi}{\partial x^2}, \qquad \psi(x, 0) = -\phi(x) - \varphi(0),$$

$$\frac{\partial \psi(0, t)}{\partial x} = 0, \qquad \frac{\partial \psi(L, t)}{\partial x} = 0, \qquad (5\text{-}22)$$

and

$$\frac{d\varphi}{dt} = a \frac{d^2 \phi}{dx^2}, \qquad (5\text{-}23)$$

$$\frac{d\phi(0)}{dx} = 0, \qquad +k \frac{d\phi(L)}{dx} = q''. \qquad (5\text{-}24)$$

Since $\varphi(t)$ and $\phi(x)$ can vary independently, Eq. (5-23) holds when it is equal to a constant, say C. Then the general solution of Eq. (5-23) is obtained in the form

$$\varphi(t) = aCt + C_1, \qquad (5\text{-}25)$$

$$\phi(x) = \tfrac{1}{2} C x^2 + C_2 x + C_3. \qquad (5\text{-}26)$$

Here C_2 and C may readily be evaluated by introducing Eq. (5-26) into Eq. (5-24). The result is $C_2 = 0$, $C = q''/kL$. Hence Eqs. (5-25) and (5-26) become $\varphi(t) = (q''t/\rho cL) + C_1$, $\phi(x) = (q''x^2/2kL) + C_3$, where C_1 and C_3 are the remaining constants. However, noting that the solution of $\psi(x, t)$ depends on $\varphi(t)$ and $\phi(x)$, we may arbitrarily set these constants equal to zero. Thus

$$\varphi(t) = \frac{q''t}{\rho cL}, \qquad \phi(x) = \frac{q''x^2}{2kL}. \qquad (5\text{-}27)$$

On the other hand, the product solution $\psi(x, t) = X(x)\tau(t)$ applied to Eq. (5-22) results in

$$\frac{d^2 X}{dx^2} + \lambda^2 X = 0; \qquad \frac{dX(0)}{dx} = 0, \qquad \frac{dX(L)}{dx} = 0, \qquad (5\text{-}28)$$

$$\frac{d\tau}{dt} + a\lambda^2 \tau = 0. \qquad (5\text{-}29)$$

---

† A similar superposition which has the form $\theta(x, y) = \psi(x, y) + \phi(y) + \varphi(x)$ should be used for Problem 4-14.

The solution of Eq. (5–28) is

$$X_n(x) = A_n \varphi_n(x), \quad \varphi_n(x) = \cos \lambda_n x, \quad \text{characteristic functions,}$$
$$\lambda_n L = n\pi, \quad n = 0, 1, 2, 3, \ldots, \quad \text{characteristic values.}$$

The general solution of Eq. (5–29) is

$$\tau_n(t) = C_n e^{-a\lambda_n^2 t}.$$

Thus the product solution of $\psi(x, t)$ yields

$$\psi(x, t) = a_0 + \sum_{n=1}^{\infty} a_n e^{-a\lambda_n^2 t} \cos \lambda_n x, \tag{5–30}$$

where $a_0 = A_0 C_0$ and $a_n = A_n C_n$.

Finally, the initial value of Eq. (5–30), which is equal to $-\phi(x)$, gives

$$-\frac{q''x^2}{2kL} = a_0 + \sum_{n=1}^{\infty} a_n \cos \lambda_n x; \tag{5–31}$$

the coefficients a_0, a_n are

$$a_0 = -\frac{q''L}{6k}, \quad a_n = -(-1)^n \frac{2q''L}{k(\lambda_n L)^2}. \tag{5–32}$$

Hence, combining Eqs. (5–21), (5–27), (5–30), and (5–32), we find the unsteady temperature of the plate to be

$$\frac{\theta(x, t)}{q''L/k} = \frac{at}{L^2} + \frac{1}{2}\left(\frac{x}{L}\right)^2 - \frac{1}{6} - 2\sum_{n=1}^{\infty} \frac{(-1)^n}{(\lambda_n L)^2} e^{-a\lambda_n^2 t} \cos \lambda_n x. \tag{5–33}$$

(How does the temperature vary for large values of time? Can this variation be guessed in advance of the solution?)

Thus Examples 5–4 and 5–5 teach us the possibility of two procedures for handling the nonhomogeneities of unsteady problems. *If an unsteady problem approaches a steady solution as $t \to \infty$, then the assumption of Eq. (5–14) becomes convenient, whereas if there is no steady solution as $t \to \infty$, then Eq. (5–21) is suitable.*

We now return to the solution of the differential formulation of Example 2–3. Since the problem has a steady solution as $t \to \infty$, it may be solved starting with the superposition given by Eq. (5–14). The result is

$$\frac{T - T_\infty}{q''L/k} = 1 - \left(\frac{x}{L}\right) + \frac{1}{\text{Bi}} - 2\sum_{n=1}^{\infty} \frac{e^{-a\lambda_n^2 t} \cos \lambda_n x}{(\lambda_n L)(\lambda_n L + \sin \lambda_n L \cos \lambda_n L)}, \tag{5–34}$$

where x is measured from the lower surface of the plate, $\text{Bi} = hL/k$, and the characteristic values are the zeros of $(\lambda_n L) \sin \lambda_n L = \text{Bi} \cos \lambda_n L$ (see Fig. 4–13). The details of this problem are left to the reader.

Now let us suppose that we wish to find the periodic solution of Example 5-1 corresponding to the distributed formulation. In this connection, let us first investigate the procedure of steady periodic solutions by means of a simpler problem.

Example 5-6. The surface temperature of a semi-infinite body oscillates as $\theta_0 \cos \omega t$ (Fig. 5-6). The initial temperature of the body is uniform, say zero. We wish to find the steady periodic temperature of the body.

The formulation of the problem is

$$\frac{\partial \theta}{\partial t} = a \frac{\partial^2 \theta}{\partial x^2}, \qquad (5\text{-}35)$$

$$\theta(x, 0) = 0, \qquad (5\text{-}36)$$

$$\theta(0, t) = \theta_0 \cos \omega t, \qquad (5\text{-}37)$$

$$\theta(\infty, t) = 0. \qquad (5\text{-}38)$$

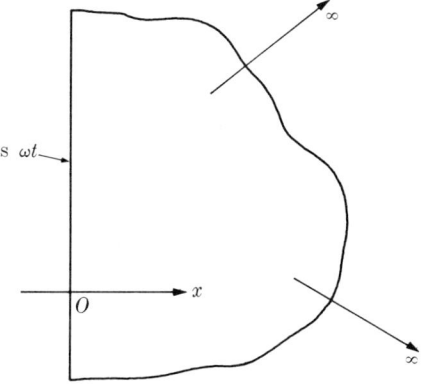

FIG. 5-6

Since the solution is to be valid for large values of time only, it is independent of the initial condition given by Eq. (5-36), and is periodic in time. Due to the linearity of the problem this periodicity is equal to that of the disturbance, as will be seen in the solution. The periodic time dependence of the temperature may be satisfied by an imaginary separation parameter. This gives exponential solutions in time, with imaginary arguments which are expressible in terms of circular functions. If we then apply the product solution $\theta(x, t) = X(x)\tau(t)$ to Eq. (5-35), using appropriate separation parameters, we obtain

$$\frac{1}{a\tau} \frac{d\tau}{dt} = \frac{1}{X} \frac{d^2 X}{dx^2} = \pm i\lambda^2. \qquad (5\text{-}39)$$

It is clear that both $+i\lambda^2$ and $-i\lambda^2$ satisfy the foregoing argument.

Defining the temperatures $\theta_+(x, t)$ and $\theta_-(x, t)$ corresponding to $+i\lambda^2$ and $-i\lambda^2$, respectively, we have

$$\begin{aligned} \theta_+(x, t) &= e^{ia\lambda^2 t}(C_1 e^{\lambda \sqrt{i}\, x} + C_2 e^{-\lambda \sqrt{i}\, x}), \\ \theta_-(x, t) &= e^{-ia\lambda^2 t}(C_3 e^{\lambda \sqrt{-i}\, x} + C_4 e^{-\lambda \sqrt{-i}\, x}). \end{aligned} \qquad (5\text{-}40)$$

Noting Fig. 5-7, we may rearrange Eq. (5-40) in the form

$$\begin{aligned} \theta(x, t) &= \theta_+(x, t) + \theta_-(x, t) \\ &= C_1 e^{\lambda x/\sqrt{2}} e^{i(a\lambda^2 t + \lambda x/\sqrt{2})} + C_2 e^{-\lambda x/\sqrt{2}} e^{i(a\lambda^2 t - \lambda x/\sqrt{2})} \\ &\quad + C_3 e^{-\lambda x/\sqrt{2}} e^{-i(a\lambda^2 t - \lambda x/\sqrt{2})} + C_4 e^{\lambda x/\sqrt{2}} e^{-i(a\lambda^2 t + \lambda x/\sqrt{2})}. \end{aligned} \qquad (5\text{-}41)$$

The first and fourth terms of Eq. (5–41) violate the boundary condition given by Eq. (5–38) and therefore are not admissible. The remaining terms may be expressed in terms of circular functions to give

$$\theta(x, t) = e^{-\lambda x/\sqrt{2}}[A \cos(a\lambda^2 t - \lambda x/\sqrt{2}) + B \sin(a\lambda^2 t - \lambda x/\sqrt{2})]. \tag{5–42}$$

Equation (5–42), employed with the boundary condition given by Eq. (5–37), yields

$$\theta_0 \cos \omega t = A \cos a\lambda^2 t + B \sin a\lambda^2 t. \tag{5–43}$$

Equation (5–43) can hold for all values of t only if $\theta_0 = A$, $0 = B$, and $\omega = a\lambda^2$; this confirms the fact that the solution has the period ω.

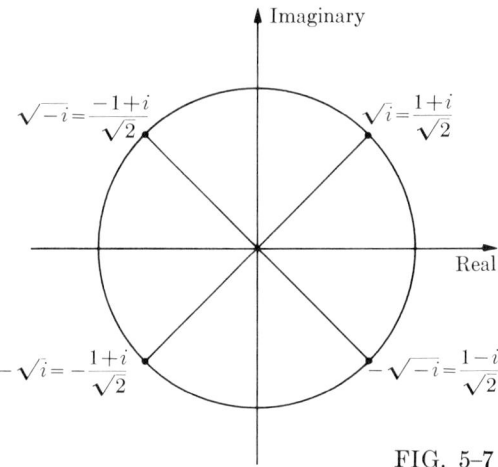

FIG. 5–7

Hence the steady periodic temperature of the semi-infinite body is found to be

$$\frac{\theta(x, t)}{\theta_0} = e^{-(\omega/2a)^{1/2}x} \cos\left[\omega t - \left(\frac{\omega}{2a}\right)^{1/2} x\right]. \tag{5–44}$$

Equation (5–44) is shown by means of a perspective diagram in Fig. 5–8. At a given depth x, the temperature oscillates with the amplitude $e^{-(\omega/2a)^{1/2}}$ relative to that of the disturbance, and has the period of and the time lag $(x/2a\omega)^{1/2}$ relative to the disturbance.

Note that even the steady periodic solution of a simple problem like the one of the present example requires an involved procedure in terms of the separation of variables. The methods suitable to steady and unsteady periodic problems will be developed in Chapters 6 and 7, respectively.

Let us now turn to two examples of transient problems in one-dimensional cylindrical geometry.

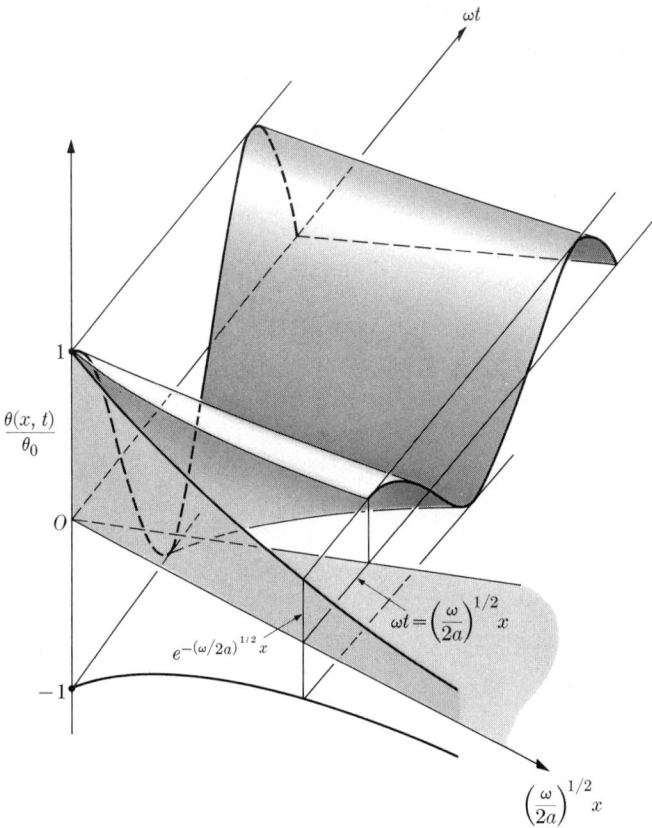

FIG. 5-8

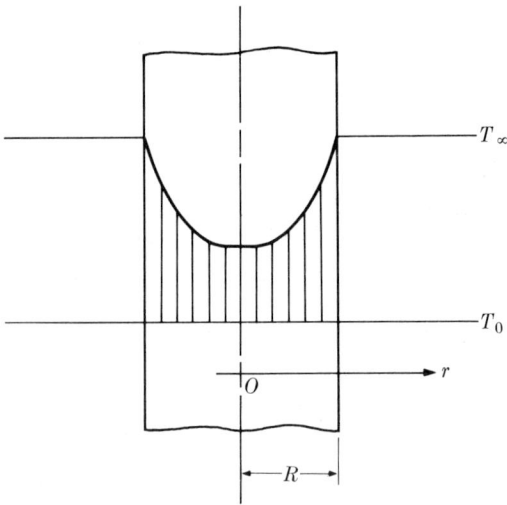

FIG. 5-9

Example 5–7. An infinitely long rod of radius R having the uniform initial temperature T_0 is plunged suddenly into a bath at temperature T_∞. The heat transfer coefficient is large (Fig. 5–9). We wish to find the unsteady temperature of the rod.

The formulation of the problem in terms of $\theta = T - T_\infty$ is

$$\frac{\partial \theta}{\partial t} = \frac{a}{r}\frac{\partial}{\partial r}\left(r\frac{\partial \theta}{\partial r}\right), \qquad \theta(r, 0) = \theta_0 = T_0 - T_\infty,$$

$$\frac{\partial \theta(0, t)}{\partial r} = 0, \qquad \theta(R, t) = 0. \tag{5-45}$$

The product solution $\theta(r, t) = \mathcal{R}(r)\tau(t)$ introduced into Eq. (5–45) results in

$$\frac{d}{dr}\left(r\frac{d\mathcal{R}}{dr}\right) + \lambda^2 r\mathcal{R} = 0; \qquad \frac{d\mathcal{R}(0)}{dr} = 0, \quad \mathcal{R}(R) = 0, \tag{5-46}$$

$$\frac{d\tau}{dt} + a\lambda^2 \tau = 0. \tag{5-47}$$

The solution of Eq. (5–46) is

$$\mathcal{R}_n(r) = A_n \varphi_n(r), \qquad \varphi_n(r) = J_0(\lambda_n r), \qquad \text{characteristic functions,}$$

and the zeros of $J_0(\lambda_n R) = 0$ are the characteristic values. The solution of Eq. (5–47) is $\tau_n(t) = C_n e^{-a\lambda_n^2 t}$. Hence the product solution becomes

$$\theta(r, t) = \sum_{n=1}^{\infty} a_n e^{-a\lambda_n^2 t} J_0(\lambda_n r), \tag{5-48}$$

where $a_n = A_n C_n$.

The initial value of Eq. (5–48) is

$$\theta_0 = \sum_{n=1}^{\infty} a_n J_0(\lambda_n r). \tag{5-49}$$

Equation (5–49) is the Fourier-Bessel series expansion of θ_0. Here the coefficient a_n may be calculated in the usual manner.* The result is

$$a_n = \frac{2\theta_0}{(\lambda_n R) J_1(\lambda_n R)}. \tag{5-50}$$

Thus, introducing Eq. (5–50) into Eq. (5–48), we find that the unsteady temperature of the rod is

$$\frac{T(r, t) - T_\infty}{T_0 - T_\infty} = 2\sum_{n=1}^{\infty} \frac{e^{-a\lambda_n^2 t} J_0(\lambda_n r)}{(\lambda_n R) J_1(\lambda_n R)}. \tag{5-51}$$

The only difference between Examples 5–2 and 5–7 is the effect of curvature included in the latter.

---

* This is the special case of Eq. (4–203) corresponding to $F(r) = \theta_0$.

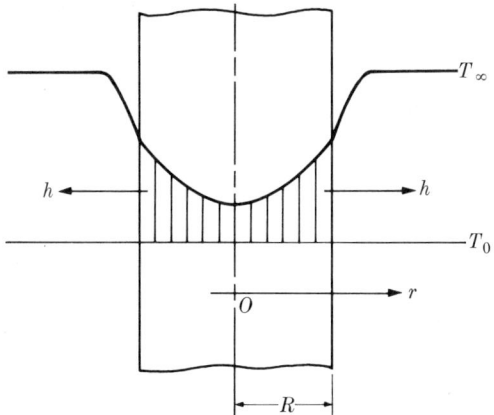

FIG. 5–10

Example 5–8. Let us re-solve Example 5–7, given a moderate heat transfer coefficient h (Fig. 5–10).

The formulation of the problem now becomes

$$\frac{\partial \theta}{\partial t} = \frac{a}{r}\frac{\partial}{\partial r}\left(r\frac{\partial \theta}{\partial r}\right), \qquad \theta(r, 0) = \theta_0,$$

$$\frac{\partial \theta(0, t)}{\partial r} = 0, \qquad -k\frac{\partial \theta(R, t)}{\partial r} = h\theta(R, t). \tag{5-52}$$

The product solution yields in the r-direction

$$\frac{d}{dr}\left(r\frac{d\mathcal{R}}{dr}\right) + \lambda^2 r\mathcal{R} = 0; \qquad \frac{d\mathcal{R}(0)}{dr} = 0, \qquad \frac{d\mathcal{R}(R)}{dr} + \frac{h}{k}\mathcal{R}(R) = 0, \tag{5-53}$$

and in the time-direction an equation identical to Eq. (5–47) of the previous example. The solution of Eq. (5–53) is

$$\mathcal{R}_n(r) = A_n\varphi_n(r), \qquad \varphi_n(r) = J_0(\lambda_n r), \qquad \text{characteristic functions,}$$

and the zeros of $(\lambda_n R)J_1(\lambda_n R) + \text{Bi}\, J_0(\lambda_n R) = 0$, where $\text{Bi} = hR/k$, are the characteristic values.

The form of the product solution is the same as that of the previous problem, Eq. (5–48). However, the coefficient a_n [see Eq. (4–214)] now becomes

$$a_n = \frac{2\theta_0\, \text{Bi}}{(\lambda_n^2 R^2 + \text{Bi}^2)J_0(\lambda_n R)}. \tag{5-54}$$

Thus the unsteady temperature of the rod is

$$\frac{T(r, t) - T_\infty}{T_0 - T_\infty} = 2\,\text{Bi}\sum_{n=1}^{\infty}\frac{e^{-a\lambda_n^2 t}J_0(\lambda_n r)}{(\lambda_n^2 R^2 + \text{Bi}^2)J_0(\lambda_n R)}. \tag{5-55}$$

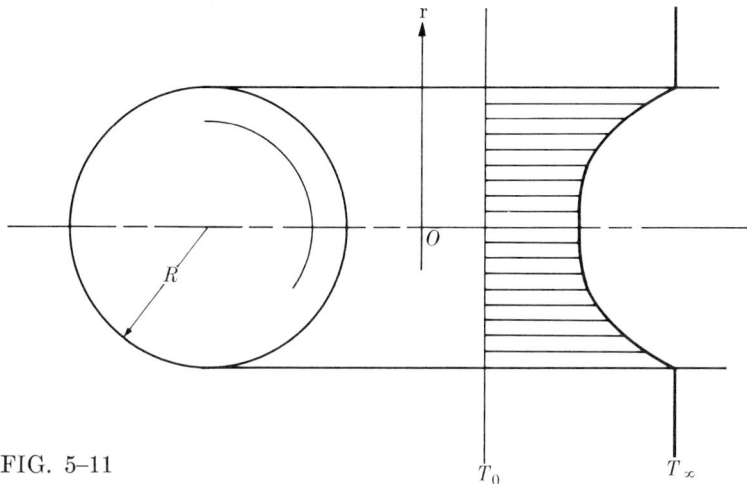

FIG. 5-11

The physics of this cylindrical problem is identical to that of the cartesian problem of Example 5-3.

The next two examples illustrate transient problems in one-dimensional spherical geometry.

Example 5-9. A solid sphere of radius R having the uniform initial temperature T_0 is plunged suddenly into a bath at temperature T_∞. The heat transfer coefficient is large (Fig. 5-11). We wish to find the unsteady temperature of the sphere.

The formulation of the problem in terms of $\theta = T - T_\infty$ is

$$\frac{\partial \theta}{\partial t} = \frac{a}{r^2} \frac{\partial}{\partial r}\left(r^2 \frac{\partial \theta}{\partial r}\right), \qquad \theta(r, 0) = \theta_0,$$

$$\theta(0, t) = \text{finite} \quad \text{or} \quad \frac{\partial \theta(0, t)}{\partial r} = 0, \qquad \theta(R, t) = 0. \tag{5-56}$$

The product solution $\theta(r, t) = \Re(r)\tau(t)$ introduced into Eq. (5-56) yields

$$\frac{d}{dr}\left(r^2 \frac{d\Re}{dr}\right) + \lambda^2 r^2 \Re = 0; \qquad \frac{d\Re(0)}{dr} = 0, \quad \Re(R) = 0, \tag{5-57}$$

$$\frac{d\tau}{dt} + a\lambda^2 \tau = 0. \tag{5-58}$$

Particular solutions corresponding to the differential equation of Eq. (5-57) (see Table 3-1) are

$$J_{1/2}(\lambda r)/r^{1/2}, \qquad J_{-1/2}(\lambda r)/r^{1/2}. \tag{5-59}$$

Furthermore, noting from Eq. (3-140) that

$$J_{1/2}(\lambda r) \sim \sin \lambda r / r^{1/2}, \qquad J_{-1/2}(\lambda r) \sim \cos \lambda r / r^{1/2},$$

we may rearrange the solutions (5–59) to give

$$\sin \lambda r / r, \qquad \cos \lambda r / r.$$

This result explains the use of the well-known transformation

$$\theta(r, t) = \psi(r, t)/r \qquad (5\text{–}60)$$

for problems of spherical geometry: Eq. (5–60) reduces the spherical Laplacian to the cartesian Laplacian whose solution is expressible in terms of circular functions.

We may now express Eq. (5–56) in terms of ψ by using Eq. (5–60) and the condition of finite center temperature rather than that of temperature symmetry. The result is

$$\frac{\partial \psi}{\partial t} = a \frac{\partial^2 \psi}{\partial r^2}, \qquad \psi(r, 0) = r\theta_0,$$
$$\psi(0, t) = 0, \qquad \psi(R, t) = 0. \qquad (5\text{–}61)$$

Hence the problem is reduced to a problem of cartesian geometry.

The product solution $\psi(r, t) = \mathcal{R}(r)\tau(t)$ yields

$$\frac{d^2 \mathcal{R}}{dr^2} + \lambda^2 \mathcal{R} = 0; \qquad \mathcal{R}(0) = 0, \quad \mathcal{R}(R) = 0, \qquad (5\text{–}62)$$

$$\frac{d\tau}{dt} + a\lambda^2 \tau = 0. \qquad (5\text{–}63)$$

The solution of Eq. (5–62) is

$$\mathcal{R}_n(r) = A_n \varphi_n(r), \qquad \varphi_n(r) = \sin \lambda_n r, \qquad \text{characteristic functions,}$$
$$\lambda_n R = n\pi, \qquad n = 1, 2, 3, \ldots, \qquad \text{characteristic values,}$$

and the solution of Eq. (5–63) is $\tau_n(t) = C_n e^{-a\lambda_n^2 t}$.

Thus the product solution becomes

$$\psi(r, t) = \sum_{n=1}^{\infty} a_n e^{-a\lambda_n^2 t} \sin \lambda_n r. \qquad (5\text{–}64)$$

The initial value of Eq. (5–64) is

$$r\theta_0 = \sum_{n=1}^{\infty} a_n \sin \lambda_n r.$$

The coefficient a_n is

$$a_n = (-1)^{n+1} \frac{2\theta_0}{\lambda_n}. \qquad (5\text{–}65)$$

[5-1]   DISTRIBUTED SYSTEMS HAVING STEPWISE DISTURBANCES   287

Finally, the unsteady temperature of the sphere is found to be

$$\frac{T(r, t) - T_\infty}{T_0 - T_\infty} = 2 \sum_{n=1}^{\infty} (-1)^{n+1} e^{-a\lambda_n^2 t} \frac{\sin \lambda_n r}{\lambda_n r}. \tag{5-66}$$

Note that Examples 5–2, 5–7, and 5–9 represent the same physical situation in cartesian, cylindrical, and spherical geometries, respectively.

Example 5–10. Let us modify Example 5–9 by assuming a moderate heat transfer coefficient h (Fig. 5–12).

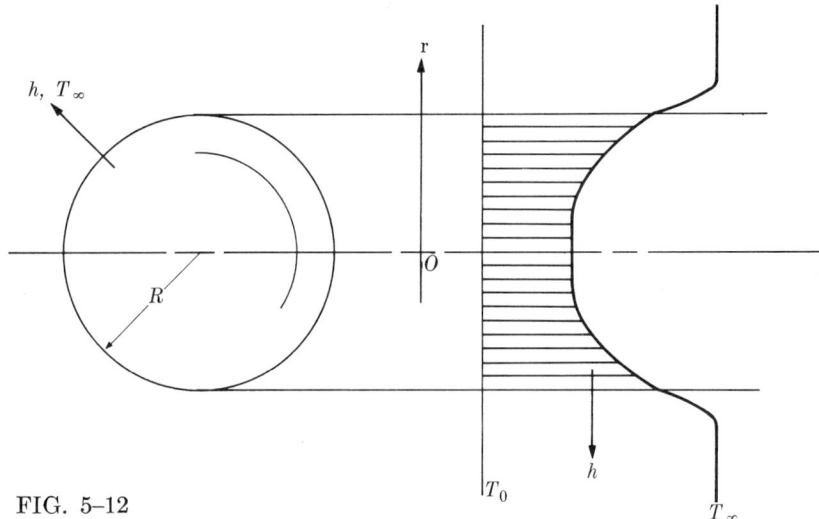

FIG. 5–12

The formulation of the problem is identical to that of the preceding example, Eq. (5–56), except for the surface boundary condition $\theta(R, t) = 0$, which must now be replaced by $-k [\partial \theta(R, t)/\partial r] = h\theta(R, t)$. Thus in terms of ψ we have

$$\frac{\partial \psi}{\partial t} = a \frac{\partial^2 \psi}{\partial r^2}, \qquad \psi(r, 0) = r\theta_0,$$

$$\psi(0, t) = 0, \qquad -k \frac{\partial \psi(R, t)}{\partial r} = \left(h - \frac{k}{R}\right) \psi(R, t). \tag{5-67}$$

The product solution gives

$$\frac{d^2 \Re}{dr^2} + \lambda^2 \Re = 0; \qquad \Re(0) = 0, \qquad \frac{d\Re(R)}{dr} + \left(\frac{h}{k} - \frac{1}{R}\right) \Re(R) = 0, \tag{5-68}$$

in r and Eq. (5–63) in time. The solution of Eq. (5–68) is

$$\Re_n(r) = A_n \varphi_n(r), \qquad \varphi_n(r) = \sin \lambda_n r, \qquad \text{characteristic functions,}$$

and the zeros of $(\lambda_n R) \cos \lambda_n R = (1 - \text{Bi}) \sin \lambda_n R$, where $\text{Bi} = hR/k$, are the characteristic values.*

The product solution of the previous example, Eq. (5–64), and its initial value apply to the present problem also. However, the coefficient a_n is different:

$$a_n = \frac{2\theta_0(\sin \lambda_n R - \lambda_n R \cos \lambda_n R)}{\lambda_n(\lambda_n R - \sin \lambda_n R \cos \lambda_n R)}. \tag{5–69}$$

Thus the unsteady temperature of the sphere is

$$\frac{T(r,t) - T_\infty}{T_0 - T_\infty} = 2 \sum_{n=1}^{\infty} \left(\frac{\sin \lambda_n R - \lambda_n R \cos \lambda_n R}{\lambda_n R - \sin \lambda_n R \cos \lambda_n R}\right) e^{-a\lambda_n^2 t} \frac{\sin \lambda_n r}{\lambda_n r}. \tag{5–70}$$

Note that the physics of the cartesian, cylindrical, and spherical problems considered in Examples 5–3, 5–8, and 5–10 is the same. Since two- and three-dimensional problems require no further mathematics, they may be treated in a manner similar to one-dimensional problems. One example taken from two-dimensional cartesian geometry may serve to illustrate this similarity.

Example 5–11. Let us consider a transient problem associated with Example 4–8. Assume that the initial temperature of the fin is T_∞, and that the temperature of one surface in ξ is suddenly changed to T_0 (see Fig. 4–24). We wish to find the unsteady temperature of the fin.

The formulation of the problem in terms of $\theta = T - T_\infty$ is

$$\frac{\partial \theta}{\partial t} = a \left(\frac{\partial^2 \theta}{\partial \xi^2} + \frac{\partial^2 \theta}{\partial y^2}\right), \qquad \theta(\xi, y, 0) = 0,$$

$$\theta(0, y, t) = 0, \qquad \theta(L, y, t) = \theta_0 = T_0 - T_\infty, \tag{5–71}$$

$$\frac{\partial \theta(\xi, 0, t)}{\partial y} = 0, \qquad \theta(\xi, l, t) = 0.$$

The nonhomogeneity of one boundary condition suggests that the principle of superposition be used. Noting that the problem has a steady solution as $t \to \infty$, and generalizing Eq. (5–14), we assume

$$\theta(\xi, y, t) = \psi(\xi, y, t) + \phi(\xi, y) \tag{5–72}$$

such that $\phi(\xi, y)$ and $\psi(\xi, y, t)$ satisfy

$$\frac{\partial^2 \phi}{\partial \xi^2} + \frac{\partial^2 \phi}{\partial y^2} = 0,$$

$$\phi(0, y) = 0, \qquad \phi(L, y) = \theta_0, \tag{5–73}$$

$$\frac{\partial \phi(\xi, 0)}{\partial y} = 0, \qquad \phi(\xi, l) = 0,$$

---

* Show these characteristic values on a sketch as intersections of two different curves for the values of $\text{Bi} > 1$ and $\text{Bi} < 1$.

and

$$\frac{\partial \psi}{\partial t} = a\left(\frac{\partial^2 \psi}{\partial \xi^2} + \frac{\partial^2 \psi}{\partial y^2}\right), \qquad \psi(\xi, y, 0) = -\phi(\xi, y),$$

$$\psi(0, y, t) = 0, \qquad \psi(L, y, t) = 0, \qquad (5\text{-}74)$$

$$\frac{\partial \psi(\xi, 0, t)}{\partial y} = 0, \qquad \psi(\xi, l, t) = 0.$$

The steady problem of Eq. (5–73) is identical to that of Example 4–8 provided $\theta(\xi, y)$ of that example is replaced by $\phi(\xi, y)$. It follows then that

$$\frac{\phi(\xi, y)}{\theta_0} = 2 \sum_{n=0}^{\infty} \frac{(-1)^n}{(\lambda_n l)} \left(\frac{\sinh \lambda_n \xi}{\sinh \lambda_n L}\right) \cos \lambda_n y, \qquad (4\text{-}123)$$

where $\lambda_n l = (2n+1)\pi/2$, $n = 0, 1, 2, 3, \ldots$

On the other hand, the product solution $\psi(\xi, y, t) = \Xi(\xi)Y(y)\tau(t)$ introduced into Eq. (5–74) yields

$$\frac{d^2 \Xi}{d\xi^2} + \mu^2 \Xi = 0; \qquad \Xi(0) = 0, \quad \Xi(L) = 0, \qquad (5\text{-}75)$$

$$\frac{d^2 Y}{dy^2} + \nu^2 Y = 0; \qquad \frac{dY(0)}{dy} = 0, \quad Y(l) = 0, \qquad (5\text{-}76)$$

$$\frac{d\tau}{dt} + a(\mu^2 + \nu^2)\tau = 0. \qquad (5\text{-}77)$$

The solution of Eq. (5–75) is

$$\Xi_m(\xi) = A_m \varphi_m(\xi), \qquad \varphi_m(\xi) = \sin \mu_m \xi, \qquad \text{characteristic functions,}$$
$$\mu_m L = m\pi, \qquad m = 1, 2, 3, \ldots, \qquad \text{characteristic values,}$$

and the solution of Eq. (5–76) is

$$Y_k(y) = B_k \varphi_k(y), \qquad \varphi_k(y) = \cos \nu_k y, \qquad \text{characteristic functions,}$$
$$\nu_k l = (2k+1)\pi/2, \qquad k = 0, 1, 2, 3, \ldots, \qquad \text{characteristic values.}$$

The general solution of Eq. (5–77) is

$$\tau_{mk}(t) = C_{mk} e^{-a(\mu_m^2 + \nu_k^2)t}.$$

Thus the product solution becomes

$$\psi(\xi, y, t) = \sum_{m=1}^{\infty} \sum_{k=0}^{\infty} a_{mk} e^{-a(\mu_m^2 + \nu_k^2)t} \sin \mu_m \xi \cos \nu_k y, \qquad (5\text{-}78)$$

where $a_{mk} = A_m B_k C_{mk}$.

Using the initial condition, from Eq. (5–78) we obtain

$$-\phi(\xi, y) = \sum_{m=1}^{\infty} \sum_{k=0}^{\infty} a_{mk} \sin \mu_m \xi \cos \nu_k y,$$

which is the expansion of $-\phi(\xi, y)$ into a *double Fourier series*. Here the coefficient a_{mk} is*

$$a_{mk} = \frac{4\theta_0(-1)^{k+m}(\mu_m L)}{(\mu_m^2 + \nu_k^2)L^2(\nu_k l)}. \tag{5–79}$$

Hence the unsteady part of the problem, expressed by $\psi(\xi, y, t)$, is

$$\frac{\psi(\xi, y, t)}{\theta_0} = 4 \sum_{k=0}^{\infty} \sum_{m=1}^{\infty} (-1)^{k+m} \left(\frac{\mu_m L}{\nu_k l}\right) \frac{e^{-a(\nu_k^2 + \mu_m^2)t}}{(\nu_k^2 + \mu_m^2)L^2} \sin \mu_m \xi \cos \nu_k y. \tag{5–80}$$

The sum of Eqs. (4–123) and (5–80) gives the complete solution of the problem.

5–2. Multidimensional Problems Expressible in Terms of One-Dimensional Ones. Use of One-Dimensional Charts

In this section we consider a class of multidimensional problems whose solution can be found by expressing the problem in terms of two or more one-dimensional problems. First an example will be taken from two-dimensional cartesian geometry, then the results will be generalized to three-dimensional cartesian and other geometries.

Example 5–12. An infinitely long rod of rectangular cross section ($2L \times 2l$) having the uniform initial temperature T_0 is plunged suddenly into a bath at constant temperature T_∞. The heat transfer coefficient is h (Fig. 5–13). We wish to find the unsteady temperature of the rod.

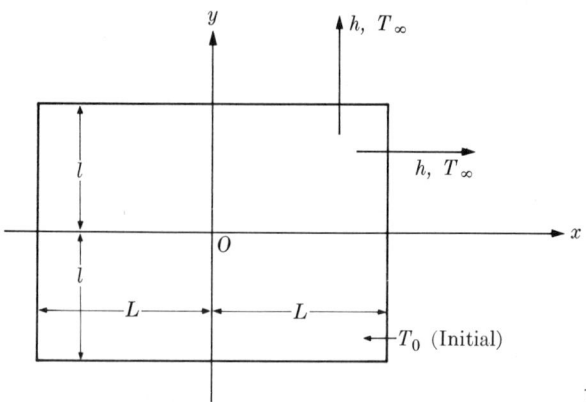

FIG. 5–13

---

* Use the argument which follows Eq. (4–277).

The formulation of the problem in terms of the dimensionless temperature*
$\vartheta = (T - T_\infty)/(T_0 - T_\infty)$ is

$$\frac{\partial \vartheta}{\partial t} = a\left(\frac{\partial^2 \vartheta}{\partial x^2} + \frac{\partial^2 \vartheta}{\partial y^2}\right), \qquad \vartheta(x, y, 0) = 1,$$

$$\frac{\partial \vartheta(0, y, t)}{\partial x} = 0, \qquad -k\frac{\partial \vartheta(L, y, t)}{\partial x} = h\vartheta(L, y, t), \qquad (5\text{-}81)$$

$$\frac{\partial \vartheta(x, 0, t)}{\partial y} = 0, \qquad -k\frac{\partial \vartheta(x, l, t)}{\partial y} = h\vartheta(x, l, t).$$

The problem could have been solved by the usual separation $\vartheta(x, y, t) = X(x)Y(y)\tau(t)$. Here, however, a less restrictive form,

$$\vartheta(x, y, t) = X(x, t)Y(y, t), \qquad (5\text{-}82)$$

will be *assumed*. If Eq. (5-82) works, we are led to the important conclusion that *it is possible to express an unsteady two-dimensional problem as the product of two unsteady one-dimensional problems*.

Introducing Eq. (5-82) into the differential equation of Eq. (5-81) and rearranging gives

$$\frac{1}{X}\left(\frac{\partial X}{\partial t} - a\frac{\partial^2 X}{\partial x^2}\right) = -\frac{1}{Y}\left(\frac{\partial Y}{\partial t} - a\frac{\partial^2 Y}{\partial y^2}\right). \qquad (5\text{-}83)$$

Since x and y can vary independently, both sides of Eq. (5-83) must be independent of either variable, and equal to a parameter, say $\pm\lambda^2(t)$, which now may depend on the common variable, time. However, because of the geometric as well as thermal symmetry of the problem, the characteristic-value problems in the x- and y-directions must be similar. This can occur only with $\lambda^2(t) = 0$. Employing this value of $\lambda^2(t)$, and introducing Eq. (5-82) into the initial and boundary conditions of Eq. (5-81), we have

$$\frac{\partial X}{\partial t} = a\frac{\partial^2 X}{\partial x^2}, \qquad\qquad \frac{\partial Y}{\partial t} = a\frac{\partial^2 Y}{\partial y^2},$$

$$X(x, 0) = 1, \qquad\qquad Y(y, 0) = 1,$$

$$\frac{\partial X(0, t)}{\partial x} = 0, \qquad\qquad \frac{\partial Y(0, t)}{\partial y} = 0,$$

$$-k\frac{\partial X(L, t)}{\partial x} = hX(L, t), \qquad -k\frac{\partial Y(l, t)}{\partial y} = hY(l, t).$$

Thus the problem becomes expressible as the product of two one-dimensional unsteady problems. These are identical to each other, and to the formulation

---

* We could have solved the previous problems in terms of the same dimensionless temperature, but did not need to do so. As we shall see in the separation of the initial condition, the present problem *requires* the use of this temperature.

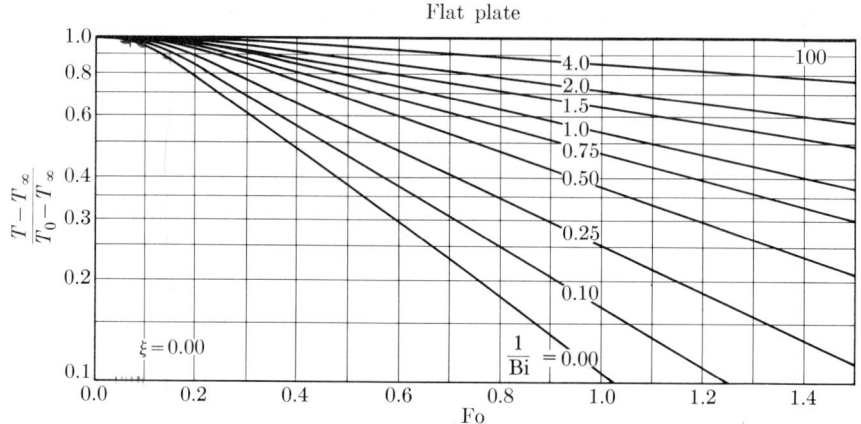
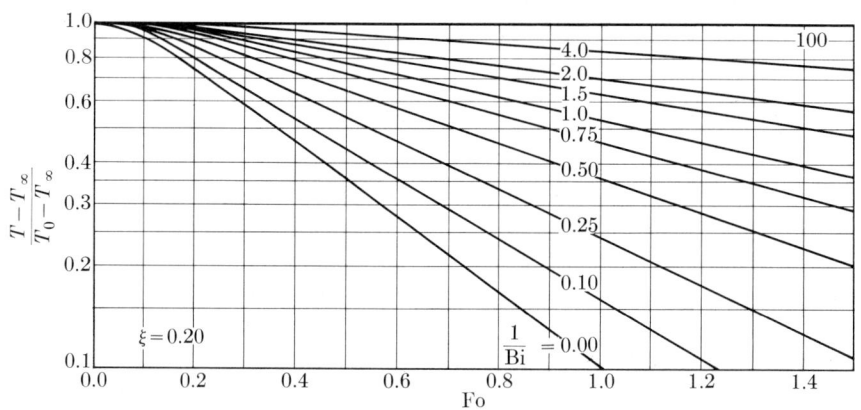
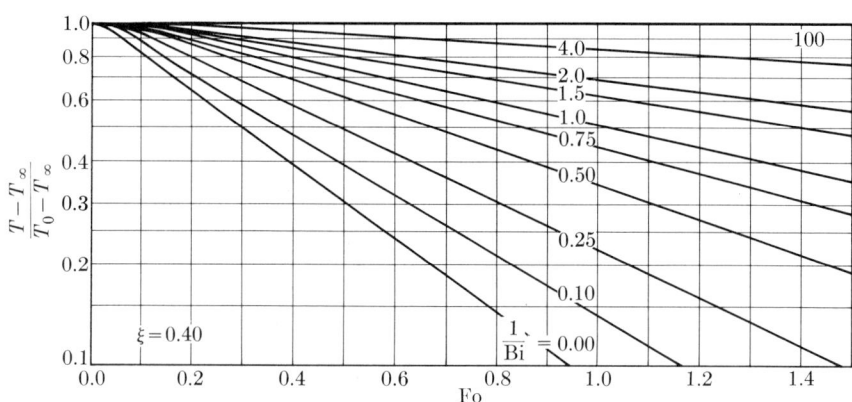

FIG. 5-14

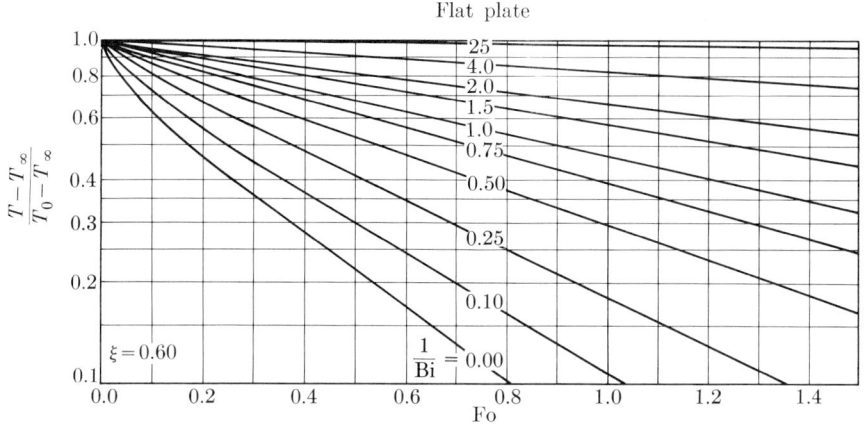
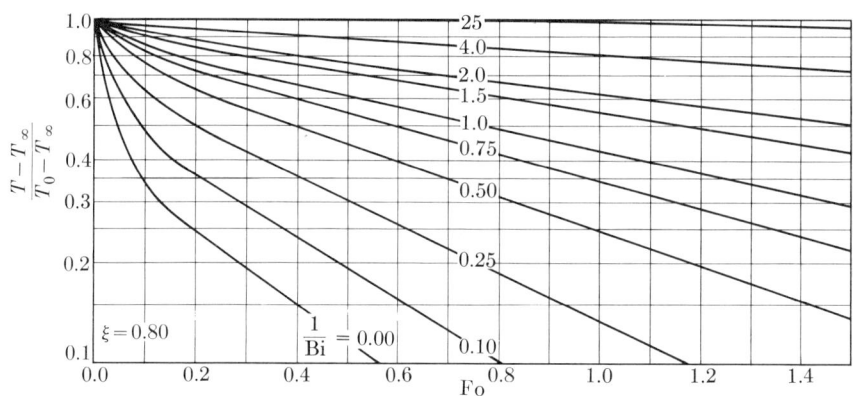
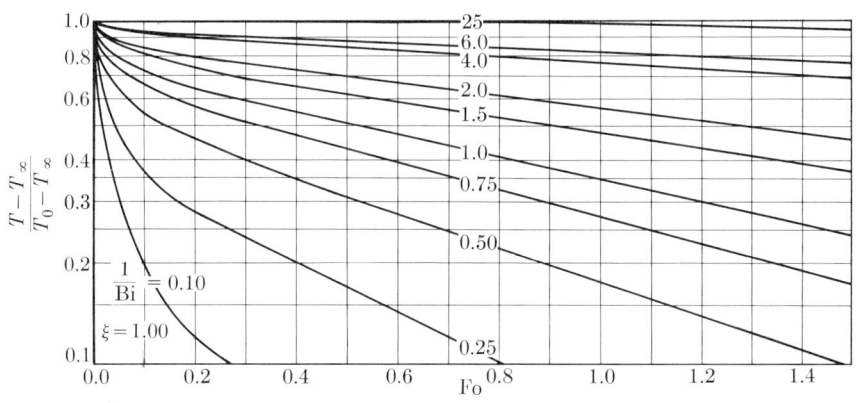

FIG. 5-14 (cont.)

of Example 5–3, whose solution is given by Eq. (5–13). The dimensionless form of Eq. (5–13) is

$$\left(\frac{T - T_\infty}{T_0 - T_\infty}\right)_{\substack{2L\,or\,2l \\ \text{Plate}}} = 2 \sum_{n=1}^{\infty} \left(\frac{\sin \mu_n}{\mu_n + \sin \mu_n \cos \mu_n}\right) e^{-\mu_n^2 \text{Fo}} \cos \mu_n \xi, \qquad (5\text{--}84)$$

where $\xi = x/L$ (or y/l), Fo $= at/L^2$ (Fourier number), $\mu_n = \lambda_n L$, and μ_n are the zeros of $\mu_n \sin \mu_n =$ Bi $\cos \mu_n$. In Fig. 5–14,* $[(T - T_\infty)/(T_0 - T_\infty)]_{2L\,\text{Plate}}$ is plotted against Fo for the values $\xi = 0.0, 0.2, 0.4, 0.6, 0.8, 1.0$, with Bi as parameter.

In the time interval $(0, t)$ the ratio between the total heat transfer q from (or to) the plate and its initial internal energy $q_i = AL\rho c(T_0 - T_\infty)$† is

$$\left(\frac{q}{q_i}\right)_{2L\,\text{Plate}} = \frac{-kA \int_0^t (\partial/\partial x)(T - T_\infty)_{x=L}\,dt}{AL\rho c(T_0 - T_\infty)}. \qquad (5\text{--}85)$$

Introducing Eq. (5–84) into the dimensionless form of Eq. (5–85), we have

$$\left(\frac{q}{q_i}\right)_{\substack{2L\,or\,2l \\ \text{Plate}}} = 2 \sum_{n=1}^{\infty} \frac{\sin^2 \mu_n}{\mu_n(\mu_n + \sin \mu_n \cos \mu_n)} (1 - e^{-\mu_n^2 \text{Fo}}). \qquad (5\text{--}86)$$

In Fig. 5–15,‡ $(q/q_i)_{2L\,\text{Plate}}$ is plotted against $(\text{Bi})^2\text{Fo}$ for some values of Bi.§

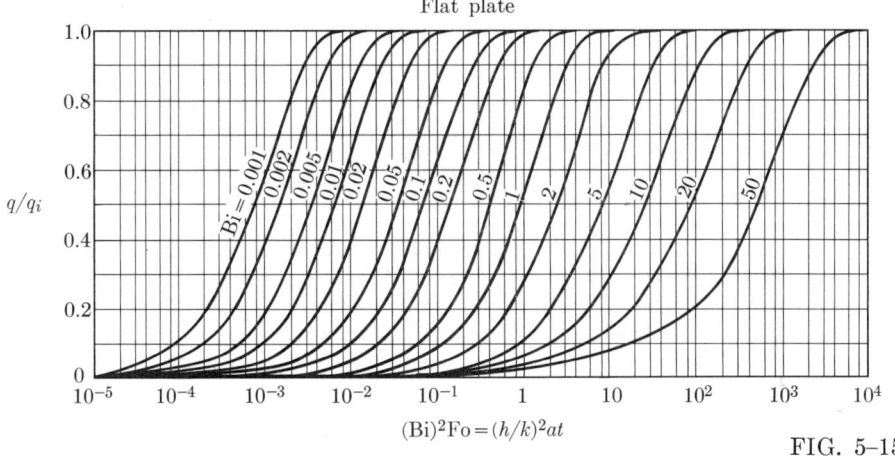

FIG. 5–15

---

* Figures 5–14, 5–16, 5–18, 5–22, 5–23 adapted from L. M. K. Boelter, V. H. Cherry, H. A. Johnson, and R. C. Martinelli, *Heat Transfer Notes*. New York: McGraw-Hill, 1965. Used by permission.
† Here A denotes the heat transfer area of one surface of the plate.
‡ Figures 5–15, 5–17, 5–19 adapted from H. Gröber, S. Erk, and U. Grigull, *Fundamentals of Heat Transfer*. New York: McGraw-Hill, 1961. Used by permission.
§ The selection of coordinates and parameters in Figs. (5–14) and (5–15) is to a large extent arbitrary.

Thus, noting that Eq. (5-82) may be written in the form

$$\left(\frac{T-T_\infty}{T_0-T_\infty}\right)_{\substack{2L,2l \\ \text{Rod}}} = \left(\frac{T-T_\infty}{T_0-T_\infty}\right)_{\substack{2L \\ \text{Plate}}} \left(\frac{T-T_\infty}{T_0-T_\infty}\right)_{\substack{2l \\ \text{Plate}}}, \quad (5\text{-}87)$$

and using Eq. (5-87) with the one-dimensional temperature charts given by Fig. 5-14, we may readily find the instantaneous temperature of an infinitely long rod of rectangular cross section $(2L \times 2l)$.

The foregoing procedure may now be extended to three-dimensional cartesian and two-dimensional cylindrical geometries. The result for the cartesian case is

$$\left(\frac{T-T_\infty}{T_0-T_\infty}\right)_{\substack{2\mathcal{L},2L,2l \\ \text{Parallelepiped}}} = \left(\frac{T-T_\infty}{T_0-T_\infty}\right)_{\substack{2\mathcal{L} \\ \text{Plate}}} \left(\frac{T-T_\infty}{T_0-T_\infty}\right)_{\substack{2L \\ \text{Plate}}} \left(\frac{T-T_\infty}{T_0-T_\infty}\right)_{\substack{2l \\ \text{Plate}}},$$

$$(5\text{-}88)$$

and that for a cylindrical rod of radius R and height $2L$ is

$$\left(\frac{T-T_\infty}{T_0-T_\infty}\right)_{\substack{2R,2L \\ \text{Rod}}} = \left(\frac{T-T_\infty}{T_0-T_\infty}\right)_{\substack{\text{Infinite} \\ 2R \text{ Rod}}} \left(\frac{T-T_\infty}{T_0-T_\infty}\right)_{\substack{2L \\ \text{Plate}}} \quad (5\text{-}89)$$

Equation (5-89) requires that we know the temperature of an infinite rod of radius R whose solution is given in Example 5-8 by Eq. (5-55). The dimensionless form of Eq. (5-55) is

$$\left(\frac{T-T_\infty}{T_0-T_\infty}\right)_{2R\,\text{Rod}} = 2\sum_{n=1}^{\infty} \frac{\text{Bi}\, e^{-\mu_n^2 \text{Fo}} J_0(\mu_n \rho)}{(\mu_n^2 + \text{Bi}^2)J_0(\mu_n)}, \quad (5\text{-}90)$$

where

$$\rho = r/R, \quad \text{Fo} = at/R^2, \quad \text{Bi} = hR/k, \quad \mu_n = \lambda_n R,$$

and μ_n are the zeros of $\mu_n J_1(\mu_n) + \text{Bi}\, J_0(\mu_n) = 0$. In Fig. 5-16,

$$[(T-T_\infty)/(T_0-T_\infty)]_{2R\,\text{Rod}}$$

is plotted against Fo for the values $\rho = 0.0, 0.2, 0.4, 0.6, 0.8, 1.0$, with Bi as parameter.

In the time interval $(0, t)$, the ratio between the total heat transfer q from a length L of the infinitely long rod and its initial internal energy

$$q_i = \pi R^2 L \rho c (T_0 - T_\infty)$$

is

$$\left(\frac{q}{q_i}\right)_{2R\,\text{Rod}} = \frac{-k 2\pi R L \int_0^t (\partial/\partial r)(T-T_\infty)_{r=R}\, dt}{\pi R^2 L \rho c (T_0 - T_\infty)}. \quad (5\text{-}91)$$

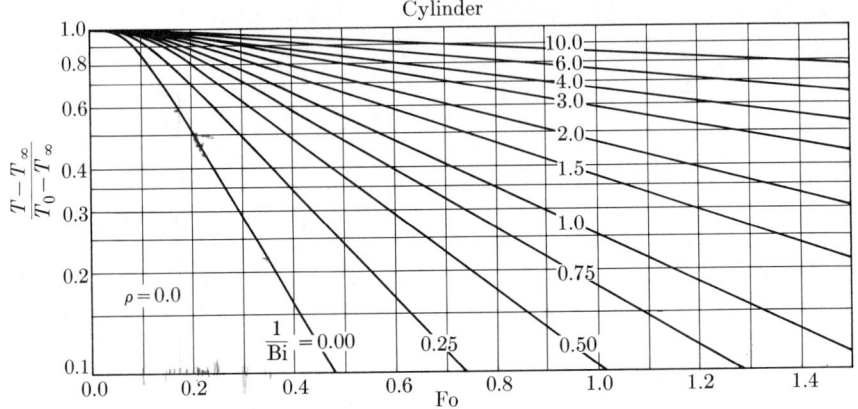
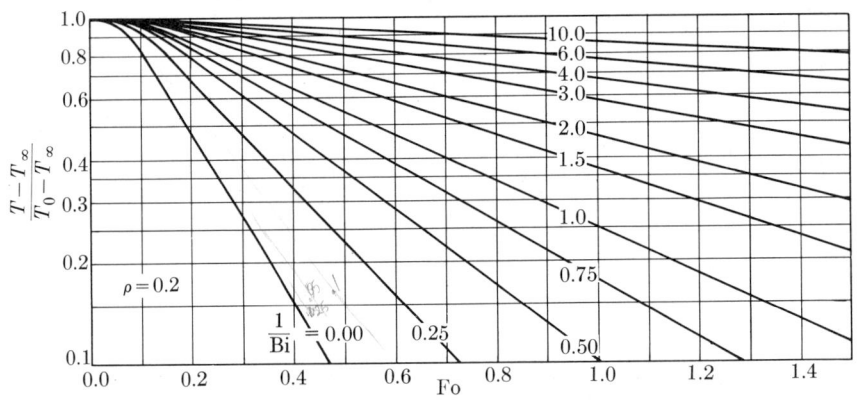
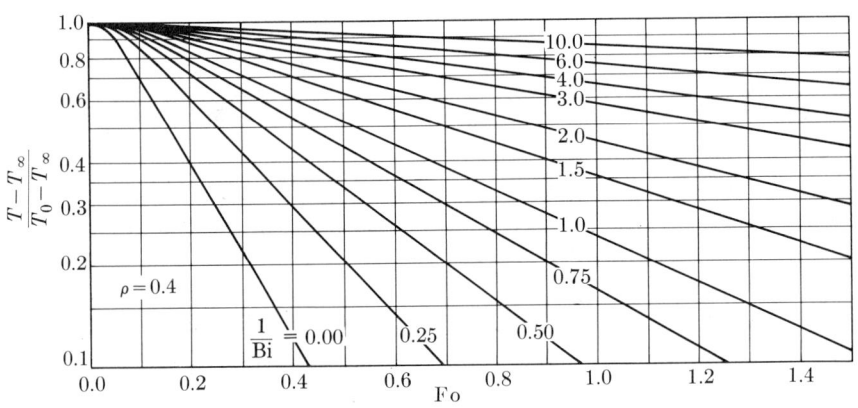

FIG. 5-16

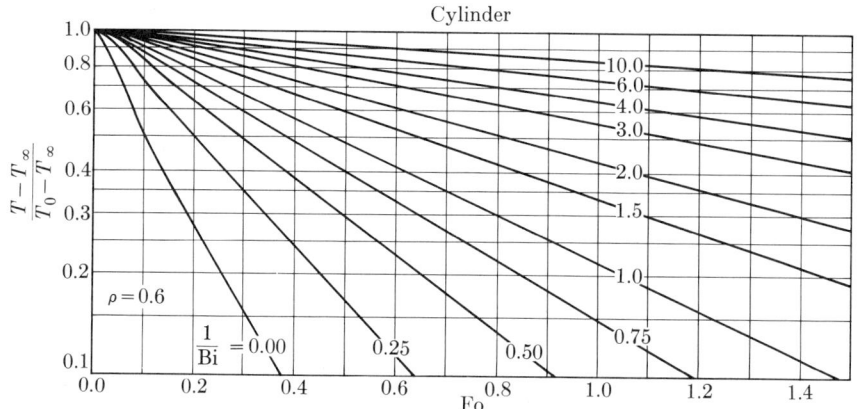

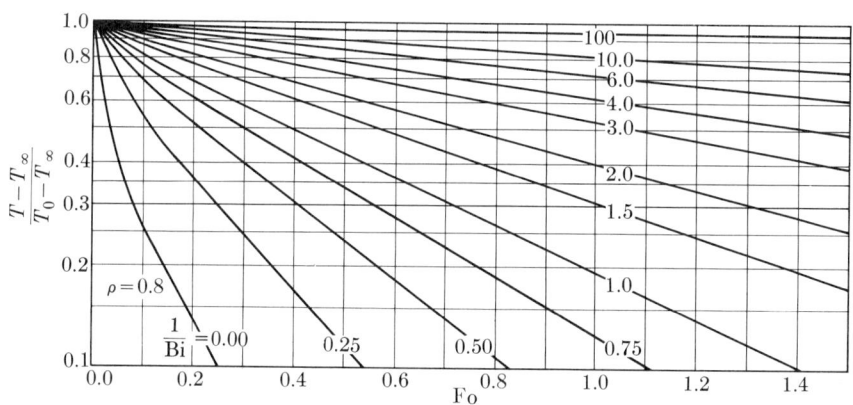

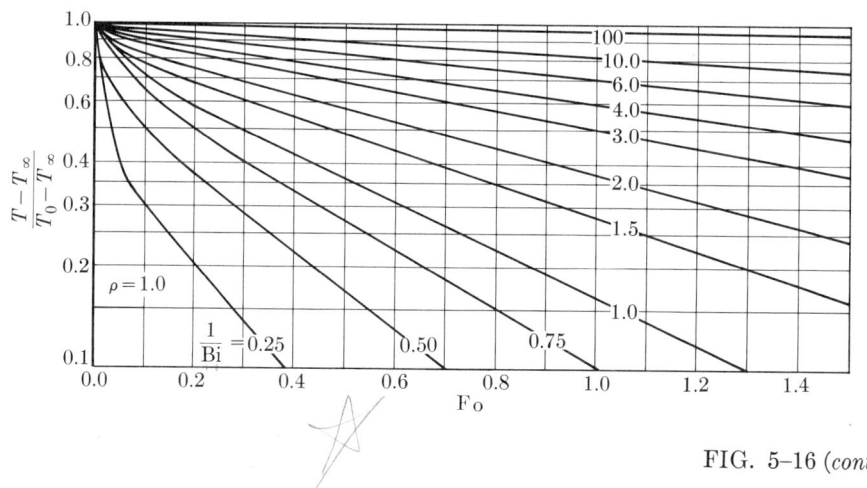

FIG. 5–16 (cont.)

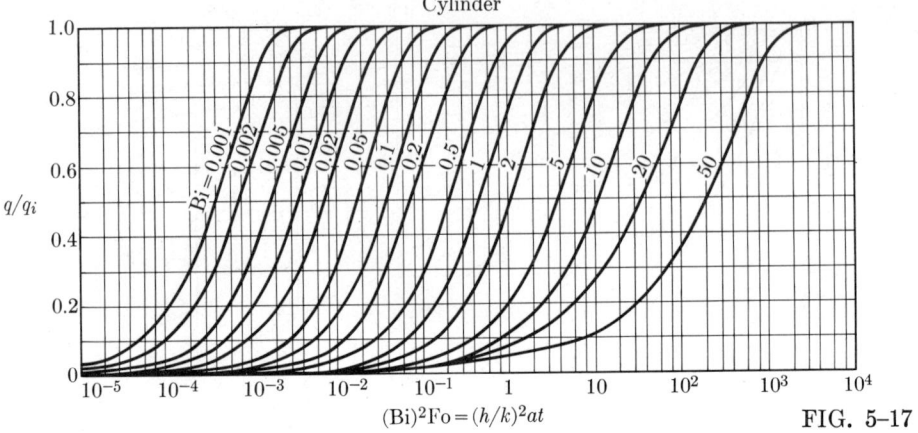

FIG. 5-17

Equation (5–90), introduced into the dimensionless form of Eq. (5–91), gives

$$\left(\frac{q}{q_i}\right)_{2R\,\text{Rod}} = 4 \sum_{n=1}^{\infty} \frac{\text{Bi}^2}{\mu_n^2(\mu_n^2 + \text{Bi}^2)} (1 - e^{-\mu_n^2 \text{Fo}}). \qquad (5\text{–}92)$$

In Fig. 5–17, $(q/q_i)_{2R\,\text{Rod}}$ is plotted against $(\text{Bi})^2 \text{Fo}$ for some values of Bi.

To complete our study of unsteady one-dimensional charts we now consider spherical geometry. The formulation and the solution of an example of this case are given by Example 5–10 and Eq. (5–70), respectively. Equation (5–70), expressed in dimensionless variables for numerical convenience, gives the instantaneous temperature of the sphere in the form

$$\left(\frac{T - T_\infty}{T_0 - T_\infty}\right)_{\text{Sphere}} = 2 \sum_{n=1}^{\infty} \left(\frac{\text{Bi} \sin \mu_n}{\mu_n - \sin \mu_n \cos \mu_n}\right) e^{-\mu_n^2 \text{Fo}} \frac{\sin \mu_n \rho}{\mu_n \rho}, \qquad (5\text{–}93)$$

where $\rho = r/R$, $\text{Fo} = at/R^2$, $\text{Bi} = hR/k$, $\mu_n = \lambda_n R$, and μ_n are the zeros of $\mu_n \cos \mu_n = (1 - \text{Bi}) \sin \mu_n$. In Fig. 5–18, $[(T - T_\infty)/(T_0 - T_\infty)]_{\text{Sphere}}$ is plotted against Fo for the values $\rho = 0.0, 0.2, 0.4, 0.6, 0.8, 1.0$, with Bi as parameter.

In the time interval $(0, t)$, the ratio between the total heat transfer q from (or to) the sphere and its initial internal energy $q_i = \frac{4}{3}\pi R^3 \rho c (T_0 - T_\infty)$ is

$$\left(\frac{q}{q_i}\right)_{\text{Sphere}} = \frac{-k 4\pi R^2 \int_0^t (\partial/\partial r)(T - T_\infty)_{r=R}\, dt}{\frac{4}{3}\pi R^3 \rho c (T_0 - T_\infty)}. \qquad (5\text{–}94)$$

Equation (5–93), introduced into the dimensionless form of Eq. (5–94), yields

$$\left(\frac{q}{q_i}\right)_{\text{Sphere}} = 6 \sum_{n=1}^{\infty} \frac{(\sin \mu_n - \mu_n \cos \mu_n)}{\mu_n^3(\mu_n - \sin \mu_n \cos \mu_n)} (1 - e^{-\mu_n^2 \text{Fo}}). \qquad (5\text{–}95)$$

[5–2] USE OF ONE-DIMENSIONAL CHARTS 299

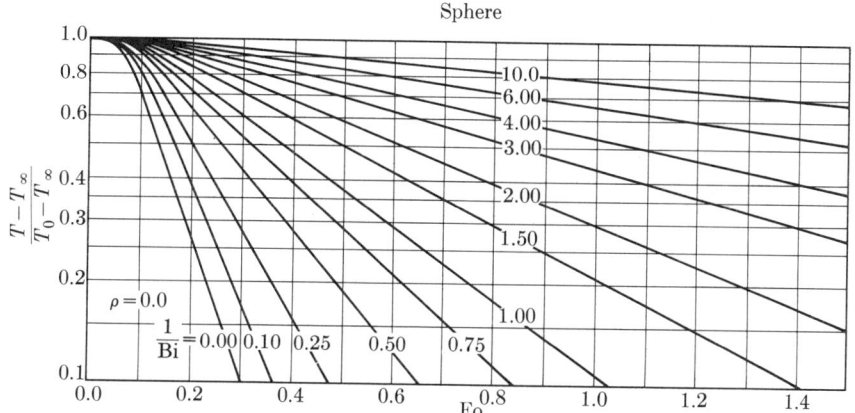

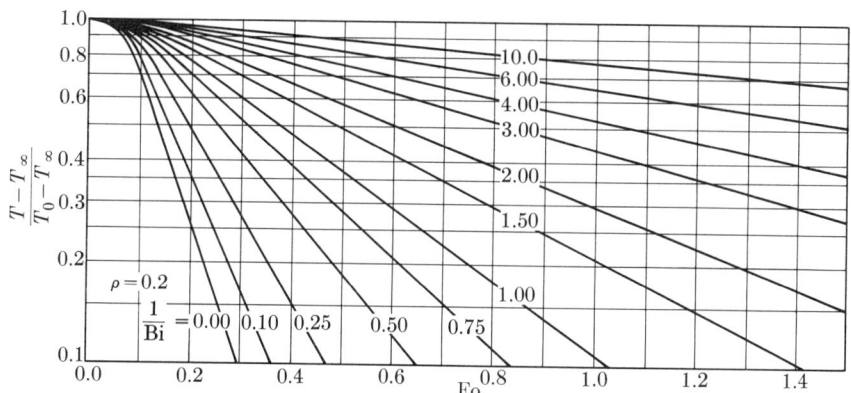

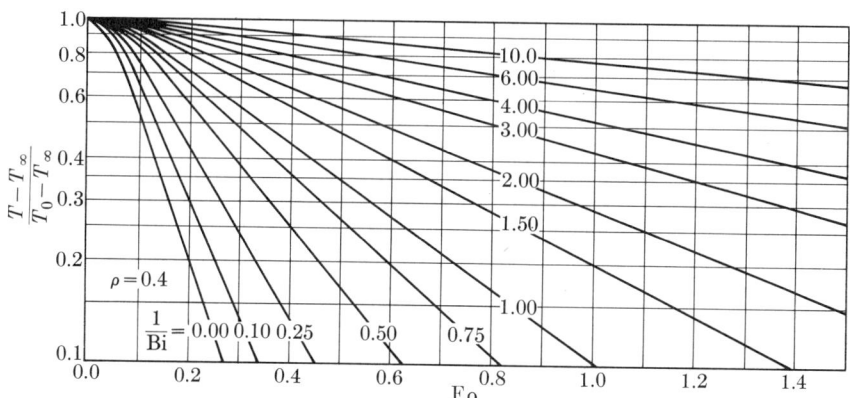

FIG. 5–18

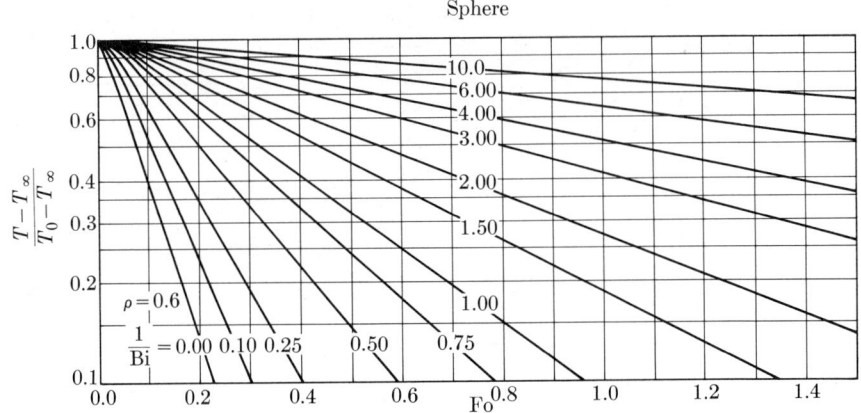
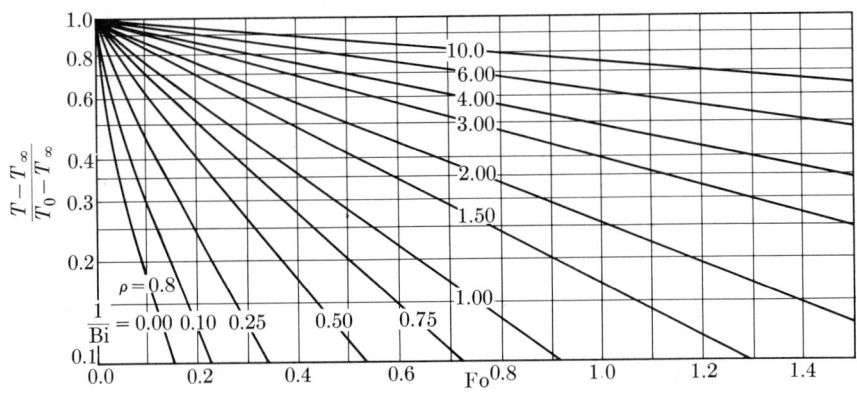
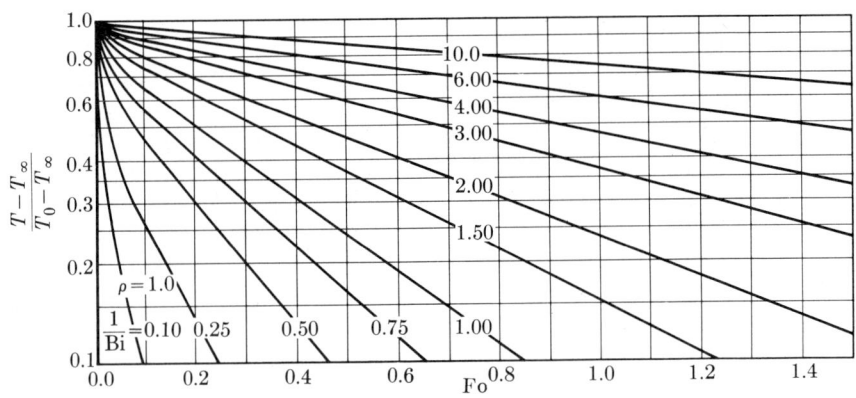

FIG. 5–18 (*cont.*)

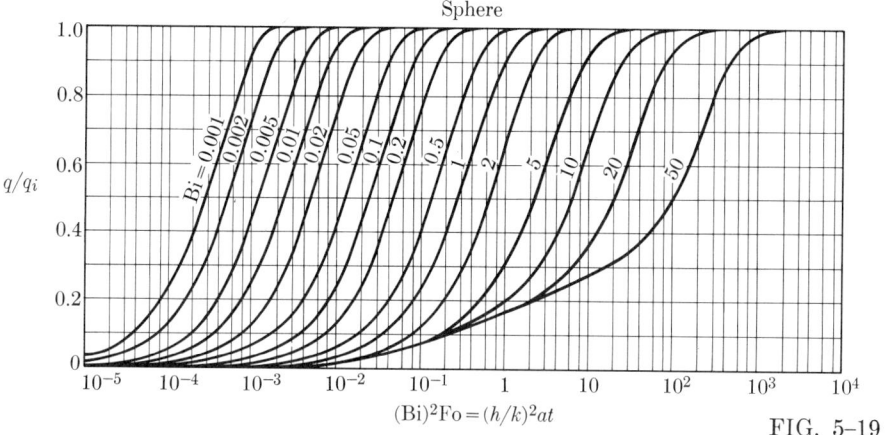

FIG. 5-19

In Fig. 5-19, we have plotted $(q/q_i)_{\text{Sphere}}$ against $(\text{Bi})^2\text{Fo}$, with Bi as parameter.

In this section we have discovered an important property of a certain class of unsteady problems, now generalized as follows: *Any unsteady three-dimensional cartesian problem $T(x, y, z, t)$ or two-dimensional cylindrical problem $T(r, z, t)$ subject to a homogeneous differential equation and homogeneous boundary conditions can be expressed as the product of unsteady one-dimensional problems if its initial condition is constant or expressible in the product form $X_i(x)Y_i(y)Z_i(z)$ or $R_i(r)Z_i(z)$, respectively.*

This procedure can also be extended to cases of semi-infinite geometry. For example, consider an infinitely thick flat plate at the uniform initial temperature T_0. The temperature of the ambient is changed suddenly to the temperature T_∞. The heat transfer coefficient is h (Fig. 5-20).

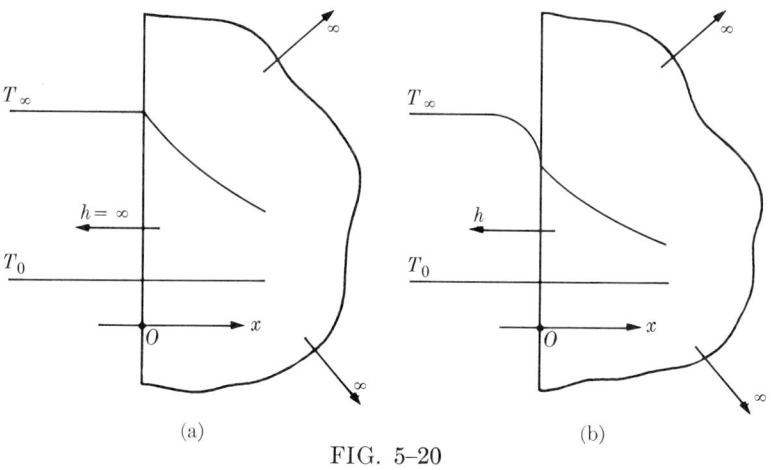

FIG. 5-20

The unsteady temperature of the plate corresponding to an infinite heat transfer coefficient is*

$$\left(\frac{T-T_\infty}{T_0-T_\infty}\right)_x = \text{erf}\left[\frac{x}{2(at)^{1/2}}\right] \qquad (5\text{-}96)$$

and that corresponding to a finite heat transfer coefficient is

$$\left(\frac{T-T_\infty}{T_0-T_\infty}\right)_x = \text{erf}\left[\frac{x}{2(at)^{1/2}}\right]$$

$$- e^{hx/k+(h/k)^2 at}\,\text{erf}\left[\frac{x}{2(at)^{1/2}} + \frac{h}{k}(at)^{1/2}\right]. \qquad (5\text{-}97)$$

In Fig. 5-21, Eq. (5-96) is plotted against $x/2(at)^{1/2}$. In Figs. 5-22 and 5-23, Eq. (5-97) is plotted against different values of the local Biot modulus, $\text{Bi}_x = hx/k$, with $(h/k)^2 at$ as parameter.

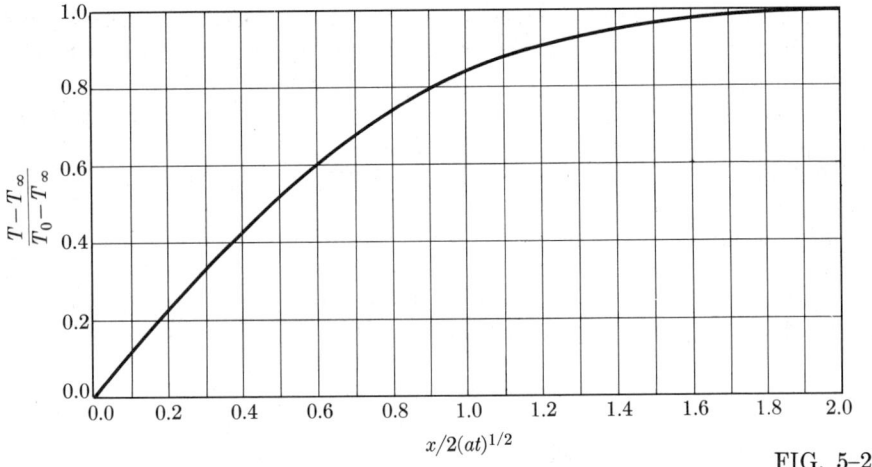

FIG. 5-21

By referring to Eqs. (5-96) and (5-97), we may write the temperature of two- and three-dimensional corners (Fig. 5-24) in the form

$$\left(\frac{T-T_\infty}{T_0-T_\infty}\right)_{x,y} = \left(\frac{T-T_\infty}{T_0-T_\infty}\right)_x \left(\frac{T-T_\infty}{T_0-T_\infty}\right)_y \qquad (5\text{-}98)$$

and

$$\left(\frac{T-T_\infty}{T_0-T_\infty}\right)_{x,y,z} = \left(\frac{T-T_\infty}{T_0-T_\infty}\right)_x \left(\frac{T-T_\infty}{T_0-T_\infty}\right)_y \left(\frac{T-T_\infty}{T_0-T_\infty}\right)_z. \qquad (5\text{-}99)$$

---

* Equations (5-96) and (5-97) cannot be obtained by the techniques of Chapters 4 and 5, and are left to Chapter 7. See Example 7-3 for Eq. (5-96).

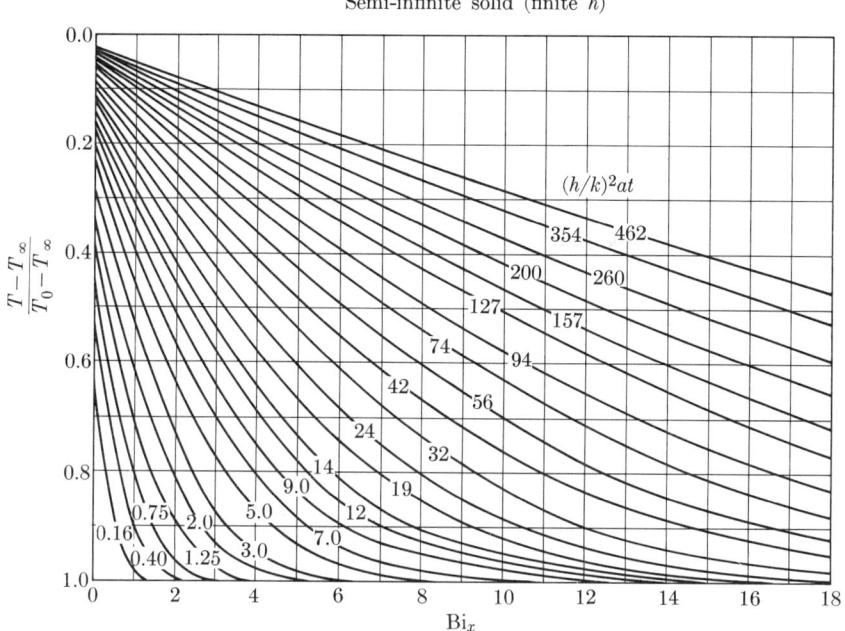

FIG. 5-22

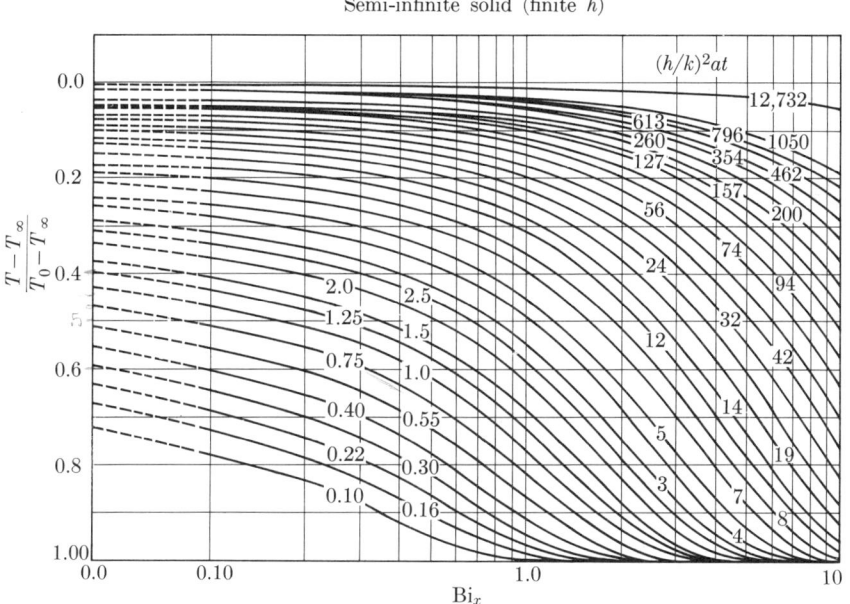

FIG. 5-23

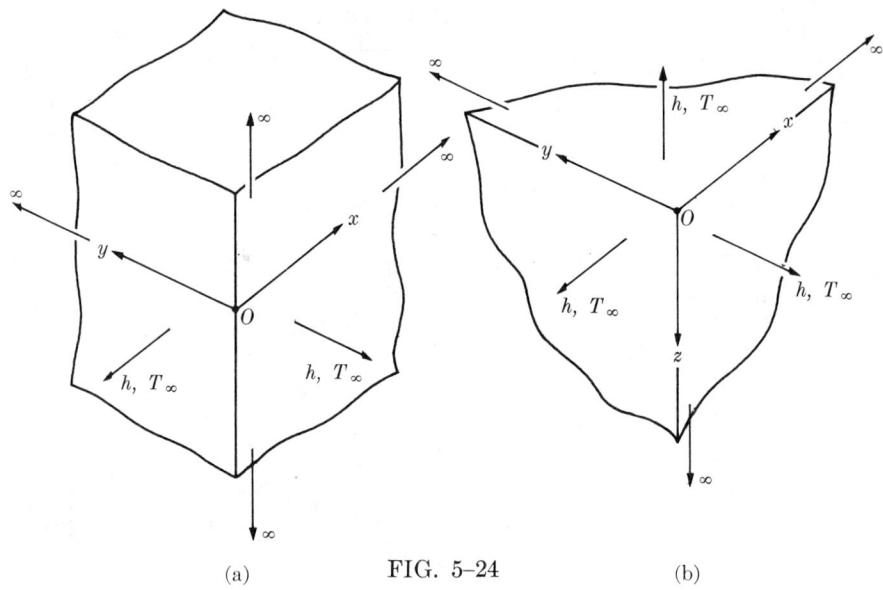

Similarly, the temperature of a semi-infinite rod of radius R (Fig. 5–25) is found to be

$$\left(\frac{T - T_\infty}{T_0 - T_\infty}\right)_{r,z} = \left(\frac{T - T_\infty}{T_0 - T_\infty}\right)_r \left(\frac{T - T_\infty}{T_0 - T_\infty}\right)_z, \qquad (5\text{--}100)$$

where $[(T - T_\infty)/(T_0 - T_\infty)]_r$, which depends on the heat transfer coefficient, is given by Eq. (5–51) or Eq. (5–55). The details of the semi-infinite geometry are left to the reader.

The use of one-dimensional charts can best be illustrated by a numerical problem.

Example 5–13. A heat treatment process requires that a cylindrical steel rod of diameter $D = 2''$ and length $2L = 4''$ heated to $T_0 = 1700°F$ in a furnace be quenched in an oil bath at $T_\infty = 100°F$. The heat transfer co-

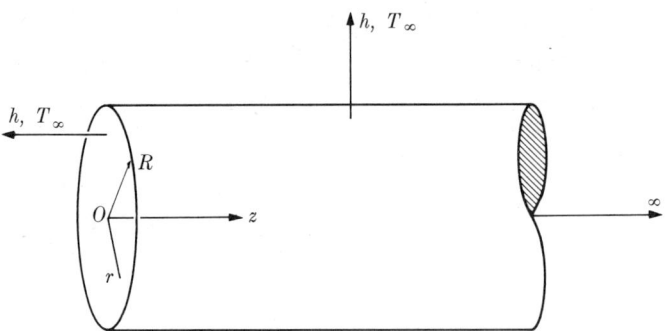

FIG. 5–25

efficient h is 100 Btu/ft²·hr·°F, and the density, specific heat, and thermal conductivity of the rod are $\rho = 500$ lbm/ft³, $c_p = 0.10$ Btu/lbm·°F and $k = 12.5$ Btu/ft·hr·°F. (a) How long will it take for the center M of the cylinder (Fig. 5–26) to reach a temperature of 500°F? (b) What are the temperature differences between the locations M–P and M–N when the center M assumes the temperature of part (a)?

(a) The two-dimensional temperature of the rod, expressed in terms of two one-dimensional problems, is

$$\left(\frac{T - T_\infty}{T_0 - T_\infty}\right)_{\substack{2R,2L \\ \text{Rod}}} = \left(\frac{T - T_\infty}{T_0 - T_\infty}\right)_{\substack{\text{Infinite} \\ 2R \text{ Rod}}} \left(\frac{T - T_\infty}{T_0 - T_\infty}\right)_{\substack{2L \\ \text{Plate}}} \tag{5–89}$$

The desired dimensionless temperature at $M(r/R = 0, x/L = 0)$,

$$\left(\frac{T - T_\infty}{T_0 - T_\infty}\right)_{\substack{2R,2L \\ \text{Rod}}} = \frac{500 - 100}{1700 - 100} = 0.25,$$

must be equal to the product of the temperatures of the $2R$ rod and the $2L$ plate. This requires a trial-and-error procedure. However, noting that the heat transfer from the ends of the cylinder is much smaller than that from the peripheral area, we may neglect the effect of the plate in a first approximation.

We calculate the Biot number

$$\text{Bi}_R = \frac{hR}{k} = \frac{100 \times 1/12}{12.5} = \frac{1}{1.5}, \quad \frac{1}{\text{Bi}_R} = 1.5,$$

and from the charts of Fig. 5–16 for $r/R = 0$ and $(T - T_\infty)/(T_0 - T_\infty) = 0.25$ we obtain the Fourier number

$$\text{Fo}_R = at/R^2 \cong 1.36,$$

which corresponds to the time

$$t = 1.36\frac{R^2}{a} = 1.36\frac{(1/12)^2}{12.5/(500 \times 0.1)} = 0.038 \text{ hr} \quad (2 \text{ min } 17 \text{ sec}).$$

The heat transfer from the ends slightly shortens this time. A trial-and-error procedure yields $t = 0.035$ hr (2 min 6 sec). The Fourier numbers based on this time are

$$\text{Fo}_R = \frac{at}{R^2} = \frac{0.25 \times 0.035}{(1/12)^2} = 1.26,$$

$$\text{Fo}_L = \frac{at}{L^2} = \frac{0.25 \times 0.035}{(2/12)^2} = 0.315.$$

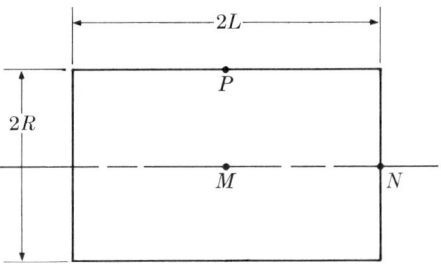

FIG. 5–26

Using Fig. 5–16 for $r/R = 0$, $1/\text{Bi}_R = 1.5$, and $\text{Fo}_R = 1.26$, we have

$$\left(\frac{T - T_\infty}{T_0 - T_\infty}\right)_{\substack{\text{Infinite} \\ 2R \text{ Rod}}} \cong 0.28, \qquad r/R = 0,$$

and Figure 5–14 yields for $x/L = 0$, $1/\text{Bi}_L = 1/2\text{Bi}_R = 0.75$, and $\text{Fo}_L = 0.315$

$$\left(\frac{T - T_\infty}{T_0 - T_\infty}\right)_{2L \text{ Plate}} \cong 0.88, \qquad x/L = 0.$$

Hence

$$\left(\frac{T - T_\infty}{T_0 - T_\infty}\right)_{\substack{2R, 2L \\ \text{Rod}}} \cong 0.28 \times 0.88 \cong 0.246,$$

which is close to the actual value, 0.250. Note that the effect of finite length, as seen from the value 0.88 obtained for the plate, lowers the center temperature by only 12%.

(b) Figure 5–16 gives for $r/R = 1$, $1/\text{Bi}_R = 1.5$, and $\text{Fo}_R = 1.26$

$$\left(\frac{T - T_\infty}{T_0 - T_\infty}\right)_{\substack{\text{Infinite} \\ 2R \text{ Rod}}} \cong 0.21, \qquad r/R = 1.$$

The temperature at $P(r/R = 1, x/L = 0)$ is then

$$\left(\frac{T - T_\infty}{T_0 - T_\infty}\right)_{\substack{2R, 2L \\ \text{Rod}}} \cong 0.21 \times 0.88 = 0.18,$$

and the temperature difference between the locations M–P is found to be

$$\Delta T_{M-P} = 1600 \times (0.25 - 0.18) = 112°\text{F}.$$

Similarly, Figure 5–14 yields for $x/L = 1$, $1/\text{Bi}_L = 0.75$, and $\text{Fo}_L = 0.315$

$$\left(\frac{T - T_\infty}{T_0 - T_\infty}\right)_{2L \text{ Plate}} \cong 0.50, \qquad x/L = 1,$$

and the temperature at $N(r/R = 0, x/L = 1)$ is

$$\left(\frac{T - T_\infty}{T_0 - T_\infty}\right)_{\substack{2R, 2L \\ \text{Rod}}} = 0.28 \times 0.50 = 0.14.$$

Thus the temperature difference between the locations M–N is

$$\Delta T_{M-N} = 1600 \times (0.25 - 0.14) \cong 176°\text{F}.$$

5–3. Time-Dependent Boundary Conditions. Duhamel's Superposition Integral

In preceding sections of this chapter we have examined unsteady problems arising from stepwise disturbances, such as sudden internal energy generation in a body or sudden change in temperature, heat flux, or ambient temperature along part or all of a boundary. When such disturbances vary with time instead, the procedure employed in Sections 5–1 and 5–2 fails to yield a solution. Two other methods of solution are, however, available.* The first is due to Duhamel, who showed that the problem of a time-dependent disturbance could be reduced to that of a stepwise disturbance. The second involves the use of Laplace transforms. Here we consider only Duhamel's method, leaving Laplace transforms to Chapter 7. Our investigation will be restricted to systems initially at zero temperature, for in practice most problems may be reduced to this case by the method of superposition.

Let the unsteady temperature of a problem resulting from a stepwise *unit* disturbance be $\psi(\mathbf{r}, t)$ relative to a uniform initial temperature, say zero. If the disturbance is kept zero until a certain time $t = s$, and at that instant is raised to unity and maintained constant thereafter, the new temperature $\phi(\mathbf{r}, t)$ may readily be expressed in terms of $\psi(\mathbf{r}, t)$ as

$$\phi(\mathbf{r}, t) = \begin{cases} 0, & t < s \\ \psi(\mathbf{r}, t - s), & t > s. \end{cases} \quad (5\text{-}101)$$

If the disturbance of the same problem is time-dependent rather than stepwise (Fig. 5–27), the following procedure is adopted. First this disturbance is approximated by assuming that it is increased suddenly to $D(0)$ when $t = 0$

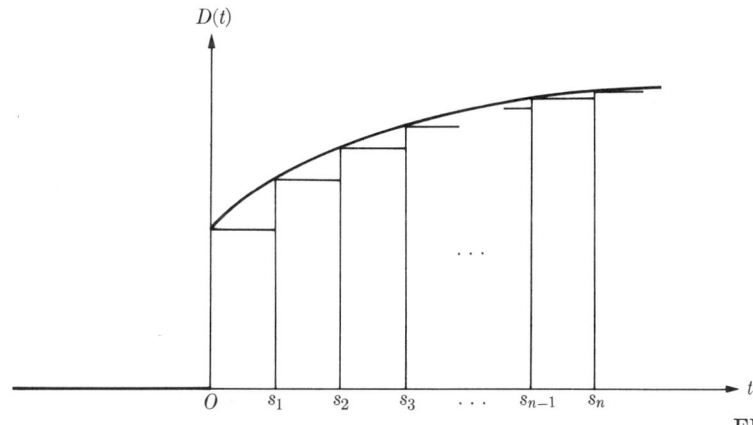

FIG. 5–27

---

* A third possible technique is the use of Green's function, which we shall not discuss in this text.

and is kept at this value until $t = s_1$, at which time it is raised suddenly by an amount $D(s_1) - D(0)$ and held at that value until $t = s_2$, and so on. Next, noting the linearity of the problem, we write the temperature variation $\phi(\mathbf{r}, t)$ after the time $t = s_n$ in the form

$$\begin{aligned}\phi(\mathbf{r}, t) = D(0)\psi(\mathbf{r}, t) &+ [D(s_1) - D(0)]\psi(\mathbf{r}, t - s_1) \\ &+ [D(s_2) - D(s_1)]\psi(\mathbf{r}, t - s_2) \\ &\vdots \\ &+ [D(s_n) - D(s_{n-1})]\psi(\mathbf{r}, t - s_n).\end{aligned} \quad (5\text{-}102)$$

Introducing the definitions

$$D(s_m) - D(s_{m-1}) = \Delta D_m, \qquad s_m - s_{m-1} = \Delta s_m,$$

we rearrange Eq. (5-102) to give

$$\phi(\mathbf{r}, t) = D(0)\psi(\mathbf{r}, t) + \sum_{m=1}^{n} \psi(\mathbf{r}, t - s_m) \left(\frac{\Delta D}{\Delta s}\right)_m \Delta s_m. \quad (5\text{-}103)$$

Thus in the limit as $n \to \infty$ all intervals between successive steps tend to zero, and the definition of integral yields

$$\phi(\mathbf{r}, t) = D(0)\psi(\mathbf{r}, t) + \int_0^t \psi(\mathbf{r}, t - s) \frac{dD(s)}{ds} ds. \quad (5\text{-}104)$$

This is *Duhamel's superposition integral*, which gives the solution $\phi(\mathbf{r}, t)$ corresponding to the time-dependent disturbance $D(t)$ in terms of the solution $\psi(\mathbf{r}, t)$ corresponding to the stepwise disturbance of the same kind.

An alternative form of Eq. (5-104) may be obtained by an integration by parts as

$$\phi(\mathbf{r}, t) = D(0)\psi(\mathbf{r}, t) + D(s)\psi(\mathbf{r}, t - s)\big|_{s=0}^{s=t} - \int_0^t D(s) \frac{\partial \psi(\mathbf{r}, t - s)}{\partial s} ds,$$

which, since

$$\partial \psi(\mathbf{r}, t - s)/\partial s = -\partial \psi(\mathbf{r}, t - s)/\partial t,$$

may be rearranged in the form

$$\phi(\mathbf{r}, t) = D(t)\psi(\mathbf{r}, 0) + \int_0^t D(s) \frac{\partial \psi(\mathbf{r}, t - s)}{\partial t} ds. \quad (5\text{-}105)$$

Furthermore, since $\psi(\mathbf{r}, 0) = 0$ according to the statement of the problem, Eq. (5-105) reduces to

$$\phi(\mathbf{r}, t) = \int_0^t D(s) \frac{\partial \psi(\mathbf{r}, t - s)}{\partial t} ds. \quad (5\text{-}106)$$

The use of the foregoing procedure will now be illustrated by three examples. First we consider a lumped problem whose solution is also available by standard methods.

Example 5–14. We wish to re-solve the unsteady periodic case of Example 5–1 by using its transient solution and Duhamel's integral.

According to the notation of this section, the solution of the transient case is

$$\frac{\psi(t)}{q''/h} = 1 - e^{-mt}, \tag{2-172}$$

and the time-dependent disturbance is

$$D(t) = \cos \omega t. \tag{5-107}$$

Introducing Eqs. (2–172) and (5–107) into Duhamel's integral, Eq. (5–104), we have

$$\frac{\phi(t)}{q''/h} = 1 - e^{-mt} - \omega \int_0^t [1 - e^{-m(t-s)}] \sin \omega s \, ds. \tag{5-108}$$

Evaluating the integral of Eq. (5–108) and rearranging yields the solution of the problem,

$$\frac{\phi(t)}{q''/h} = \frac{m}{(m^2 + \omega^2)^{1/2}} \cos(\omega t - \alpha) - \frac{m^2 e^{-mt}}{(m^2 + \omega^2)},$$

which is identical to Eq. (5–3).

Sometimes it is more convenient to use the alternative form of Duhamel's integral given by Eq. (5–106). For example, Eqs. (2–172) and (5–107) introduced into Eq. (5–106) would yield

$$\frac{\phi(t)}{q''/h} = m \int_0^t e^{-m(t-s)} \cos \omega s \, ds, \tag{5-109}$$

which also results in Eq. (5–3). Although the integration of Eq. (5–109) is similar to the integration of Eq. (5–108), the evaluation of the latter is simpler than that of the former.

The foregoing procedure may readily be extended to the distributed case of the same problem. Thus, introducing Eqs. (5–34) and (5–107) into Eq. (5–106), we obtain

$$\frac{\phi(x,t)}{q''L/k} = 2 \sum_{n=1}^{\infty} \frac{a\lambda_n^2 (a\lambda_n^2 \cos \omega t + \omega \sin \omega t - a\lambda_n^2 e^{-a\lambda_n^2 t})}{(\lambda_n L)(\lambda_n L + \sin \lambda_n L \cos \lambda_n L)(a^2 \lambda_n^4 + \omega^2)} \cos \lambda_n x. \tag{5-110}$$

The steps of this solution are left to the reader.

Example 5-15. Suppose that the ambient temperature of Example 5-2 now varies linearly in time as measured from its initial value. The heat transfer coefficient remains large. We wish to find the unsteady temperature of the plate.

The solution of Example 5-2 written in terms of $\psi = (T - T_0)/(T_\infty - T_0)$ can be utilized for the solution ϕ of the present problem. Here ψ, obtained from Eq. (5-10), satisfies

$$\psi(x, t) = 1 - 2 \sum_{n=0}^{\infty} \frac{(-1)^n}{(\lambda_n L)} e^{-a\lambda_n^2 t} \cos \lambda_n x, \qquad (5\text{-}111)$$

and the time-dependent disturbance (surface temperature) is

$$D(t) = (\phi_0/t_0)t, \qquad (5\text{-}112)$$

where $\phi_0 = T_\infty - T_0$ and t_0 is a time constant. Replacing r by x and introducing Eqs. (5-111) and (5-112) into Eq. (5-104), we have

$$\phi(x, t) = \frac{\phi_0}{t_0} \int_0^t \left[1 - 2 \sum_{n=0}^{\infty} \frac{(-1)^n}{(\lambda_n L)} e^{-a\lambda_n^2(t-s)} \cos \lambda_n x \right] ds. \qquad (5\text{-}113)$$

The integration of Eq. (5-113) gives the solution of the problem in the form

$$\frac{\phi(x, t)}{\phi_0} = \left(\frac{t}{t_0}\right) - 2 \sum_{n=0}^{\infty} \frac{(-1)^n (1 - e^{-a\lambda_n^2 t})}{(\lambda_n L)(a\lambda_n^2 t_0)} \cos \lambda_n x. \qquad (5\text{-}114)$$

Similarly, inserting Eqs. (5-111) and (5-112) into Eq. (5-106), we have

$$\phi(x, t) = 2 \frac{\phi_0}{t_0} \sum_{n=0}^{\infty} \frac{(-1)^n (a\lambda_n^2)}{(\lambda_n L)} e^{-a\lambda_n^2 t} \int_0^t s e^{a\lambda_n^2 s} \, ds. \qquad (5\text{-}115)$$

The integration of Eq. (5-115) yields the alternative form of the solution

$$\frac{\phi(x, t)}{\phi_0} = 2 \sum_{n=0}^{\infty} \frac{(-1)^n}{(\lambda_n L)} \left[\left(\frac{t}{t_0}\right) - \left(\frac{1 - e^{-a\lambda_n^2 t}}{a\lambda_n^2 t_0}\right) \right] \cos \lambda_n x. \qquad (5\text{-}116)$$

Since $\psi(x, 0) = 0$, we have from Eq. (5-111)

$$1 = 2 \sum_{n=0}^{\infty} \frac{(-1)^n}{(\lambda_n L)} \cos \lambda_n x.$$

Using this result, we may show that Eq. (5-114) is identical to Eq. (5-116), as expected.

Example 5-16. Let us reconsider the empty skillet of Example 3–12. Assume that the initial temperature of the skillet is equal to the ambient temperature T_∞, and that the uniform heat flux q'' is suddenly applied to the bottom of the skillet. We wish to find the unsteady temperature of the skillet.

Here we consider only the simple case for which the bottom temperature of the skillet can be lumped. (See the discussion in Example 3–12.) The lumped analysis of the bottom readily gives

$$\rho c A \delta \frac{dT_2}{dt} = q''A - h_2 A(T_2 - T_\infty), \qquad (5\text{-}117)$$

where A is the total surface area of the bottom.

The solution of Eq. (5–117), subject to the initial condition $T_2(0) = T_\infty$, is

$$\theta_2(t) = \frac{q''}{h_2}(1 - e^{-bt}), \qquad (5\text{-}118)$$

where $\theta_2(t) = T_2(t) - T_\infty$ and $b = h_2/\rho c \delta$. Now, employing Eq. (5–118) as a boundary condition, we may write for the side walls

$$\frac{\partial^2 \theta_1}{\partial x^2} - m_1^2 \theta_1 = \frac{1}{a}\frac{\partial \theta_1}{\partial t}, \qquad m_1^2 = 2h_1/k\delta,$$

$$\theta_1(x, 0) = 0, \qquad (5\text{-}119)$$

$$\frac{\partial \theta_1(0, t)}{\partial x} = 0, \qquad \theta_1(L, t) = \frac{q''}{h_2}(1 - e^{-bt}).$$

As a step toward the solution of Eq. (5–119), we first consider the auxiliary problem $\psi(x, t)$, whose formulation is identical to that of Eq. (5–119) except for the last boundary condition, now replaced by the step disturbance $\theta_1(L, t) = q''/h_2$. The nonhomogeneity of this problem suggests the use of the superposition $\psi(x, t) = \phi(x, t) + \varphi(x)$, where $\varphi(x)$ and $\phi(x, t)$ satisfy

$$\frac{d^2\varphi}{dx^2} - m_1^2\varphi = 0, \qquad \frac{d\varphi(0)}{dx} = 0, \qquad \varphi(L) = \frac{q''}{h_2}, \qquad (5\text{-}120)$$

and

$$\frac{\partial^2 \phi}{\partial x^2} - m_1^2 \phi = \frac{1}{a}\frac{\partial \phi}{\partial t}, \qquad \phi(x, 0) = -\varphi(x),$$

$$\frac{\partial \phi(0, t)}{\partial x} = 0, \qquad \phi(L, t) = 0. \qquad (5\text{-}121)$$

The steady problem of Eq. (5–120) is identical to Example 3–10 provided θ_0

is replaced by q''/h_2, and m_1 is interpreted properly. Hence the solution of Eq. (5–120) is

$$\frac{\varphi(x)}{q''/h_2} = \frac{\cosh m_1 x}{\cosh m_1 L}. \tag{5-122}$$

The method of separation of variables applied to Eq. (5–121) yields

$$\frac{d^2 X}{dx^2} + \lambda^2 X = 0,$$

$$\frac{dX(0)}{dx} = 0, \quad X(L) = 0; \tag{5-123}$$

$$\frac{d\tau}{dt} + a(\lambda^2 + m^2)\tau = 0. \tag{5-124}$$

The solution of Eq. (5–123) is

$X_n(x) = A_n \cos \lambda_n x, \quad \phi_n(x) = \cos \lambda_n x, \quad$ characteristic functions,

$\lambda_n L = (2n+1)\pi/2, \quad n = 0, 1, 2, 3, \ldots, \quad$ characteristic values,

and the solution of Eq. (5–124) is

$$\tau_n(t) = C_n e^{-a(\lambda_n^2 + m^2)t}.$$

Hence the product solution of ϕ becomes

$$\phi(x, t) = \sum_{n=0}^{\infty} a_n e^{-a(\lambda_n^2 + m^2)t} \cos \lambda_n x. \tag{5-125}$$

The initial value of Eq. (5–125), which is equal to $-\varphi(x)$, gives

$$-\frac{q''}{h_2} \frac{\cosh m_1 x}{\cosh m_1 L} = \sum_{n=0}^{\infty} a_n \cos \lambda_n x. \tag{5-126}$$

The coefficient a_n of Eq. (5–126) may be evaluated in the usual manner. The result is

$$a_n = -(-1)^n \frac{2}{L} \left(\frac{q''}{h_2}\right) \frac{\lambda_n}{m_1^2 + \lambda_n^2}. \tag{5-127}$$

Introducing Eq. (5–127) into Eq. (5–125) and adding the result to Eq. (5–122) gives the solution of $\psi(x, t)$ in the form

$$\frac{\psi(x, t)}{q''/h_2} = \frac{\cosh m_1 x}{\cosh m_1 L} - 2 \sum_{n=0}^{\infty} \frac{(-1)^n (\lambda_n L) e^{-a(m_1^2 + \lambda_n^2)t}}{(m_1^2 + \lambda_n^2) L^2} \cos \lambda_n x. \tag{5-128}$$

Finally, noting that $D(s) = 1 - e^{-bs}$ and using Eq. (5–104), we find the solution of $\theta_1(x, t)$ to be

$$\frac{\theta_1(x, t)}{q''/h_2} = b \int_0^t \left[\frac{\cosh m_1 x}{\cosh m_1 L} - 2 \sum_{n=0}^{\infty} \frac{(-1)^n (\lambda_n L) e^{-a(m_1^2 + \lambda_n^2)(t-s)}}{(m_1^2 + \lambda_n^2) L^2} \cos \lambda_n x \right] e^{-bs} \, ds. \tag{5–129}$$

The integration of Eq. (5–129) yields the unsteady temperature of the side walls as

$$\frac{\theta_1(x, t)}{q''/h_2} = \frac{\cosh m_1 x}{\cosh m_1 L} (1 - e^{-bt})$$

$$+ 2 \left(\frac{b}{a} \right) \sum_{n=0}^{\infty} \frac{(-1)^n (\lambda_n L)[e^{-a(m_1^2 + \lambda_n^2)t} - e^{-bt}]}{(m_1^2 + \lambda_n^2) L^2 (m_1^2 + \lambda_n^2 - b/a)} \cos \lambda_n x. \tag{5–130}$$

Equation (5–130) may also be given an alternative form. Thus, combining Eqs. (5–126) and (5–127), we obtain

$$\frac{\cosh m_1 x}{\cosh m_1 L} = 2 \sum_{n=0}^{\infty} \frac{(-1)^n (\lambda_n L)}{(m_1^2 + \lambda_n^2) L^2} \cos \lambda_n x, \tag{5–131}$$

and replacing m_1 by $(m_1^2 - b/a)^{1/2}$ in Eq. (5–131), we have

$$\frac{\cosh (m_1^2 - b/a)^{1/2} x}{\cosh (m_1^2 - b/a)^{1/2} L} = 2 \sum_{n=0}^{\infty} \frac{(-1)^n (\lambda_n L)}{(m_1^2 + \lambda_n^2 - b/a) L^2} \cos \lambda_n x. \tag{5–132}$$

Then, from the difference of Eqs. (5–132) and (5–131), we have

$$\frac{\cosh (m_1^2 - b/a)^{1/2} x}{\cosh (m_1^2 - b/a)^{1/2} L} - \frac{\cosh m_1 x}{\cosh m_1 L} = 2 \left(\frac{b}{a} \right) \sum_{n=0}^{\infty} \frac{(-1)^n (\lambda_n L) \cos \lambda_n x}{(m_1^2 + \lambda_n^2) L^2 (m_1^2 + \lambda_n^2 - b/a)}. \tag{5–133}$$

Multiplying Eq. (5–133) by e^{-bt} and introducing the result into Eq. (5–130), we may rearrange the solution of $\theta_1(x, t)$ in the form

$$\frac{\theta_1(x, t)}{q''/h_2} = \frac{\cosh m_1 x}{\cosh m_1 L} - \frac{e^{-bt} \cosh (m_1^2 - b/a)^{1/2} x}{\cosh (m_1^2 - b/a)^{1/2} L}$$

$$+ 2 \left(\frac{b}{a} \right) \sum_{n=0}^{\infty} \frac{(-1)^n (\lambda_n L) e^{-a(m_1^2 + \lambda_n^2)t}}{(m_1^2 + \lambda_n^2) L^2 (m_1^2 + \lambda_n^2 - b/a)} \cos \lambda_n x. \tag{5–134}$$

Equation (5–134) is identical to the solution to be obtained by Laplace transforms in Chapter 7 (see Example 7–14).

References

1. H. S. CARSLAW and J. C. JAEGER, *Conduction of Heat in Solids.* Oxford: Clarendon Press, 1959.
2. R. V. CHURCHILL, *Fourier Series and Boundary Value Problems.* New York: McGraw-Hill, 1963.
3. F. B. HILDEBRAND, *Advanced Calculus for Engineers.* Englewood Cliffs: Prentice-Hall, 1956.
4. L. M. K. BOELTER, V. H. CHERRY, H. A. JOHNSON, and R. C. MARTINELLI, *Heat Transfer Notes.* New York: McGraw-Hill, 1965.
5. H. GRÖBER, S. ERK, and U. GRIGULL, *Fundamentals of Heat Transfer.* New York: McGraw-Hill, 1961.

Problems

Problems 5-1 through 5-6 deal with lumped systems.

5-1. An electron tube with steel casing is shown in Fig. 5-28. The thermal conductivity, mean diameter, length, and thickness of the tube are k, D, L, and δ, respectively. The heat transfer coefficient is h, and the ambient temperature T_∞. When the filament is heated, we may assume that the inner surface of the tube is suddenly subjected to a net radiant heat flux q''. Find the unsteady temperature of the tube.

5-2. An empty vertical tube is initially at the ambient temperature T_∞; then a fluid having the temperature T_0 starts filling the tube with the uniform velocity V (Fig. 5-29). The axial conduction in and the radial resistance of the tube may be neglected. The density, specific heat, and cross-sectional area of the tube and fluid are (ρ_w, ρ_f), (c_{p_w}, c_{p_f}), and (A_w, A_f), respectively. It is known that $(\rho c_p A)_f \gg (\rho c_p A)_w$. The inner and outer heat transfer coefficients are h_i, h_o. The heat transfer between the fluid and ambient can be neglected compared to that between the fluid and tube.

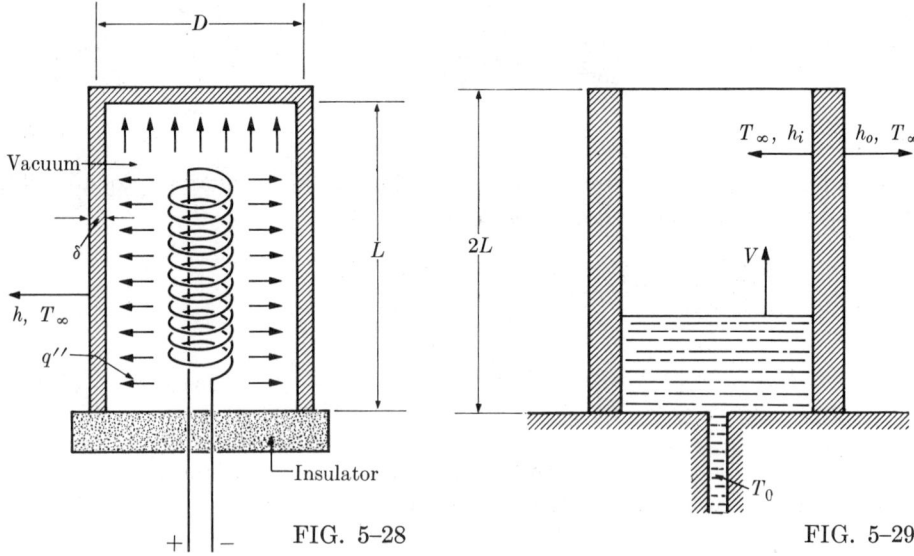

FIG. 5-28   FIG. 5-29

Calculate the temperature of the tube halfway between the ends when the fluid reaches the top of the tube.

5-3. A wall divides two fluids as shown in Fig. 5-30. The density, heat capacity, and thickness of the wall and the finite fluid are (ρ_w, ρ_f), (c_{p_w}, c_{p_f}), and (δ_w, δ_f), respectively. The system is initially at the uniform temperature T_0; then the temperature of the infinite fluid is changed suddenly to the temperature T_∞. The heat transfer coefficients are h_1, h_2. Find the unsteady temperature of the wall and of the finite fluid. [*Hint:* Introduce the operator $D \equiv d/dt$ into the formulation of the problem. Solve the resulting simultaneous algebraic equations by elimination or by determinants. Replace D by d/dt, and solve by classical means the two separate linear second-order differential equations with constant coefficients thus obtained. Eliminate the extra integration constants by inserting these solutions into the governing equation of the wall or that of the fluid. Evaluate the remaining integration constants by considering the initial condition of the problem.]

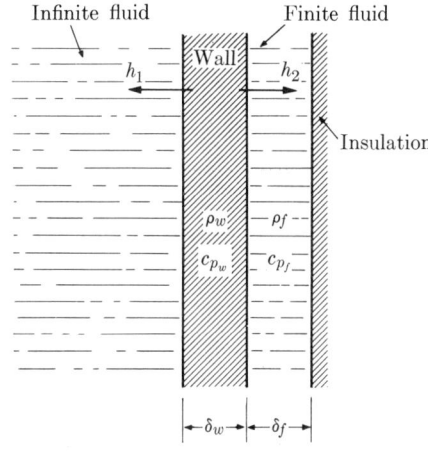

FIG. 5-30

5-4. The brake system of a car is shown in Fig. 5-31. The density, heat capacity, and thickness of the brake shoes and drum are (ρ_1, ρ_2), (c_{p_1}, c_{p_2}), and (δ_1, δ_2), respectively. The system has the uniform initial temperature T_0. During a stopping period Δt, the pressure p between the shoes and drum, as well as the coefficient of dry friction μ, are constant and the angular velocity $\omega(t)$ is specified. The effect of curvature is negligible. (a) Formulate the problem. (b) Calculate the temperature rise in the shoes and drum during the time interval Δt.

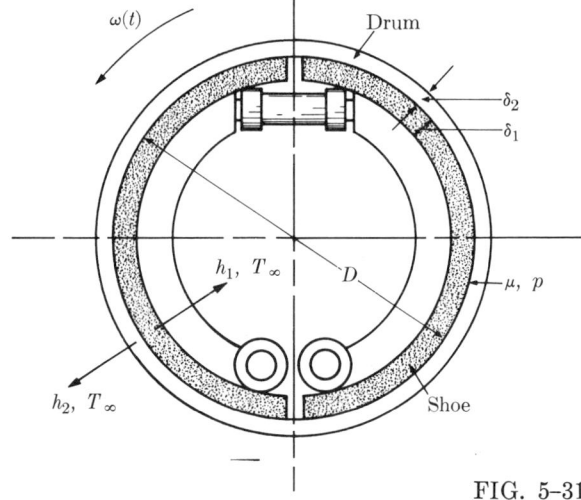

FIG. 5-31

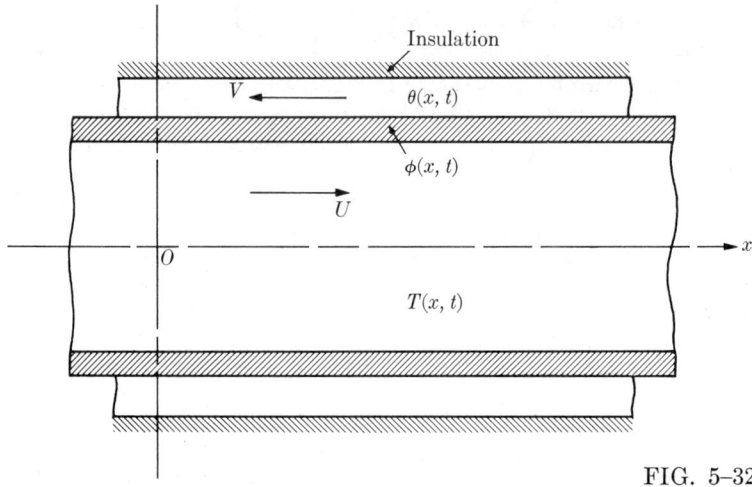

FIG. 5-32

5-5. A two-fluid counterflow heat exchanger is assumed to be made up by two coaxial tubes (Fig. 5-32). *The heat capacity of the inner tube is not negligible.* The system has the uniform initial temperature T_0; then the inlet temperature of one fluid is changed suddenly to the temperature T_∞. Formulate the problem on the basis of a radially lumped analysis.

5-6. Consider an ordinary light bulb (Fig. 5-33). (a) Calculate the steady surface temperature of the lighted bulb. (b) Find the time required for the surface temperature of the bulb to reach this steady value. Data:

$$P = 200 \text{ watts}, \quad \delta = \tfrac{1}{64} \text{ in.,}$$
$$\rho = 170 \text{ Btu/lbm·°F}, \quad \alpha = 0.10 \text{ (emissivity)},$$
$$c_p = 0.2 \text{ lbm/ft}^3, \quad T_\infty = 70°F,$$
$$D = 3 \text{ in.}$$

Problems 5-7 through 5-32 deal with distributed systems.

5-7. The classical form of the unsteady one-dimensional conduction equation

$$\frac{\partial T}{\partial t} = a \frac{\partial^2 T}{\partial x^2}$$

is the first-order approximation of the first law of thermodynamics combined with the conduction law. (a) Show that this equation is obtained by neglecting the timewise variation of conduction compared to its spatial variation, and by neglecting the spacewise variation of internal energy compared to its change in time. (b) Show that part (a) is the common approximation in the formulation of field theories (hydrodynamics, gas dynamics, heat transfer, elasticity, plasticity, electromagnetism, etc.) and can be generalized as follows: *In expressing a general law [combined with particular law(s)] in terms of a system or control volume, the timewise variation of a surface term is neglected compared to its spatial variation, and the spacewise variation of a volume term is neglected compared to its change in time.*

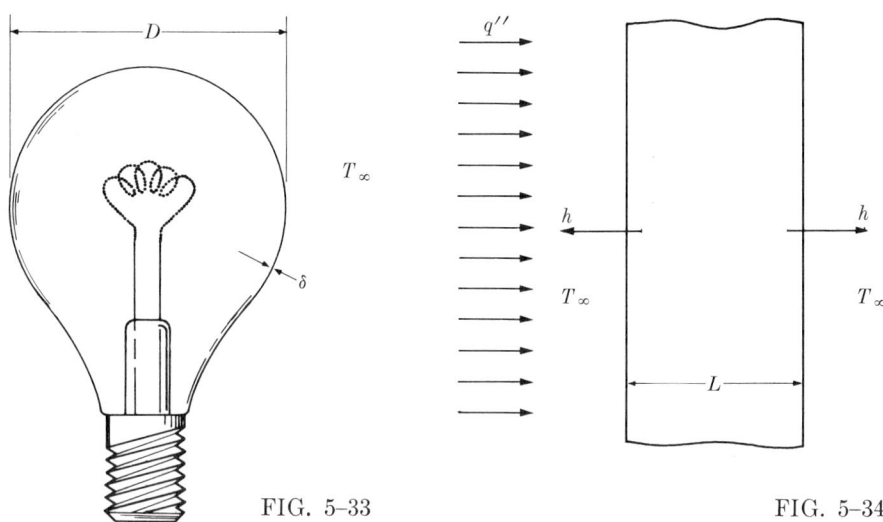

FIG. 5-33   FIG. 5-34

5-8. (a) Re-solve Example 5-4 (the differential formulation of Example 2-2) for a moderate heat transfer coefficient. (b) Re-solve Example 5-4 by the integral method and the use of second-order profiles. (See Example 2-2 for the first-order profiles.)

5-9. One surface of a flat plate of thickness L is suddenly subjected to the net radiant heat flux q'' (Fig. 5-34). The initial temperature of the plate is equal to the ambient temperature T_∞. The heat transfer coefficient h is the same for both surfaces of the plate. Find the unsteady temperature of the plate.

5-10. Find the solution of the unsteady problem associated with Example 3-10. (See Example 3-14 for an approximate first-order solution.) Assume that the initial temperature is equal to the ambient temperature T_∞, and that the base temperature is suddenly changed from T_∞ to T_0.

5-11. Repeat the preceding problem for Example 3-11.

5-12. A ring-shaped electrical resistor is made by rounding and welding a straight wire (Fig. 5-35). Assume that the electrical resistance of the wire is considerably

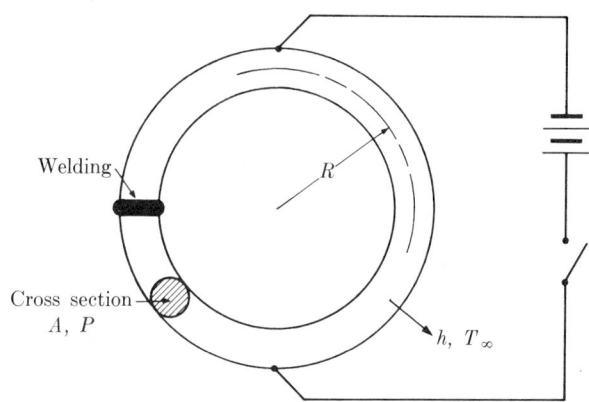

FIG. 5-35

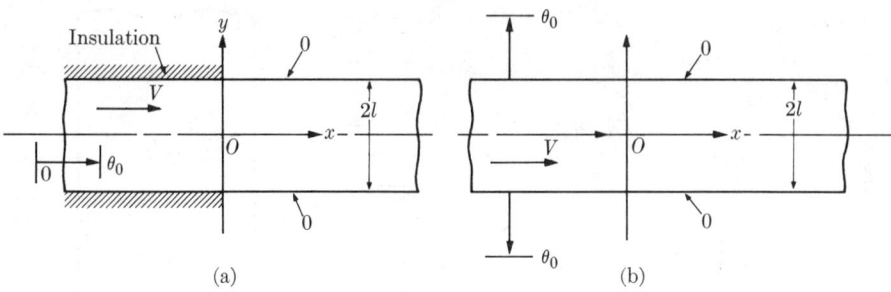

FIG. 5-36

increased at the welding point during the welding process. Initially the resistor is at the ambient temperature T_∞; then electrical power $\mathcal{P}$ is drawn to the resistor. Find the temperature variation in the resistor.

5-13. Interpret the physics of Example 4-5 as the steady slug flow of an ideal fluid between two parallel plates of negligible thickness. The plates extend from $-\infty$ to $+\infty$ in the direction of flow, say x, and are a distance $2l$ apart. The axial conduction is negligible. The initial temperature of the system is uniform, say 0. Distinguish two cases: (i) while the plates are kept at zero temperature in $0 \leq x < +\infty$ and insulated in $-\infty < x \leq 0$, the upstream temperature of the fluid is suddenly changed to θ_0 (Fig. 5-36a); (ii) while the plates are kept at zero temperature in $0 \leq x < +\infty$, they are suddenly subjected to temperature θ_0 in $-\infty < x \leq 0$ (Fig. 5-36b). Find the temperature variation in the fluid corresponding to these two cases.

5-14. Find the melting velocity of the welding electrode shown in Fig. 5-37. The diameter and initial length of the electrode are d and L, respectively. The heat transfer coefficient is h, the ambient temperature T_∞, and the electric power drawn $\mathcal{P}$. It may be assumed that the tip of the electrode instantaneously reaches the melting temperature, and that the radial temperature variation is negligible.

5-15. Re-solve Example 5-4 and Problem 5-8(a) for an infinitely long solid cylinder of radius R.

5-16. Re-solve Example 5-5 for an infinitely long solid cylinder of radius R.

5-17. Reconsider Problem 3-30 (Fig. 3-64). Both disks are initially stationary, and their temperature is equal to the ambient temperature T_∞. Then the upper disk

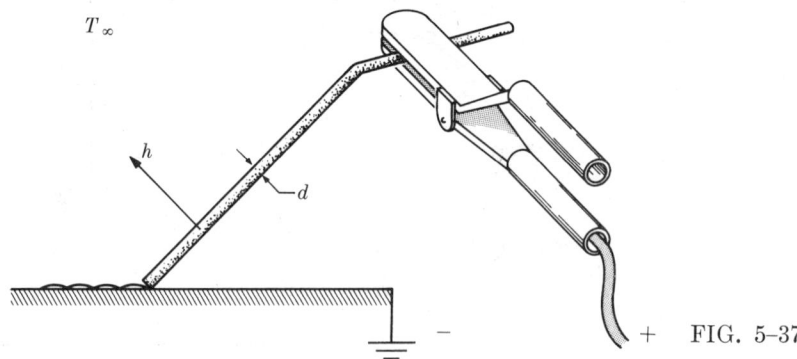

FIG. 5-37

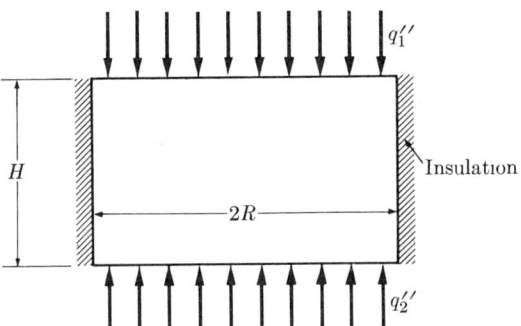

FIG. 5-38

suddenly assumes the constant angular velocity ω. Find the unsteady temperature of the system, assuming constant wear.

5–18. Reconsider Problem 3–31 (Fig. 3–65). Assume that the initial temperature of the pipe is equal to the ambient temperature T_∞, and that the pipe is suddenly filled with condensing steam at the temperature T_0. Find the unsteady temperature of the fins, assuming that the thermal conductivity of the pipe walls is large compared with that of the fins.

5–19. A solid cylinder of radius R and height H is insulated peripherally (Fig. 5–38). The initial temperature of the cylinder is T_0; then the top and the bottom of the cylinder are subjected to heat fluxes q_1'' and q_2'', respectively. Find the temperature variation in the cylinder.

5–20. Reconsider the pool reactor of Problem 2–11, assuming now that the thickness of the cladding is significant, that the internal energy generation is uniform, and that the heat transfer coefficient is finite (Fig. 5–39). Find the unsteady temperature of the fuel elements by the integral technique.

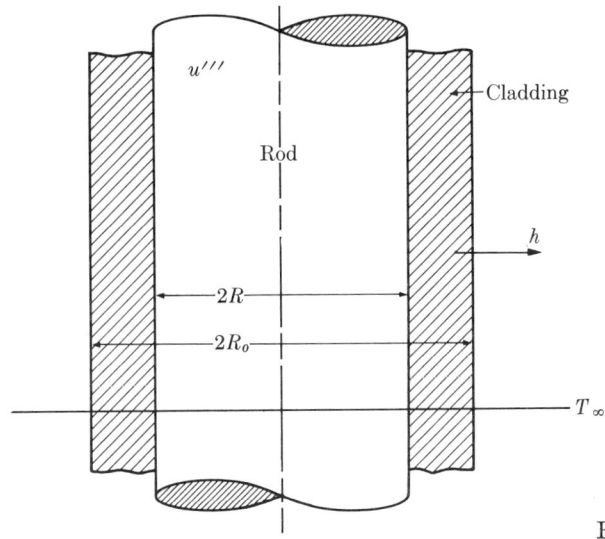

FIG. 5-39

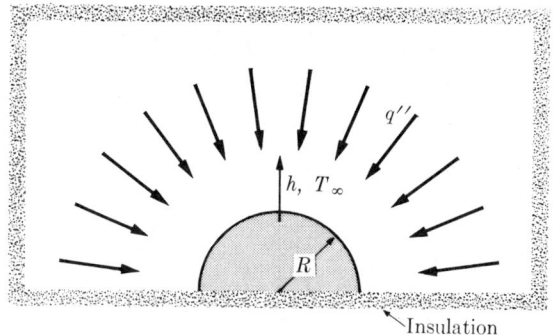

FIG. 5-40

5-21. Re-solve Example 5-4 and Problem 5-8(a) for a solid sphere of radius R.

5-22. Re-solve Example 5-5 for a solid sphere of radius R.

5-23. A biscuit at the initial temperature T_0 is placed in an oven heated to the uniform temperature T_∞ (Fig. 5-40). Assume that the biscuit is a solid hemisphere of radius R, and that its properties remain constant. Find the temperature variation in the biscuit.

5-24. The spherical container of a space vehicle initially contains gaseous and liquid oxygen, GOX and LOX, at the uniform temperature T_s (Fig. 5-41). Assume now that the container is subjected to a zero gravitational field and the constant heat flux q'' during the launch and flight of the vehicle. A venting process results in constant pressure and density of the GOX, which obeys the perfect gas law. Recent experiments with these containers indicate that the GOX forms a concentric sphere under the effect of zero gravity. To simplify the problem, the interfacial mass transfer between GOX and LOX is neglected. Find the unsteady temperature of the LOX.

5-25. A spine of variable cross section is built into a wall of thickness L (Fig. 5-42). The transversal and axial cross sections of the spine are a circle and a parabola, respectively. The base radius and the length of the spine are R and $2L$. The thermal conductivity of the wall is negligibly small compared with that of the spine. Initially the system is at the ambient temperature T_∞; then the periphery of the spine is suddenly subjected to the uniform heat flux q''. The base heat transfer coefficient is h. Find the unsteady temperature of the spine.

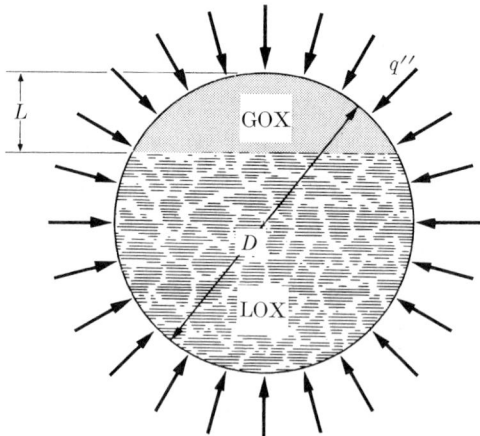

FIG. 5-41

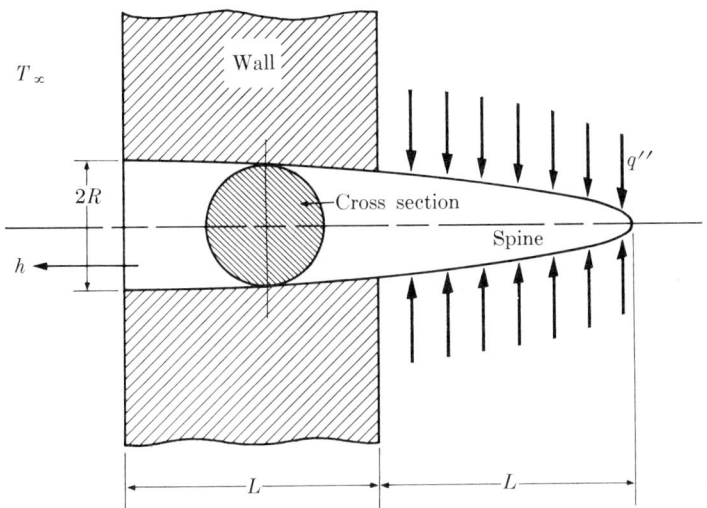

FIG. 5-42

5-26. A solid sphere of radius R_0 and at the initial temperature T_0 is suddenly placed in a large room whose walls are kept at the uniform temperature $T_w(\gg T_0)$. The ambient temperature in the room may be assumed to be zero. The heat transfer coefficient between the sphere and the ambient is h. The emissivity of the sphere and that of the walls are ϵ_s and ϵ_w, respectively. The latent heat of melting and the melting temperature of the sphere are h_{sl} and T_m. (a) Find the unsteady temperature of the sphere on the basis of a lumped analysis. Sketch this temperature as a function of time, showing its dependence on the relative magnitudes of the radiant and convective heat transfers. (b) Repeat part (a), using a distributed-integral analysis.

5-27. Consider the unsteady problem associated with the electric heater of Example 2-1. (See Problem 2-13 for a first-order approximate solution based on the integral formulation of the problem.) Initially the heater is at the ambient temperature T_∞; then the constant internal energy u''' is uniformly generated in the heater. Find the unsteady temperature of the heater corresponding to (i) a large heat transfer coefficient, (ii) a moderate heat transfer coefficient.

5-28. Consider a long rod of triangular cross section (Fig. 5-43). The initial temperature of the rod is uniform, say T_0, and two of its surfaces are insulated. The temperature of the third surface, the hypotenuse, is suddenly changed to the tempera-

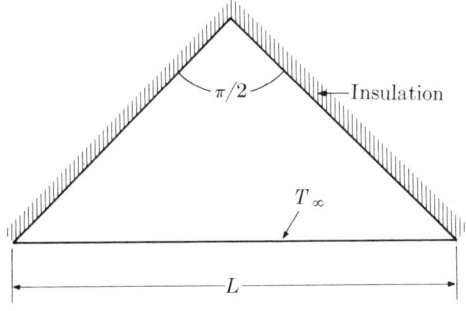

FIG. 5-43

ture T_∞. Can this problem be expressed as the product of two unsteady one-dimensional problems?

5–29. Consider a solid cylinder of radius R and height H. The initial temperature of the cylinder is T_∞. The surfaces of the cylinder are insulated except for a circular part of the bottom of radius R_0 which is suddenly subjected to the uniform heat flux q'' (Fig. 5–44). Find the unsteady temperature of the cylinder.

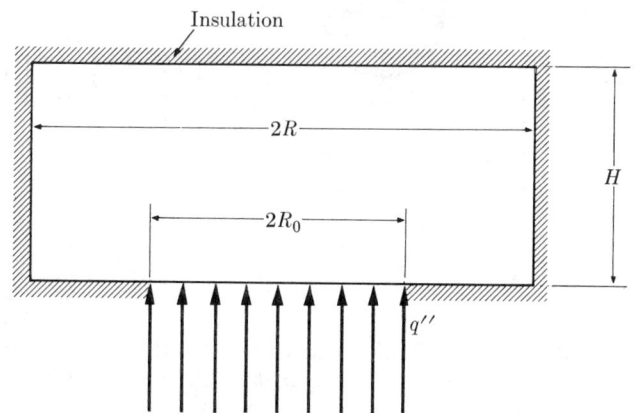

FIG. 5–44

5–30. Consider a solid cylinder of radius $R = 1''$ and height $H = 1''$ (Fig. 5–45). The peripheral surface and the bottom of the cylinder are insulated. The density, specific heat, and thermal conductivity of the cylinder are $\rho = 500$ lbm/ft^3, $c_p = 0.1$ Btu/lbm·°F, and $k = 10$ Btu/ft hr·°F. The initial temperature of the cylinder is $T_0 = 50°$F; then the upper surface of the cylinder is covered by a fluid at the temperature $T_\infty = 200°$F. The heat transfer coefficient is $h = 30$ Btu/ft^2·hr·°F. (a) Find the temperature at locations M and P after 40 minutes have elapsed. (b) Determine the amount of heat transferred to the cylinder in this same time interval.

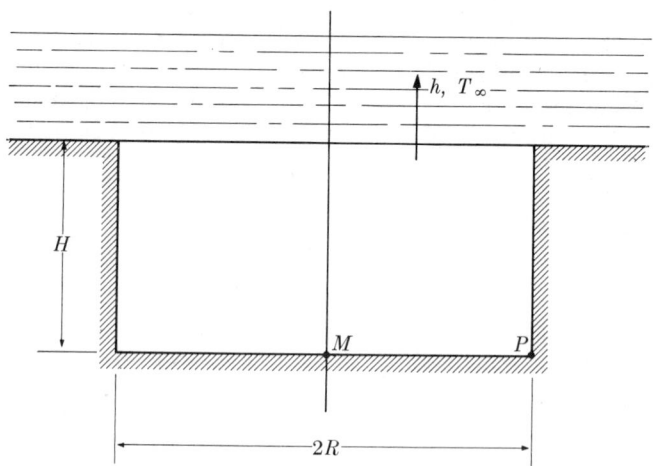

FIG. 5–45

5-31. Reconsider the lumped formulation of Example 2-2. Assume that the internal energy generation in the fuel plates (a) increases linearly as $q = q_0(t/t_0)$; (b) increases exponentially as $q = q_0(1 - e^{-bt})$; (c) oscillates as $q = q_0(1 + \epsilon \sin \omega t)$. Find the unsteady temperature of the plates on the basis of each of these assumptions.

5-32. Repeat the preceding problem for the distributed formulation of the same example.

CHAPTER 6

STEADY PERIODIC PROBLEMS. COMPLEX TEMPERATURE

When we wish to find the response of a body to a periodic disturbance for large values of time only, that is, when we are seeking the steady periodic solution of a problem, the method of complex temperature proves to be convenient. This method will be investigated here for its application to conduction problems, although it is equally useful for other linear problems of engineering and may also be used for approximate solutions of nonlinear problems.

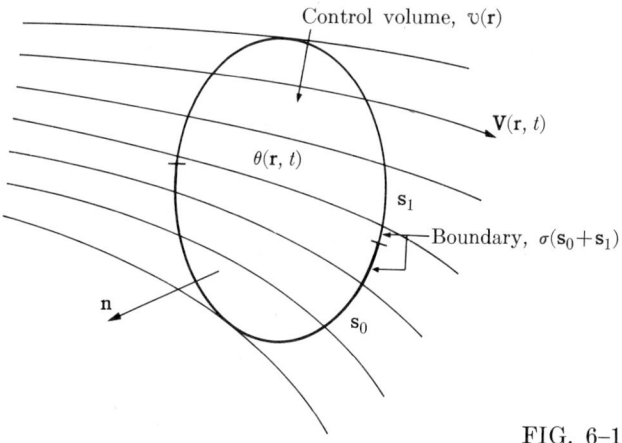

FIG. 6–1

Consider a control volume fixed in space (Fig. 6–1) through which a constant-property medium flows with the velocity $\mathbf{V}$. The temperature distribution $\theta(\mathbf{r}, t)$ of this control volume satisfies the differential form of the conduction equation

$$\frac{\partial \theta}{\partial t} + \mathbf{V} \cdot \boldsymbol{\nabla} \theta = a \nabla^2 \theta + \frac{u'''}{\rho c}, \qquad (6\text{--}1)$$

where $\mathbf{r}$ is the space vector prescribing the control volume $\mathcal{U}$. Initially, the temperature throughout the control volume is

$$\theta(\mathbf{r}, 0) = \theta_i(\mathbf{r}). \qquad (6\text{--}2)$$

Equation (6–2) has no importance for the present problem, however, since its effect disappears for large values of time.

Before stating the boundary conditions, let us note that a periodic disturbance may be introduced on the surfaces and/or in the volume of a system (or control volume). The *volumetric disturbances* are the *fluctuating velocity* and/or the *internal energy generation*. The velocity oscillations lead to a nonlinear problem and hence will not be considered here. Leaving the linear case of oscillating energy generation to the reader,† we shall discuss in the text only the *surface disturbances*.

Returning now to the boundary conditions of the problem, we assume that part of the boundary, say s_0, is subjected to one of the following steady homogeneous boundary conditions:

$$\theta(s_0, t) = 0, \qquad \frac{\partial \theta(s_0, t)}{\partial n} = 0, \qquad \pm k \frac{\partial \theta(s_0, t)}{\partial n} = h\theta(s_0, t), \qquad (6\text{-}3)$$

and that the remaining part of the boundary, s_1,‡ is subjected to one of the periodic nonhomogeneous boundary conditions

$$\theta(s_1, t) = \theta_0 \cos m\omega t, \qquad \pm k \frac{\partial \theta(s_1, t)}{\partial n} = q_0'' \cos m\omega t,$$

$$\pm k \frac{\partial \theta(s_1, t)}{\partial n} = h[\theta(s_1, t) - \theta_\infty \cos m\omega t], \qquad (6\text{-}4)$$

where n is the outward normal to the boundary, k the thermal conductivity of the medium, h the heat transfer coefficient, q_0'' the heat flux on the boundary, and $m\omega$ the angular frequency of the periodic disturbance. A harmonic disturbance is employed for algebraic simplicity. Otherwise the procedure applies to any kind of periodic disturbance (see Problem 6–1).

The method of complex temperature is based on the construction of another conduction problem $\theta^*(\mathbf{r}, t)$ subject to the differential equation and to the initial and homogeneous boundary conditions of the original problem, but subject also to a periodic nonhomogeneous boundary condition having a phase angle of $\pi/2$ relative to the original nonhomogeneous boundary condition. Thus the new problem, which will hereafter be referred to as the *supplementary problem*, satisfies the differential equation

$$\frac{\partial \theta^*}{\partial t} + \mathbf{V} \cdot \nabla \theta^* = a \nabla^2 \theta^*, \qquad (6\text{-}5)$$

the initial condition

$$\theta^*(\mathbf{r}, 0) = \theta_i(\mathbf{r}), \qquad (6\text{-}6)$$

one of the following boundary conditions on s_0,

$$\theta^*(s_0, t) = 0, \qquad \frac{\partial \theta^*(s_0, t)}{\partial n} = 0, \qquad \pm k \frac{\partial \theta^*(s_0, t)}{\partial n} = h\theta^*(s_0, t), \qquad (6\text{-}7)$$

---

† See Problems 6–4, 6–7, and 6–12.
‡ s_0 and s_1 can further be subdivided such that each part assumes one boundary condition from Eqs. (6–3) and (6–4), respectively.

and one of the following nonhomogeneous conditions on s_1:

$$\theta^*(s_1, t) = \theta_0 \sin m\omega t, \qquad \pm k \frac{\partial \theta^*(s_1, t)}{\partial n} = q_0'' \sin m\omega t, \qquad (6\text{-}8)$$

$$\pm k \frac{\partial \theta^*(s_1, t)}{\partial n} = h[\theta^*(s_1, t) - \theta_\infty \sin m\omega t].$$

The disturbance of the supplementary problem has been so arranged that it has a phase lag of $\pi/2$ with respect to the disturbance of the original problem.

The complex temperature $\psi(r, t)$ is now defined as

$$\psi(r, t) = \theta(r, t) + i\theta^*(r, t), \qquad (6\text{-}9)$$

where i is the conventional imaginary unit, $i = (-1)^{1/2}$. The differential equation and the initial and boundary conditions satisfied by $\psi(r, t)$ may readily be obtained by multiplying the original problem by i and adding the result to the supplementary problem. It follows then that

$$\frac{\partial \psi}{\partial t} + \mathbf{V} \cdot \nabla \psi = a \nabla^2 \psi, \qquad (6\text{-}10)$$

$$\psi(r, 0) = (1 + i)\theta_i(r), \qquad (6\text{-}11)$$

$$\psi(s_0, t) = 0, \qquad \frac{\partial \psi(s_0, t)}{\partial n} = 0, \qquad \pm k \frac{\partial \psi(s_0, t)}{\partial n} = h\psi(s_0, t), \qquad (6\text{-}12)$$

$$\psi(s_1, t) = \theta_0 e^{im\omega t}, \qquad \pm k \frac{\partial \psi(s_1, t)}{\partial n} = q_0'' e^{im\omega t},$$

$$\pm k \frac{\partial \psi(s_1, t)}{\partial n} = h[\psi(s_1, t) - \theta_\infty e^{im\omega t}]. \qquad (6\text{-}13)$$

Now we may try the solution of $\psi(r, t)$ by the method of separation of variables, which suggests the product solution

$$\psi(r, t) = \phi(r)\tau(t). \qquad (6\text{-}14)$$

However, since the desired solution is valid only for large values of time, it is periodic (with the same period as that of the disturbance) and, by further specifying Eq. (6–14), can be expressed in the form

$$\psi(r, t) = \phi(r) e^{im\omega t}. \qquad (6\text{-}15)$$

Equation (6–15) no longer satisfies the initial condition given by Eq. (6–11). The differential equation and boundary conditions satisfied by $\phi(r)$ are readily obtained by introducing Eq. (6–15) into Eqs. (6–10), (6–12), and (6–13). The result is

$$im\omega\phi + \mathbf{V}\cdot\nabla\phi = a\nabla^2\phi, \qquad (6\text{--}16)$$

$$\phi(\mathbf{s}_0) = 0, \quad \frac{\partial\phi(\mathbf{s}_0)}{\partial n} = 0, \quad \pm k\frac{\partial\phi(\mathbf{s}_0)}{\partial n} = h\phi(\mathbf{s}_0), \qquad (6\text{--}17)$$

$$\phi(\mathbf{s}_1) = \theta_0, \quad \pm k\frac{\partial\phi(\mathbf{s}_1)}{\partial n} = q_0'', \quad \pm k\frac{\partial\phi(\mathbf{s}_1)}{\partial n} = h[\phi(\mathbf{s}_1) - \theta_\infty]. \qquad (6\text{--}18)$$

Thus the method of complex temperature reduces the original problem $\theta(\mathbf{r}, t)$ *to the solution of the steady problem* $\phi(\mathbf{r})$. *The solution of* $\phi(\mathbf{r})$ *may be obtained by known methods. The result inserted into Eq.* (6–15) *yields the complex temperature* $\psi(\mathbf{r}, t)$. *The real part of* $\psi(\mathbf{r}, t)$ *is the solution of the original problem* $\theta(\mathbf{r}, t)$, *and the imaginary part is the solution of the supplementary problem* $\theta^*(\mathbf{r}, t)$ *for large values of time.*

Example 6–1. Let us find the steady periodic solution of Example 5–1. The formulation of the original problem is

$$\frac{d\theta}{dt} + m\theta = n\cos\omega t, \qquad (5\text{--}1)$$

$$\theta(0) = 0, \qquad (5\text{--}2)$$

which implies the supplementary problem

$$\frac{d\theta^*}{dt} + m\theta^* = n\sin\omega t, \qquad (6\text{--}19)$$

$$\theta^*(0) = 0. \qquad (6\text{--}20)$$

The complex temperature

$$\psi(t) = \theta(t) + i\theta^*(t)$$

satisfies the differential equation and the initial condition obtained by multiplying the supplementary problem of Eqs. (6–19) and (6–20) by i, and then adding the result to the original problem of Eqs. (5–1) and (5–2). Thus we have

$$\frac{d\psi}{dt} + m\psi = ne^{i\omega t}, \qquad (6\text{--}21)$$

$$\psi(0) = 0. \qquad (6\text{--}22)$$

The steady periodic solution of $\psi(t)$, which no longer satisfies the initial condition given by Eq. (6–22), has the form

$$\psi(t) = \phi e^{i\omega t}. \qquad (6\text{--}23)$$

Since $\psi(t)$ is lumped, now ϕ is only an unknown parameter to be determined. Introducing Eq. (6–23) into Eq. (6–21) readily gives

$$\phi = \frac{n}{m + i\omega}. \qquad (6\text{–}24)$$

Multiplying and dividing Eq. (6–24) by the conjugate of its denominator we obtain

$$\phi = \frac{n}{m^2 + \omega^2}(m - i\omega). \qquad (6\text{–}25)$$

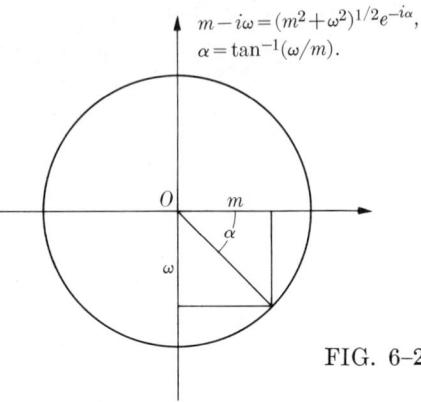

FIG. 6-2

By use of the polar form of $(m - i\omega)$, which is shown in Fig. (6–2), Eq. (6–25) may be rearranged in the form

$$\phi = \frac{n}{(m^2 + \omega^2)^{1/2}} e^{-i\alpha}. \qquad (6\text{–}26)$$

Introducing Eq. (6–26) into Eq. (6–23), we find the complex temperature of the plate to be

$$\psi(t) = \frac{n}{(m^2 + \omega^2)^{1/2}} e^{i(\omega t - \alpha)}. \qquad (6\text{–}27)$$

The real and imaginary parts of Eq. (6–27) are the steady periodic solutions of the original and supplementary problems, respectively. The real part of Eq. (6–27),

$$\frac{T - T_\infty}{q''/h} = \frac{m}{(m^2 + \omega^2)^{1/2}} \cos(\omega t - \alpha),$$

$$\alpha = \tan^{-1}(\omega/m),$$

is identical to the periodic part of Eq. (5–3).

Example 6–2. Let us find the steady periodic solution of Example 5–6 by the method of complex temperature.

The formulation of the problem is

$$\frac{\partial \theta}{\partial t} = a \frac{\partial^2 \theta}{\partial x^2}, \qquad (5\text{–}35)$$

$$\theta(x, 0) = 0, \qquad (5\text{–}36)$$

$$\theta(0, t) = \theta_0 \cos \omega t, \qquad (5\text{–}37)$$

$$\theta(\infty, t) = 0. \qquad (5\text{–}38)$$

The supplementary problem multiplied by i and then added to the original problem results in the complex temperature $\psi(x, t)$, which satisfies

$$\frac{\partial \psi}{\partial t} = a \frac{\partial^2 \psi}{\partial x^2}, \tag{6-28}$$

$$\psi(x, 0) = 0, \tag{6-29}$$

$$\psi(0, t) = \theta_0 e^{i\omega t}, \tag{6-30}$$

$$\psi(\infty, t) = 0. \tag{6-31}$$

The steady periodic solution,

$$\psi(x, t) = \phi(x) e^{i\omega t}, \tag{6-32}$$

introduced into Eqs. (6-28), (6-30), and (6-31) yields

$$\frac{d^2\phi}{dx^2} - \left(\frac{i\omega}{a}\right)\phi = 0, \tag{6-33}$$

$$\phi(0) = \theta_0, \tag{6-34}$$

$$\phi(\infty) = 0. \tag{6-35}$$

The solution of Eq. (6-33) which satisfies Eqs. (6-34) and (6-35) is

$$\frac{\phi(x)}{\theta_0} = e^{-(i\omega/a)^{1/2} x}. \tag{6-36}$$

Finally, combining Eqs. (6-32) and (6-36), and noting that $i^{1/2} = (1 + i)/2^{1/2}$, we find that the complex temperature is

$$\frac{\psi(x, t)}{\theta_0} = e^{-(\omega/2a)^{1/2} x} e^{i[\omega t - (\omega/2a)^{1/2} x]}. \tag{6-37}$$

The real part of Eq. (6-37),

$$\frac{\theta(x, t)}{\theta_0} = e^{-(\omega/2a)^{1/2} x} \cos\left[\omega t - \left(\frac{\omega}{2a}\right)^{1/2} x\right],$$

which is identical to Eq. (5-44), gives the steady periodic solution of the problem. (Note the simplicity of the foregoing procedure as compared with the method of separation of variables employed in Example 5-6.)

Example 6-3. An incompressible fluid at uniform velocity V flows steadily through a tube (Fig. 6-3). The thermal conductivity, density, specific heat, and cross-sectional area of the tube and fluid are (k', k), (ρ', ρ), (c', c), and (A', A), respectively. The inner and outer peripheries of the tube are P_i, P_o,

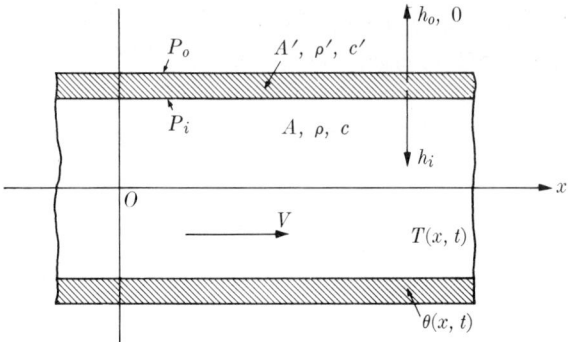

FIG. 6-3

and the inside and outside heat transfer coefficients h_i, h_o. Initially the system is at the ambient temperature, say zero; then the inlet temperature starts to oscillate as $T_0 \cos \omega t$. Using a radially lumped analysis, find the steady periodic temperature of the system.

The formulation of the problem is

$$\frac{\partial T}{\partial t} + V \frac{\partial T}{\partial x} + b_1(T - \theta) = 0, \tag{6-38}$$

$$\frac{\partial \theta}{\partial t} - b_2(T - \theta) + b_3 \theta = 0, \tag{6-39}$$

$$T(x, 0) = 0, \tag{6-40}$$

$$\theta(x, 0) = 0, \tag{6-41}$$

$$T(0, t) = T_0 \cos \omega t, \tag{6-42}$$

where $b_1 = (h_i P_i / \rho c A)$, $b_2 = (h_i P_i / \rho' c' A')$, and $b_3 = (h_o P_o / \rho' c' A')$.

The complex temperatures of the fluid and of the wall,

$$\psi = T + iT^*, \qquad \phi = \theta + i\theta^*,$$

satisfy the equations

$$\frac{\partial \psi}{\partial t} + V \frac{\partial \psi}{\partial x} + b_1(\psi - \phi) = 0, \tag{6-43}$$

$$\frac{\partial \phi}{\partial t} - b_2 \psi + b_4 \phi = 0, \tag{6-44}$$

$$\psi(x, 0) = 0, \tag{6-45}$$

$$\phi(x, 0) = 0, \tag{6-46}$$

$$\psi(0, t) = T_0 e^{i\omega t}, \tag{6-47}$$

where $b_4 = b_2 + b_3$. The complex steady periodic temperatures of the fluid

and of the wall,
$$\psi(x, t) = \Lambda(x)e^{i\omega t}, \qquad \phi(x, t) = \Gamma(x)e^{i\omega t}, \tag{6-48}$$

introduced into Eqs. (6–43), (6–44), and (6–47) give
$$i\omega\Lambda + V\frac{d\Lambda}{dx} + b_1(\Lambda - \Gamma) = 0, \tag{6-49}$$

$$i\omega\Gamma - b_2\Lambda + b_4\Gamma = 0, \tag{6-50}$$

$$\Lambda(0) = T_0. \tag{6-51}$$

Solving Eq. (6–50) for $\Gamma(x)$ and inserting the result into Eq. (6–49) yields
$$\frac{d\Lambda}{dx} + \frac{1}{V}\left[i\omega + b_1\left(\frac{b_3 + i\omega}{b_4 + i\omega}\right)\right]\Lambda = 0. \tag{6-52}$$

The solution of Eq. (6–52) which satisfies Eq. (6–51) is
$$\Lambda(x) = T_0 \exp\left[-\left(\frac{b_3 b_4 + \omega^2}{b_4^2 + \omega^2}\right)\left(\frac{b_1 x}{V}\right)\right]\exp\left[-i\left(1 + \frac{b_1 b_2}{b_4^2 + \omega^2}\right)\left(\frac{\omega x}{V}\right)\right]. \tag{6-53}$$

Using this result and Eq. (6–50), we obtain
$$\Gamma(x) = T_0 \left(\frac{b_2}{b_4^2 + \omega^2}\right)\exp\left[-\left(\frac{b_3 b_4 + \omega^2}{b_4^2 + \omega^2}\right)\left(\frac{b_1 x}{V}\right)\right]$$

$$\times \exp\left\{-i\left[\left(1 + \frac{b_1 b_2}{b_4^2 + \omega^2}\right)\left(\frac{\omega x}{V}\right) + \beta\right]\right\}, \tag{6-54}$$

where $\beta = \tan^{-1}(\omega/b_4)$.

Thus from Eqs. (6–48), (6–53), and (6–54) the complex steady periodic temperatures of the fluid and wall are readily obtained. The real parts of these temperatures,
$$\frac{T(x, t)}{T_0} = \exp\left[-\left(\frac{b_3 b_4 + \omega^2}{b_4^2 + \omega^2}\right)\left(\frac{b_1 x}{V}\right)\right]\cos\left[\omega t - \left(1 + \frac{b_1 b_2}{b_4^2 + \omega^2}\right)\left(\frac{\omega x}{V}\right)\right], \tag{6-55}$$

$$\frac{\theta(x, t)}{T_0} = \left(\frac{b_2}{b_4^2 + \omega^2}\right)\exp\left[-\left(\frac{b_3 b_4 + \omega^2}{b_4^2 + \omega^2}\right)\left(\frac{b_1 x}{V}\right)\right]$$

$$\times \cos\left[\omega t - \left(1 + \frac{b_1 b_2}{b_4^2 + \omega^2}\right)\left(\frac{\omega x}{V}\right) - \beta\right], \tag{6-56}$$

are the steady periodic temperatures of the fluid and wall, respectively.

Example 6–4. From the initial condition of zero temperature the surface temperature of an infinitely long rod of radius R starts to oscillate as $\theta_0 \cos \omega t$. We wish to find the steady periodic temperature of the rod.

The complex temperature of the rod satisfies

$$\frac{\partial \psi}{\partial t} = a\left(\frac{\partial^2 \psi}{\partial r^2} + \frac{1}{r}\frac{\partial \psi}{\partial r}\right), \tag{6-57}$$

$$\psi(r, 0) = 0, \tag{6-58}$$

$$\frac{\partial \psi(0, t)}{\partial r} = 0, \tag{6-59}$$

$$\psi(R, t) = \theta_0 e^{i\omega t}. \tag{6-60}$$

The complex steady periodic temperature,

$$\psi(r, t) = \phi(r) e^{i\omega t}, \tag{6-61}$$

introduced into Eqs. (6–57), (6–59), and (6–60) gives

$$\frac{d^2\phi}{dr^2} + \frac{1}{r}\frac{d\phi}{dr} - \left(\frac{i\omega}{a}\right)\phi = 0, \tag{6-62}$$

$$\frac{d\phi(0)}{dr} = 0, \tag{6-63}$$

$$\phi(R) = \theta_0. \tag{6-64}$$

The solution of Eq. (6–62) which satisfies Eqs. (6–63) and (6–64) is

$$\frac{\phi(r)}{\theta_0} = \frac{I_0[(i\omega/a)^{1/2} r]}{I_0[(i\omega/a)^{1/2} R]} = \frac{J_0[i^{3/2}(\omega/a)^{1/2} r]}{J_0[i^{3/2}(\omega/a)^{1/2} R]}, \tag{6-65}$$

where the second form of Eq. (6–65) has been obtained by using Eq. (3–119).

The Bessel functions for complex arguments are complex numbers. These, by contrast with the trigonometric functions, cannot be separated into real and imaginary parts expressed in terms of known functions. However, the real functions U_0, V_0, U_1, V_1 exist,[†] and are defined for the polar argument $z = \rho e^{i\varphi}$ in the form

$$J_0(\rho e^{i\varphi}) = U_0(\rho, \varphi) + iV_0(\rho, \varphi), \qquad J_1(\rho e^{i\varphi}) = U_1(\rho, \varphi) + iV_1(\rho, \varphi),$$

tabulated for the first quadrant of the complex plane. The values of these functions for other quadrants are deduced from the tabulated entries for the first

---

[†] See *Table of Bessel Functions $J_0(z)$ and $J_1(z)$ for Complex Arguments*. New York: Columbia University Press, 1943.

quadrant. Thus

$$J_0[i^{3/2}(\omega/a)^{1/2}r] = J_0[(\omega/a)^{1/2}re^{i3\pi/4}]$$
$$= U_0[(\omega/a)^{1/2}r, \pi/4] - iV_0[(\omega/a)^{1/2}r, \pi/4]. \quad (6\text{-}66)$$

By use of the shortened notation

$$U_0[(\omega/a)^{1/2}r, \pi/4] = U_0(r), \qquad V_0[(\omega/a)^{1/2}r, \pi/4] = V_0(r),$$
$$U_0[(\omega/a)^{1/2}R, \pi/4] = U_0, \qquad V_0[(\omega/a)^{1/2}R, \pi/4] = V_0, \quad (6\text{-}67)$$

Eq. (6-65) may be rearranged as

$$\frac{\phi(r)}{\theta_0} = \frac{U_0(r) - iV_0(r)}{U_0 - iV_0}. \quad (6\text{-}68)$$

Introducing Eq. (6-68) into Eq. (6-61) and considering the real part of the result, we obtain the solution of the problem

$$\frac{\theta(r,t)}{\theta_0} = \frac{([U_0 U_0(r) + V_0 V_0(r)]^2 + [U_0 V_0(r) - V_0 U_0(r)]^2)^{1/2}}{U_0^2 + V_0^2} \cos(\omega t - \alpha), \quad (6\text{-}69)$$

where

$$\alpha = \tan^{-1}\left[\frac{U_0 V_0(r) - V_0 U_0(r)}{U_0 U_0(r) + V_0 V_0(r)}\right].$$

A simpler but only approximate solution of the problem may be obtained by using the power series expansions of Bessel functions for small and large arguments. This is left to the reader (see Problem 6-8).

Note that the Bessel functions of the foregoing problem are also defined in the form†

$$J_\nu(xe^{\pm i3\pi/4}) = \text{ber}_\nu x \pm i \, \text{bei}_\nu x.$$

However, here we employ the notation U_0, V_0 because it is more convenient for algebraic manipulation and is more frequently used in numerical tables.

Problems

6-1. Using the Fourier analysis and the principle of superposition, extend the general procedure of complex temperature to an arbitrary periodic disturbance.

6-2. A flat plate divides an ambient into two regions (Fig. 6-4). The thickness of the plate is δ, the heat transfer coefficient h. The initial temperature of the plate

---

† See Eq. (3-146).

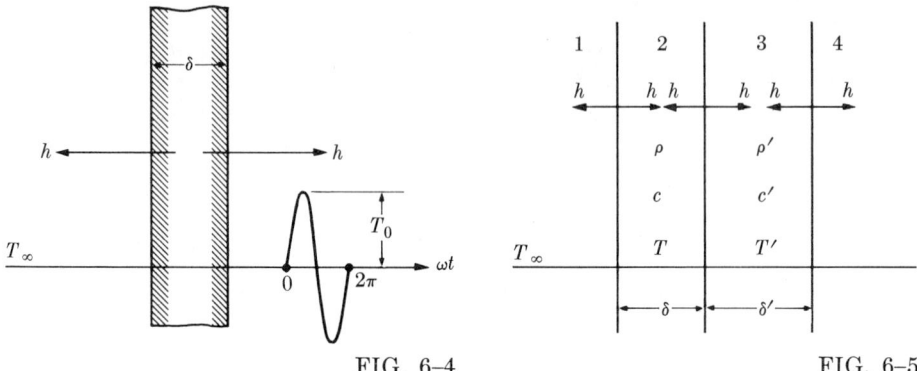

FIG. 6-4   FIG. 6-5

is uniform, say T_∞; then the temperature of one region starts to oscillate as $T_\infty + T_0 \sin \omega t$. Assuming that lumped system conditions prevail, find the steady periodic temperature of the plate.

6-3. Three parallel plates of negligible heat capacity divide a fluid into four regions, 1, 2, 3, 4 (Fig. 6-5). The initial temperature of the system is T_∞; then the temperatures of regions 1 and 4 oscillate as $T_\infty(1 + \sin \omega t)$ and $T_\infty(1 + \sin 2\omega t)$, respectively. Find the steady periodic temperatures of regions 2 and 3 using a lumped analysis.

6-4. Reconsider the problem of Example 6-3, assuming now that the outer surface of the tube is insulated and that the disturbance is introduced by the oscillating energy generation $u_0''' \cos \omega t$ in the tube walls. Find the steady periodic temperature of the system.†

6-5. Reconsider the problem associated with Example 5-2. Assume that the plate is plunged into a bath whose temperature oscillates as $T(t) = T_\infty + T_0 \sin \omega t$. The heat transfer coefficient remains large. Find the steady periodic temperature of the plate.

6-6. Reconsider the differential formulation of Example 2-3. This time assume that the heat flux applied to the bottom of the plate oscillates as $q'' \cos \omega t$. Find the steady periodic temperature of the plate. (See Example 5-1 for the solution of the lumped formulation of the problem.)

6-7. Reconsider the part of Problem 5-15 associated with Example 5-4, assuming now that the energy generation in the rod is $u''' + u_0''' \sin \omega t$. Find the steady periodic temperature of the rod.

6-8. Reconsider Example 6-4. Find the steady periodic temperature of the rod by expanding the Bessel functions appearing in Eq. (6-65) into series for small and large arguments. (For small- and large-argument behavior of Bessel functions, see Section 3-7.)

---

† See W. J. Yang, "The Dynamic Response of Heat Exchangers with Sinusoidal Time Dependent Internal Heat Generation," Ph.D. Thesis, University of Michigan, 1960; also W. J. Yang, J. A. Clark, and V. S. Arpacı, "Dynamic Response of Heat Exchangers Having Internal Heat Sources, Part IV," *Trans. ASME, C, Journal of Heat Transfer*, **83**, 321 (1961).

6-9. Reconsider the problem associated with Example 5-10. Modify this problem by letting the ambient temperature oscillate as $T_\infty + T_0 \cos \omega t$. Find the steady periodic temperature of the sphere.

6-10. Reconsider the problem associated with Example 6-3. Find the steady periodic temperature of the system on the basis of a radially lumped fluid temperature but a radially distributed tube-wall temperature.†

6-11. Reconsider Example 4-2, assuming now that the base temperature oscillates as $\theta_0 \sin \omega t$. Find the steady periodic temperature of the fin.

6-12. Reconsider the differential formulation of Example 2-1, assuming now that the energy generation is $u_0''' \cos \omega t$. The heat transfer coefficient is large. Find the steady periodic temperature of the heater.

† See J. R. Cairns, "Periodic Thermal Stresses in Thick-Walled Tubes," Ph.D. Thesis, University of Michigan, 1963.

CHAPTER 7

UNSTEADY PROBLEMS.
LAPLACE TRANSFORMS

In Chapters 4 and 5 we learned the method of separation of variables may be used provided the geometry of a problem is finite in orthogonal directions. In this chapter we shall study the solution of unsteady one- or multi-dimensional problems by the method of Laplace transforms, and we shall see that the condition of finite geometry in orthogonal directions is not required when the Laplace transforms are employed. Section 7–1 will be devoted to an introduction to transform calculus, Section 7–2 to a typical example of the use of Laplace transforms, Section 7–3 to the properties of Laplace transforms, and Section 7–4 to two short tables of Laplace transforms and to the use of these tables in the solution of a number of illustrative problems. In Sections 7–5, 7–6, and 7–7, the necessary mathematics will be briefly developed for the general theorem on the inversion of Laplace transforms. In Section 7–8, this theorem will be reduced to a contour integration in the complex domain, and two special cases of this contour suitable to our problems will be given. In Section 7–9, a number of problems will be solved by employing these special contours. Section 7–10 will be devoted to solution methods valid for small and large values of time.

7–1. Transform Calculus

Let us first consider the two basic operations of calculus, differentiation and integration. The operation of differentiation consists in transforming a function, say $f(t)$, to the function $f'(t)$. If the differential operator is denoted by the letter D, the transformation can be written

$$D\{f(t)\} = f'(t).$$

Thus the function $f'(t)$ is the *transform* of $f(t)$ under the operation of differentiation. Similarly, representing the operation of integration by the letter I, we have

$$I\{f(t)\} = \int_0^x f(t)\, dt. \qquad (7\text{–}1)$$

Equation (7–1) is the transform of $f(t)$ under the operation of integration.

The class of functions to which a given transformation applies is limited to some extent. As we know, the transformation $D\{f(t)\}$ applies only to differentiable functions, and the transformation $I\{f(t)\}$ only to integrable functions. Within these limitations, there exists an inverse transformation corresponding to each of the foregoing examples; that is, when the transform is given, a function $f(t)$ may be found which has that transform.

A transformation $T\{f(t)\}$ is *linear* if, for every pair of functions $f(t)$ and $g(t)$ and for each pair of constants C_1 and C_2, it satisfies the relation

$$T\{C_1 f(t) + C_2 g(t)\} = C_1 T\{f(t)\} + C_2 T\{g(t)\}. \tag{7-2}$$

Two frequent special cases of Eq. (7-2) are $C_2 = 0$ and $C_1 = C_2 = 1$. Differentiation and integration are linear transformations.

The linear *integral transform* defined by Eq. (7-1) may be extended in an obvious way. When $K(p, t)$ is a known function of the variable t and a parameter p, and the definite integral

$$T\{f(t)\} = \int_a^b K(p, t) f(t)\, dt \tag{7-3}$$

exists, then Eq. (7-3) defines a function of the parameter p. This function is called the integral transform of the function $f(t)$ by the *kernel* $K(p, t)$. A special case of Eq. (7-3), corresponding to $a = 0$, $b = \infty$, and $K(p, t) = e^{-pt}$, is the *Laplace transform*

$$L\{f(t)\} = \bar{f}(p) = \int_0^\infty e^{-pt} f(t)\, dt. \tag{7-4}$$

The integral of Eq. (7-4) exists if (i) $f(t)$ is *piecewise continuous*[†] in any interval $t_1 \leq t \leq t_0$ where $t_1 > 0$; (ii) $t^n |f(t)|$ is bounded near $t = 0$ for $n < 1$; (iii) $f(t)$ is of exponential order, that is, $e^{-\gamma t}|f(t)|$ is bounded for a positive number γ as $t \to \infty$. Of the two notations given by Eq. (7-4), $L\{f(t)\}$ is convenient when stating properties of the Laplace transforms, whereas $\bar{f}(p)$ is a convenient short notation to employ in the algebra of the solution. Because of its definition, the Laplace transform is most suitable to semi-infinite domains; thus it is often (but not necessarily) applied to the time variable of unsteady problems.[‡] Other well-known particular cases of Eq. (7-3) are the Fourier, Hankel, and Mellin transforms. In this text we consider only Laplace transforms.

7–2. An Introductory Example

Before studying the properties of Laplace transforms, let us demonstrate one of their most useful applications by considering a simple problem, say the lumped formulation of Example 2–2. Assume that the nuclear internal energy is generated in the plates at the exponential decay rate, that is,

$$u''' = u_0''' e^{-bt}.$$

[†] A function $f(t)$ is said to be piecewise continuous over a finite range if it is possible to divide that range into a finite number of intervals in each of which $f(t)$ is continuous.
[‡] For the application of Laplace transforms to a steady problem of finite geometry, see Problem 7–13.

The lumped formulation of the problem in terms of the temperature above the ambient, $\theta = T - T_\infty$, is

$$\frac{d\theta}{dt} + m\theta = ne^{-bt} \tag{7-5}$$

and

$$\theta(0) = 0, \tag{7-6}$$

where $m = h/\rho c L$ and $n = u_0'''/\rho c$.

Instead of integrating Eq. (7–5) by the classical methods of linear first-order differential equations and determining the resulting arbitrary constant by satisfying Eq. (7–6), let us proceed as follows. First we obtain the Laplace transform of Eq. (7–5) by multiplying both sides of Eq. (7–5) by e^{-pt} and integrating the results with respect to t from 0 to ∞. Thus we have

$$\int_0^\infty \frac{d\theta}{dt} e^{-pt}\, dt + m\int_0^\infty \theta e^{-pt}\, dt = n\int_0^\infty e^{-(b+p)t}\, dt. \tag{7-7}$$

Let us assume, for the time being, that the separate integrals exist at least for some range of values of p. The integral on the right of Eq. (7–7) readily yields

$$\int_0^\infty e^{-(b+p)t}\, dt = \frac{1}{p+b}, \tag{7-8}$$

provided $p + b > 0$. The first integral on the left of Eq. (7–7) may be rearranged by integration by parts to give

$$\int_0^\infty \frac{d\theta}{dt} e^{-pt}\, dt = -\theta(0) + \int_0^\infty \theta e^{-pt}\, dt, \tag{7-9}$$

with the assumption that $\theta(t)e^{-pt} \to 0$ as $t \to \infty$ for sufficiently large values of p. Equation (7–9) shows the fundamental operational property of the Laplace transforms, the property that makes it possible to express the transform of $d\theta/dt$ in terms of the transform and the initial value of θ itself.

Now, inserting Eq. (7–6) into Eq. (7–9), and then the result, together with Eq. (7–8), into Eq. (7–7) yields

$$\int_0^\infty \theta e^{-pt}\, dt = \frac{n}{(p+b)(p+m)}. \tag{7-10}$$

The original problem is thus reduced to finding a temperature $\theta(t)$ whose Laplace transform is the right-hand side of Eq. (7–10). To determine such a function, we rearrange Eq. (7–10) by expansion in partial fractions as

$$\int_0^\infty \theta e^{-pt}\, dt = \frac{n}{m-b}\left(\frac{1}{p+b} - \frac{1}{p+m}\right). \tag{7-11}$$

Noting from Eq. (7–8) that $1/(p+b)$ and $1/(p+m)$ are the transforms of e^{-bt} and e^{-mt}, respectively, we find that the temperature $\theta(t)$ that satisfies Eq. (7–10) is†

$$\theta(t) = \frac{n}{m-b}(e^{-bt} - e^{-mt}). \tag{7-12}$$

It is of course desirable to shorten the foregoing procedure by eliminating the lengthy integration process in each case. This may be done by reference to the properties of Laplace transforms (Section 7–3). The use of these would transform Eqs. (7–5) and (7–6) directly into Eq. (7–10). Once the transform of a function is obtained, the simplest method of finding its inverse is to look up the transform in a previously prepared *table of Laplace transforms* (Section 7–4). If the transform does not appear in tables, the inverse function is determined from its transform by the use of the *inversion theorem for Laplace transforms* (Section 7–6).

7–3. Properties of Laplace Transforms

The most frequently needed properties are given below; after the statement of each property, we give a brief proof without mentioning the conditions for its validity.

I. $$L\{C_1 f(t) + C_2 g(t)\} = C_1 \bar{f}(p) + C_2 \bar{g}(p). \tag{7-13}$$

Equation (7–13) expresses the linearity of Laplace transformation. Its proof follows directly from the linear property of the integral operation involved in the definition of Laplace transforms.

II. $$L\left\{\frac{df(t)}{dt}\right\} = p\bar{f}(p) - f(0+), \tag{7-14}$$

where $f(0+)$ denotes the right-hand limit of $f(t)$ at $t = 0$. Equation (7–14) follows immediately from integration by parts, since

$$\int_0^\infty \frac{df(t)}{dt} e^{-pt}\, dt = f(t)e^{-pt}\Big|_0^\infty + p\int_0^\infty f(t)e^{-pt}\, dt$$
$$= -f(0+) + p\bar{f}(p).$$

III. $$L\left\{\frac{\partial^n f(x_i, t)}{\partial x_i^n}\right\} = \frac{\partial^n \bar{f}(x_i, p)}{\partial x_i^n}, \tag{7-15}$$

---

† The uniqueness of the solution given by Eq. (7–12), being beyond the scope of the text, is not discussed here.

where x_i is a variable independent of t. Equation (7–15) is equivalent to

$$\int_0^\infty \frac{\partial^n f(x_i, t)}{\partial x_i^n} e^{-pt}\, dt = \frac{\partial^n}{\partial x_i^n} \int_0^\infty f(x_i, t) e^{-pt}\, dt,$$

which is true when the order of integration and differentiation can be interchanged.

The foregoing three properties are the ones of greatest importance. Other properties occasionally employed are as follows:

IV.
$$L\left\{\int_0^t f(\tau)\, d\tau\right\} = \frac{1}{p}\bar{f}(p). \tag{7–16}$$

Like Eq. (7–14), Eq. (7–16) follows from an integration by parts which yields

$$\int_0^\infty \left(\int_0^t f(\tau)\, d\tau\right) e^{-pt}\, dt = -\frac{1}{p}\left(\int_0^t f(\tau)\, d\tau\right) e^{-pt}\bigg|_0^\infty$$

$$+ \frac{1}{p}\int_0^\infty f(t) e^{-pt}\, dt = \frac{1}{p}\bar{f}(p).$$

V. If α is a positive constant and $L\{f(t)\} = \bar{f}(p)$, then

$$L\{f(\alpha t)\} = \frac{1}{\alpha}\bar{f}\left(\frac{p}{\alpha}\right). \tag{7–17}$$

Introducing a new variable $\tau = \alpha t$, we may rearrange the definition of Laplace transforms to give

$$\int_0^\infty f(\alpha t) e^{-pt}\, dt = \frac{1}{\alpha}\int_0^\infty f(\tau) e^{-(p/\alpha)\tau}\, d\tau = \frac{1}{\alpha}\bar{f}\left(\frac{p}{\alpha}\right).$$

VI. If β is any constant and $L\{f(t)\} = \bar{f}(p)$, then

$$L\{e^{-\beta t}f(t)\} = \bar{f}(p + \beta), \tag{7–18}$$

which also follows from the definition of Laplace transforms,

$$\int_0^\infty [e^{-\beta t}f(t)] e^{-pt}\, dt = \int_0^\infty f(t) e^{-(p+\beta)t}\, dt = \bar{f}(p + \beta).$$

Equation (7–18) states that the transform of $e^{-\beta t}f(t)$ is the transform of $f(t)$ shifted through a distance β in the negative direction of p.

VII. If α is a positive constant and $L\{f(t)\} = \bar{f}(p)$, then

$$L\left\{\begin{matrix}0, & t < \alpha \\ f(t - \alpha), & t \geq \alpha\end{matrix}\right\} = e^{-\alpha p}\bar{f}(p). \tag{7–19}$$

Note first that if $f(t)$ is defined for $t \geq 0$ and is zero for $t < 0$, then $f(t - \alpha)$ denotes the displacement of $f(t)$ through a distance α in the positive direction of t. Then the definition of Laplace transforms rearranged by a change of the variable, $t = \tau + \alpha$, gives

$$\int_\alpha^\infty f(t - \alpha)e^{-pt}\,dt = \int_0^\infty f(\tau)e^{-p(\tau+\alpha)}\,d\tau$$
$$= e^{-\alpha p}\int_0^\infty f(\tau)e^{-p\tau}\,d\tau = e^{-\alpha p}\bar{f}(p).$$

VIII. $\qquad L\left\{\int_0^t f(t - \tau)g(\tau)\,d\tau\right\} = \bar{f}(p)\bar{g}(p).$ \hfill (7–20)

The integral of Eq. (7–20) is called the *convolution* of $f(t)$ and $g(t)$. Because of the symmetry in $\bar{f}(p)\bar{g}(p)$, the integrand of this integral can be replaced by $f(\tau)g(t - \tau)$. Equation (7–20) becomes indispensable when $\bar{F}(p)$ is not the transform of a known function but can be expressed as the product of two functions, $\bar{F}(p) = \bar{f}(p)\bar{g}(p)$, each of which is the transform of a known function.

To establish Eq. (7–20), we refer to the definition of Laplace transforms, and express the right-hand side of this equation in terms of different dummy variables,

$$\bar{f}(p)\bar{g}(p) = \left(\int_0^\infty f(s)e^{-ps}\,ds\right)\left(\int_0^\infty g(\tau)e^{-p\tau}\,d\tau\right),$$

which may also be written in the form

$$\int_0^\infty g(\tau)\left\{\int_0^\infty f(s)e^{-p(s+\tau)}\,ds\right\}d\tau. \qquad (7\text{–}21)$$

Introducing the variable $t = s + \tau$, and rearranging Eq. (7–21), we obtain

$$\int_0^\infty \left\{\int_\tau^\infty f(t - \tau)g(\tau)e^{-pt}\,dt\right\}d\tau. \qquad (7\text{–}22)$$

The domain to which Eq. (7–22) applies and the integration limits are shown in Fig. 7–1(a); first t is varied from τ to ∞, then τ is varied from 0 to ∞. This,

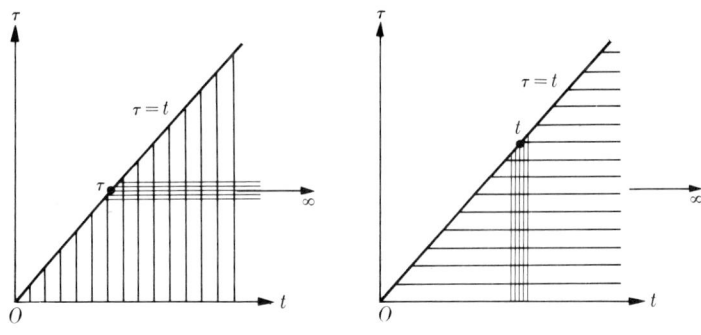

FIG. 7–1

of course, is identical to first letting τ vary from 0 to t, then letting t vary from 0 to ∞ Fig. 7–1(b). Now, interchanging the order of integration in Eq. (7–22) and expressing the limits according to Fig. 7–1(b) rather than Fig. 7–1(a), we obtain

$$\int_0^\infty \left\{ \int_0^t f(t - \tau) g(\tau) \, d\tau \right\} e^{-pt} \, dt,$$

which is identical to the left-hand side of Eq. (7–20).

It may be shown by Leibnitz's integral formula (see Problem 2–8) that the relation

$$\phi(\mathbf{r}, t) = \frac{\partial}{\partial t} \int_0^t D(\tau) \psi(\mathbf{r}, t - \tau) \, d\tau$$

is identical to Eq. (5–108). Thus *the differentiation of the convolution integral with respect to the parameter t gives Duhamel's superposition integral.*

Having established the properties of Laplace transforms, let us return to the solution of conduction problems by means of Laplace transforms. The solution procedure outlined in Section 7–2 may be broken into two steps:

(i) Application of Laplace transforms to the problem, that is, multiplication of its formulation by e^{-pt} and integration of the result with respect to t from 0 to ∞. The appropriate properties of the Laplace transforms are then employed to obtain the transform function or a relation involving this function, say a differential equation whose solution gives the transform function.

(ii) Inversion of the transform function by the use of either a table of transforms or the inversion theorem for Laplace transforms. Tables of transforms give transform pairs which have generally been established in advance by the use of the inversion theorem. Nevertheless, a large number of simple cases, for example,

$$L\{1\} = \int_0^\infty e^{-pt} \, dt = \frac{1}{p}, \qquad p > 0, \qquad (7\text{–}23)$$

$$L\{e^{-\alpha t}\} = \int_0^\infty e^{-(p+\alpha)t} \, dt = \frac{1}{p + \alpha}, \qquad p + \alpha > 0, \qquad (7\text{–}24)$$

$$L\{\sin \alpha t\} = \int_0^\infty \sin \alpha t \, e^{-pt} \, dt = \frac{\alpha}{p^2 + \alpha^2}, \qquad p > 0, \qquad (7\text{–}25)$$

and many others, may easily be found from the Laplace transforms of simple functions, obtained by direct integration of the definition of Laplace transforms rather than by using the inversion theorem. [Prove Eqs. (7–23), (7–24), and (7–25) by direct integration.] Thus, for example, the transform pairs (1), (12), and (24) (see Table 7–2) are simply the reverse statements of Eqs. (7–23), (7–24), and (7–25).

TABLE 7-1

PROPERTIES OF LAPLACE TRANSFORMS

No.	Function	Transform
I	$C_1 f(t) + C_2 g(t)$	$C_1 \bar{f}(p) + C_2 \bar{g}(p)$
II	$\dfrac{df(t)}{dt}$	$p\bar{f}(p) - f(0+)$
III	$\dfrac{\partial^n f(x_i, t)}{\partial x_i^n}$	$\dfrac{\partial^n \bar{f}(x_i, p)}{\partial x_i^n}$
IV	$\int_0^t f(\tau)\, d\tau$	$\dfrac{1}{p} \bar{f}(p)$
V	$f(\alpha t)$	$\dfrac{1}{\alpha} \bar{f}\left(\dfrac{p}{\alpha}\right)$
VI	$e^{-\beta t} f(t)$	$\bar{f}(p + \beta)$
VII	$\begin{cases} 0, & t < \alpha \\ f(t - \alpha), & t \geq \alpha \end{cases}$	$e^{-\alpha p} \bar{f}(p)$
VIII	$\int_0^t f(t - \tau) g(\tau)\, d\tau$	$\bar{f}(p) \bar{g}(p)$

7–4. Solutions Obtainable by the Table of Transforms

A collection of Laplace transforms particularly suitable to the problems of this text is given in Tables 7–1 and 7–2.† The first of these tables summarizes the list of properties established in Section 7–3. This table is employed in obtaining the transform function of a given problem. The inverse transforms of the simple functions of frequent occurrence are tabulated in Table 7–2. The more complicated inverse transforms of a large number of problems may be deduced from those given, by referring to the appropriate properties in Table 7–1. The manipulations involved in obtaining the transform and the inverse transform of a problem are illustrated by the examples below.

Example 7–1. Let us reconsider Example 5–1, which deals with a flat plate whose bottom is subjected to an oscillating heat flux. We wish to find the unsteady lumped temperature of the plate.

---

† For more extensive tables see References 10 and 11.

TABLE 7-2

INVERSION OF LAPLACE TRANSFORMS

In this table $q = (p/a)^{1/2}$; a and x are positive real; α, β, γ are unrestricted; k is a finite integer; n is a finite integer or zero; ν is a fractional number; $1 \cdot 2 \cdot 3 \cdots n = n!$; $1 \cdot 3 \cdot 5 \cdots (2n-1) = (2n-1)!!$; $n\Gamma(n) = \Gamma(n+1) = n!$; $\Gamma(1) = 0! = 1$; $\Gamma(\nu)\Gamma(1-\nu) = \pi/\sin \nu\pi$; $\Gamma(\frac{1}{2}) = \pi^{1/2}$.

No.	Transform	Function
1	$\dfrac{1}{p}$	1
2	$\dfrac{1}{p^2}$	t
3	$\dfrac{1}{p^k}$	$\dfrac{t^{k-1}}{(k-1)!}$
4	$\dfrac{1}{p^{1/2}}$	$\dfrac{1}{(\pi t)^{1/2}}$
5	$\dfrac{1}{p^{3/2}}$	$2\left(\dfrac{t}{\pi}\right)^{1/2}$
6	$\dfrac{1}{p^{k+1/2}}$	$\dfrac{2^k}{\pi^{1/2}(2k-1)!!} t^{k-1/2}$
7	$\dfrac{1}{p^\nu}$	$\dfrac{t^{\nu-1}}{\Gamma(\nu)}$
8	$p^{1/2}$	$-\dfrac{1}{2\pi^{1/2}t^{3/2}}$
9	$p^{3/2}$	$\dfrac{3}{4\pi^{1/2}t^{5/2}}$
10	$p^{k-1/2}$	$\dfrac{(-1)^k(2k-1)!!}{2^k\pi^{1/2}t^{k+1/2}}$
11	$p^{n-\nu}$	$\dfrac{t^{\nu-n-1}}{\Gamma(\nu-n)}$
12	$\dfrac{1}{p+\alpha}$	$e^{-\alpha t}$
13	$\dfrac{1}{(p+\alpha)(p+\beta)}$	$\dfrac{e^{-\beta t} - e^{-\alpha t}}{\alpha - \beta}$
14	$\dfrac{1}{(p+\alpha)^2}$	$te^{-\alpha t}$

TABLE 7-2 (Continued)

No.	Transform	Function
15	$\dfrac{1}{(p+\alpha)(p+\beta)(p+\gamma)}$	$\dfrac{(\gamma-\beta)e^{-\alpha t}+(\alpha-\gamma)e^{-\beta t}+(\beta-\alpha)e^{-\gamma t}}{(\alpha-\beta)(\beta-\gamma)(\gamma-\alpha)}$
16	$\dfrac{1}{(p+\alpha)^2(p+\beta)}$	$\dfrac{e^{-\beta t}-e^{-\alpha t}[1-(\beta-\alpha)t]}{(\beta-\alpha)^2}$
17	$\dfrac{1}{(p+\alpha)^3}$	$\tfrac{1}{2}t^2 e^{-\alpha t}$
18	$\dfrac{1}{(p+\alpha)^k}$	$\dfrac{t^{k-1}e^{-\alpha t}}{(k-1)!}$
19	$\dfrac{p}{(p+\alpha)(p+\beta)}$	$\dfrac{\alpha e^{-\alpha t}-\beta e^{-\beta t}}{\alpha-\beta}$
20	$\dfrac{p}{(p+\alpha)^2}$	$(1-\alpha t)e^{-\alpha t}$
21	$\dfrac{p}{(p+\alpha)(p+\beta)(p+\gamma)}$	$\dfrac{\alpha(\beta-\gamma)e^{-\alpha t}+\beta(\gamma-\alpha)e^{-\beta t}+\gamma(\alpha-\beta)e^{-\gamma t}}{(\alpha-\beta)(\beta-\gamma)(\gamma-\alpha)}$
22	$\dfrac{p}{(p+\alpha)^2(p+\beta)}$	$\dfrac{[\beta-\alpha(\beta-\alpha)t]e^{-\alpha t}-\beta e^{-\beta t}}{(\beta-\alpha)^2}$
23	$\dfrac{p}{(p+\alpha)^3}$	$t(1-\tfrac{1}{2}\alpha t)e^{-\alpha t}$
24	$\dfrac{\alpha}{p^2+\alpha^2}$	$\sin\alpha t$
25	$\dfrac{p}{p^2+\alpha^2}$	$\cos\alpha t$
26	$\dfrac{\alpha}{p^2-\alpha^2}$	$\sinh\alpha t$
27	$\dfrac{p}{p^2-\alpha^2}$	$\cosh\alpha t$
28	e^{-qx}	$\dfrac{x}{2(\pi a t^3)^{1/2}}e^{-x^2/4at}$
29	$\dfrac{e^{-qx}}{q}$	$\left(\dfrac{a}{\pi t}\right)^{1/2}e^{-x^2/4at}$
30	$\dfrac{e^{-qx}}{p}$	$\operatorname{erfc}\left[\dfrac{x}{2(at)^{1/2}}\right]$
31	$\dfrac{e^{-qx}}{qp}$	$2\left(\dfrac{at}{\pi}\right)^{1/2}e^{-x^2/4at}-x\operatorname{erfc}\left[\dfrac{x}{2(at)^{1/2}}\right]$
32	$\dfrac{e^{-qx}}{p^2}$	$\left(t+\dfrac{x^2}{2a}\right)\operatorname{erfc}\left[\dfrac{x}{2(at)^{1/2}}\right]-x\left(\dfrac{t}{\pi a}\right)^{1/2}e^{-x^2/4at}$

TABLE 7-2 (*Continued*)

No.	Transform	Function
33	$\dfrac{e^{-qx}}{p^{1+n/2}}$	$(4t)^{n/2} i^n \operatorname{erfc}\left[\dfrac{x}{2(at)^{1/2}}\right]$
34	$\dfrac{e^{-qx}}{p^{3/4}}$	$\dfrac{1}{\pi}\left(\dfrac{x}{2ta^{1/2}}\right)^{1/2} e^{-x^2/8at} K_{1/4}\left(\dfrac{x^2}{8at}\right)$
35	$\dfrac{e^{-qx}}{q+\beta}$	$\left(\dfrac{a}{\pi t}\right)^{1/2} e^{-x^2/4at} - a\beta e^{\beta x + a\beta^2 t} \operatorname{erfc}\left[\dfrac{x}{2(at)^{1/2}} + \beta(at)^{1/2}\right]$
36	$\dfrac{e^{-qx}}{q(q+\beta)}$	$a e^{\beta x + a\beta^2 t} \operatorname{erfc}\left[\dfrac{x}{2(at)^{1/2}} + \beta(at)^{1/2}\right]$
37	$\dfrac{e^{-qx}}{p(q+\beta)}$	$\dfrac{1}{\beta}\operatorname{erfc}\left[\dfrac{x}{2(at)^{1/2}}\right] - \dfrac{1}{\beta} e^{\beta x + a\beta^2 t} \operatorname{erfc}\left[\dfrac{x}{2(at)^{1/2}} + \beta(at)^{1/2}\right]$
38	$\dfrac{e^{-qx}}{qp(q+\beta)}$	$\dfrac{2}{\beta}\left(\dfrac{at}{\pi}\right)^{1/2} e^{-x^2/4at} - \dfrac{(1+\beta x)}{\beta^2}\operatorname{erfc}\left[\dfrac{x}{2(at)^{1/2}}\right]$ $+ \dfrac{1}{\beta^2} e^{\beta x + a\beta^2 t} \operatorname{erfc}\left[\dfrac{x}{2(at)^{1/2}} + \beta(at)^{1/2}\right]$
39	$\dfrac{e^{-qx}}{q^{n+1}(q+\beta)}$	$\dfrac{a}{(-\beta)^n} e^{\beta x + a\beta^2 t} \operatorname{erfc}\left[\dfrac{x}{2(at)^{1/2}} + \beta(at)^{1/2}\right]$ $- \dfrac{a}{(-\beta)^n} \sum\limits_{r=0}^{n-1} [-2\beta(at)^{1/2}]^r i^r \operatorname{erfc}\left[\dfrac{x}{2(at)^{1/2}}\right]$
40	$\dfrac{e^{-qx}}{(q+\beta)^2}$	$-2\beta\left(\dfrac{a^3 t}{\pi}\right)^{1/2} e^{-x^2/4at}$ $+ a(1 + \beta x + 2a\beta^2 t) e^{\beta x + a\beta^2 t} \operatorname{erfc}\left[\dfrac{x}{2(at)^{1/2}} + \beta(at)^{1/2}\right]$
41	$\dfrac{e^{-qx}}{p(q+\beta)^2}$	$\dfrac{1}{\beta^2}\operatorname{erfc}\left[\dfrac{x}{2(at)^{1/2}}\right] - \dfrac{2}{\beta}\left(\dfrac{at}{\pi}\right)^{1/2} e^{-x^2/4at}$ $- \dfrac{1}{\beta^2}(1 - \beta x - 2a\beta^2 t) e^{\beta x + a\beta^2 t} \operatorname{erfc}\left[\dfrac{x}{2(at)^{1/2}} + \beta(at)^{1/2}\right]$
42	$\dfrac{e^{-qx}}{p-\gamma}$	$\tfrac{1}{2} e^{\gamma t}\left\{ e^{-x(\gamma/a)^{1/2}} \operatorname{erfc}\left[\dfrac{x}{2(at)^{1/2}} - (\gamma t)^{1/2}\right]\right.$ $\left. + e^{x(\gamma/a)^{1/2}} \operatorname{erfc}\left[\dfrac{x}{2(at)^{1/2}} + (\gamma t)^{1/2}\right]\right\}$

TABLE 7-2 (Continued)

No.	Transform	Function
43	$\dfrac{e^{-qx}}{q(p-\gamma)}$	$\frac{1}{2}e^{\gamma t}\left(\dfrac{a}{\gamma}\right)^{1/2}\left\{e^{-x(\gamma/a)^{1/2}}\operatorname{erfc}\left[\dfrac{x}{2(at)^{1/2}}-(\gamma t)^{1/2}\right]\right.$ $\left. - e^{x(\gamma/a)^{1/2}}\operatorname{erfc}\left[\dfrac{x}{2(at)^{1/2}}+(\gamma t)^{1/2}\right]\right\}$
44	$\dfrac{e^{-qx}}{(p-\gamma)^2}$	$\frac{1}{2}e^{\gamma t}\left\{\left[t-\dfrac{x}{2(at)^{1/2}}\right]e^{-x(\gamma/a)^{1/2}}\operatorname{erfc}\left[\dfrac{x}{2(at)^{1/2}}-(\gamma t)^{1/2}\right]\right.$ $\left. +\left[t+\dfrac{x}{2(at)^{1/2}}\right]e^{x(\gamma/a)^{1/2}}\operatorname{erfc}\left[\dfrac{x}{2(at)^{1/2}}+(\gamma t)^{1/2}\right]\right\}$
45	$\dfrac{e^{-qx}}{(p-\gamma)(q+\beta)},$ $\gamma \neq a\beta^2$	$\frac{1}{2}e^{\gamma t}\left\{\dfrac{a^{1/2}}{a^{1/2}\beta+\gamma^{1/2}}e^{-x(\gamma/a)^{1/2}}\operatorname{erfc}\left[\dfrac{x}{2(at)^{1/2}}-(\gamma t)^{1/2}\right]\right.$ $\left. +\dfrac{a^{1/2}}{a^{1/2}\beta-\gamma^{1/2}}e^{x(\gamma/a)^{1/2}}\operatorname{erfc}\left[\dfrac{x}{2(at)^{1/2}}+(\gamma t)^{1/2}\right]\right\}$ $-\dfrac{a\beta}{a\beta^2-\gamma}e^{\beta x+a\beta^2 t}\operatorname{erfc}\left[\dfrac{x}{2(at)^{1/2}}+\beta(at)^{1/2}\right]$
46	$e^{x/p}-1$	$\left(\dfrac{x}{t}\right)^{1/2}I_1[2(xt)^{1/2}]$
47	$\dfrac{1}{p}e^{x/p}$	$I_0[2(xt)^{1/2}]$
48	$\dfrac{1}{p^\nu}e^{x/p}$	$\left(\dfrac{t}{x}\right)^{(\nu-1)/2}I_{\nu-1}[2(xt)^{1/2}]$
49	$K_0(qx)$	$\dfrac{1}{2t}e^{-x^2/4at}$
50	$\dfrac{1}{p^{1/2}}K_{2\nu}(qx)$	$\dfrac{1}{2(\pi t)^{1/2}}e^{-x^2/8at}K_\nu\left(\dfrac{x^2}{8at}\right)$
51	$p^{\nu/2-1}K_\nu(qx)$	$x^{-\nu}a^{\nu/2}2^{\nu-1}\displaystyle\int_{x^2/4at}^\infty e^{-u}u^{\nu-1}\,du$
52	$p^{\nu/2}K_\nu(qx)$	$\dfrac{x^\nu}{a^{\nu/2}(2t)^{\nu+1}}e^{-x^2/4at}$

TABLE 7-2 (Continued)

No.	Transform	Function
53	$[p - (p^2 - x^2)^{1/2}]^\nu$	$\nu \dfrac{x^\nu}{t} I_\nu(xt)$
54	$e^{x[(p+\alpha)^{1/2}-(p+\beta)^{1/2}]^2} - 1$	$\dfrac{x(\alpha - \beta)e^{-(\alpha+\beta)t/2} I_1[\tfrac{1}{2}(\alpha - \beta)t^{\frac{1}{2}}(t + 4x)^{\frac{1}{2}}]}{t^{\frac{1}{2}}(t + 4x)^{\frac{1}{2}}}$
55	$\dfrac{e^{x[p-(p+\alpha)^{1/2}(p+\beta)^{1/2}]}}{(p+\alpha)^{\frac{1}{2}}(p+\beta)^{\frac{1}{2}}}$	$e^{-(\alpha+\beta)(t+x)/2} I_0[\tfrac{1}{2}(\alpha - \beta)t^{\frac{1}{2}}(t + 2x)^{\frac{1}{2}}]$
56	$\dfrac{e^{x[(p+\alpha)^{1/2}-(p+\beta)^{1/2}]^2}}{(p+\alpha)^{\frac{1}{2}}(p+\beta)^{\frac{1}{2}}[(p+\alpha)^{\frac{1}{2}} + (p+\beta)^{\frac{1}{2}}]^{2\nu}}$	$\dfrac{t^{\nu/2} e^{-(\alpha+\beta)t/2} I_\nu[\tfrac{1}{2}(\alpha - \beta)t^{\frac{1}{2}}(t + 4x)^{\frac{1}{2}}]}{(\alpha - \beta)^\nu (t + 4x)^{\nu/2}}$

The formulation of the problem is†

$$\frac{d\theta}{dt} + m\theta = n \cos \omega t, \tag{5-1}$$

$$\theta(0) = 0, \tag{5-2}$$

where n and m are defined as in Example 5–1.

Let us apply the Laplace transforms to Eq. (5–1) by considering the transform pair (II) from Table 7–1 with Eq. (5–2) and the pair (25) from Table 7–2. (Hereafter the initials TP will be used in place of "transform pair" for convenience.) The transform of Eq. (5–1) then becomes

$$(p\bar{\theta} - 0) + m\bar{\theta} = n \frac{p}{p^2 + \omega^2},$$

which may be rearranged to give

$$\bar{\theta}(p) = \frac{np}{(p + m)(p^2 + \omega^2)}. \tag{7-26}$$

In terms of the conventional methods of partial fractions, Eq. (7–26) may be expanded as

$$\frac{np}{(p + m)(p^2 + \omega^2)} = \frac{A}{p + m} + \frac{Bp + C}{p^2 + \omega^2}. \tag{7-27}$$

After clearing fractions, we require Eq. (7–27) to be an identity, and obtain

$$\bar{\theta}(p) = \left(\frac{n}{m^2 + \omega^2}\right)\left(-\frac{m}{p + m} + \frac{mp}{p^2 + \omega^2} + \frac{\omega^2}{p^2 + \omega^2}\right). \tag{7-28}$$

---

† See Example 5–1.

The use of TP(12), TP(24), and TP(25) readily gives the inverse transform of Eq. (7–28) as

$$\theta(t) = \left(\frac{n}{m^2 + \omega^2}\right)(-me^{-mt} + m\cos\omega t + \omega\sin\omega t),$$

which can be rearranged in the form

$$\frac{\theta(t)}{q''/h} = \frac{m}{(m^2 + \omega^2)^{1/2}}\cos(\omega t - \alpha) - \frac{m^2 e^{-mt}}{(m^2 + \omega^2)}, \tag{5-3}$$

where $\alpha = \tan^{-1}(\omega/m)$. This result is identical to that obtained by the classical methods of solution for ordinary differential equations (see Example 5–1).

Example 7–2. Let us reconsider Problem 5–3, in which a wall divides two fluids as shown in Fig. 5–30. We wish to find the unsteady lumped temperature of the wall and of the finite fluid.

The formulation of the problem in terms of $\psi = T - T_\infty$ and $\phi = \theta - T_\infty$ is

$$\frac{d\psi}{dt} = -b_1(\psi - \phi), \tag{7-29}$$

$$\frac{d\phi}{dt} = b_2(\psi - \phi) - b_3\phi, \tag{7-30}$$

$$\psi(0) = \phi(0) = T_0 - T_\infty = \psi_0, \tag{7-31}$$

where $b_1 = h_2/\rho_f c_f \delta_f$, $b_2 = h_2/\rho_w c_w \delta_w$, and $b_3 = h_1/\rho_w c_w \delta_w$ (see Fig. 5–30 for the notation).

Using TP(II) and noting Eq. (7–31), we reduce Eqs. (7–29) and (7–30) to a set of two algebraic equations which can be rearranged in the form

$$(p + b_1)\bar{\psi} - b_1\bar{\phi} = \psi_0, \tag{7-32}$$

$$-b_2\bar{\psi} + (p + b_2 + b_3)\bar{\phi} = \psi_0. \tag{7-33}$$

Solving $\bar{\psi}$ and $\bar{\phi}$ from Eqs. (7–32) and (7–33), we obtain

$$\frac{\bar{\psi}(p)}{\psi_0} = \frac{p + (b_1 + b_2 + b_3)}{p^2 + (b_1 + b_2 + b_3)p + b_1 b_3}, \tag{7-34}$$

$$\frac{\bar{\phi}(p)}{\psi_0} = \frac{p + (b_1 + b_2)}{p^2 + (b_1 + b_2 + b_3)p + b_1 b_3}. \tag{7-35}$$

To permit its expansion in partial fractions, the common denominator of Eqs. (7–34) and (7–35) is first factored in the form

$$p^2 + (b_1 + b_2 + b_3)p + b_1 b_3 = (p + \lambda_1)(p + \lambda_2). \tag{7-36}$$

It follows from Eq. (7–36) that $\lambda_1 + \lambda_2 = b_1 + b_2 + b_3$ and $\lambda_1 \lambda_2 = b_1 b_3$.

Now, rearranging $\bar{\psi}$ and $\bar{\phi}$ in terms of λ_1 and λ_2 as

$$\frac{\bar{\psi}(p)}{\psi_0} = \frac{p + (\lambda_1 + \lambda_2)}{(p + \lambda_1)(p + \lambda_2)}, \qquad (7\text{--}37)$$

$$\frac{\bar{\phi}(p)}{\psi_0} = \frac{p + (b_1 + b_2)}{(p + \lambda_1)(p + \lambda_2)}, \qquad (7\text{--}38)$$

and expanding Eqs. (7–37) and (7–38), we have

$$\frac{\bar{\psi}(p)}{\psi_0} = \left(\frac{1}{\lambda_1 - \lambda_2}\right)\left(\frac{\lambda_1}{p + \lambda_2} - \frac{\lambda_2}{p + \lambda_1}\right), \qquad (7\text{--}39)$$

$$\frac{\bar{\phi}(p)}{\psi_0} = \left(\frac{1}{\lambda_1 - \lambda_2}\right)\left(\frac{\lambda_1 - b_3}{p + \lambda_2} - \frac{\lambda_2 - b_3}{p + \lambda_1}\right). \qquad (7\text{--}40)$$

Reference to TP(12) readily gives the inverse of Eqs. (7–39) and (7–40), and hence the temperatures of the wall and the finite fluid,

$$\frac{T(t) - T_\infty}{T_0 - T_\infty} = \frac{\lambda_1 e^{-\lambda_2 t} - \lambda_2 e^{-\lambda_1 t}}{\lambda_1 - \lambda_2}, \qquad (7\text{--}41)$$

$$\frac{\theta(t) - T_\infty}{T_0 - T_\infty} = \frac{T(t) - T_\infty}{T_0 - T_\infty} - b_3 \frac{e^{-\lambda_2 t} - e^{-\lambda_1 t}}{\lambda_1 - \lambda_2}. \qquad (7\text{--}42)$$

It is important to note that when the transform table in hand permits, part or all of the procedure of expanding in partial fractions may be eliminated. Thus, for example, the inversion of Eq. (7–37) may be obtained directly by TP(13) and TP(19) without an expansion. (Compare the algebra of this problem with that of the differential operator method, suggested in Chapter 5 for the same problem.)

Example 7–3. Consider a semi-infinite solid that is initially at temperature T_0 (Fig. 7–2). Assume that the surface temperature of the solid is suddenly changed to T_∞. We wish to find the unsteady temperature in the solid.

Note that the problem cannot be solved in terms of Fourier series† (why?). Thus the Laplace transform procedure, which readily yields a solution, becomes indispensable.

The formulation of the problem in terms of $\theta = T - T_\infty$ is

$$\frac{\partial \theta}{\partial t} = a \frac{\partial^2 \theta}{\partial x^2}, \qquad (7\text{--}43)$$

$$\theta(x, 0) = T_0 - T_\infty = \theta_0, \qquad (7\text{--}44)$$

$$\theta(0, t) = 0, \qquad (7\text{--}45)$$

$$\theta(\infty, t) = \theta_0. \qquad (7\text{--}46)$$

---

† In Section 7–5 the solution of the problem will be given in terms of Fourier integrals.

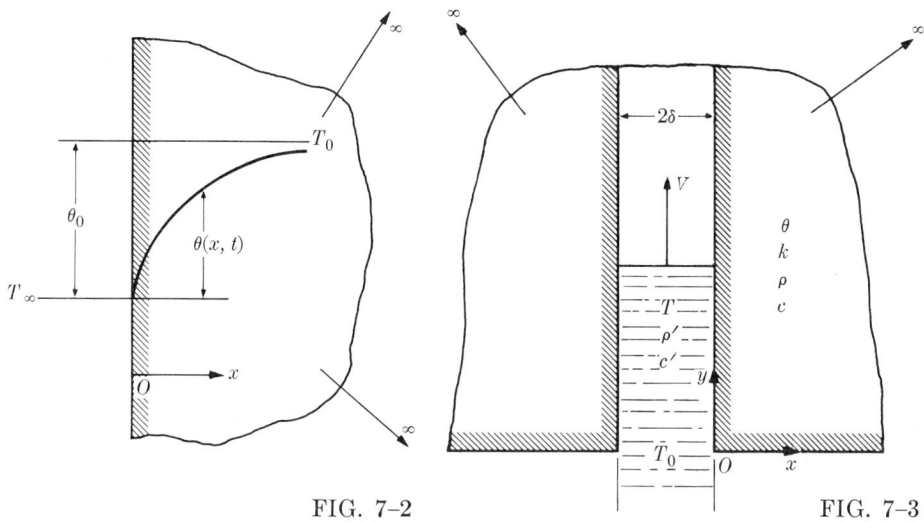

FIG. 7-2 FIG. 7-3

Employing TP(II), Eq. (7–44), and TP(III), we may write the transform of Eq. (7–43) in time,

$$\frac{d^2\bar{\theta}}{dx^2} - q^2\bar{\theta} = \frac{\theta_0}{a}, \tag{7-47}$$

subject to the transforms of Eqs. (7–45) and (7–46),

$$\bar{\theta}(0, p) = 0, \tag{7-48}$$

$$\bar{\theta}(\infty, p) = \theta_0/p. \tag{7-49}$$

[In Eq. (7–47), we use $q^2 = p/a$ for convenience in the algebra of the solution.]

The general solution of Eq. (7–47), suitably expressed in terms of exponentials, is

$$\bar{\theta}(x, p) = \theta_0/p + Ae^{-qx} + Be^{qx}. \tag{7-50}$$

The integration constants A and B are determined by considering Eqs. (7–42) and (7–43). There then follows

$$\frac{\bar{\theta}(x, p)}{\theta_0} = \frac{1}{p} - \frac{1}{p}e^{-qx}. \tag{7-51}$$

The use of TP(1) and TP(30) results in the inversion of Eq. (7–51) and hence the desired temperature of the solid,

$$\frac{T(x, t) - T_\infty}{T_0 - T_\infty} = \text{erf}\left[\frac{x}{2(at)^{1/2}}\right]. \tag{7-52}$$

Example 7–4. Consider the semi-infinite solid of Fig. 7–3. The solid has a slit of thickness 2δ extending to infinity in the z-direction. Initially, the temperature of the system is assumed to be zero; then a liquid having the uniform velocity V begins to fill up the slit. The inlet temperature of the liquid is T_0. The heat transfer coefficient between the liquid and the solid is large. Conduction in the y-direction can be neglected both in the liquid and in the solid. We wish to find the unsteady temperature in the liquid and in the solid.

The geometry of the problem suggests that a lumped liquid temperature be considered in the x-direction. The first law of thermodynamics written for the liquid then becomes a boundary condition for the solid at $x = 0 (y > 0)$. In view of this fact the formulation of the problem may be written in the form[†]

$$\frac{\partial \theta}{\partial t} = a \frac{\partial^2 \theta}{\partial x^2}, \tag{7-53}$$

$$\theta(x, y, 0) = 0, \tag{7-54}$$

$$\theta(\infty, y, t) = 0, \tag{7-55}$$

$$\left(\frac{\partial \theta}{\partial t} + V \frac{\partial \theta}{\partial y} = \epsilon \frac{\partial \theta}{\partial x}\right)_{x=0}, \tag{7-56}$$

where $\epsilon = k/\rho' c' \delta$. Since conduction in the y-direction is neglected, the only source of y-dependency is the term $V(\partial \theta/\partial y)_{x=0}$ of Eq. (7-56). Because of this term, the formulation requires one more boundary condition, to be considered in the y-direction. The inlet temperature of the liquid expressed in terms of the solid temperature gives this condition as

$$\theta(0, 0, t) = T_0. \tag{7-57}$$

The transformed differential equation and the transformed boundary conditions in time are

$$\frac{\partial^2 \bar{\theta}}{\partial x^2} - q^2 \bar{\theta} = 0, \tag{7-58}$$

$$\bar{\theta}(\infty, y, p) = 0, \tag{7-59}$$

$$\left(p\bar{\theta} + V \frac{\partial \bar{\theta}}{\partial y} = \epsilon \frac{\partial \bar{\theta}}{\partial x}\right)_{x=0}, \tag{7-60}$$

$$\bar{\theta}(0, 0, p) = \frac{T_0}{p}, \tag{7-61}$$

where $q^2 = p/a$ as before. The solution of Eq. (7-58) satisfying Eq. (7-59) can

---

[†] For a clear understanding of the formulation above, we suggest that the reader formulate the problem in terms of the fluid temperature $T(y, t)$ and the solid temperature $\theta(x, y, t)$, then eliminate $T(y, t)$ by considering the equality of interface temperatures, $T(y, t) = \theta(0, y, t)$.

[7-4] SOLUTIONS OBTAINABLE BY THE TABLE OF TRANSFORMS   353

be written as
$$\bar{\theta}(x, y, p) = \bar{\theta}(0, y, p)e^{-qx}. \qquad (7\text{-}62)$$

Evaluating the right-hand side of Eq. (7-60) in terms of Eq. (7-62), we obtain
$$\left[\frac{\partial \bar{\theta}}{\partial y} + \frac{1}{V}(p + q\epsilon)\bar{\theta}\right]_{x=0} = 0. \qquad (7\text{-}63)$$

The solution of Eq. (7-63) subject to Eq. (7-61) is
$$\frac{\bar{\theta}(0, y, p)}{T_0} = e^{-(y/V)p} \frac{e^{-(\epsilon y/V)q}}{p}. \qquad (7\text{-}64)$$

Finally, inserting Eq. (7-64) into Eq. (7-62), we get
$$\frac{\bar{\theta}(x, y, p)}{T_0} = e^{-(y/V)p} \frac{e^{-(x+\epsilon y/V)q}}{p}. \qquad (7\text{-}65)$$

The inversion of Eq. (7-65) in terms of TP(VII) and TP(30) gives the unsteady solid temperature
$$\frac{\theta(x, y, t^*)}{T_0} = \begin{cases} 0, & t < y/V \\ \mathrm{erfc}\left[\dfrac{xV + y\epsilon}{2V(at^*)^{1/2}}\right], & t \geq y/V, \end{cases} \qquad (7\text{-}66)$$

where $t^* = t - y/V$. Note that for $x = 0$ Eq. (7-66) denotes the unsteady liquid temperature. [Interpret Eq. (7-66) with respect to an observer located at a fixed y.]

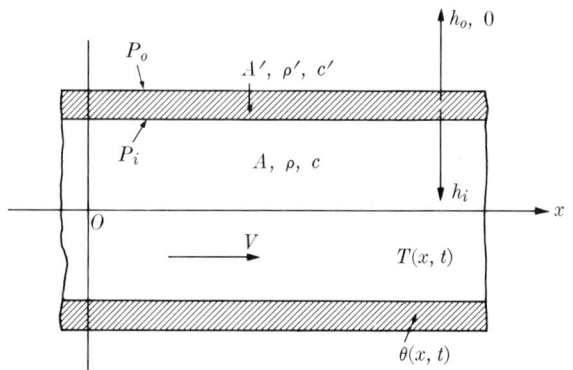

FIG. 7-4

Example 7-5. An incompressible fluid at uniform velocity V flows steadily through a tube (Fig. 7-4). The axial conduction both in the fluid and in the tube walls is negligible. The thermal conductivity, density, specific heat, and cross-sectional area of the fluid and tube are (k, k'), (ρ, ρ'), (c, c'), and (A, A'), respectively. The inner and outer peripheries of the tube are P_i, P_o, and the inside and outside heat transfer coefficients h_i, h_o. Initially the system is at the ambient temperature, say zero; then the inlet temperature of the fluid is suddenly changed to temperature T_0.† We wish to find the unsteady temperature in the system, using a radially lumped analysis.

The formulation of the problem is

$$\frac{\partial T}{\partial t} + V \frac{\partial T}{\partial x} + b_1(T - \theta) = 0, \tag{7-67}$$

$$\frac{\partial \theta}{\partial t} - b_2(T - \theta) + b_3\theta = 0, \tag{7-68}$$

$$T(x, 0) = 0, \tag{7-69}$$

$$\theta(x, 0) = 0, \tag{7-70}$$

$$T(0, t) = T_0, \tag{7-71}$$

where $b_1 = h_i P_i / \rho c A$, $b_2 = h_i P_i / \rho' c' A'$, and $b_3 = h_o P_o / \rho' c' A'$.

The transforms of Eqs. (7–67) and (7–68), including Eqs. (7–69) and (7–70), respectively, are

$$p\bar{T} + V \frac{d\bar{T}}{dx} + b_1(\bar{T} - \bar{\theta}) = 0, \tag{7-72}$$

$$p\bar{\theta} - b_2(\bar{T} - \bar{\theta}) + b_3\bar{\theta} = 0, \tag{7-73}$$

subject to the transform of Eq. (7–71),

$$\bar{T}(0, p) = T_0/p. \tag{7-74}$$

Solving for $\bar{\theta}$ from Eq. (7–73) and inserting the result into Eq. (7–72) yields

$$\frac{d\bar{T}}{dx} + \frac{1}{V}\left(p + b_1 - \frac{b_1 b_2}{p + b_4}\right)\bar{T} = 0, \tag{7-75}$$

where $b_4 = b_2 + b_3$. The solution of Eq. (7–75) that satisfies Eq. (7–74) is

$$\frac{\bar{T}(x, p)}{T_0} = e^{-b_1 x/V} e^{-(x/V)p} \frac{1}{p} e^{b_1 b_2 x/V(p+b_4)}. \tag{7-76}$$

Introducing Eq. (7–76) into Eq. (7–73) and rearranging gives

$$\frac{\bar{\theta}(x, p)}{T_0} = b_2 e^{-b_1 x/V} \underbrace{[e^{-(x/V)p}]}_{\text{(i)}} \underbrace{\left[\frac{e^{b_1 b_2 x/V(p+b_4)}}{p(p + b_4)}\right]}_{\text{(ii)}}. \tag{7-77}$$

TP(46) and TP(47), first combined with TP(VI), then applied to Eqs. (7–76) and (7–77), respectively, and later operated on by TP(IV) and TP(VII) may

---

† The only difference between Example 6–3 and the present example is the variation of the inlet fluid temperature.

give the fluid and the tube-wall temperatures. The result would be in terms of two integrals involving the combination of an exponential function with modified Bessel functions (of the first kind, of order zero and one). However, since these integrals require elaborate tabulations for their use in numerical computations, it is more convenient to express the solution in terms of one integral rather than two. This can be done by rearranging the last term of Eq. (7–76) according to the last term of Eq. (7–77), or vice versa. Selecting the former procedure, for example, and multiplying and dividing the last term of Eq. (7–76) by $(p + b_4)$, we obtain

$$\frac{\overline{T}(x, p)}{T_0} = e^{-b_1 x/V}[e^{-(x/V)p}]\left\{\underbrace{\left[\frac{e^{b_1 b_2 x/V(p+b_4)}}{(p+b_4)}\right]}_{\text{(iii)}} + b_4\underbrace{\left[\frac{e^{b_1 b_2 x/V(p+b_4)}}{p(p+b_4)}\right]}_{\text{(ii)}}\right\}. \quad (7\text{–}78)$$

$\underbrace{\phantom{e^{-b_1 x/V}}}_{\text{(i)}}$

To find the inversions of Eqs. (7–77) and (7–78), let us first consider TP(47),

$$e^{x/p}/p = L\{I_0[2(xt)^{1/2}]\}. \quad (7\text{–}79)$$

Replacing x by $b_1 b_2 x/V$ and employing TP(VI) with $\beta = b_4$, we may rearrange Eq. (7–79) in the form

$$\bar{g}(p) = \frac{e^{b_1 b_2 x/V(p+b_4)}}{(p+b_4)} = L\left\{e^{-b_4 t}I_0\left[2\left(\frac{b_1 b_2 xt}{V}\right)^{1/2}\right]\right\}. \quad (7\text{–}80)$$

Equation (7–80) gives the inversion of (iii) of Eq. (7–78). Furthermore, according to TP(VIII), the convolution of Eq. (7–80) with $\bar{f}(p) = 1/p$ yields the inversion of (ii) of Eqs. (7–77) and (7–78). Finally, noting that (i) of both equations implies a shift in time according to TP(VII) with $\alpha = t/V$, we obtain the desired temperatures of the tube wall and the fluid,

$$\theta(x, t) = T(x, t) = 0, \qquad \text{when} \quad t < x/V \quad (7\text{–}81)$$

and

$$\frac{\theta(x, t)}{T_0} = b_2 e^{-b_1 x/V}\int_0^{t^*} e^{-b_4 s}I_0\left[2\left(\frac{b_1 b_2 x s}{V}\right)^{1/2}\right]ds, \qquad \text{when} \quad t \geq \frac{x}{V}, \quad (7\text{–}82)$$

$$\frac{T(x, t)}{T_0} = e^{-b_1 x/V}\left\{e^{-b_4 t^*}I_0\left[2\left(\frac{b_1 b_2 x t^*}{V}\right)^{1/2}\right]\right.$$

$$\left. + b_4\int_0^{t^*} e^{-b_4 s}I_0\left[2\left(\frac{b_1 b_2 x s}{V}\right)^{1/2}\right]ds\right\}, \qquad \text{when} \quad t \geq \frac{x}{V}, \quad (7\text{–}83)$$

where $t^* = t - x/V$.

Equations (7–82) and (7–83) may further be rearranged to make them suitable for numerical computations. Thus, employing the series expansion of the

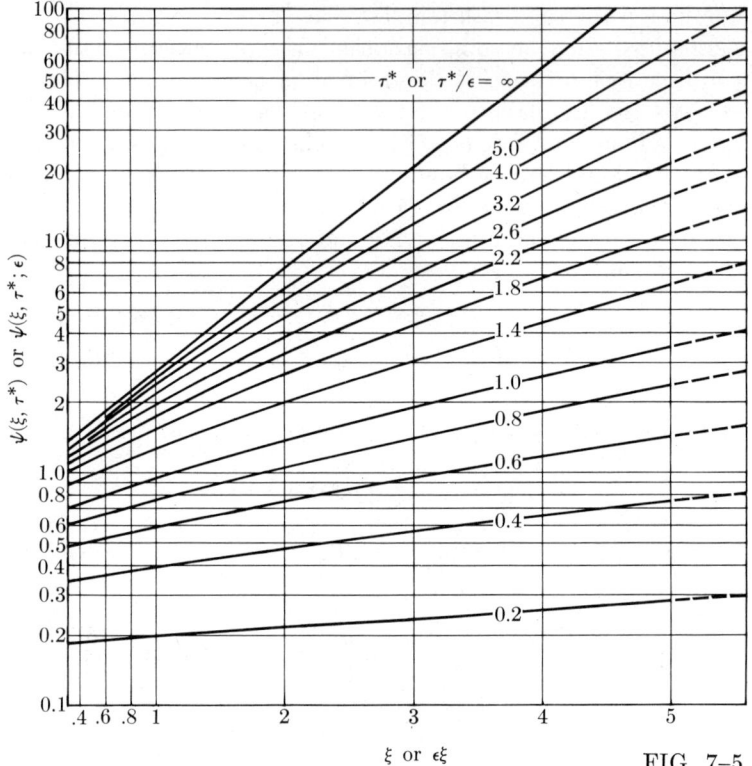

FIG. 7-5

modified Bessel function for small values of the argument,

$$I_0(x) = \sum_{k=0}^{\infty} \frac{(x/2)^{2k}}{(k!)^2}, \qquad (7\text{-}84)$$

we may write the integral appearing in these equations as

$$b_4 \int_0^{t^*} e^{-b_4 s} I_0\left[2\left(\frac{b_1 b_2 x s}{V}\right)^{1/2}\right] ds = \sum_{k=0}^{\infty} \frac{(\epsilon \xi)^k}{(k!)^2} \int_0^{\tau^*/\epsilon} \zeta^k e^{-\zeta} d\zeta = \psi(\xi, \tau^*; \epsilon), \qquad (7\text{-}85)$$

where $\xi = b_1 x/V$, $\tau^* = b_2 t^*$, and $\epsilon = b_2/b_4$. Then, in terms of the function $\psi(\xi, \tau^*; \epsilon)$, Eqs. (7-82) and (7-83) become

$$\frac{\theta(\xi, \tau^*; \epsilon)}{T_0} = \epsilon e^{-\xi} \psi(\xi, \tau^*; \epsilon), \qquad \tau^* \geq 0, \qquad (7\text{-}86)$$

$$\frac{T(\xi, \tau^*; \epsilon)}{T_0} = e^{-\xi}\{e^{-\tau^*/\epsilon} I_0[2(\xi \tau^*)^{1/2}] + \psi(\xi, \tau^*; \epsilon)\}, \qquad \tau^* \geq 0. \qquad (7\text{-}87)$$

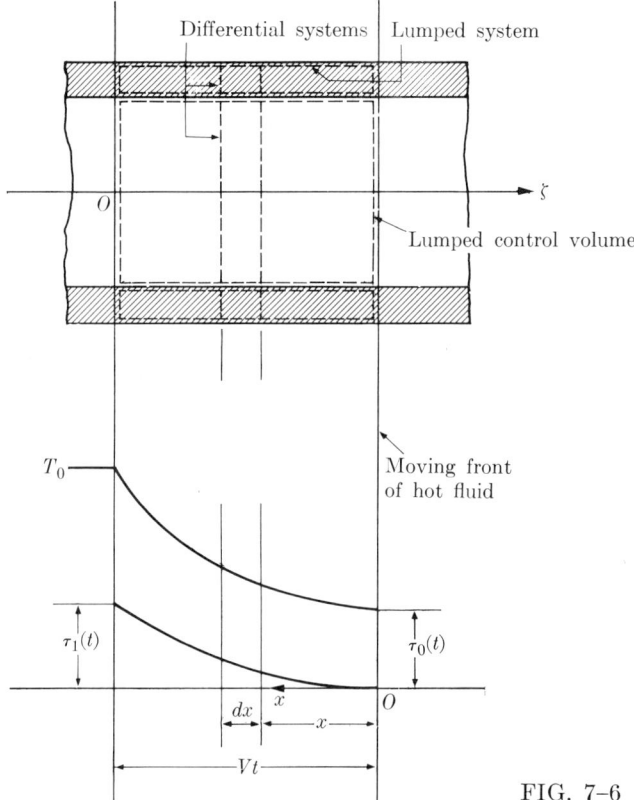

FIG. 7-6

The special case of this problem corresponding to $\epsilon = 1$ may be found in Carslaw and Jaeger† in terms of two integrals, tabulated later by Rizika.‡ $\psi(\xi, \tau^*; \epsilon)$ becomes identical to Rizika's second table provided the space and time variables of this table are replaced by $\epsilon\xi$ and τ^*/ϵ, respectively (Fig. 7-5). (What is the physical significance of $\epsilon = 1$?)

In practice, an approximate solution may be more convenient than the foregoing exact but somewhat complicated solution. Such an approximate solution is obtained in the next example.

Example 7-6. Let us obtain the integral formulation of Example 7-5 and an approximate solution to this formulation.

Consider the lumped control volume and differential system for the fluid, and the lumped and differential systems for the tube walls as shown in Fig. 7-6. Evaluating the terms of the first law of thermodynamics for the lumped control volume and for the lumped system by means of the corresponding differen-

---

† See Reference 1, p. 394, Eqs. (16) and (17).
‡ See Reference 13, from which our Fig. 7-5 is reproduced by permission.

tial systems, we obtain the first law for the fluid,

$$\frac{d}{dt}\int_0^{Vt} \rho c A T\, dx = \rho c A V T_0 - \int_0^{Vt} h_i P_i (T - \theta)\, dx, \qquad (7\text{--}88)$$

and for the tube walls,

$$\frac{d}{dt}\int_0^{Vt} \rho' c' A' \theta\, dx = \int_0^{Vt} h_i P_i (T - \theta)\, dx - \int_0^{Vt} h_o P_o \theta\, dx. \qquad (7\text{--}89)$$

Equations (7–88) and (7–89) may conveniently be rearranged in terms of the constants b_1, b_2, and b_3 as defined in Example 7–5. It follows then that

$$\frac{d}{dt}\int_0^{Vt} T\, dx = V T_0 - b_1 \int_0^{Vt} (T - \theta)\, dx, \qquad (7\text{--}90)$$

$$\frac{d}{dt}\int_0^{Vt} \theta\, dx = b_2 \int_0^{Vt} (T - \theta)\, dx - b_3 \int_0^{Vt} \theta\, dx. \qquad (7\text{--}91)$$

Before selecting temperature profiles, let us simplify Eqs. (7–90) and (7–91) in terms of the spacewise averaged values of the fluid and tube-wall temperatures,

$$T_m(t) = \frac{1}{V T_0}\int_0^{Vt} T(x, t)\, dx, \qquad (7\text{--}92)$$

$$\theta_m(t) = \frac{1}{V T_0}\int_0^{Vt} \theta(x, t)\, dx. \qquad (7\text{--}93)$$

Inserting Eqs. (7–92) and (7–93) into Eqs. (7–90) and (7–91) gives

$$\frac{dT_m}{dt} = 1 - b_1 (T_m - \theta_m), \qquad (7\text{--}94)$$

$$\frac{d\theta_m}{dt} = b_2 (T_m - \theta_m) - b_3 \theta_m, \qquad (7\text{--}95)$$

subject to the initial conditions

$$T_m(0) = 0, \qquad \theta_m(0) = 0, \qquad (7\text{--}96)$$

which are evident from Eqs. (7–92) and (7–93).

Delaying for the moment the discussion of the spacewise distribution of the temperature profiles, we may proceed here to the solution of the spacewise averaged formulation. Taking the Laplace transforms of Eqs. (7–94) and (7–95), and noting Eq. (7–96), we obtain

$$(p + b_1)\overline{T}_m - b_1 \overline{\theta}_m = 1/p, \qquad (7\text{--}97)$$

$$b_2 \overline{T}_m - (p + b_2 + b_3)\overline{\theta}_m = 0. \qquad (7\text{--}98)$$

Equations (7-97) and (7-98) yield

$$\overline{T}_m(p) = \frac{p + b_2 + b_3}{p[p^2 + (b_1 + b_2 + b_3)p + b_1 b_3]}, \qquad (7\text{-}99)$$

$$\overline{\theta}_m(p) = \frac{b_2}{p[p^2 + (b_1 + b_2 + b_3)p + b_1 b_3]}, \qquad (7\text{-}100)$$

which may be rearranged by introducing

$$b_1 + b_2 + b_3 = \lambda_1 + \lambda_2, \qquad b_1 b_3 = \lambda_1 \lambda_2.$$

The result is

$$\overline{T}_m(p) = \frac{1}{(p + \lambda_1)(p + \lambda_2)} + \frac{b_4}{p(p + \lambda_1)(p + \lambda_2)}, \qquad (7\text{-}101)$$

$$\overline{\theta}_m(p) = \frac{b_2}{p(p + \lambda_1)(p + \lambda_2)}, \qquad (7\text{-}102)$$

where $b_4 = b_2 + b_3$. Either by first expanding Eqs. (7-101) and (7-102) in partial fractions and then employing TP(12), or by appropriately applying TP(13) and TP(15) to Eqs. (7-101) and (7-102), we obtain

$$T_m(t) = \frac{(b_4/\lambda_2 - 1)(1 - e^{-\lambda_2 t}) - (b_4/\lambda_1 - 1)(1 - e^{-\lambda_1 t})}{\lambda_1 - \lambda_2}, \qquad (7\text{-}103)$$

$$\theta_m(t) = \frac{(b_2/\lambda_2)(1 - e^{-\lambda_2 t}) - (b_2/\lambda_1)(1 - e^{-\lambda_1 t})}{\lambda_1 - \lambda_2}. \qquad (7\text{-}104)$$

Finally, let us consider the spacewise distribution of the temperature profiles in terms of the simplest physically admissible polynomials, that is, in terms of parabolas. Referring to Fig. 7-6, and for simplicity of integration selecting as the origin of the coordinate axis x the moving front of the hot fluid, we have the first-order Kantorovich profiles in terms of the temperatures $\tau_0(t)$ and $\tau_1(t)$,

$$T(x, t) = \tau_0(t) + [T_0 - \tau_0(t)](x/Vt)^2, \qquad (7\text{-}105)$$

$$\theta(x, t) = \tau_1(t)(x/Vt)^2. \qquad (7\text{-}106)$$

The right-hand sides of Eqs. (7-92) and (7-93) integrated by means of Eqs. (7-105) and (7-106) give the relation between $T_m(t)$ and $\tau_0(t)$, and $\theta_m(t)$ and $\tau_1(t)$,

$$\tau_0(t)/T_0 = \tfrac{3}{2}[T_m(t)/t] - \tfrac{1}{2}, \qquad (7\text{-}107)$$

$$\tau_1(t)/T_0 = 3\theta_m(t)/t. \qquad (7\text{-}108)$$

Inserting Eqs. (7-103) and (7-104) into Eqs. (7-107) and (7-108), and the result into Eqs. (7-105) and (7-106), we obtain the approximate temperatures of the fluid and the tube walls,

$$\frac{T(x,t)}{T_0} = \left(\frac{x}{Vt}\right)^2 + \frac{1}{2}\left\{\left(\frac{3b_4}{\lambda_1 - \lambda_2}\right)\left[\left(1 - \frac{\lambda_2}{b_4}\right)\left(\frac{1 - e^{-\lambda_2 t}}{\lambda_2 t}\right)\right.\right.$$
$$\left.\left. - \left(1 - \frac{\lambda_1}{b_4}\right)\left(\frac{1 - e^{-\lambda_1 t}}{\lambda_1 t}\right)\right] - 1\right\}\left[1 - \left(\frac{x}{Vt}\right)^2\right] \quad (7\text{--}109)$$

and

$$\frac{\theta(x,t)}{T_0} = \left(\frac{3b_2}{\lambda_1 - \lambda_2}\right)\left[\left(\frac{1 - e^{-\lambda_2 t}}{\lambda_2 t}\right) - \left(\frac{1 - e^{-\lambda_1 t}}{\lambda_1 t}\right)\right]\left(\frac{x}{Vt}\right)^2, \quad (7\text{--}110)$$

both valid when $\zeta \leq Vt$. According to the physics of the problem and Fig. 7–6, the fluid and the tube-wall temperature are both zero when $\zeta > Vt$.

Note the simplicity of the functions involved in Eqs. (7–109) and (7–110), as compared with those involved in the exact solution of the problem, Eqs. (7–86) and (7–87). The detailed comparison of the exact and approximate solutions, although important, requires an involved parametric study, and hence is not considered here.

Example 7–7. Let us reconsider the problem of Examples 7–5 and 7–6, but assume that the outer surface of the tube is insulated. Also, instead of there being a sudden change in the inlet temperature of the fluid, assume that the constant internal energy u''' begins to be generated uniformly in the tube walls (Fig. 7–7). We wish to obtain the radially lumped and axially differential formulation of the problem, and its solution.

The formulation of the problem is

$$\frac{\partial T}{\partial t} + V\frac{\partial T}{\partial x} + b_1(T - \theta) = 0, \quad (7\text{--}67)$$

$$\frac{\partial \theta}{\partial t} - b_2(T - \theta) = u, \quad (7\text{--}111)$$

$$T(x, 0) = 0, \quad (7\text{--}69)$$

$$\theta(x, 0) = 0, \quad (7\text{--}70)$$

$$T(0, t) = 0, \quad (7\text{--}112)$$

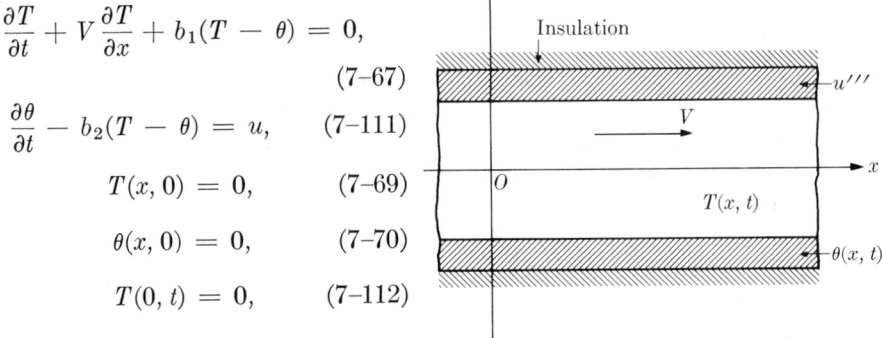

FIG. 7–7

where $u = u'''/\rho' c'$, and b_1 and b_2 are defined as in Example 7–5.

Taking the Laplace transforms of Eqs. (7–67) and (7–111), and employing Eqs. (7–69) and (7–70), we have

$$p\bar{T} + V\frac{d\bar{T}}{dx} + b_1(\bar{T} - \bar{\theta}) = 0, \quad (7\text{--}72)$$

$$p\bar{\theta} - b_2(\bar{T} - \bar{\theta}) = u/p, \quad (7\text{--}113)$$

subject to the transform of Eq. (7–112),

$$\bar{T}(0, p) = 0. \qquad (7\text{–}114)$$

Solving $\bar{\theta}$ from Eq. (7–113), inserting the result into Eq. (7–72), and rearranging gives

$$\frac{d\bar{T}}{dx} + \frac{1}{V}\left(p + b_1 - \frac{b_1 b_2}{p + b_2}\right)\bar{T} = \frac{ub_1/V}{p(p + b_2)}. \qquad (7\text{–}115)$$

The general solution of Eq. (7–115) satisfied by Eq. (7–114) is

$$\frac{\bar{T}(x, p)}{u} = \frac{b_1}{p^2(p + b)}\{1 - e^{-b_1 x/V} e^{-(x/V)p} e^{b_1 b_2 x/V(p+b_2)}\}, \qquad (7\text{–}116)$$

where $b = b_1 + b_2$. Introducing Eq. (7–116) into Eq. (7–113), we obtain

$$\frac{\bar{\theta}(x, p)}{u} = \left[\frac{1}{p(p + b)} + \frac{b_1}{p^2(p + b)}\right]$$
$$\phantom{\frac{\bar{\theta}(x, p)}{u} = }\text{(iv)}$$
$$- b_1 b_2 e^{-b_1 x/V}[e^{-(x/V)p}]\left[\frac{1}{p^2(p + b)}\right]\left[\frac{e^{b_1 b_2 x/V(p+b_2)}}{(p + b_2)}\right]. \qquad (7\text{–}117)$$
$$\text{(i)}\text{(iii)}\text{(ii)}$$

For convenience in tabulations and numerical computations, we may follow the steps of Example 7–5 and express the solutions in terms of integrals involving the modified Bessel function of the first kind, of order zero. Equation (7–116) may then be rearranged accordingly by multiplying and dividing the last of its terms by $(p + b_2)$. Thus we have

$$\frac{\bar{T}(x, p)}{u} = b_1\left[\frac{1}{p^2(p + b)}\right]$$
$$\text{(iii)}$$
$$- b_1 e^{-b_1 x/V}[e^{-(x/V)p}]\left[\frac{1}{p(p + b)} + \frac{b_2}{p^2(p + b)}\right]\left[\frac{e^{b_1 b_2 x/V(p+b_2)}}{(p + b_2)}\right].$$
$$\text{(i)}\text{(v)}\text{(ii)}$$
$$\qquad (7\text{–}118)$$

Before the inversion process, let us first note that the terms in the first brackets of Eqs. (7–117) and (7–118) do not involve the factor $e^{-(x/V)p}$, and yield inversions valid for $t \geq 0$. Thus, inserting $\alpha = 0$, $\beta = b$ into TP(13) and TP(16) for (iv) of Eq. (7–117), and again $\alpha = 0$, $\beta = b$ into TP(16) for (iii) of Eq. (7–118), we obtain

$$\frac{T(t)}{ub_1/b^2} = bt - (1 - e^{-bt}), \qquad 0 \leq t \leq x/V, \qquad (7\text{–}119)$$

$$\frac{\theta(t)}{ub_1/b^2} = bt + \frac{b_2}{b_1}(1 - e^{-bt}), \qquad 0 \leq t \leq x/V. \qquad (7\text{–}120)$$

[Why are Eqs. (7–119) and (7–120) independent of x? Write the particular formulation of the problem which yields these solutions. What is the physical significance of this formulation?] Introducing the dimensionless quantities $\xi = b_1 x/V$, $\tau = b_2 t$, and $\lambda = 1 + b_1/b_2$, we may rearrange Eqs. (7–119) and (7–120) in the form

$$\frac{T(\tau;\lambda)}{(u'''A'/h_iP_i)(\lambda-1)/\lambda^2} = \lambda\tau - (1 - e^{-\lambda\tau}), \qquad 0 \leq \tau \leq (\lambda-1)\xi, \tag{7-121}$$

$$\frac{\theta(\tau;\lambda)}{(u'''A'/h_iP_i)(\lambda-1)/\lambda^2} = \lambda\tau + \left(\frac{1}{\lambda-1}\right)(1 - e^{-\lambda\tau}),$$

$$0 \leq \tau \leq (\lambda-1)\xi. \tag{7-122}$$

When $t \geq x/V$, the solutions given by Eqs. (7–119) and (7–120) must be supplemented by the inversion of the terms involving $e^{-(x/V)p}$. The procedure of Example 7–5 may again be followed. First replacing b_4 by b_2, we have from Eq. (7–80)

$$\bar{g}(p) = \frac{e^{b_1 b_2 x/V(p+b_2)}}{(p+b_2)} = L\left\{e^{-b_2 t}I_0\left[2\left(\frac{b_1 b_2 xt}{V}\right)^{1/2}\right]\right\}. \tag{7-123}$$

Let $\bar{f}_1(p)$ and $\bar{f}_2(p)$ denote (iii) of Eq. (7–117) and (v) of Eq. (7–118), respectively. Noting that the inverse of (iii) has already been found, and the inverse of (v) can be obtained from that of (iv) by simply replacing b_1 of the latter by b_2, we may write

$$f_1(t) = \frac{1}{b^2}[bt + (1 - e^{-bt})], \tag{7-124}$$

$$f_2(t) = \frac{1}{b^2}[bb_2 t - b_1(1 - e^{-bt})]. \tag{7-125}$$

Then the convolution of Eqs. (7–123) and (7–124), and that of Eqs. (7–123) and (7–125) according to TP(VIII), and the use of TP(VII) with $\alpha = x/V$ give the fluid and tube-wall temperatures

$$\frac{T(x,t)}{ub_1/b^2} = bt - (1 - e^{-bt}) - e^{-b_1 x/V}$$

$$\times \left\{(bb_2 t^* + b_1)\int_0^{t^*} e^{-b_2 s} I_0\left[2\left(\frac{b_1 b_2 xs}{V}\right)^{1/2}\right]ds\right.$$

$$- bb_2 \int_0^{t^*} s e^{-b_2 s} I_0\left[2\left(\frac{b_1 b_2 xs}{V}\right)^{1/2}\right]ds$$

$$\left. - b_1 e^{-bt^*}\int_0^{t^*} e^{b_1 s} I_0\left[2\left(\frac{b_1 b_2 xs}{V}\right)^{1/2}\right]ds\right\}, \qquad t \geq \frac{x}{V}, \tag{7-126}$$

$$\frac{\theta(x,t)}{ub_1/b^2} = bt + \frac{b_2}{b_1}(1 - e^{-bt}) - e^{-b_1 x/V}$$

$$\times \left\{ b_2(bt^* - 1) \int_0^{t^*} e^{-b_2 s} I_0\left[2\left(\frac{b_1 b_2 xs}{V}\right)^{1/2}\right] ds \right.$$

$$- bb_2 \int_0^{t^*} s e^{-b_2 s} I_0\left[2\left(\frac{b_1 b_2 xs}{V}\right)^{1/2}\right] ds$$

$$\left. + b_2 e^{-bt^*} \int_0^{t^*} e^{b_1 s} I_0\left[2\left(\frac{b_1 b_2 xs}{V}\right)^{1/2}\right] ds \right\}, \quad t \geq \frac{x}{V}. \quad (7\text{-}127)$$

The integrals of Eqs. (7-126) and (7-127), like those of Example 7-5, may be expressed in terms of the series expansion of the appropriate Bessel function, Eq. (7-84). It follows then that

$$\frac{T(\xi, \tau; \lambda)}{(u'''A'/h_i P_i)(\lambda - 1)/\lambda^2} = \lambda \tau - (1 - e^{-\lambda \tau})$$

$$- e^{-\xi}[(\lambda \tau^* + \lambda - 1)\Psi(\xi, \tau^*) - \lambda \Phi(\xi, \tau^*) - e^{-\lambda \tau^*}\Lambda(\xi, \tau^*; \lambda)],$$

$$\tau \geq (\lambda - 1)\xi, \quad (7\text{-}128)$$

$$\frac{\theta(\xi, \tau; \lambda)}{(u'''A'/h_i P_i)(\lambda - 1)/\lambda^2} = \lambda \tau + \left(\frac{1}{\lambda - 1}\right)(1 - e^{-\lambda \tau})$$

$$- e^{-\xi}\left[(\lambda \tau^* - 1)\Psi(\xi, \tau^*) - \lambda \Phi(\xi, \tau^*) + \left(\frac{1}{\lambda - 1}\right)e^{-\lambda \tau^*}\Lambda(\xi, \tau^*; \lambda)\right],$$

$$\tau \geq (\lambda - 1)\xi, \quad (7\text{-}129)$$

where

$$\Psi(\xi, \tau^*) = \sum_{k=0}^{\infty} \frac{\xi^k}{(k!)^2} \int_0^{\tau^*} \zeta^k e^{-\zeta} d\zeta, \quad (7\text{-}130)$$

$$\Phi(\xi, \tau^*) = \sum_{k=0}^{\infty} \frac{\xi^k}{(k!)^2} \int_0^{\tau^*} \zeta^{k+1} e^{-\zeta} d\zeta, \quad (7\text{-}131)$$

$$\Lambda(\xi, \tau^*; \lambda) = \sum_{k=0}^{\infty} \frac{[\xi/(\lambda - 1)]^k}{(k!)^2} \int_0^{(\lambda-1)\tau^*} \zeta^k e^{\zeta} d\zeta. \quad (7\text{-}132)$$

Equation (7-130) is a special case of Eq. (7-85) for $\epsilon = 1$. Equations (7-131) and (7-132) have been tabulated in References 14 and 15;† the results are re-

---

† The functions Ψ_4 and Ψ_5 of References 14 and 15 correspond to Λ and Φ, respectively. Figures 7-8 and 7-9 reproduced by permission.

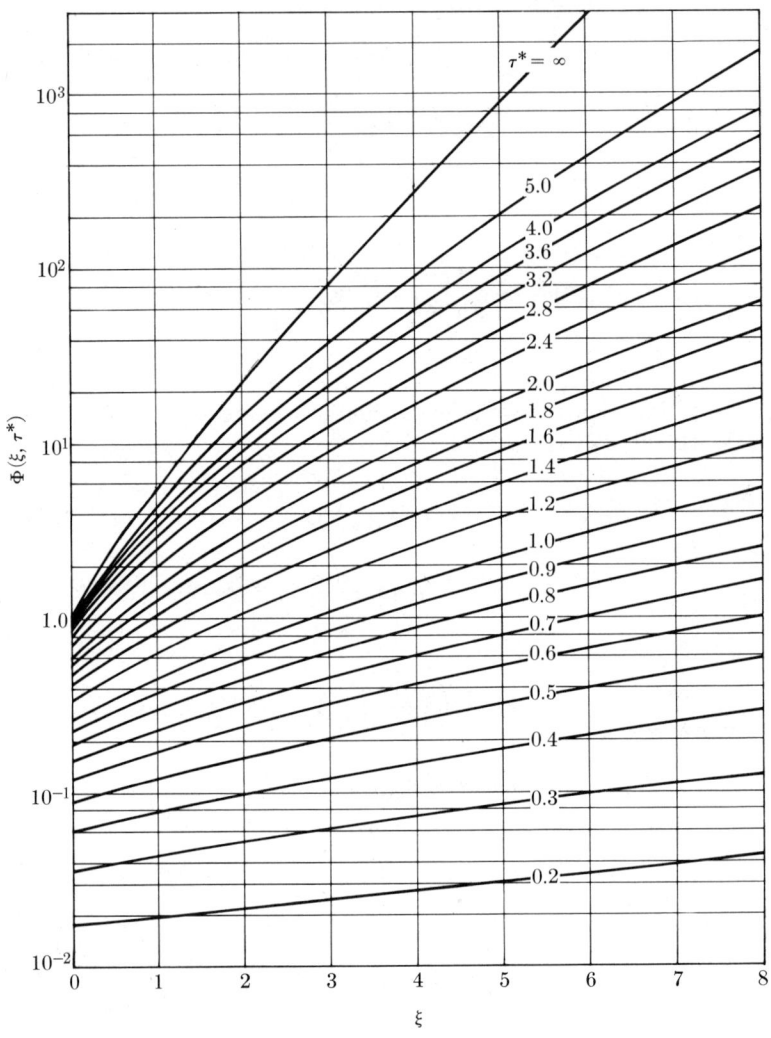

FIG. 7-8

produced here in Figs. 7-8 and 7-9, respectively. A parametric study of the fluid and tube-wall temperatures for various values of λ, although useful, will not be elaborated here.

As we saw in the foregoing examples, the inverse transforms of a large number of problems can be obtained directly by the use of transform tables. There remain, however, many problems whose transforms are not suitable to tables and which thus require the use of the inversion theorem for Laplace transforms. The next three sections are devoted to topics related to this theorem. In Section 7-5, the Fourier integrals and the Fourier transforms are briefly reviewed.

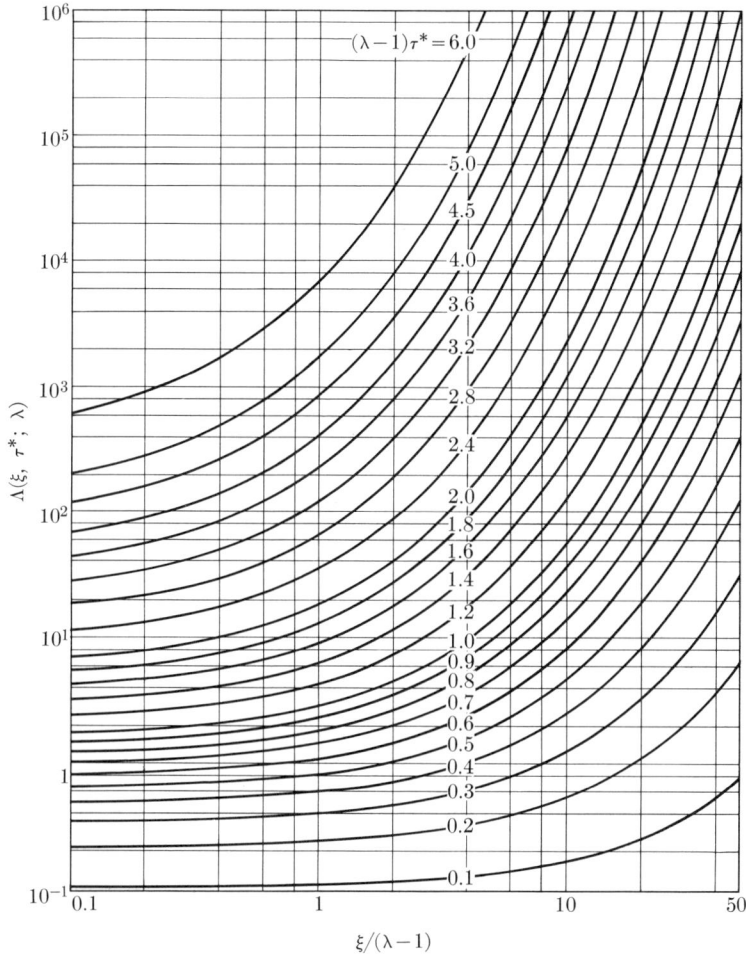

FIG. 7-9

From the definition of Fourier transforms, the inversion theorem is developed in Section 7-6. Then, in Section 7-7, a short treatment is given to functions of complex variables and the theory of residues, which are useful for evaluating the integral of the inversion theorem.

7-5. Fourier Integrals

In Chapters 4 and 5 we solved a number of problems converting them to characteristic-value problems by the technique of separation of variables. In this connection, we showed that linear homogeneous second-order differential equations subject to homogeneous boundary conditions may have an infinite number of particular solutions, the characteristic functions $\varphi_n(x)$, and that to each of these functions there corresponds a discrete (characteristic) value of a parameter λ_n involved in the differential equation. We then established that any two characteristic functions from a given set are orthogonal with respect to a weighting function $w(x)$ over an interval (a, b), that is,

$$\int_a^b w(x)\varphi_m(x)\varphi_n(x)\,dx = 0, \qquad m \neq n. \tag{4-14}$$

For every problem of Chapters 4 and 5, we proceeded by expanding a function $f(x)$, which was related to the nonhomogeneous boundary condition, in terms of a given set of orthogonal functions,

$$f(x) = \sum_{n=0}^{\infty} b_n \varphi_n(x), \tag{4-21}$$

then employed the orthogonality given by Eq. (4–14) to obtain the unknown coefficients

$$b_n = \frac{\int_a^b w(x)f(x)\varphi_n(x)\,dx}{\int_a^b w(x)\varphi_n^2(x)\,dx}. \tag{4-23}$$

We used the method summarized above for problems of finite geometry, and for those of semi-infinite geometry provided the latter were or could be made *homogeneous* in *finite* directions of their geometry. Some problems, however, are *nonhomogeneous* in *infinite* directions of their geometry and require an expansion in terms of an appropriate set of functions over a semi-infinite or infinite interval of a nonhomogeneous direction. In such a case, one or both of the integration limits of Eq. (4–14) extend to infinity, the functions $\varphi_n(x)$ vanish, and Eq. (4–21) fails to represent $f(x)$ over the interval (a, b). It is the purpose of this section to develop a method, that of the so-called Fourier integrals, by which a given function $f(x)$ can be represented in terms of a set of functions over the interval $0 < x < \infty$ or $-\infty < x < \infty$.

Let us first reconsider the problem of the finite interval L,

$$\frac{d^2y}{dx^2} + \lambda^2 y = 0, \tag{4-7}$$

subject to the boundary conditions,

$$y(0) = 0, \tag{4-8}$$

$$y(L) = 0, \tag{4-9}$$

and then see what happens when we let $L \to \infty$. The characteristic functions corresponding to Eqs. (4–7), (4–8), and (4–9) are $\varphi_n(x) = \sin \lambda_n x$, with the characteristic values $\lambda_n = n\pi/L$, $n = 1, 2, 3, \ldots$ These characteristic functions are orthogonal with respect to the weighting function 1 over the interval $(0, L)$ according to the general proof given by Eqs. (4–16) through (4–20). Now recall from Chapter 4 that for each particular case the orthogonality of the characteristic functions can be shown by direct integration. In the present case, for instance, the use of appropriate trigonometric properties readily gives

$$\frac{2}{L}\int_0^L \sin \lambda_m x \sin \lambda_n x \, dx = \frac{\sin(\lambda_m - \lambda_n)L}{(\lambda_m - \lambda_n)L} - \frac{\sin(\lambda_m + \lambda_n)L}{(\lambda_m + \lambda_n)L}. \quad (7\text{--}133)$$

The right-hand side of Eq. (7–133) vanishes if λ_m and λ_n are different integral multiples of π/L, and is equal to unity if λ_m and λ_n both assume the same integral multiple of π/L. When $L \to \infty$, however, the right-hand side approaches zero for any rational values of λ_m and λ_n if $\lambda_m \neq \lambda_n$, but becomes unity if $\lambda_m = \lambda_n$. Hence the limit of Eq. (7–133) becomes

$$\lim_{L \to \infty} \frac{2}{L} \int_0^L \sin \lambda_m x \sin \lambda_n x \, dx = \begin{cases} 0, & \lambda_m \neq \lambda_n, \\ 1, & \lambda_m = \lambda_n. \end{cases} \quad (7\text{--}134)$$

Equation (7–134) implies that $\sin \lambda x$, corresponding to any two positive rational but different values of λ, has a sort of orthogonality property in the semi-infinite interval $(0, \infty)$. Hence the representation of a given function $f(x)$ in terms of $\sin \lambda x$ over a semi-infinite interval involves all positive rational values of λ and is not restricted, as in the finite case, to a set of discrete values.† Thus in the limiting case $L \to \infty$, the series representation of an arbitrary function $f(x)$ by the sum of $b_n \sin \lambda_n x$ corresponding to discrete characteristic values should be replaced by the infinite integral of $b(\lambda) \sin \lambda x$ to include all contributions of continuous characteristic values. There then follows

$$f(x) = \int_0^\infty b(\lambda) \sin \lambda x \, d\lambda, \quad 0 < x < \infty. \quad (7\text{--}135)$$

Assuming that such a representation exists, for a sufficiently well-behaved function $f(x)$, we have yet to determine $b(\lambda)$ for all positive values of λ. Following a similar procedure to that employed for finite intervals, that is, multiplying both sides of the Eq. (7–135) by $\sin \lambda^* x$, then integrating the result with respect to x over an interval $(0, L)$, we have

$$\int_0^L f(x) \sin \lambda^* x \, dx = \int_0^L \sin \lambda^* x \left(\int_0^\infty b(\lambda) \sin \lambda x \, d\lambda \right) dx. \quad (7\text{--}136)$$

---

† As with a finite interval, the negative values need not be considered.

Assuming that the order of integration may be interchanged, we may rearrange Eq. (7–136) as

$$\int_0^L f(x) \sin \lambda^* x \, dx = \int_0^\infty b(\lambda) \left(\int_0^L \sin \lambda x \sin \lambda^* x \, dx \right) d\lambda,$$

which by the use of Eq. (7–133) becomes

$$\int_0^L f(x) \sin \lambda^* x \, dx = \frac{L}{2} \int_0^\infty b(\lambda) \left[\frac{\sin (\lambda - \lambda^*)L}{(\lambda - \lambda^*)L} - \frac{\sin (\lambda + \lambda^*)L}{(\lambda + \lambda^*)L} \right] d\lambda. \tag{7–137}$$

Now we introduce the new dummy variables $(\lambda - \lambda^*)L = \beta$ and $(\lambda + \lambda^*)L = \gamma$ for the first and second terms on the right-hand side of Eq. (7–137), respectively, and obtain

$$\int_0^L f(x) \sin \lambda^* x \, dx = \frac{1}{2} \int_{-\lambda^* L}^\infty b\left(\lambda^* + \frac{\beta}{L}\right) \frac{\sin \beta}{\beta} d\beta$$

$$- \frac{1}{2} \int_{\lambda^* L}^\infty b\left(-\lambda^* + \frac{\gamma}{L}\right) \frac{\sin \gamma}{\gamma} d\gamma. \tag{7–138}$$

Since $\lambda^* > 0$, when $L \to \infty$ the second integral on the right of Eq. (7–138) approaches zero, whereas the first one yields

$$\frac{1}{2} \int_{-\infty}^\infty b(\lambda^*) \frac{\sin \beta}{\beta} d\beta = \frac{b(\lambda^*)}{2} \int_{-\infty}^\infty \frac{\sin \beta}{\beta} d\beta. \tag{7–139}$$

The value of the integral on the right of Eq. (7–139) is available in integral tables. It follows then that

$$\int_0^\infty f(x) \sin \lambda^* x \, dx = \frac{\pi}{2} b(\lambda^*), \qquad \lambda^* > 0.$$

To distinguish the dummy variable used here from the independent variable in $f(x)$, x may be replaced by ξ to give

$$b(\lambda) = \frac{2}{\pi} \int_0^\infty f(\xi) \sin \lambda \xi \, d\xi, \tag{7–140}$$

where, for convenience, the parameter λ^* is also replaced by λ.

Inserting Eq. (7–140) into Eq. (7–135), we get

$$f(x) = \frac{2}{\pi} \int_0^\infty \sin \lambda x \int_0^\infty f(\xi) \sin \lambda \xi \, d\xi \, d\lambda, \qquad 0 < x < \infty. \tag{7–141}$$

Equation (7–141) is the *Fourier sine integral* representation of $f(x)$ and can be shown to be valid when $x > 0$, if $f(x)$ is piecewise differentiable in any positive interval and if the integral $\int_0^\infty |f(\xi)|\,d\xi$ exists. Like the Fourier series, Eq. (7–141) represents $f(x)$ at points of continuity, and gives $[f(x+) + f(x-)]/2$ at points where $f(x)$ has finite jumps.

The *Fourier cosine integral* representation of $f(x)$ can be obtained in a manner similar to the development of Eq. (7–141). Thus, from the definition of this representation,

$$f(x) = \int_0^\infty a(\lambda) \cos \lambda x\, d\lambda, \qquad (7\text{–}142)$$

we obtain

$$a(\lambda) = \frac{2}{\pi} \int_0^\infty f(\xi) \cos \lambda \xi\, d\xi. \qquad (7\text{–}143)$$

Introducing Eq. (7–143) into Eq. (7–142) gives

$$f(x) = \frac{2}{\pi} \int_0^\infty \cos \lambda x \int_0^\infty f(\xi) \cos \lambda \xi\, d\xi\, d\lambda, \qquad 0 < x < \infty. \qquad (7\text{–}144)$$

As in the finite case, Eq. (7–141) is valid for all values of x when $f(x)$ is an odd function; similarly, Eq. (7–144) is valid for all values of x when $f(x)$ is an even function. Using these facts, we may obtain a representation of an arbitrary function $f(x)$ for all values of x in terms of the sine and cosine integrals. First, writing $f(x)$ in the form†

$$f(x) = \tfrac{1}{2}[f(x) + f(-x)] + \tfrac{1}{2}[f(x) - f(-x)],$$

we get

$$f(x) = f_e(x) + f_o(x),$$

where $f_e(x)$ and $f_o(x)$ are even and odd functions, respectively. It follows then that

$$\frac{1}{\pi} \int_0^\infty \cos \lambda x \int_{-\infty}^\infty f(\xi) \cos \lambda \xi\, d\xi\, d\lambda$$

$$= \frac{1}{\pi} \int_0^\infty \cos \lambda x \int_{-\infty}^\infty f_e(\xi) \cos \lambda \xi\, d\xi\, d\lambda$$

$$= \frac{2}{\pi} \int_0^\infty \cos \lambda x \int_0^\infty f_e(\xi) \cos \lambda \xi\, d\xi\, d\lambda = f_e(x), \qquad -\infty < x < \infty, \qquad (7\text{–}145)$$

and similarly,

$$\frac{1}{\pi} \int_0^\infty \sin \lambda x \int_{-\infty}^\infty f(\xi) \sin \lambda \xi\, d\xi\, d\lambda = f_o(x), \qquad -\infty < x < \infty. \qquad (7\text{–}146)$$

---

† See Fig. 4–1.

Finally, adding Eqs. (7–145) and (7–146), we obtain the representation

$$f(x) = \int_0^\infty [a(\lambda) \cos \lambda x + b(\lambda) \sin \lambda x] \, d\lambda, \qquad -\infty < x < \infty, \qquad (7\text{–}147)$$

where

$$a(\lambda) = \frac{1}{\pi} \int_{-\infty}^\infty f(\xi) \cos \lambda \xi \, d\xi, \qquad (7\text{–}148)$$

$$b(\lambda) = \frac{1}{\pi} \int_{-\infty}^\infty f(\xi) \sin \lambda \xi \, d\xi. \qquad (7\text{–}149)$$

Equation (7–147) combined with Eqs. (7–148) and (7–149) becomes

$$f(x) = \frac{1}{\pi} \int_0^\infty \left[\int_{-\infty}^\infty f(\xi) \cos \lambda(x - \xi) \, d\xi \right] d\lambda, \qquad -\infty < x < \infty. \qquad (7\text{–}150)$$

Furthermore, noting that its integrand is an even function with respect to λ, we may rearrange Eq. (7–150) in the form

$$f(x) = \frac{1}{2\pi} \int_{-\infty}^\infty \int_{-\infty}^\infty f(\xi) \cos \lambda(x - \xi) \, d\xi \, d\lambda, \qquad -\infty < x < \infty. \qquad (7\text{–}151)$$

This expression is the *complete Fourier integral* representation of $f(x)$ and represents $f(x)$ for all values of x in the usual sense, if $f(x)$ is piecewise differentiable in every finite interval, and if the integral $\int_{-\infty}^\infty |f(\xi)| \, d\xi$ exists. By expressing the cosine in terms of exponentials, we may rearrange Eq. (7–151) in a *complex* form as

$$f(x) = \frac{1}{2\pi} \int_0^\infty \int_{-\infty}^\infty [f(\xi) e^{i\lambda(x-\xi)} + f(\xi) e^{-i\lambda(x-\xi)}] \, d\xi \, d\lambda.$$

If λ is replaced by $-\lambda$ for the second term of this integrand, then it follows that

$$f(x) = \frac{1}{2\pi} \int_0^\infty \int_{-\infty}^\infty f(\xi) e^{i\lambda(x-\xi)} \, d\xi \, d\lambda - \frac{1}{2\pi} \int_0^{-\infty} \int_{-\infty}^\infty f(\xi) e^{i\lambda(x-\xi)} \, d\xi \, d\lambda.$$

Interchanging the limits of first integration of the second term, we obtain

$$f(x) = \frac{1}{2\pi} \int_{-\infty}^\infty e^{i\lambda x} \left(\int_{-\infty}^\infty f(\xi) e^{-i\lambda \xi} \, d\xi \right) d\lambda, \qquad -\infty < x < \infty. \qquad (7\text{–}152)$$

Equation (7–152) is known as the *complex Fourier integral* representation of $f(x)$. Denoting the inner integral of Eq. (7–152) by $f(\lambda)$, we have

$$f(\lambda) = \int_{-\infty}^\infty f(\xi) e^{-i\lambda \xi} \, d\xi, \qquad -\infty < \lambda < \infty, \qquad (7\text{–}153)$$

$$f(x) = \frac{1}{2\pi} \int_{-\infty}^\infty f(\lambda) e^{i\lambda x} \, d\lambda, \qquad -\infty < x < \infty. \qquad (7\text{–}154)$$

The function $f(\lambda)$ defined by Eq. (7–153) is called the *Fourier transform* of $f(x)$. For a function $f(x)$ which vanishes when $x < 0$, the Fourier transform becomes identical with the Laplace transform provided $i\lambda$ is replaced by p. Analogous integral representations involving cylindrical (Bessel) functions rather than circular functions can be obtained in a similar manner. The expressions thus obtained are known as the *Fourier-Bessel integral* representations; these will not be considered here.

Before utilizing the Fourier integrals in the development of the inversion theorem for Laplace transforms, let us demonstrate their use in the solution of problems involving a nonhomogeneous direction to be made orthogonal over a semi-infinite domain.

Example 7–8. Let us re-solve Example 7–3 by means of Fourier integrals. The formulation of the problem is

$$\frac{\partial \theta}{\partial t} = a \frac{\partial^2 \theta}{\partial x^2}, \tag{7-43}$$

$$\theta(x, 0) = \theta_0, \tag{7-44}$$

$$\theta(0, t) = 0, \tag{7-45}$$

$$\theta(\infty, t) = \theta_0. \tag{7-46}$$

A product solution of Eq. (7–43) satisfied by Eq. (7–45) has the form

$$\theta_\lambda(x, t) = b(\lambda) e^{-a\lambda^2 t} \sin \lambda x,$$

where λ is the separation parameter to be determined. Since the space interval (the only possible orthogonal direction of the problem) extends to infinity, we seek an integral solution involving all positive values of λ, that is

$$\theta(x, t) = \int_0^\infty b(\lambda) e^{-a\lambda^2 t} \sin \lambda x \, d\lambda, \qquad x > 0. \tag{7-155}$$

Employing Eq. (7–44), we obtain from Eq. (7–155)

$$\theta_0 = \int_0^\infty b(\lambda) \sin \lambda x \, d\lambda,$$

which is the Fourier sine integral representation of θ_0. Here $b(\lambda)$ may be evaluated by means of Eq. (7–140). This gives

$$b(\lambda) = 2 \frac{\theta_0}{\pi} \int_0^\infty \sin \lambda \xi \, d\xi. \tag{7-156}$$

Inserting Eq. (7–156) into Eq. (7–155), we have

$$\frac{\theta(x, t)}{\theta_0} = \frac{2}{\pi} \int_0^\infty \int_0^\infty e^{-a\lambda^2 t} \sin \lambda x \sin \lambda \xi \, d\lambda \, d\xi,$$

which can be rearranged according to Eq. (7–133). The result is

$$\frac{\theta(x,t)}{\theta_0} = \frac{1}{\pi} \int_0^\infty \int_0^\infty e^{-a\lambda^2 t} [\cos \lambda(x-\xi) - \cos \lambda(x+\xi)] \, d\lambda \, d\xi. \qquad (7\text{--}157)$$

Furthermore, referring to the value of the integral

$$\int_0^\infty e^{-a\lambda^2 t} \cos \lambda x \, d\lambda = \frac{1}{2} \left(\frac{\pi}{at}\right)^{1/2} e^{-x^2/4at},$$

which can be found in any integral tables, we may write Eq. (7–157) in the form

$$\frac{\theta(x,t)}{\theta_0} = \frac{1}{2(\pi at)^{1/2}} \int_0^\infty [e^{-(x-\xi)^2/4at} - e^{-(x+\xi)^2/4at}] \, d\xi. \qquad (7\text{--}158)$$

Introducing the variable η^2 for $(x-\xi)^2/4at$ in the first integral of Eq. (7–158) and for $(x+\xi)^2/4at$ in the second integral gives

$$\frac{\theta(x,t)}{\theta_0} = \frac{1}{\pi^{1/2}} \left(\int_{-x/2(at)^{1/2}}^\infty e^{-\eta^2} \, d\eta - \int_{x/2(at)^{1/2}}^\infty e^{-\eta^2} \, d\eta \right),$$

or equivalently,†

$$\frac{\theta(x,t)}{\theta_0} = \frac{2}{\pi^{1/2}} \int_0^{x/2(at)^{1/2}} e^{-\eta^2} \, d\eta = \text{erf}\left[\frac{x}{2(at)^{1/2}}\right]. \qquad (7\text{--}52)$$

Equation (7–52) is identical to the result obtained by means of Laplace transforms.

Comparing the algebra of Examples 7–3 and 7–8 we see that the use of Laplace transforms proved to be more convenient than the use of Fourier integrals, because with Fourier integrals we had to go through the integration process, whereas with Laplace transforms we could refer to the result of integration which was tabulated in advance.

In the next section we develop the inversion theorem for Laplace transforms from Fourier integrals. As we shall see, the use of this theorem, reducing the solution of a given problem to an integration in the complex plane, may not be easier than the use of Fourier integrals.

7–6. Inversion Theorem for Laplace Transforms

Let us consider a function $f(x)$ which vanishes when $x < 0$. According to Eq. (7–152) the complex form of the Fourier integral representation of $e^{-\gamma x} f(x)$‡ is

$$e^{-\gamma x} f(x) = \frac{1}{2\pi} \lim_{L \to \infty} \int_{-L}^L e^{i\lambda x} \left(\int_0^\infty e^{-i\lambda \xi} e^{-\gamma \xi} f(\xi) \, d\xi \right) d\lambda, \qquad (7\text{--}159)$$

---

† Note that the problem has been solved without employing the boundary condition at infinity, Eq. (7–46). This condition has already been included in Eq. (7–44), however, and need not be satisfied explicitly.

‡ See the sentence following Eq. (7–160) for the reason for introducing $e^{-\gamma x}$.

where the limit form is used for convenience. [See the integration limits of Eq. (7–161).] Noting that x is a parameter for the right-hand side and multiplying both sides by $e^{\gamma x}$, we may rearrange Eq. (7–159) in the form

$$f(x) = \frac{1}{2\pi} \lim_{L \to \infty} \int_{-L}^{L} e^{(\gamma + i\lambda)x} \left(\int_{0}^{\infty} e^{-(\gamma + i\lambda)\xi} f(\xi) \, d\xi \right) d\lambda.$$

Furthermore, introducing the variable $p = \gamma + i\lambda$, we have

$$f(x) = \frac{1}{2\pi i} \lim_{L \to \infty} \int_{\gamma - iL}^{\gamma + iL} e^{px} \left(\int_{0}^{\infty} e^{-p\xi} f(\xi) \, d\xi \right) dp. \qquad (7\text{–}160)$$

The transformation given by Eq. (7–160) can be shown to be legitimate if $f(x)$ is piecewise differentiable and of exponential order,† when γ is sufficiently large that the integral $\int_{0}^{\infty} e^{-\gamma \xi} |f(\xi)| \, d\xi$ exists.

By making appropriate changes in the notation, replacing x by t and p by z, and showing the Laplace transform of $f(t)$ by $\bar{f}(p)$,

$$\bar{f}(p) = \int_{0}^{\infty} e^{-pt} f(t) \, dt,$$

where p is now complex, we can express $f(t)$ in terms of its transform $\bar{f}(p)$ by the complex integral

$$f(t) = \frac{1}{2\pi i} \lim_{L \to \infty} \int_{\gamma - iL}^{\gamma + iL} e^{tz} \bar{f}(z) \, dz, \qquad (7\text{–}161)$$

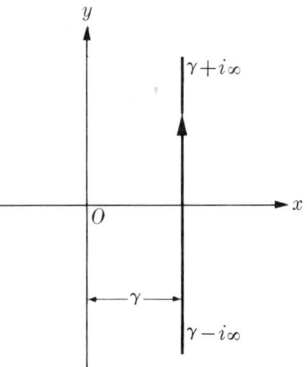

FIG. 7–10

which is known as the *inversion theorem for Laplace transforms*. This integral, as shown in the complex plane of Fig. 7–10, is taken along the infinite line $x = \gamma$, and can usually be evaluated by the contour integration to be explained in Section 7–8.

The inversion theorem given by Eq. (7–161) is also valid for multidimensional problems. If the Laplace transform of the function $f(\mathbf{r}, t)$ with respect to time is $\bar{f}(\mathbf{r}, p)$, then the inversion theorem obtained by rearranging Eq. (7–161) has the form

$$f(\mathbf{r}, t) = \frac{1}{2\pi i} \lim_{L \to \infty} \int_{\gamma - iL}^{\gamma + iL} e^{tz} \bar{f}(\mathbf{r}, z) \, dz, \qquad (7\text{–}162)$$

where $\mathbf{r}$ denotes the position (space) vector. The inversion theorem may also be used for multidimensional problems to be (Laplace) transformed in a suitable space direction.

---

† Note the existence conditions of Eq. (7–4).

In the next section we review the functions of a complex variable as preparation for the application of the inversion theorem.

7–7. Functions of a Complex Variable

A complete treatment of the functions of a complex variable is beyond the scope of the text. Only the definitions and theorems that are needed in the evaluation of the integral appearing in the inversion theorem are given below, and in most cases we indicate the proofs briefly without careful statement of the conditions. For a more extensive study, the reader may refer to any text on the theory of functions of a complex variable.

(1) *Complex variables.* The complex variable z is an expression of the form

$$z = x + iy, \qquad (7\text{–}163)$$

where x and y are real variables (the so-called *real* and *imaginary* parts of z), and i is the *imaginary unit* defined by $i^2 = -1$.

It is convenient to represent z by the point $P(x, y)$ in the cartesian coordinate system known as the *complex plane* (Fig. 7–11). By definition, the real and imaginary numbers constitute the x- and y-axes, respectively, of this plane. Then a complex number may also be considered as a vector from the origin to the point $P(x, y)$. The length of this vector is the *absolute value* or the *modulus* of z and is denoted by $|z|$. Thus

$$|z| = |x + iy| = \sqrt{x^2 + y^2}.$$

From Fig. 7–11 it also follows that

$$x = \rho \cos \theta,$$
$$y = \rho \sin \theta. \qquad (7\text{–}164)$$

Combining Eqs. (7–163) and (7–164), we obtain the polar form of z as

$$z = \rho(\cos \theta + i \sin \theta), \qquad (7\text{–}165)$$

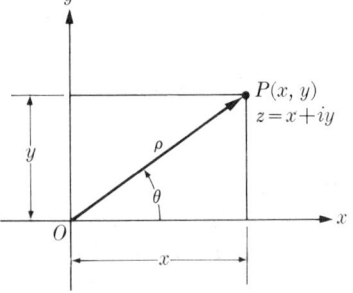

FIG. 7–11

where ρ is the modulus and θ is the so-called *argument* of z. Note that θ is not unique, and it can only be determined within an integral multiple of 2π.

Two complex numbers are equal if and only if their corresponding real and imaginary parts are equal.

The *conjugate* of z is denoted by $\bar{z}$, and is defined to be

$$\bar{z} = x - iy.$$

The rules for adding, subtracting, multiplying, and dividing complex numbers are the same as those for real numbers.

(2) *Elementary functions.* The elementary functions of a complex variable, such as *polynomials, power series, exponential functions,* and *trigonometric* (circular and hyperbolic) *functions,* are so defined that they reduce to those of real variables when z is replaced by a real variable. The *complex logarithmic function* Ln z and the *generalized power function* z^a, where a may be real or complex, will be defined later, in (6).

From the definitions of sin z and cos z we obtain

$$e^{iz} = \cos z + i \sin z, \qquad (7\text{-}166)$$

which is known as *Euler's formula*. The polar form of z can be rearranged by the use of Eq. (7-166). Thus, by combining Eqs. (7-165) and (7-166), we have the form

$$z = \rho e^{i\theta}, \qquad (7\text{-}167)$$

which will prove useful in our later discussions.

Comparing the definitions of the circular and hyperbolic functions, we find the following relations:

$$\begin{aligned} \sinh iz &= i \sin z, & \cosh iz &= \cos z, \\ \sin iz &= i \sinh z, & \cos iz &= \cosh z. \end{aligned} \qquad (7\text{-}168)$$

(3) *Single-valued functions.* If a function $f(z)$ and the complex variable z are so related that to each value of z in a given region R of the complex plane there corresponds one value of $f(z)$, then $f(z)$ is said to be a *single-valued* function of z. On the other hand, if more than one value of $f(z)$ corresponds to each value of z, then $f(z)$ is *multiple-valued*. Polynomials, power series, the exponential functions, and the trigonometric functions are single-valued functions. The complex logarithmic function and the generalized power function, as we shall see in (6), are multiple-valued functions.

(4) *Existence of $df(z)/dz$.* The first derivative of $f(z)$ at a point exists if the limit

$$\frac{df}{dz} = \lim_{\Delta z \to 0} \frac{\Delta f}{\Delta z} = \lim_{\Delta z \to 0} \frac{f(z + \Delta z) - f(z)}{\Delta z} \qquad (7\text{-}169)$$

is finite and unique as $\Delta z \to 0$ from any direction in the complex plane.

Suppose that the limit given by Eq. (7-169) exists. Then denoting the real and imaginary parts of $f(z)$ by u and v, respectively, we have

$$\frac{df}{dz} = \lim_{\substack{\Delta x \to 0 \\ \Delta y \to 0}} \left(\frac{\Delta u + i \Delta v}{\Delta x + i \Delta y} \right). \qquad (7\text{-}170)$$

Since Eq. (7-170) must exist regardless of the direction in which $\Delta z \to 0$, let us first approach this limit along a line parallel to the x-axis; thus

$$\frac{df}{dz} = \lim_{\Delta x \to 0} \left(\frac{\Delta u + i\,\Delta v}{\Delta x} \right) = \frac{\partial u}{\partial x} + i\,\frac{\partial v}{\partial y}. \qquad (7\text{--}171)$$

Similarly, an approach parallel to the y-axis gives

$$\frac{df}{dz} = \lim_{\Delta y \to 0} \left(\frac{\Delta u + i\,\Delta v}{i\,\Delta y} \right) = \frac{\partial v}{\partial y} - i\,\frac{\partial u}{\partial y}. \qquad (7\text{--}172)$$

The assumed existence (finiteness and uniqueness) of df/dz requires the equality of Eqs. (7–171) and (7–172); this equality obtains when the respective real and imaginary parts of these equations are equal. Therefore,

$$\frac{\partial u}{\partial x} = \frac{\partial v}{\partial y}, \qquad \frac{\partial u}{\partial y} = -\frac{\partial v}{\partial x}. \qquad (7\text{--}173)$$

These equations are known as the *Cauchy-Riemann conditions*. Thus *the derivative df/dz of a single-valued function $f(z) = u + iv$ exists in a region R of the complex plane if and only if the Cauchy-Riemann conditions are satisfied in R.*

(5) *Analytic functions.* A function $f(z)$ is said to be *analytic* in a region R if $f(z)$ is single-valued in R and its derivative df/dz exists at each point of R.

(6) *Singularities of analytic functions.* The points at which $f(z)$ is not analytic are called the *singular points* of $f(z)$. Then from the definition of an analytic function, the point $z = a$ is a singular point of $f(z)$ if (i) $f(a)$ is not single-valued, or (ii) $df(a)/dz$ does not exist.

(i) The points at which $f(z)$ is not single-valued are called the *branch points* of $f(z)$. When a point $P(z)$ traces out a closed path around $z = a$ without enclosing any other singular point, and the value assumed by the function $f(z)$ differs from its initial value, then, by definition, $z = a$ is a branch point of $f(z)$. A function having branch points is a *multiple-valued* function; each particular value is called a *branch* of this function. A function having branch points in a region R can be made analytic in this region by excluding these points from R by proper *cuts*. Let us illustrate the multiple-valued functions by considering two elementary functions.

The *complex logarithmic function*, represented temporarily by the notation

$$w = \operatorname{Ln} z, \qquad (7\text{--}174)$$

is defined as the inverse of the exponential function

$$z = e^w. \qquad (7\text{--}175)$$

Denoting the real and imaginary parts of Eq. (7–174) by u and v, respectively, and employing Eq. (7–166), we may rearrange Eq. (7–175) as

$$z = x + iy = e^{u+iv} = e^u(\cos v + i \sin v). \tag{7-176}$$

It follows from the real and imaginary parts of Eq. (7–176) that

$$x = e^u \cos v, \quad y = e^u \sin v. \tag{7-177}$$

Then solving Eq. (7–177) for u and v, we obtain

$$e^{2u} = x^2 + y^2 = |z|^2 = \rho^2, \tag{7-178}$$

$$\tan v = x/y = \tan \theta. \tag{7-179}$$

Since Eq. (7–178) is in terms of real variables, the ordinary *real logarithm* of this equation gives

$$u = \ln |z| = \ln \rho. \tag{7-180}$$

Furthermore, we have from Eq. (7–179)

$$v = \theta \mp 2k\pi, \quad k = 0, 1, 2, \ldots, \tag{7-181}$$

where θ lies in the range $0 \leq \theta < 2\pi$. Inserting Eqs. (7–180) and (7–181) into Eq. (7–174), we get the complex logarithmic function

$$\text{Ln } z = \ln |z| + i(\theta \mp 2k\pi), \quad k = 0, 1, 2, \ldots \tag{7-182}$$

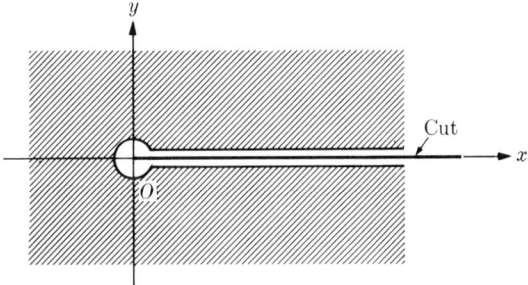

FIG. 7–12

Equation (7–182) clearly shows that Ln z is a single-valued function for all values of z except the branch point $z = 0$, where it becomes *infinitely many-valued*. To prevent the transition from one branch to another we may introduce a cut, say along the entire positive real axis (Fig. 7–12), thus making Ln z single-valued on the *cut complex plane*. For convenience in a given case, the cut of Fig. 7–12 can be replaced by a cut along any ray or curve extending from $z = 0$ to infinity.

The generalized power function z^a, where a may be real or complex, is defined in terms of the complex logarithm by the relation

$$z^a = e^{a \operatorname{Ln} z}. \tag{7-183}$$

If z is real and positive and a is real, Eq. (7–183) gives the usual definition of the real logarithm. If z is complex and a is a positive integer, say $a = n$, then inserting Eq. (7–182) into Eq. (7–183) and noting that $|z| = \rho$, we have

$$z^n = e^{n[\ln\rho + i(\theta \mp 2k\pi)]}, \qquad k = 0, 1, 2, \ldots, \tag{7–184}$$

which may be rearranged as

$$z^n = e^{\ln\rho^n} e^{in\theta} e^{\mp 2ink\pi}. \tag{7–185}$$

Since both n and k are integers, Eq. (7–166) gives

$$e^{\mp 2ink\pi} = \cos 2nk\pi \mp i \sin 2nk\pi = 1.$$

Then Eq. (7–185) becomes

$$z^n = \rho^n e^{in\theta}, \qquad n \text{ an integer.} \tag{7–186}$$

Inspection of Eq. (7–186) reveals that z^n is a single-valued function. More generally, if a is a real rational number,† say $a = n/m$ and n, m are integers having no common factor, then Eq. (7–183) takes the form

$$z^{n/m} = \rho^{n/m} e^{i(n/m)\theta} e^{\mp 2i(n/m)k\pi}, \qquad k = 0, 1, 2, \ldots \tag{7–187}$$

Since n and m are given integers, whereas k assumes all integer values, the factor $e^{\mp 2i(n/m)k\pi}$ takes on m different values for each sign of the exponent when $k < m$, but $2m$ values so obtained are repeated periodically when $k \geq m$. Hence $z^{n/m}$ is exactly a $2m$-valued function. Since $z = 0$ is again a branch point, the cut shown in Fig. 7–12 for Ln z may also be used for $z^{n/m}$.

(ii) Let us consider next the singularity of a function $f(z)$ at a point $z = a$ arising from the nonexistence of $df(a)/dz$. In this case if $(z - a)^k f(z)$ is analytic at $z = a$, where k is a positive integer, then $z = a$ is said to be a *pole* of $f(z)$. The smallest value of k which satisfies this condition is the *order* of the pole. For example, $f(z) = 1/(z - a)^k$, where k is a positive integer, has a pole of order k at $z = a$. Similarly the kth term of $f(z) = e^{1/z} = \sum_{k=0}^{\infty} (1/k!)(1/z^k)$ has a pole of order k at $z = 0$.

(7) *The complex line integral. The Cauchy integral theorem.* Let C be any continuous curve of finite length joining the points P_1 and P_2, and let $f(z) = u + iv$ be a continuous function defined at all points of C (Fig. 7–13). The line integral of $f(z)$ along C is defined by

$$\int_C f(z)\,dz = \int_C (u + iv)(dx + i\,dy)$$

$$= \int_C (u\,dx - v\,dy) + i \int_C (v\,dx + u\,dy). \tag{7–188}$$

---

† The most general case, corresponding to a complex a, is not considered here.

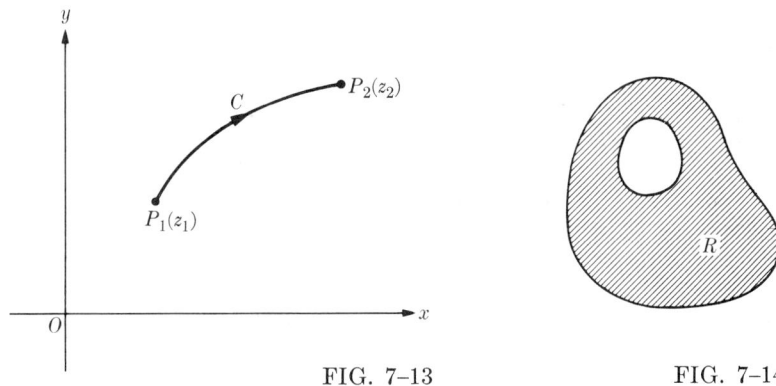

FIG. 7-13   FIG. 7-14

The real and imaginary parts of Eq. (7–188) are ordinary real line integrals. If the integrands of these integrals are exact differentials, say dF and dG, then

$$dF = u\, dx - v\, dy = \frac{\partial F}{\partial x} dx + \frac{\partial F}{\partial y} dy,$$

$$dG = v\, dx + u\, dy = \frac{\partial G}{\partial x} dx + \frac{\partial G}{\partial y} dy. \qquad (7\text{–}189)$$

Furthermore, if the partial derivatives involved in Eq. (7–189) are continuous, then the order of differentiation is irrelevant, and we obtain

$$\frac{\partial^2 F}{\partial x\, \partial y}\left(\text{or } -\frac{\partial v}{\partial x}\right) = \frac{\partial^2 F}{\partial y\, \partial x}\left(\text{or } \frac{\partial u}{\partial y}\right),$$

$$\frac{\partial^2 G}{\partial x\, \partial y}\left(\text{or } \frac{\partial u}{\partial x}\right) = \frac{\partial^2 G}{\partial y\, \partial x}\left(\text{or } \frac{\partial v}{\partial y}\right),$$

which are precisely the Cauchy-Riemann conditions. It follows that the *line integral* $\int_C f(z)\, dz$ *is independent of the path C joining the end points* P_1 *and* P_2 *if C can be enclosed in a simply connected region R in which* $f(z)$ *is analytic.*

A *simply connected region* R is so defined that any closed curve lying in R can be continuously shrunk to a point without crossing the boundaries of R. If the curve cannot be so shrunk, we have a *multiply connected region* as demonstrated by Fig. 7-14.

Denoting then $f(z)\, dz$ by an exact differential function $F(z)$, we can evaluate the *line integral of the analytic function* $f(z)$ in the usual manner, according to

$$\int_{z_1}^{z_2} f(z)\, dz = F(z_2) - F(z_1).$$

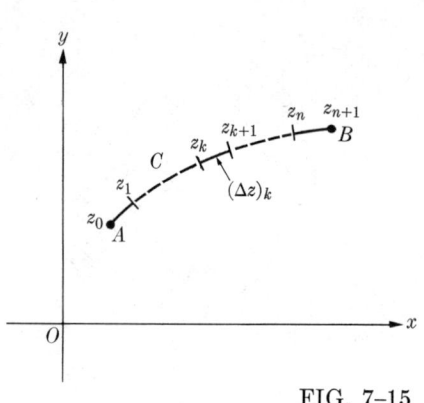

FIG. 7–15   FIG. 7–16

Particularly, if the end points coincide, we have

$$\oint_C f(z)\, dz = 0. \tag{7-190}$$

This result is known as *Cauchy's integral theorem*.

(8) *An upper bound for complex line integrals.* By definition, the complex line integral of an analytic function $f(z)$ along a path C (Fig. 7–15) is the limit of the sum

$$\int_C f(z)\, dz = \lim_{n\to\infty} \sum_{k=0}^{n} f(z_k)(\Delta z)_k. \tag{7-191}$$

On the other hand, remembering that a complex number can be represented by a vector on the complex plane, we have from Fig. 7–16

$$|\zeta_0 + \zeta_1| \leq |\zeta_0| + |\zeta_1|,$$

which can be readily generalized to the sum of n complex numbers as

$$\left|\sum_{k=0}^{n} \zeta_k\right| \leq \sum_{k=0}^{n} |\zeta_k|. \tag{7-192}$$

Employing Eq. (7–192) with the consideration that $\zeta_k = f(z_k)(\Delta z)_k$, we may rearrange the absolute value of the right-hand side of Eq. (7–191) to give

$$\left|\sum_{k=0}^{n} f(z_k)(\Delta z)_k\right| \leq \sum_{k=0}^{n} |f(z_k)(\Delta z)_k| = \sum_{k=0}^{n} |f(z_k)|\,|(\Delta z)_k|. \tag{7-193}$$

Then the absolute value of Eq. (7–191) becomes

$$\left|\int_C f(z)\, dz\right| \leq \lim_{n\to\infty} \sum_{k=0}^{n} |f(z_k)|\,|(\Delta z)_k|. \tag{7-194}$$

Proceeding to the limit for the right-hand side of Eq. (7-194) as $n \to \infty$, we obtain

$$\left| \int_C f(z) \, dz \right| \leq \int_C |f(z)| \, |dz|. \qquad (7\text{-}195)$$

Furthermore, noting that $|dz| = ds$ is the differential arc length along C, we may write Eq. (7-195) in the form

$$\left| \int_C f(z) \, dz \right| \leq \int_C |f(z)| \, ds. \qquad (7\text{-}196)$$

If $|f(z)|$ is bounded, say $|f(z)| \leq M$ on C, then Eq. (7-196) yields

$$\left| \int_C f(z) \, dz \right| \leq ML, \qquad (7\text{-}197)$$

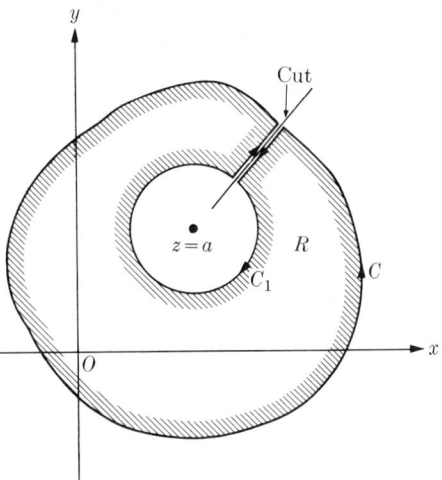

FIG. 7-17

where L is the length of path C between the points A and B.

(9) *The value of $\int_C (z - a)^n \, dz$ for integer values of n.* If n is zero or a positive integer, then according to the discussion on the generalized power function,† $(z - a)^n$ is analytic everywhere in the complex plane, and Cauchy's integral theorem, Eq. (7-190), gives

$$\oint_C (z - a)^n \, dz = 0, \qquad n \text{ zero or a positive integer}, \qquad (7\text{-}198)$$

where the region enclosed by the contour C may or may not include the point $z = a$. If n is a negative integer, then $(z - a)^n$, having a pole of order n, is not analytic at the point $z = a$. For a region excluding $z = a$, $(z - a)^n$ is analytic and Eq. (7-198) still holds; however, when the region includes $z = a$, $(z - a)^n$ is no longer analytic and Eq. (7-198) ceases to be valid. In this case we proceed as follows. Let C_1 be the unit circle around $z = a$ and let C be an arbitrary contour enclosing C_1. The region R between C and C_1 is not simply connected but can be made so by introducing an arbitrary cut as shown in Fig. 7-17. Since $(z - a)^n$ is analytic in the cut R region, employing Cauchy's integral theorem, we obtain

$$\oint_C (z - a)^n \, dz - \oint_{C_1} (z - a)^n \, dz = 0, \qquad (7\text{-}199)$$

where the first and second integrals are taken in the positive and negative directions, respectively.‡ The integrals along the cut, being considered in oppo-

---
† See Eqs. (7-183) through (7-187).
‡ Hereafter the counterclockwise and clockwise directions are assumed to be the positive and negative directions, respectively.

site directions, cancel as the width of the cut approaches zero. The first integral of Eq. (7–199) may now be evaluated in terms of the second integral. Noting that $\rho = 1$, we find that the polar expression of C_1 is $z = a + e^{i\theta}$. Then $dz = ie^{i\theta}\,d\theta$, and

$$\oint_{C_1} (z-a)^n\,dz = i\int_0^{2\pi} e^{i(n+1)\theta}\,d\theta$$

$$= i\int_0^{2\pi} [\cos(n+1)\theta + i\sin(n+1)\theta]\,d\theta. \qquad (7\text{–}200)$$

Thus, by direct integration of Eq. (7–200), we find that

$$\oint_{C_1} (z-a)^n\,dz = \begin{cases} 0, & n \neq -1, \text{ but a negative integer,} \\ 2\pi i, & n = -1. \end{cases} \qquad (7\text{–}201)$$

Inserting Eq. (7–201) into Eq. (7–199), and considering Eq. (7–198), we obtain the important relation†

$$\oint_C (z-a)^n\,dz = \begin{cases} 0, & n \neq -1, \text{ but any other integer or zero,} \\ 2\pi i, & n = -1, \end{cases} \qquad (7\text{–}202)$$

where C is an arbitrary contour enclosing the point $z = a$.

(10) *The Cauchy integral formula.* Let $f(z)$ be an analytic function in a simply connected region R bounded by a contour C, and let a be a point inside C such that $f(a) \neq 0$. Then the function

$$f(z)/(z-a) \qquad (7\text{–}203)$$

is analytic at all points of R except $z = a$, where it has a simple pole. Let C_ρ be a small circle of radius ρ with center at $z = a$. Now the function defined by Eq. (7–203) is analytic between the contours C and C_ρ (Fig. 7–18), and according to Cauchy's integral theorem we have

$$\oint_C \frac{f(z)\,dz}{(z-a)} = \oint_{C_\rho} \frac{f(z)\,dz}{(z-a)}. \qquad (7\text{–}204)$$

Employing the polar representation $z = a + \rho e^{i\theta}$ of C_ρ and noting that $dz = i\rho e^{i\theta}\,d\theta$, we may rearrange the right-hand side of Eq. (7–204) as follows:

$$\oint_{C_\rho} \frac{f(z)\,dz}{(z-a)} = i\int_0^{2\pi} f(a + \rho e^{i\theta})\,d\theta. \qquad (7\text{–}205)$$

Then the combination of Eq. (7–204) with the limiting value of Eq. (7–205) as

---

† Note the conceptual similarity between Eq. (7–202) and the condition of orthogonality over a finite or infinite interval.

$\rho \to 0$ gives

$$\oint_C \frac{f(z)\,dz}{(z-a)} = 2\pi i f(a). \qquad (7\text{-}206)$$

Equation (7–206) is known as *Cauchy's integral formula;* it gives the value of a function $f(z)$ at any point a within a contour C in terms of an integral taken along C in which $f(z)$ is analytic.

Assuming that differentiation under the integral sign is permissible, and differentiating both sides of Eq. (7–206) with respect to a, we obtain

$$f'(a) = \frac{1}{2\pi i}\oint_C \frac{f(z)\,dz}{(z-a)^2}.$$

In a similar manner it can be shown that

$$f^{(n)}(a) = \frac{n!}{2\pi i}\oint_C \frac{f(z)\,dz}{(z-a)^{n+1}}. \qquad (7\text{-}207)$$

We therefore conclude that *if $f(z)$ is analytic at a point $z = a$, then derivatives of $f(z)$ of all orders exist at that point.*†

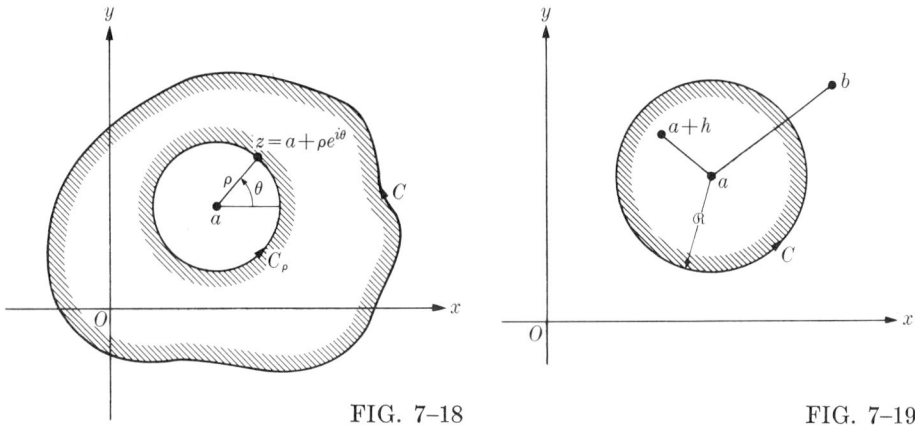

FIG. 7–18   FIG. 7–19

(11) *The Taylor series.* Let $z = a$ be any point at which $f(z)$ is analytic, and let $z = b$ be the nearest singularity of $f(z)$ to $z = a$. Then $f(z)$ is analytic at all points on and in the circle C of radius $\Re < |b - a|$ and with center at $z = a$ (Fig. 7–19). If $z = a + h$ is a point within C, the value of $f(z)$ at this point can be evaluated by Cauchy's integral formula, Eq. (7–206). Thus we have

$$f(a+h) = \frac{1}{2\pi i}\oint_C \frac{f(z)\,dz}{(z-a-h)}. \qquad (7\text{-}208)$$

---

† Since no branch point is assumed, all derivatives are single-valued. Therefore, if a derivative exists, then the previous derivative is analytic. This implies that the derivatives of all orders are analytic.

Noting that by algebraic manipulation

$$\frac{1}{(z-a-h)} = \frac{1}{(z-a)[1-h/(z-a)]} = \sum_{n=0}^{\infty} \frac{h^n}{(z-a)^{n+1}}, \quad (7\text{-}209)$$

and inserting Eq. (7–209) into Eq. (7–208), we obtain

$$f(a+h) = \sum_{n=0}^{\infty} \frac{1}{2\pi i} \left[\oint_C \frac{f(z)\,dz}{(z-a)^{n+1}} \right] h^n. \quad (7\text{-}210)$$

Finally, combination of Eqs. (7–207) and (7–210) gives

$$f(a+h) = \sum_{n=0}^{\infty} \frac{f^{(n)}(a)}{n!} h^n. \quad (7\text{-}211)$$

Equation (7–211) is identical in form to the *Taylor series* for functions of a real variable. The transformation $z = a + h$ gives a more frequently used representation of this series as follows:

$$f(z) = \sum_{n=0}^{\infty} \frac{f^{(n)}(a)}{n!} (z-a)^n. \quad (7\text{-}212)$$

Thus *if $f(z)$ is analytic at a point $z = a$, then derivatives of $f(z)$ of all orders exist, and $f(z)$ can be expanded into a Taylor series; the circle of convergence of this series is the largest circle (with center at $z = a$) in which $f(z)$ is analytic.*

(12) *The Cauchy residue theorem.* Suppose that $z = a$ is a pole of order k and is the only singularity of a function $f(z)$ in a region R. Then, according to the definition of the pole, $(z-a)^k f(z)$ is analytic and, by use of Eq. (7–212), can be expanded into a Taylor series

$$(z-a)^k f(z) = \sum_{n=0}^{\infty} A_n (z-a)^n, \quad (7\text{-}213)$$

where

$$A_n = \frac{1}{n!} \left\{ \frac{d^n}{dz^n} [(z-a)^k f(z)] \right\}_{z=a}. \quad (7\text{-}214)$$

If $z \neq a$, then dividing both sides of Eq. (7–213) by $(z-a)^k$, we get

$$f(z) = \sum_{n=0}^{\infty} A_n (z-a)^{n-k}. \quad (7\text{-}215)$$

Integration of Eq. (7–215) around a contour C enclosing $z = a$ gives

$$\oint_C f(z)\,dz = \sum_{n=0}^{\infty} A_n \oint_C (z-a)^{n-k}\,dz. \quad (7\text{-}216)$$

However, according to Eq. (7–202) we have for any integer value of $(n - k)$

$$\oint_C (z - a)^{n-k} \, dz = \begin{cases} 0, & n - k \neq -1, \\ 2\pi i, & n - k = -1. \end{cases} \quad (7\text{–}217)$$

Finally, inserting Eq. (7–217) into the right-hand side of Eq. (7–216), we obtain

$$\oint_C f(z) \, dz = 2\pi i A_{k-1}. \quad (7\text{–}218)$$

Equation (7–218) is known as *Cauchy's residue theorem;* the coefficient A_{k-1} is called the *residue* of $f(z)$ at $z = a$, and is usually denoted by Res (a) or, more explicitly, by Res $[f(z); a]$. Hence, if $f(z)$ has a pole of order k at $z = a$, then

$$\oint_C f(z) \, dz = 2\pi i \text{ Res } (a), \quad (7\text{–}219)$$

where

$$\text{Res } (a) = \frac{1}{(k-1)!} \left\{ \frac{d^{k-1}}{dz^{k-1}} [(z - a)^k f(z)] \right\}_{z=a}, \quad (7\text{–}220)$$

and C is a contour surrounding $z = a$ but excluding all other singularities of $f(z)$. It is clear from Eq. (7–215) that Res (a) is the coefficient of $1/(z - a)$ in the expansion of $f(z)$ in powers of $(z - a)$, and can be either calculated from Eq. (7–220) or obtained directly from the expansion of $f(z)$. The latter proves useful when the order of the pole cannot be readily determined.

The special case of Eq. (7–220) that corresponds to a *simple* pole ($k = 1$) at $z = a$ gives

$$\text{Res } (a) = \lim_{z \to a} [(z - a)f(z)], \quad (7\text{–}221)$$

where the use of the limit becomes necessary, since $f(z) \to \infty$ and the product $(z - a)f(z)$ takes the indeterminate form $0 \times \infty$ as $z \to a$.

Particularly, if $f(z)$ takes the frequently encountered form

$$f(z) = \frac{P(z)}{z^k Q(z)}, \quad (7\text{–}222)$$

with a_n, $n = 1, 2, \ldots, N$ as the *simple* roots of $Q(z)$, and $Q(0) \neq 0$, $P(0) \neq 0$, and $P(a_n) \neq 0$, then for the residues at $z = a_n$, we have from Eq. (7–221)

$$\sum_{n=1}^{N} \text{Res } (a_n) = \sum_{n=1}^{N} \lim_{z \to a_n} \frac{(z - a_n) P(z)}{z^k Q(z)}. \quad (7\text{–}223)$$

Now L'Hôpital's rule applied to Eq. (7–223) yields

$$\sum_{n=1}^{N} \text{Res } (a_n) = \sum_{n=1}^{N} \lim_{z \to a_n} \frac{P(z) + (z - a_n)(dP/dz)}{k z^{k-1} Q(z) + z^k (dQ/dz)}. \quad (7\text{–}224)$$

It follows from Eq. (7–224) that

$$\sum_{n=1}^{N} \text{Res}\,(a_n) = \sum_{n=1}^{N} \frac{P(a_n)}{a_n^k (dQ/dz)_{z=a_n}}. \tag{7-225}$$

The evaluation of the residue for the multiple pole at $z = 0$ may be obtained either from the coefficient of $1/z$ in the expansion of $f(z)$ or by employing Eq. (7–220).

As another particular case, if $f(z)$ has the form

$$f(z) = \frac{P(z)}{Q(z)\,R(z)}, \tag{7-226}$$

with a_n, $n = 1, 2, \ldots, N$ and b_m, $m = 1, 2, \ldots, M$ as the *simple* roots of $Q(z)$ and $R(z)$, respectively, and $a_n \neq b_m$, $P(a_n) \neq 0$, and $P(b_m) \neq 0$, then following a procedure similar to that of the previous particular case, we may easily show the sum of the residues at all the poles to be

$$\sum_{n=1}^{N} \text{Res}\,(a_n) + \sum_{m=1}^{M} \text{Res}\,(b_m)$$

$$= \sum_{n=1}^{N} \frac{P(a_n)}{(dQ/dz)_{z=a_n} R(a_n)} + \sum_{m=1}^{M} \frac{P(b_m)}{Q(b_m)(dR/dz)_{z=b_m}}. \tag{7-227}$$

This completes our review of the functions of a complex variable. In the next section we express the inversion integral given by Eq. (7–161) or by Eq. (7–162) in terms of two particular contours which are suitable to our problems.

7–8. Evaluation of the Inversion Theorem in Terms of Two Particular Contours

Let us return now to the solution of conduction problems by Laplace transforms. Suppose that the solution of a given problem has been reduced to the inversion of its transform, $\overline{T}(\mathbf{r}, p)$. If $\overline{T}(\mathbf{r}, p)$ does not appear in or cannot be made suitable to the direct use of tables of transforms, the inversion theorem given by Eq. (7–162) may be employed to write the solution $T(\mathbf{r}, t)$ in terms of a contour integral in the complex plane. Two standard procedures are available for evaluating this contour integral.

(i) If $\overline{T}(\mathbf{r}, z)$ is analytic except for a number (say N) of poles, all of which are to the left of some line $x = \gamma$, we complete the contour of Eq. (7–162) and Fig. 7–20(a) by a large circle C of radius $\mathcal{R}$, enclosing all poles of the integrand (Fig. 7–20b). (Hereafter we shall refer to the closed contour of Fig. 7–20(b) as Contour I.) In the limit as $N \to \infty$, it follows from Cauchy's residue theorem, Eq. (7–219), that

$$\lim_{L\to\infty} \int_{\gamma-iL}^{\gamma+iL} e^{tz}\overline{T}(\mathbf{r},z)\,dz + \lim_{\mathfrak{R}\to\infty} \int_C e^{tz}\overline{T}(\mathbf{r},z)\,dz = 2\pi i \lim_{N\to\infty} \sum_{n=1}^{N} \text{Res}\,(a_n).$$

(7–228)

In all the problems we discuss here it can be shown by employing Eq. (7–197) that the integral over the circle C vanishes as $\mathfrak{R} \to \infty$. Thus, in the limit, we have

$$T(\mathbf{r},t) = \frac{1}{2\pi i} \int_{\gamma-i\infty}^{\gamma+i\infty} e^{tz}\overline{T}(\mathbf{r},z)\,dz = \sum_{n=1}^{\infty} \text{Res}\,(a_n). \qquad (7\text{–}229)$$

This case usually arises in problems involving conduction of heat in *finite regions*.

Let us illustrate the use of Eq. (7–229) by the inversion of

$$\frac{e^{x/p}}{p} = \sum_{n=0}^{\infty} \frac{x^n}{n!\,p^{n+1}}. \qquad (7\text{–}230)$$

Since the $(n+1)$th term of Eq. (7–230) has only a pole of order $(n+1)$ at $z = 0$, it follows from Eq. (7–220) that

$$\text{Res}\,(0) = \sum_{n=0}^{\infty} \frac{x^n}{n!}\left[\frac{1}{n!}\frac{d^n(e^{tz})}{dz^n}\right]_{z=0} = \sum_{n=0}^{\infty} \frac{(xt)^n}{(n!)^2}. \qquad (7\text{–}231)$$

Furthermore, comparing Eq. (7–231) with the series expansion of the modified

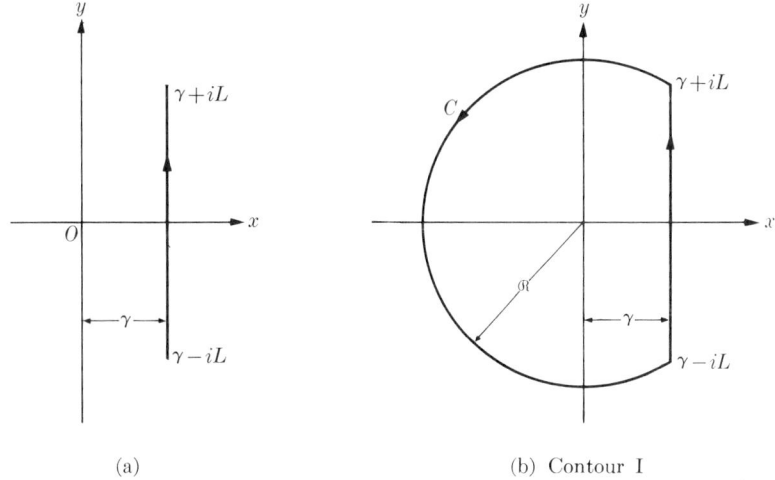

FIG. 7–20

Bessel function of the first kind, of order zero [see Eq. (7–84)], we have

$$e^{x/p}/p = L\{I_0[2(xt)^{1/2}]\}, \tag{7-79}$$

which is identical to TP(47) in Table 7–2. This transform pair plays an important role in the solution of unsteady problems related to single-fluid heat exchangers (see Examples 7–5 and 7–7).

(ii) *If $f(z)$ is analytic except for a number of poles*, all of which are to the left of some line $x = \gamma$, and if it also has a branch point at $z = 0$, then by introducing a cut along the negative real axis,† we complete the contour of the inversion integral by a large circle of radius $\mathcal{R}$ which encloses all poles of the integrand, and a loop $\mathcal{L}$ along the cut and around the branch point (Fig. 7–21).

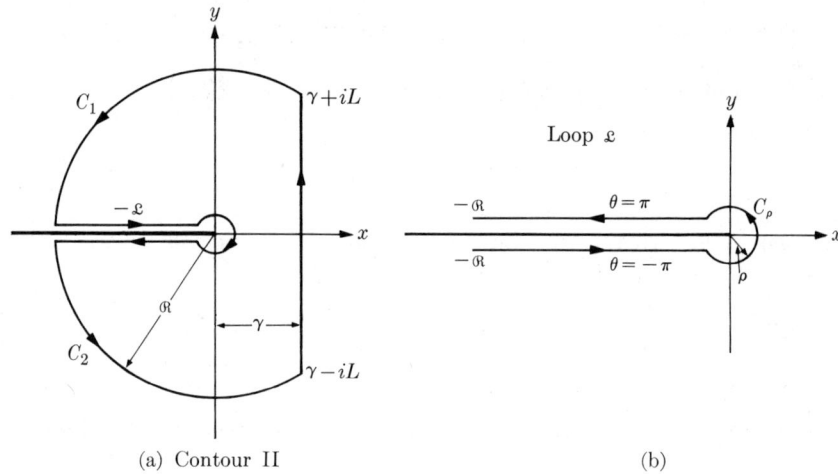

(a) Contour II   (b)

FIG. 7–21

Then Cauchy's residue theorem applied to the closed contour of Fig. 7–21(a), say *Contour II*, gives

$$\lim_{L\to\infty} \int_{\gamma-iL}^{\gamma+iL} e^{tz}\overline{T}(\mathbf{r}, z)\, dz + \lim_{\mathcal{R}\to\infty} \int_{C_1+C_2} e^{tz}\overline{T}(\mathbf{r}, z)\, dz - \lim_{\substack{\mathcal{R}\to\infty \\ \rho\to 0}} \int_{\mathcal{L}} e^{tz}\overline{T}(\mathbf{r}, z)\, dz$$

$$= 2\pi i \lim_{N\to\infty} \sum_{n=1}^{N} \text{Res}\,(a_n). \tag{7-232}$$

Again, by use of Eq. (7–197), the integral over the circle $C_1 + C_2$ may be shown

---

† The selection of the negative real axis as the cut is arbitrary. As long as we stay on the left of $x = \gamma$, the direction of any radius or any curve connecting the origin to the circle may be used for this purpose.

to vanish as $\Re \to \infty$. Thus in the limit we get

$$T(\mathbf{r}, t) = \frac{1}{2\pi i} \int_{\gamma-i\infty}^{\gamma-i\infty} e^{tz} \overline{T}(\mathbf{r}, z)\, dz = \sum_{n=1}^{\infty} \text{Res}\,(a_n) + \frac{1}{2\pi i} \lim_{\substack{\Re \to \infty \\ \rho \to 0}} \int_{\mathcal{L}} e^{tz} \overline{T}(\mathbf{r}, z)\, dz.$$
(7-233)

For convenience, the loop integral of Eq. (7-233) may be expressed more explicitly; noting that $z = -\rho$ on both the lower and upper branches of the loop, we obtain

$$\int_{\mathcal{L}} e^{tz}\overline{T}(\mathbf{r}, z)\, dz = -\int_{\Re}^{\rho} e^{-t\rho}\overline{T}(\mathbf{r}, \rho e^{-i\pi})\, dz + \lim_{\rho \to 0} \int_{C_\rho} e^{tz}\overline{T}(\mathbf{r}, z)\, dz$$

$$- \int_{\rho}^{\Re} e^{-t\rho}\overline{T}(\mathbf{r}, \rho e^{i\pi})\, dz. \qquad (7\text{-}234)$$

Since $\overline{T}(\mathbf{r}, z)$ is considered implicitly, its argument along the lower and upper branches of the loop must be kept as $z = \rho e^{-i\pi}$ and $z = \rho e^{i\pi}$, respectively. As $\Re \to \infty$ and $\rho \to 0$ Eq. (7-234) tends to

$$\lim_{\substack{\Re \to \infty \\ \rho \to 0}} \int_{\mathcal{L}} e^{tz}\overline{T}(\mathbf{r}, z)\, dz = \int_{0}^{\infty} e^{-t\rho}[\overline{T}(\mathbf{r}, \rho e^{-i\pi}) - \overline{T}(\mathbf{r}, \rho e^{i\pi})]\, d\rho$$

$$+ \lim_{\rho \to 0} \int_{C_\rho} e^{tz}\overline{T}(\mathbf{r}, z)\, dz. \qquad (7\text{-}235)$$

In Eq. (7-235) the integral around the small circle C_ρ is left complex, to be evaluated individually for each particular problem. Note that ρ remains constant and the differential of $z = \rho e^{i\theta}$ is $dz = i\rho e^{i\theta}\, d\theta$ for this integral. Now, inserting Eq. (7-235) into Eq. (7-233), we get finally†

$$T(\mathbf{r}, t) = \frac{1}{2\pi i} \int_{\gamma-i\infty}^{\gamma+i\infty} e^{tz}\overline{T}(\mathbf{r}, z)\, dz = \sum_{n=1}^{\infty} \text{Res}\,(a_n) + \frac{1}{2\pi i} \lim_{\rho \to 0} \int_{C_\rho} e^{tz}\overline{T}(\mathbf{r}, z)\, dz$$

$$+ \frac{1}{2\pi i} \int_{0}^{\infty} e^{-t\rho}[\overline{T}(\mathbf{r}, \rho e^{-i\pi}) - \overline{T}(\mathbf{r}, \rho e^{i\pi})]\, d\rho. \qquad (7\text{-}236)$$

This case usually arises in problems involving conduction of heat in *semi-infinite regions*.

Let us illustrate the use of Eq. (7-236) by the inversion of

$$e^{-qx}/p, \qquad (7\text{-}237)$$

where $q = (p/a)^{1/2}$. According to Definition (6) of Section 7-7, Eq. (7-237) is multiple-valued at $z = 0$. This may readily be seen from every other term

---

† If $\overline{T}(\mathbf{r}, z)$ does not have a branch point at $z = 0$, then Eq. (7-236) reduces to Eq. (7-229), as expected.

of its Maclaurin expansion when the argument of these terms is changed by 2π. Thus Eq. (7–237) has a branch point at $z = 0$. Since $z = 0$ must be excluded by a cut from the region in which the function is analytic, it cannot be a simple pole even though this makes the denominator of Eq. (7–237) zero. Thus the branch point $z = 0$ is the only singularity of Eq. (7–237). It follows from Eq. (7–236) that the inversion of Eq. (7–237) is

$$\frac{1}{2\pi i}\lim_{\rho \to 0}\int_{C_\rho}[e^{tz-(z/a)^{1/2}x}]\frac{dz}{z} - \frac{1}{2\pi i}\int_0^\infty e^{-t\rho}[e^{-x(\rho/a)^{1/2}e^{-i\pi/2}} - e^{-x(\rho/a)^{1/2}e^{i\pi/2}}]\frac{d\rho}{\rho}.$$

(7–238)

Since the integrand of the first integral of Eq. (7–238) is continuous, taking the limit inside the integral sign,

$$\lim_{\rho \to 0}[e^{t\rho e^{i\theta}-x(\rho/a)^{1/2}e^{i\theta/2}}] \to 1,$$

noting that $dz/z = i\theta$, and inserting $e^{-i\pi/2} = -1$ and $e^{i\pi/2} = 1$ into the second integral, we get after rearranging

$$1 - \frac{1}{\pi}\int_0^\infty e^{-t\rho}\sin\left(\frac{\rho}{a}\right)^{1/2}x\left(\frac{d\rho}{\rho}\right) = \mathrm{erfc}\left[\frac{x}{2(at)^{1/2}}\right].$$

The equivalence of the result to the complementary error function may be seen by referring to a table of integrals. Thus we have shown by the inversion theorem that

$$\frac{e^{-qx}}{p} = L\left\{\mathrm{erfc}\left[\frac{x}{2(at)^{1/2}}\right]\right\}, \qquad (7\text{–}239)$$

which is identical to TP(30) in Table 7–2. Equation (7–239) is an important transform pair in the solution of conduction problems in semi-infinite regions (see Examples 7–3 and 7–4).

If the singularities of $\overline{T}(\mathbf{r}, z)$ belong to neither one of the foregoing two cases, the inversion process may still be carried out by a similar procedure, provided another contour suitable to the new case is chosen. However, this often requires knowledge on the techniques of contour integration that is beyond the scope of this text.

The next section is devoted to problems whose solution by the inversion theorem leads to an integration along Contours I or II.

7–9. Solutions Obtainable by the Inversion Theorem

After reconsidering an example used earlier in this chapter, we shall employ some examples from Chapter 5, to illustrate the use of the inversion theorem.

Example 7–9. Let us find the solution of Example 7–2 by means of the inversion theorem.

The transforms of the lumped wall and lumped fluid temperatures have already been obtained as

$$\frac{\bar{\psi}(p)}{\psi_0} = \frac{p + \lambda_1 + \lambda_2}{(p + \lambda_1)(p + \lambda_2)}, \qquad (7\text{–}37)$$

$$\frac{\bar{\phi}(p)}{\psi_0} = \frac{p + b_1 + b_2}{(p + \lambda_1)(p + \lambda_2)}. \qquad (7\text{–}38)$$

We now employ the inversion theorem rather than the transform tables in the inversion of Eqs. (7–37) and (7–38). Thus, from Eq. (7–161), we get

$$\frac{\psi(t)}{\psi_0} = \frac{1}{2\pi i} \int_{\gamma-i\infty}^{\gamma+i\infty} e^{tz} \frac{z + \lambda_1 + \lambda_2}{(z + \lambda_1)(z + \lambda_2)} \, dz \qquad (7\text{–}240)$$

and

$$\frac{\phi(t)}{\psi_0} = \frac{1}{2\pi i} \int_{\gamma-i\infty}^{\gamma+i\infty} e^{tz} \frac{z + b_1 + b_2}{(z + \lambda_1)(z + \lambda_2)} \, dz. \qquad (7\text{–}241)$$

The only singularities of Eqs. (7–240) and (7–241) are the simple poles at $z = -\lambda_1$ and $z = -\lambda_2$. It follows from Contour I and Eq. (7–229) that

$$\begin{Bmatrix} \psi(t)/\psi_0 \\ \phi(t)/\psi_0 \end{Bmatrix} = \text{Res}\,(-\lambda_1) + \text{Res}\,(-\lambda_2). \qquad (7\text{–}242)$$

Since the poles are simple, Eq. (7–242) may be rearranged by means of Eq. (7–221). The result is

$$\frac{\psi(t)}{\psi_0} = \left[e^{tz} \left(\frac{z + \lambda_1 + \lambda_2}{z + \lambda_2} \right) \right]_{z=-\lambda_1} + \left[e^{tz} \left(\frac{z + \lambda_1 + \lambda_2}{z + \lambda_1} \right) \right]_{z=-\lambda_2} \qquad (7\text{–}243)$$

and

$$\frac{\phi(t)}{\psi_0} = \left[e^{tz} \left(\frac{z + b_1 + b_2}{z + \lambda_2} \right) \right]_{z=-\lambda_1} + \left[e^{tz} \left(\frac{z + b_1 + b_2}{z + \lambda_1} \right) \right]_{z=-\lambda_2}. \qquad (7\text{–}244)$$

Thus from Eqs. (7–243) and (7–244) we directly obtain the solution of the problem in the expected form,

$$\frac{T(t) - T_\infty}{T_0 - T_\infty} = \frac{\lambda_1 e^{-\lambda_2 t} - \lambda_2 e^{-\lambda_1 t}}{\lambda_1 - \lambda_2} \qquad (7\text{–}41)$$

and

$$\frac{\theta(t) - T_\infty}{T_0 - T_\infty} = \frac{T(t) - T_\infty}{T_0 - T_\infty} - b_3 \frac{e^{-\lambda_2 t} - e^{-\lambda_1 t}}{\lambda_1 - \lambda_2}. \qquad (7\text{–}42)$$

Example 7-10. Let us find the solution of Example 5-2 by the use of the inversion theorem.

The formulation of the problem, taken from Example 5-2, is

$$\frac{\partial \theta}{\partial t} = a \frac{\partial^2 \theta}{\partial x^2}, \qquad \theta(x, 0) = \theta_0,$$

$$\frac{\partial \theta(0, t)}{\partial x} = 0, \qquad \theta(L, t) = 0.$$

The transformed differential equation and the transformed boundary conditions are readily found to be

$$\frac{d^2 \bar{\theta}}{dx^2} - q^2 \bar{\theta} = -\frac{\theta_0}{a}, \tag{7-245}$$

$$\frac{d\bar{\theta}(0, p)}{dx} = 0, \tag{7-246}$$

$$\bar{\theta}(L, p) = 0, \tag{7-247}$$

where, as before, $q = (p/a)^{1/2}$.

The solution of Eq. (7-245) that is satisfied by Eqs. (7-246) and (7-247) is

$$\frac{\bar{\theta}(x, p)}{\theta_0} = \frac{1}{p} - \frac{\cosh qx}{p \cosh qL}. \tag{7-248}$$

Since the second term in Eq. (7-248) does not appear in our tables of transforms, the use of the inversion theorem is necessary. Employing TP(1) for the first term and Eq. (7-162) for the second, we have

$$\frac{\theta(x, t)}{\theta_0} = 1 - \frac{1}{2\pi i} \int_{\gamma-i\infty}^{\gamma+i\infty} e^{tz} \frac{\cosh (z/a)^{1/2} x}{z \cosh (z/a)^{1/2} L} \, dz. \tag{7-249}$$

Although their arguments are double-valued at $z = 0$, $\cosh (z/a)^{1/2} x$ and $\cosh (z/a)^{1/2} L$, both being even functions, remain single-valued; they possess no branch points at $z = 0$ or elsewhere. Since z appears as a factor in the denominator of the integrand of Eq. (7-249), and $\cosh (z/a)^{1/2} L \to 1$ as $z \to 0$, $z = 0$ is a simple zero of the denominator and a simple pole of the integrand of Eq. (7-249). Furthermore, $\cosh (z/a)^{1/2} L$ appears, at first glance, to have no zeros. This is true, of course, when its argument is real. However, in the inversion theorem, z was not restricted to real values. It may be shown that $\cosh (z/a)^{1/2} L$ will continue to have no zeros for any complex argument, say $(z/a)^{1/2} = \mu + i\lambda$, but when the argument is pure imaginary, say $(z/a)^{1/2} = i\lambda$ or $z = -a\lambda^2$, noting from Eq. (7-168) that $\cosh i\lambda L = \cos \lambda L$, we find an infinite number of simple zeros for $\cosh (z/a)^{1/2} L$ corresponding to the values

$$\lambda_n L = (2n + 1)\pi/2, \qquad n = 0, 1, 2, \ldots \tag{7-250}$$

Thus the integrand of Eq. (7–249) has an infinite number of simple poles corresponding to the locations

$$z = -a[(2n+1)\pi/2L]^2, \quad n = 0, 1, 2, \ldots \tag{7-251}$$

along the negative real axis. Note that these poles are identical to the *characteristic values* of the same problem obtained by the use of separation of variables (see Example 5–2). This general fact will also be observed in the subsequent examples.

Since the integrand of Eq. (7–249) is analytic except for a number of poles, we employ Contour I and Eq. (7–229), and obtain

$$\frac{\theta(x,t)}{\theta_0} = 1 - \text{Res}(0) - \sum_{n=0}^{\infty} \text{Res}(-a\lambda_n^2). \tag{7-252}$$

It follows readily from Eq. (7–221) that the residue at the pole $z = 0$ is unity. The residue at the pole $z = -a\lambda_n^2$ may be obtained by comparing the integrand of Eq. (7–249) to Eq. (7–222); thus $P(z) = e^{tz}\cosh(z/a)^{1/2}x$, $k = 1$, and $Q(z) = \cosh(z/a)^{1/2}L$. Equation (7–252) may then be rearranged by using Eq. (7–225) for $k = 1$ to give

$$\frac{\theta(x,t)}{\theta_0} = -\sum_{n=0}^{\infty} \frac{P(-a\lambda_n^2)}{-a\lambda_n^2 (dQ/dz)_{z=-a\lambda_n^2}}. \tag{7-253}$$

Introducing the variable $\zeta = (z/a)^{1/2}L$ for convenience in the evaluation of dQ/dz, we may write Eq. (7–253) in the form

$$\frac{\theta(x,t)}{\theta_0} = \sum_{n=0}^{\infty} \frac{e^{-a\lambda_n^2 t}\cosh i\lambda_n x}{a\lambda_n^2 \left[\frac{d}{d\zeta}(\cosh \zeta)\frac{d\zeta}{dz}\right]_{z=-a\lambda_n^2}}. \tag{7-254}$$

The term in brackets in the denominator yields

$$\left\{\left[\sinh\left(\frac{z}{a}\right)^{1/2}L\right]\frac{L}{2(az)^{1/2}}\right\}_{z=-a\lambda_n^2} = (\sinh i\lambda_n L)\frac{L}{2ia\lambda_n}. \tag{7-255}$$

Now, inserting Eq. (7–255) into Eq. (7–254), and noting from Eq. (7–168) that $\cosh i\lambda_n x = \cos \lambda_n x$ and $\sinh i\lambda_n L = i\sin \lambda_n L$, and considering Eq. (7–250) to get $\sin \lambda_n L = (-1)^n$, we have finally

$$\frac{T(x,t) - T_\infty}{T_0 - T_\infty} = 2\sum_{n=0}^{\infty} \frac{(-1)^n}{(\lambda_n L)} e^{-a\lambda_n^2 t}\cos \lambda_n x, \tag{5-10}$$

which is identical to the result of Example 5–2.

Example 7–11. Let us find the solution of Example 5–3 by means of the inversion theorem.

The only difference between the formulations of Examples 5–2 and 5–3 is in their surface boundary condition at $x = L$; the former formulation is based on a large heat transfer coefficient, the latter on a moderate one. Thus the transformed formulation from Example 7–10, modified to include the effect of a moderate h, readily gives

$$\frac{d^2\bar{\theta}}{dx^2} - q^2\bar{\theta} = -\frac{\theta_0}{a}, \qquad (7\text{–}245)$$

$$\frac{d\bar{\theta}(0, p)}{dx} = 0, \qquad (7\text{–}246)$$

$$-k\frac{d\bar{\theta}(L, p)}{dx} = h\bar{\theta}(L, p). \qquad (7\text{–}256)$$

The solution of Eq. (7–245) that is satisfied by Eqs. (7–246) and (7–256) is

$$\frac{\bar{\theta}(x, p)}{\theta_0} = \frac{1}{p} - \frac{\cosh qx}{p[\cosh qL + (k/h)q \sinh qL]}. \qquad (7\text{–}257)$$

Again, the use of TP(1) for the first term and Eq. (7–162) for the second gives

$$\frac{\theta(x, t)}{\theta_0} = 1 - \frac{1}{2\pi i}\int_{\gamma-i\infty}^{\gamma+i\infty} \frac{e^{tz} \cosh (z/a)^{1/2}x}{z[\cosh (z/a)^{1/2}L + (k/h)(z/a)^{1/2} \sinh (z/a)^{1/2}L]}\, dz. \qquad (7\text{–}258)$$

Let us now follow the argument analogous to that of Example 7–10. Although $(z/a)^{1/2}$ is double-valued at $z = 0$, the functions $\cosh (z/a)^{1/2}x$, $\cosh (z/a)^{1/2}L$, and $(z/a)^{1/2} \sinh (z/a)^{1/2}L$ are single-valued, since they are all even functions; they possess no branch points at $z = 0$ or elsewhere. Since z appears as a factor in the denominator of the integrand of Eq. (7–258) and since for $z = 0$ the bracketed part of the denominator is not zero, $z = 0$ is a simple zero of the denominator and a simple pole of the integrand. It may also be shown that the bracketed part of the denominator does not have zeros for any real or complex argument except for the pure imaginary argument, say $(z/a)^{1/2} = i\lambda$ or $z = -a\lambda^2$, for which it may be rearranged as

$$(\lambda_n L) \sin \lambda_n L = \text{Bi} \cos \lambda_n L. \qquad (7\text{–}259)$$

Equation (7–259) has an infinite number of simple zeros along the negative real axis corresponding to locations $z = -a\lambda_n^2$. These zeros are the simple poles of the integrand, and are also identical to the characteristic values obtained in Example 5–3.

Like the integrand in Example 7–10, the integrand of Eq. (7–258) is analytic except for a number of simple poles, and it has the particular form of Eq. (7–222) written for $k = 1$. Thus Eq. (7-253) is also valid for the present case. Introducing again the variable $\zeta = (z/a)^{1/2}L$, we may interpret Eq. (7-253) in terms of the integrand of Eq. (7-258) to give

$$\frac{\theta(x, t)}{\theta_0} = \sum_{n=1}^{\infty} \frac{e^{-a\lambda_n^2 t} \cosh i\lambda_n x}{a\lambda_n^2 \left[\frac{d}{d\zeta}\left(\cosh \zeta + \frac{\zeta}{\text{Bi}} \sinh \zeta\right) \frac{d\zeta}{dz}\right]_{z=-a\lambda_n^2}}. \qquad (7\text{-}260)$$

The terms in brackets in the denominator of Eq. (7-260) readily yield

$$\left(\sinh i\lambda_n L + \frac{1}{\text{Bi}} \sinh i\lambda_n L + \frac{i\lambda_n L}{\text{Bi}} \cosh i\lambda_n L\right) \frac{L}{2ia\lambda_n}. \qquad (7\text{-}261)$$

Introducing Eq. (7-261) into Eq. (7-260), converting hyperbolic functions to circular functions, and eliminating the Biot modulus by means of Eq. (7-259), we have finally

$$\frac{T(x, t) - T_\infty}{T_0 - T_\infty} = 2 \sum_{n=1}^{\infty} \left(\frac{\sin \lambda_n L}{\lambda_n L + \sin \lambda_n L \cos \lambda_n L}\right) e^{-a\lambda_n^2 t} \cos \lambda_n x, \qquad (5\text{-}13)$$

which is identical to the result of Example 5-3.

Example 7-12. Let us find the solution of Example 5-4 by the inversion theorem.

The formulation of the problem, taken from Example 5-4, is

$$\frac{\partial \theta}{\partial t} = a \frac{\partial^2 \theta}{\partial x^2} + \frac{u'''}{\rho c}, \qquad \theta(x, 0) = 0,$$

$$\frac{\partial \theta(0, t)}{\partial x} = 0, \qquad \theta(L, t) = 0.$$

As we learned in Chapters 4 and 5, the use of the separation of variables in the solution of the foregoing formulation would require the homogeneity of the associated differential equation through an appropriate superposition. Here and in the following examples we shall observe that the problem of nonhomogeneity in relation to governing equations and/or boundary conditions need not be considered when we are seeking a solution by means of Laplace transforms.

The transformed formulation of the problem is

$$\frac{d^2 \bar{\theta}}{dx^2} - q^2 \bar{\theta} = -\left(\frac{u'''}{k}\right) \frac{1}{p}, \qquad (7\text{-}262)$$

$$\frac{d\bar{\theta}(0, p)}{dx} = 0, \qquad (7\text{-}263)$$

$$\bar{\theta}(L, p) = 0. \qquad (7\text{-}264)$$

The solution of Eq. (7–262) that satisfies Eqs. (7–263) and (7–264) may readily be found in the form

$$\frac{\bar{\theta}(x, p)}{u'''/\rho c} = \frac{1}{p^2} - \frac{\cosh qx}{p^2 \cosh qL}. \qquad (7\text{–}265)$$

Thus TP(2) for the first term and Eq. (7–162) for the second yield

$$\frac{\theta(x, t)}{u'''/\rho c} = t - \frac{1}{2\pi i} \int_{\gamma - i\infty}^{\gamma + i\infty} e^{tz} \frac{\cosh (z/a)^{1/2} x}{z^2 \cosh (z/a)^{1/2} L} \, dz. \qquad (7\text{–}266)$$

The integrand of Eq. (7–266) has a double pole at $z = 0$, and simple poles at $z = -a\lambda_n^2$, where $\lambda_n L = (2n + 1)\pi/2$, $n = 0, 1, 2, \ldots$ Since the integrand is analytic except for a number of poles, the use of Contour I again results in

$$\frac{\theta(x, t)}{u'''/\rho c} = t - \text{Res}(0) - \sum_{n=0}^{\infty} \text{Res}(-a\lambda_n^2). \qquad (7\text{–}267)$$

The residue at $z = 0$ may be obtained from the series expansion of the integrand.† Thus, by the Maclaurin expansion of the numerator and the denominator, we have

$$\frac{e^{tz} \cosh (z/a)^{1/2} x}{z^2 \cosh (z/a)^{1/2} L} = \frac{(1 + tz + \cdots)(1 + zx^2/2a + \cdots)}{z^2 (1 + zL^2/2a + \cdots)}.$$

Dividing now the expansion of $\cosh(z/a)^{1/2}x$ by the expansion of $\cosh(z/a)^{1/2}L$, and multiplying the result by $1/z^2$ and the expansion of e^{tz}, we get

$$\frac{1}{z^2} + \left[t - \frac{1}{2a}(L^2 - x^2) \right] \frac{1}{z} + \cdots$$

Then, according to Eqs. (7–215) and (7–218), the residue at the pole $z = 0$ is the coefficient of $1/z$, namely,

$$\text{Res}(0) = t - \frac{1}{2a}(L^2 - x^2). \qquad (7\text{–}268)$$

Noting that the integrand of Eq. (7–266) now has the particular form of Eq. (7–222) for $k = 2$, we obtain the residue at the poles $z = -a\lambda_n^2$ from Eq. (7–225) as

$$\text{Res}(-a\lambda_n^2) = \frac{P(-a\lambda_n^2)}{(-a\lambda_n^2)^2 (dQ/dz)_{z=-a\lambda_n^2}}. \qquad (7\text{–}269)$$

Further noting that $(dQ/dz)_{z=-a\lambda_n^2}$ of Eq. (7–269) is identical to that of Ex-

---

† The use of Eq. (7–220) for $k = 2$ would give the same result in a less convenient way.

ample 7–10 evaluated in Eq. (7–255), converting the hyperbolic functions to circular functions, and considering $\sin \lambda_n L = (-1)^n$, we find that

$$\text{Res}\,(-a\lambda_n^2) = \frac{2(-1)^n}{a\lambda_n^3 L} e^{-a\lambda_n^2 t} \cos \lambda_n x. \tag{7-270}$$

Introducing Eqs. (7–268) and (7–270) into Eq. (7–267), and rearranging the result, we have finally

$$\frac{\theta(x,t)}{u'''L^2/k} = \frac{1}{2}\left[1 - \left(\frac{x}{L}\right)^2\right] - 2\sum_{n=0}^{\infty} \frac{(-1)^n}{(\lambda_n L)^3} e^{-a\lambda_n^2 t} \cos \lambda_n x, \tag{5-20}$$

which is identical to the result of Example 5–4.

Example 7–13. Let us find the solution of Example 5–5 by means of the inversion theorem.

Let us first recall that Example 5–5 illustrated unsteady problems whose solution does not approach a steady value for large values of time. To take care of nonhomogeneities of this type, we were forced to employ a superposition given by Eq. (5–21) rather than Eq. (5–14). We therefore concluded that the physics of a problem should be kept in mind when we seek a solution by means of the separation of variables. Utilizing the same example, we once more demonstrate that nonhomogeneities will not be an issue when the solution is sought by means of Laplace transforms.

From Example 5–5 we have the formulation of the problem as follows:

$$\frac{\partial \theta}{\partial t} = a\frac{\partial^2 \theta}{\partial x^2}, \qquad \theta(x,0) = 0,$$

$$\frac{\partial \theta(0,t)}{\partial x} = 0, \qquad +k\frac{\partial \theta(L,t)}{\partial x} = q''.$$

The transformed formulation is readily found to be

$$\frac{d^2\bar{\theta}}{dx^2} - q^2\bar{\theta} = 0, \tag{7-271}$$

$$\frac{d\bar{\theta}(0,p)}{dx} = 0, \tag{7-272}$$

$$\frac{d\bar{\theta}(L,p)}{dx} = \left(\frac{q''}{k}\right)\frac{1}{p}. \tag{7-273}$$

The solution of Eq. (7–271) that is satisfied by Eqs. (7–272) and (7–273) has the form

$$\frac{\bar{\theta}(x,p)}{q''/k} = \frac{\cosh qx}{qp \sinh qL}. \tag{7-274}$$

Then, according to the inversion theorem, we have

$$\frac{\theta(x, t)}{q''/k} = \frac{1}{2\pi i} \int_{\gamma-i\infty}^{\gamma+i\infty} e^{tz} \frac{\cosh (z/a)^{1/2}x}{z(z/a)^{1/2} \sinh (z/a)^{1/2}L} \, dz. \quad (7\text{-}275)$$

Since $(z/a)^{1/2}L$ can be factored from the Maclaurin expansion of $\sinh (z/a)^{1/2}L$, $z(z/a)^{1/2} \sinh (z/a)^{1/2}L$ has a double zero and the integrand of Eq. (7-275) has a double pole at $z = 0$. Also, the integrand has simple poles, corresponding to the zeros of $\sinh (z/a)^{1/2}L$, at $z = -a\lambda_n^2$, where $\lambda_n L = n\pi$, $n = 1, 2, 3, \ldots$ Thus Contour I gives

$$\frac{\theta(x, t)}{q''/k} = \text{Res }(0) + \sum_{n=1}^{\infty} \text{Res }(-a\lambda_n^2). \quad (7\text{-}276)$$

In a manner similar to the evaluation of Res (0) in the previous example, we obtain from the expansion of the integrand

$$\frac{a/L}{z^2} + \left[\frac{at}{L} + \frac{1}{L}\left(\frac{x^2}{2} - \frac{L^2}{6}\right)\right]\frac{1}{z} + \cdots$$

It follows from the coefficient of $1/z$ that

$$\text{Res }(0) = \frac{at}{L} + \frac{1}{L}\left(\frac{x^2}{2} - \frac{L^2}{6}\right). \quad (7\text{-}277)$$

Comparing the integrand of Eq. (7-275) to Eq. (7-226) we note that $P(z) = e^{tz}\cosh (z/a)^{1/2}x$, $Q(z) = z(z/a)^{1/2}$, and $R(z) = \sinh (z/a)^{1/2}L$, and from Eq. (7-227) we obtain, corresponding to the zeros of $R(z)$,

$$\text{Res }(-a\lambda_n^2) = \frac{P(-a\lambda_n^2)}{Q(-a\lambda_n^2)(dR/dz)_{z=-a\lambda_n^2}}. \quad (7\text{-}278)$$

Furthermore, considering $(dR/dz)_{z=-a\lambda_n^2} = \cosh (z/a)^{1/2}L\,(L/2ia\lambda_n)$, expressing hyperbolic functions in terms of circular functions, and remembering that $\cos \lambda_n L = (-1)^n$, we may rearrange Eq. (7-278) to give

$$\text{Res }(-a\lambda_n^2) = -2L \frac{(-1)^n}{(\lambda_n L)^2} e^{-a\lambda_n^2 t} \cos \lambda_n x. \quad (7\text{-}279)$$

Introducing Eqs. (7-277) and (7-279) into Eq. (7-276), we have finally

$$\frac{\theta(x, t)}{q''L/k} = \frac{at}{L^2} + \frac{1}{2}\left(\frac{x}{L}\right)^2 - \frac{1}{6} - 2\sum_{n=1}^{\infty} \frac{(-1)^n}{(\lambda_n L)^2} e^{-a\lambda_n^2 t} \cos \lambda_n x, \quad (5\text{-}33)$$

which is identical to the result of Example 5-5.

The solution of the differential formulation of Example 2-3 by Laplace transforms requires no mathematical clarification beyond that given in previous examples. We summarize by giving the solution of the problem in terms of the inversion theorem:

$$\frac{\theta(x,t)}{q''/k} = \frac{1}{2\pi i}\int_{\gamma-i\infty}^{\gamma+i\infty} \frac{e^{tz}}{z}\left[\frac{(k/h)\cosh(z/a)^{1/2}L + (a/z)^{1/2}\sinh(z/a)^{1/2}L}{\cosh(z/a)^{1/2}L + (k/h)(z/a)^{1/2}\sinh(z/a)^{1/2}L}\right.$$

$$\left. \times \cosh\left(\frac{z}{a}\right)^{1/2}x - \left(\frac{a}{z}\right)^{1/2}\sinh\left(\frac{z}{a}\right)^{1/2}x\right]dz. \quad (7\text{-}280)$$

The integrand of Eq. (7-280) has simple poles at $z = 0$ and at the zeros of the transcendental equation $(\lambda_n L)\sin\lambda_n L = \text{Bi}\cos\lambda_n L$. Then, from Contour I, we have

$$\frac{\theta(x,t)}{q''L/k} = 1 - \left(\frac{x}{L}\right) + \frac{1}{\text{Bi}} - 2\sum_{n=1}^{\infty}\frac{e^{-a\lambda_n^2 t}\cos\lambda_n x}{\lambda_n L(\lambda_n L + \sin\lambda_n L\cos\lambda_n L)}, \quad (5\text{-}34)$$

where x is measured from the bottom of the plate upward. The details of the solution above are left to the reader.

Example 7-14. Let us find the solution of Example 5-16 by the use of the inversion theorem.

As we learned in Chapter 5, the separation of variables cannot be directly employed for problems having time-dependent nonhomogeneities. In Example 5-16, we proceeded by first applying the separation of variables to the related problem resulting from a step function. Introducing the solution thus obtained into Duhamel's integral we obtained the solution of the original problem. The Laplace transforms in time, being directly applicable to time-dependent nonhomogeneities, circumvent this lengthy procedure.

From Example 5-16 we have the formulation of the problem in the form

$$\frac{\partial^2\theta_1}{\partial x^2} - m_1^2\theta_1 = \frac{1}{a}\frac{\partial\theta_1}{\partial t}, \quad m_1^2 = 2h_1/k\delta,$$

$$\theta_1(x,0) = 0, \quad (5\text{-}119)$$

$$\frac{\partial\theta_1(0,t)}{\partial x} = 0, \quad \theta_1(L,t) = \frac{q''}{h_2}(1 - e^{-bt}).$$

Then the transformed formulation is

$$\frac{d^2\bar{\theta}_1}{dx^2} - (m_1^2 + q^2)\bar{\theta}_1 = 0, \quad (7\text{-}281)$$

$$\frac{d\bar{\theta}_1(0,p)}{dx} = 0, \quad (7\text{-}282)$$

$$\bar{\theta}_1(L,p) = \frac{q''/\rho c\delta}{p(p+b)}. \quad (7\text{-}283)$$

The solution of Eq. (7–281) that satisfies Eqs. (7–282) and (7–283) is

$$\frac{\bar{\theta}_1(x, p)}{q''/\rho c\delta} = \frac{\cosh (m_1^2 + q^2)^{1/2}x}{p(p + b) \cosh (m_1^2 + q^2)^{1/2}L} . \qquad (7\text{–}284)$$

Then, according to the inversion theorem,

$$\frac{\theta_1(x, t)}{q''/\rho c\delta} = \frac{1}{2\pi i} \int_{\gamma-i\infty}^{\gamma+i\infty} \frac{e^{tz} \cosh (m_1^2 + z/a)^{1/2}x}{z(z + b) \cosh (m_1^2 + z/a)^{1/2}L} \, dz. \qquad (7\text{–}285)$$

The integrand of Eq. (7–285) is analytic except for the simple poles at $z = 0$, $z = -b$, and $(m_1^2 + z/a)^{1/2} = i\lambda_n$ or $z = -a(m_1^2 + \lambda_n^2)$, where $\lambda_n L = (2n + 1)\pi/2$, $n = 0, 1, 2, \ldots$ It follows readily from Contour I that

$$\frac{\theta_1(x, t)}{q''/\rho c\delta} = \text{Res} (0) + \text{Res} (-b) + \sum_{n=0}^{\infty} \text{Res} [-a(m_1^2 + \lambda_n^2)]. \qquad (7\text{–}286)$$

The residues of Eq. (7–286) may be evaluated in a manner similar to the previous examples. The results are†

$$\text{Res} (0) = \frac{\cosh m_1 x}{b \cosh m_1 L}, \qquad (7\text{–}287)$$

$$\text{Res} (-b) = -\frac{e^{-bt} \cosh (m_1^2 - b/a)^{1/2}x}{b \cosh (m_1^2 - b/a)^{1/2}L}, \qquad (7\text{–}288)$$

$$\text{Res} [-a(m_1^2 + \lambda_n^2)] = -\frac{2(-1)^n(\lambda_n L)e^{-a(m_1^2+\lambda_n^2)t}}{a(m_1^2 + \lambda_n^2)L^2(m_1^2 + \lambda_n^2 - b/a)} \cos \lambda_n x. \qquad (7\text{–}289)$$

Using Eqs. (7–287), (7–288), and (7–289) in Eq. (7–286) we have finally

$$\frac{\theta_1(x, t)}{q''/h_2} = \frac{\cosh m_1 x}{\cosh m_1 L} - \frac{e^{-bt} \cosh (m_1^2 - b/a)^{1/2}x}{\cosh (m_1^2 - b/a)^{1/2}L}$$

$$+ 2\left(\frac{b}{a}\right) \sum_{n=0}^{\infty} \frac{(-1)^n(\lambda_n L)e^{-(m_1^2+\lambda_n^2)t}}{(m_1^2 + \lambda_n^2)L^2(m_1^2 + \lambda_n^2 - b/a)} \cos \lambda_n x \qquad (5\text{–}134)$$

which is identical to the result obtained in Example 5–16 by the separation of variables and Duhamel's superposition integral.

---

† To obtain Eq. (7–289) assume that the integrand of Eq. (7–285) has the form of Eq. (7–226) so that $P(z) = e^{tz} \cosh (m_1^2 + z/a)^{1/2}x$, $Q(z) = z(z + b)$, and $R(z) = \cosh (m_1^2 + z/a)^{1/2}L$, then employ the second term of Eq. (7–227).

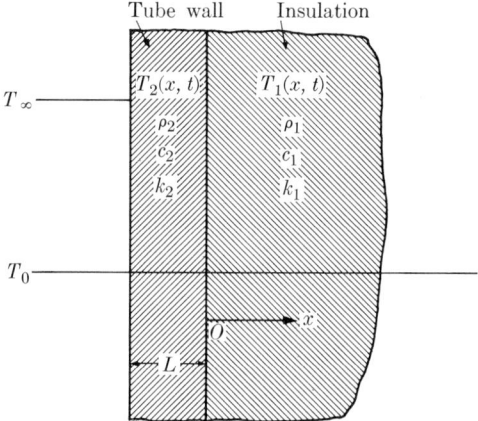

FIG. 7–22

Example 7–15. Neglecting the effect of curvature, we may treat a thick-walled insulated tube as a flat plate of thickness L attached to a semi-infinite solid (Fig. 7–22). (Here the insulation is assumed to have a large but finite thermal resistivity. Since the problem deals with the penetration of heat into the insulated tube, the effect of curvature may be neglected only in early stages of this penetration. Similarly, the assumption of infinite thickness of the insulation is valid only at early stages.) The initial temperature of the system is uniform, say T_0; then a condensing vapor at temperature T_∞ fills the tube instantaneously. The contact resistance between the plate and the solid is negligible. We wish to find the unsteady temperature of the system.

Measuring x most conveniently from the interface into the solid, we may write the formulation of the problem as follows:

$$\frac{\partial T_1}{\partial t} = a_1 \frac{\partial^2 T_1}{\partial x^2}, \quad \frac{\partial T_2}{\partial t} = a_2 \frac{\partial^2 T_2}{\partial x^2},$$

$$T_1(x, 0) = T_2(x, 0) = T_0, \quad (7\text{--}290)$$

$$T_1(0, t) = T_2(0, t), \quad k_1 \frac{\partial T_1(0, t)}{\partial x} = k_2 \frac{\partial T_2(0, t)}{\partial x},$$

$$\lim_{x \to \infty} T_1(x, t) \to T_0, \quad T_2(-L, t) = T_\infty.$$

Then in terms of $\theta_1 = T_1 - T_0$, $\theta_2 = T_2 - T_0$, $q_1^2 = p/a_1$, and $q_2^2 = p/a_2$, the transform of Eq. (7–290) becomes

$$\frac{d^2 \bar{\theta}_1}{dx^2} - q_1^2 \bar{\theta}_1 = 0, \quad \frac{d^2 \bar{\theta}_2}{dx^2} - q_2^2 \bar{\theta}_2 = 0,$$

$$\bar{\theta}_1(0, p) = \bar{\theta}_2(0, p), \quad k_1 \frac{d\bar{\theta}_1(0, p)}{dx} = k_2 \frac{d\bar{\theta}_2(0, p)}{dx}, \quad (7\text{--}291)$$

$$\lim_{x \to \infty} \bar{\theta}_1(x, p) \to 0, \quad \bar{\theta}_2(-L, p) = \theta_\infty/p,$$

where $\theta_\infty = T_\infty - T_0$. The solution of Eq. (7–291) is

$$\frac{\bar{\theta}_1(x, p)}{\theta_\infty} = \frac{e^{-q_1 x}}{p(\cosh q_2 L + \epsilon \sinh q_2 L)}, \qquad (7\text{–}292)$$

$$\frac{\bar{\theta}_2(x, p)}{\theta_\infty} = \frac{\cosh q_2 x - \epsilon \sinh q_2 x}{p(\cosh q_2 L + \epsilon \sinh q_2 L)}, \qquad (7\text{–}293)$$

where $\epsilon = (\rho_1 c_1 k_1 / \rho_2 c_2 k_2)^{1/2}$. Taking Eq. (7–292) first, for example, we have from the inversion theorem

$$\frac{\theta_1(x, t)}{\theta_\infty} = \frac{1}{2\pi i} \int_{\gamma - i\infty}^{\gamma + i\infty} \frac{e^{tz - (z/a_1)^{1/2} x}}{z[\cosh (z/a_2)^{1/2} L + \epsilon \sinh (z/a)^{1/2} L]} \, dz. \qquad (7\text{–}294)$$

Since $e^{-(z/a_1)^{1/2} x}$ and $\sinh (z/a_2)^{1/2} L$ are double-valued, the integrand of Eq. (7–294) has a branch point at $z = 0$. (Show that there are no other singularities of the integrand.) Therefore, Contour II rather than Contour I is suitable to the evaluation of Eq. (7–294). Noting that the integrand possesses no poles, and that the integral around the small circle of loop $\mathcal{L}$ (Fig. 7–21b) gives $2\pi i$ in the limit as the integrand approaches $i\theta$ when $\rho \to 0$, we have from Eq. (7–236)

$$\frac{\theta_1(x, t)}{\theta_\infty} = 1 + \frac{1}{2\pi i} \int_0^\infty e^{-t\rho}$$

$$\times \left\{ \frac{e^{-x(\rho/a_1)^{1/2} e^{-i\pi/2}}}{\cosh [(\rho/a_2)^{1/2} e^{-i\pi/2} L] + \epsilon \sinh [(\rho/a_2)^{1/2} e^{-i\pi/2} L]} \right.$$

$$\left. - \frac{e^{-x(\rho/a_1)^{1/2} e^{i\pi/2}}}{\cosh [(\rho/a_2)^{1/2} e^{i\pi/2} L] + \epsilon \sinh [(\rho/a_2)^{1/2} e^{i\pi/2} L]} \right\} \frac{d\rho}{\rho}. \qquad (7\text{–}295)$$

Employing $\xi = (\rho/a_2)^{1/2}$, $\eta = (a_2/a_1)^{1/2}$, $e^{-i\pi/2} = -i$, and $e^{i\pi/2} = i$ in Eq. (7–295) and rearranging, we get

$$\frac{\theta_1(x, t)}{\theta_\infty} = 1 - \frac{2}{\pi} \int_0^\infty e^{-a_2 \xi^2 t} \left(\frac{\cos \xi L \sin \eta \xi x + \epsilon \sin \xi L \cos \eta \xi x}{\cos^2 \xi L + \epsilon^2 \sin^2 \xi L} \right) \frac{d\xi}{\xi}. \qquad (7\text{–}296)$$

$\theta_2(x, t)$ may be obtained in a manner similar to $\theta_1(x, t)$. The details of the evaluation of $\theta_2(x, t)$ are left to the reader. The result is

$$\frac{\theta_2(x, t)}{\theta_\infty} = 1 - \epsilon \frac{2}{\pi} \int_0^\infty \frac{e^{-a_2 \xi^2 t} \sin \xi(x + L)}{\cos^2 \xi L + \epsilon^2 \sin^2 \xi L} \frac{d\xi}{\xi}. \qquad (7\text{–}297)$$

Since the integrals of Eqs. (7–296) and (7–297) cannot be integrated explicitly, reference may be made to a numerical integration procedure for a parametric study of these integrals.

Let us now proceed to illustrative examples for cylindrical geometry.

Example 7–16. Let us find the solution of Example 5–7 by means of the inversion theorem.

The formulation of the problem, taken from Example 5–7, is

$$\frac{\partial \theta}{\partial t} = \frac{a}{r}\frac{\partial}{\partial r}\left(r\frac{\partial \theta}{\partial r}\right), \qquad \theta(r, 0) = \theta_0, \qquad \frac{\partial \theta(0, t)}{\partial r} = 0, \qquad \theta(R, t) = 0. \tag{5-45}$$

The transform of Eq. (5–45) is

$$\frac{d}{dr}\left(r\frac{d\bar{\theta}}{dr}\right) - q^2 r\bar{\theta} = \frac{\theta_0}{a} r, \tag{7-298}$$

$$\frac{d\bar{\theta}(0, p)}{dr} = 0, \tag{7-299}$$

$$\bar{\theta}(R, p) = 0. \tag{7-300}$$

The solution of the homogeneous part of Eq. (7–298) may be obtained from Table 3–1. Supplementing this solution by the particular solution, θ_0/p, and satisfying the result by Eqs. (7–299) and (7–300) gives

$$\frac{\bar{\theta}(r, p)}{\theta_0} = \frac{1}{p} - \frac{I_0(qr)}{pI_0(qR)}. \tag{7-301}$$

Introducing Eq. (7–301) into the inversion theorem we have

$$\frac{\theta(r, t)}{\theta_0} = 1 - \frac{1}{2\pi i}\int_{\gamma-i\infty}^{\gamma+i\infty} \frac{e^{tz} I_0[(z/a)^{1/2} r]}{z I_0[(z/a)^{1/2} R]}\, dz. \tag{7-302}$$

Although their arguments are double-valued, $I_0[(z/a)^{1/2} r]$ and $I_0[(z/a)^{1/2} R]$, both being even functions, are single-valued; they possess no branch points at $z = 0$. Moreover, for real and complex values of its argument $I_0[(z/a)^{1/2} R]$ possesses no zeros. However, for imaginary values of this argument, say $(z/a)^{1/2} = i\lambda$ or $z = -a\lambda^2$, $I_0[(z/a)^{1/2} R]$ is reduced to $J_0(\lambda R)$, which has an infinite number of simple zeros.† These zeros are identical to the characteristic values obtained in Example 5–7. Furthermore, we note that $I_0(0) = 1$, and conclude that $zI_0[(z/a)^{1/2} R]$ has a simple zero at $z = 0$ because of the factor z. Thus $z = 0$ and the zeros of $J_0(\lambda_n R) = 0$ are simple poles of the integrand of Eq. (7–302). It follows from Contour I that

$$\frac{\theta(r, t)}{\theta_0} = 1 - \text{Res}(0) - \sum_{n=1}^{\infty} \text{Res}(-a\lambda_n^2). \tag{7-303}$$

From Eq. (7–221) we get

$$\text{Res}(0) = 1. \tag{7-304}$$

---

† Note from Eq. (3–119) that $I_0(i\lambda R) = J_0(\lambda R)$.

Comparing the integrand of Eq. (7–302) to Eq. (7–222) for $k = 1$ yields $P(z) = e^{tz}I_0[(z/a)^{1/2}r]$ and $Q(z) = I_0[(z/a)^{1/2}R]$, and we obtain from Eq. (7–225)

$$\text{Res}(-a\lambda_n^2) = \frac{P(-a\lambda_n^2)}{-a\lambda_n^2(dQ/dz)_{z=-a\lambda_n^2}}. \tag{7–305}$$

Next let us introduce the variable $\zeta = (z/a)^{1/2}R$; then, according to Eq. (3–137),

$$\left.\frac{dQ}{dz}\right|_{z=-a\lambda_n^2} = \frac{dI_0(\zeta)}{d\zeta}\left.\frac{d\zeta}{dz}\right|_{z=-a\lambda_n^2} = I_1(i\lambda_n R)\frac{R}{2ia\lambda_n}. \tag{7–306}$$

Now, inserting Eq. (7–306) into Eq. (7–305), we find the explicit form of the latter to be

$$\text{Res}(-a\lambda_n^2) = -\frac{2ie^{-a\lambda_n^2 t}I_0(i\lambda_n r)}{(\lambda_n R)I_1(i\lambda_n R)}. \tag{7–307}$$

Relating the modified Bessel functions to the Bessel functions by Eq. (3–119), and noting that the resulting Bessel functions $J_0(\lambda_n r)$ and $J_1(\lambda_n R)$ are even and odd functions, respectively, we may rearrange Eq. (7–307) to give

$$\text{Res}(-a\lambda_n^2) = -2\frac{e^{-a\lambda_n^2 t}J_0(\lambda_n r)}{(\lambda_n R)J_1(\lambda_n R)}. \tag{7–308}$$

Finally, introducing Eqs. (7–304) and (7–308) into Eq. (7–303), we have

$$\frac{T(r,t) - T_\infty}{T_0 - T_\infty} = 2\sum_{n=1}^{\infty}\frac{e^{-a\lambda_n^2 t}J_0(\lambda_n r)}{(\lambda_n R)J_1(\lambda_n R)}, \tag{5–51}$$

which is identical to the result of Example 5–7.

Example 7–17. Let us find the solution of Example 5–8 by means of Laplace transforms.

According to Example 5–8 the formulation of the problem is

$$\frac{\partial\theta}{\partial t} = \frac{a}{r}\frac{\partial}{\partial r}\left(r\frac{\partial\theta}{\partial r}\right), \quad \theta(r,0) = \theta_0, \quad \frac{\partial\theta(0,t)}{\partial r} = 0, \quad -k\frac{\partial\theta(R,t)}{\partial r} = h\theta(R,t). \tag{5–52}$$

The transform of Eq. (5–52) gives

$$\frac{d}{dr}\left(r\frac{d\bar{\theta}}{dr}\right) - q^2 r\bar{\theta} = \frac{\theta_0}{a}r, \quad \frac{d\bar{\theta}(0,p)}{dr} = 0, \quad -k\frac{d\bar{\theta}(R,p)}{dr} = h\bar{\theta}(R,p). \tag{7–309}$$

The solution of Eq. (7–309) is

$$\frac{\bar{\theta}(r,p)}{\theta_0} = \frac{1}{p} - \frac{BI_0(qr)}{p[BI_0(qR) + qI_1(qR)]}, \tag{7–310}$$

where $B = h/k$. Employing the inversion theorem, we obtain then

$$\frac{\theta(r,t)}{\theta_0} = 1 - \frac{1}{2\pi i}\int_{\gamma-i\infty}^{\gamma+i\infty} \frac{Be^{tz}I_0[(z/a)^{1/2}r]}{z\{BI_0[(z/a)^{1/2}R] + (z/a)^{1/2}I_1[(z/a)^{1/2}R]\}}\,dz. \tag{7-311}$$

The only singularities of the integrand of Eq. (7–311) are the simple poles at $z = 0$ and $z = -a\lambda_n^2$, where λ_n satisfies the transcendental equation

$$(\lambda_n R)J_1(\lambda_n R) - \text{Bi}\, J_0(\lambda_n R) = 0 \tag{7-312}$$

with $\text{Bi} = hR/k$. The use of Contour I results in Eq. (7–303) of the preceding example,

$$\frac{\theta(r,t)}{\theta_0} = 1 - \text{Res}\,(0) - \sum_{n=1}^{\infty} \text{Res}\,(-a\lambda_n^2). \tag{7-303}$$

Equation (7–305) of that example,

$$\text{Res}\,(-a\lambda_n^2) = \frac{P(-a\lambda_n^2)}{-a\lambda_n^2(dQ/dz)_{z=-a\lambda_n^2}}, \tag{7-305}$$

is also valid for the present case, provided $P(z) = Be^{tz}I_0[(z/a)^{1/2}r]$ and $Q(z) = BI_0[(z/a)^{1/2}R] + (z/a)^{1/2}I_1[(z/a)^{1/2}R]$. The residue at $z = 0$ is

$$\text{Res}\,(0) = 1. \tag{7-313}$$

For the residue at $z = -a\lambda_n^2$ we first calculate $(dQ/dz)_{z=-a\lambda_n^2}$. In terms of the variable $\zeta = (z/a)^{1/2}R$, we have

$$\left.\frac{dQ}{dz}\right|_{z=-a\lambda_n^2} = \left.\frac{dQ}{d\zeta}\frac{d\zeta}{dz}\right|_{z=-a\lambda_n^2} = [\text{Bi}\, I_1(i\lambda_n R) + i\lambda_n I_0(i\lambda_n R)]\frac{R}{2ia\lambda_n}. \tag{7-314}$$

Introducing Eq. (7–314) and the value of $P(z)$ into Eq. (7–305), rearranging the result by the transformation of the modified Bessel functions to the Bessel functions, and considering Eq. (7–312), we obtain

$$\text{Res}\,(-a\lambda_n^2) = \frac{2\text{Bi}\,e^{-a\lambda_n^2 t}J_0(\lambda_n r)}{(\lambda_n^2 R^2 + \text{Bi}^2)J_0(\lambda_n R)}. \tag{7-315}$$

Finally, the use of Eqs. (7–313) and (7–315) in Eq. (7–303) gives

$$\frac{T(r,t) - T_\infty}{T_0 - T_\infty} = 2\,\text{Bi}\sum_{n=1}^{\infty}\frac{e^{-a\lambda_n^2 t}J_0(\lambda_n r)}{(\lambda_n^2 R^2 + \text{Bi}^2)J_0(\lambda_n R)}, \tag{5-55}$$

which is identical to the result of Example 5–8.

Example 7–18. Consider a buried pipe (Fig. 7–23). The pipe and its surroundings have the same initial temperature, say T_∞. Then condensing steam at temperature T_0 suddenly fills the pipe. We wish to find the unsteady temperature of the system.

Noting that the heat transfer coefficient is large, and neglecting the temperature drop across the pipe wall because of its high thermal conductivity and small thickness, we find that the formulation of the problem in terms of $\theta = T - T_\infty$ is

$$\frac{\partial \theta}{\partial t} = \frac{a}{r}\frac{\partial}{\partial r}\left(r\frac{\partial \theta}{\partial r}\right), \qquad \theta(r,0) = 0,$$

$$\theta(R,t) = \theta_0, \qquad \lim_{r\to\infty} \theta(r,t) \to 0, \qquad (7\text{–}316)$$

where $\theta_0 = T_0 - T_\infty$. The transform of Eq. (7–316) is

$$\frac{d}{dr}\left(r\frac{d\bar\theta}{dr}\right) - q^2 r\bar\theta = 0,$$

$$\bar\theta(R,p) = \theta_0/p, \qquad \lim_{r\to\infty} \bar\theta(r,p) \to 0. \qquad (7\text{–}317)$$

The solution of Eq. (7–317) is

$$\frac{\bar\theta(r,p)}{\theta_0} = \frac{K_0(qr)}{pK_0(qR)}. \qquad (7\text{–}318)$$

Then from the inversion theorem we have

$$\frac{\theta(r,t)}{\theta_0} = \frac{1}{2\pi i}\int_{\gamma-i\infty}^{\gamma+i\infty} \frac{e^{tz}K_0[(z/a)^{1/2}r]}{zK_0[(z/a)^{1/2}R]}\,dz. \qquad (7\text{–}319)$$

Since $(z/a)^{1/2}$ is double-valued, say $\rho^{1/2}e^{\pm i\pi/2}$, we obtain from the definition of $K_\nu(ze^{\pm i\pi/2})$,†

$$K_0(\rho^{1/2}e^{i\pi/2}r) = -i\frac{\pi}{2}[J_0(\rho^{1/2}r) - iY_0(\rho^{1/2}r)],$$

$$K_0(\rho^{1/2}e^{-i\pi/2}r) = i\frac{\pi}{2}[J_0(\rho^{1/2}r) + iY_0(\rho^{1/2}r)],$$

which implies that $K_0[(z/a)^{1/2}r]$ and $K_0[(z/a)^{1/2}R]$ also are double-valued. Hence $z = 0$ is a branch point of the integrand of Eq. (7–319). The integrand has no other singularities.

Note that the integral around the small circle (Contour C_ρ of Fig. 7–21b) gives $2\pi i$ in the limit as the integrand approaches $i\theta$ when $\rho \to 0$. Then the

---

† See Eq. (3–145).

use of Contour II and Eq. (7-236) results in

$$\frac{\theta(r,t)}{\theta_0} = 1 + \frac{1}{2\pi i} \int_0^\infty e^{-t\rho} \left\{ \frac{K_0[(\rho/a)^{1/2}e^{-i\pi/2}r]}{K_0[(\rho/a)^{1/2}e^{-i\pi/2}R]} - \frac{K_0[(\rho/a)^{1/2}e^{i\pi/2}r]}{K_0[(\rho/a)^{1/2}e^{i\pi/2}R]} \right\} \frac{d\rho}{\rho}. \quad (7\text{-}320)$$

Introducing the variable $\xi = (\rho/a)^{1/2}$, and expressing K_0 in terms of J_0 and Y_0, we may rearrange Eq. (7-230) to give

$$\frac{\theta(r,t)}{\theta_0} = 1 + \frac{2}{\pi} \int_0^\infty e^{-a\xi^2 t} \left[\frac{J_0(\xi r) Y_0(\xi R) - J_0(\xi R) Y_0(\xi r)}{J_0^2(\xi R) + Y_0^2(\xi R)} \right] \frac{d\xi}{\xi}. \quad (7\text{-}321)$$

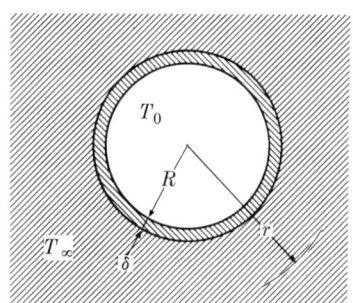

FIG. 7-23

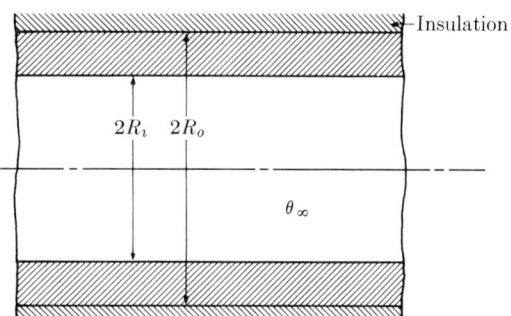

FIG. 7-24

Example 7-19. A condensing vapor at temperature θ_∞ suddenly fills an insulated pipe of inner and outer radii R_i and R_o, respectively (Fig. 7-24). The initial temperature of the pipe is zero. We wish to find the unsteady temperature of the pipe.

The formulation of the problem is

$$\frac{\partial \theta}{\partial t} = \frac{a}{r} \frac{\partial}{\partial r}\left(r \frac{\partial \theta}{\partial r}\right), \qquad \theta(r, 0) = 0,$$

$$\theta(R_i, t) = \theta_\infty, \qquad \frac{\partial \theta(R_o, t)}{\partial r} = 0. \quad (7\text{-}322)$$

The transform of Eq. (7-322) yields

$$\frac{d}{dr}\left(r \frac{d\bar{\theta}}{dr}\right) - q^2 r \bar{\theta} = 0,$$

$$\bar{\theta}(R_i, p) = \theta_\infty/p, \qquad \frac{d\bar{\theta}(R_o, p)}{dr} = 0. \quad (7\text{-}323)$$

The solution of Eq. (7-323) is found to be

$$\frac{\bar{\theta}(r,p)}{\theta_\infty} = \frac{K_1(qR_o)I_0(qr) + I_1(qR_o)K_0(qr)}{p[K_1(qR_o)I_0(qR_i) + I_1(qR_o)K_0(qR_i)]}. \quad (7\text{-}324)$$

From the inversion theorem we then have

$$\frac{\theta(r,t)}{\theta_\infty} = \frac{1}{2\pi i}$$

$$\times \int_{\gamma-i\infty}^{\gamma+i\infty} \frac{e^{tz}\{K_1[(z/a)^{1/2}R_o]I_0[(z/a)^{1/2}r] + I_1[(z/a)^{1/2}R_o]K_0[(z/a)^{1/2}r]\}}{z\{K_1[(z/a)^{1/2}R_o]I_0[(z/a)^{1/2}R_i] + I_1[(z/a)^{1/2}R_o]K_0[(z/a)^{1/2}R_i]\}}\,dz.$$

(7–325)

Although $(z/a)^{1/2}$ is double-valued, say $\rho^{1/2}e^{\pm i\pi/2}$, it can be shown by relating the modified Bessel functions to the Bessel functions through the repeated use of Eqs. (3–144) and (3–145) that the integrand of Eq. (7–325) is single-valued. The only singularities of the integrand are the simple poles corresponding to $z = 0$ and to $z = -a\lambda_n^2$, where λ_n satisfies

$$Y_1(\lambda_n R_o)J_0(\lambda_n R_i) - J_1(\lambda_n R_o)Y_0(\lambda_n R_i) = 0. \tag{7–326}$$

Then Contour I gives

$$\frac{\theta(r,t)}{\theta_\infty} = \text{Res}\,(0) + \sum_{n=1}^{\infty} \text{Res}\,(-a\lambda_n^2). \tag{7–327}$$

The residue at $z = 0$ is†

$$\text{Res}\,(0) = 1. \tag{7–328}$$

The residue at $z = -a\lambda_n^2$ may be obtained from Eq. (7–305) of Example 7–15, namely,

$$\text{Res}\,(-a\lambda_n^2) = \frac{P(-a\lambda_n^2)}{-a\lambda_n^2(dQ/dz)_{z=-a\lambda_n^2}}, \tag{7–305}$$

provided P and Q are respectively assumed to be equal to the numerator and to the term in braces in the denominator of the integrand of Eq. (7–325). Here dQ/dz may be evaluated most conveniently in terms of the variable $q = (z/a)^{1/2}$. The result is

$$\frac{dQ}{dz} = \left(\frac{dQ}{dq}\right)\frac{dq}{dz} = \left[-R_o K_0(qR_o)I_0(qR_i) - \frac{1}{q}K_1(qR_o)I_0(qR_i)\right.$$
$$+ R_i K_1(qR_o)I_1(qR_i) + R_o I_0(qR_o)K_0(qR_i)$$
$$\left. - \frac{1}{q}I_1(qR_o)K_0(qR_i) - R_i I_1(qR_o)K_1(qR_i)\right]\frac{1}{2a(z/a)^{1/2}}.$$

(7–329)

Now, relating the modified Bessel functions to the Bessel functions, inserting

---

† First note from Section 3–7 that for small arguments $I_0(z) \sim 1$, $I_1(z) \sim z/2$, $K_0(z) \sim -\ln z$, and $K_1(z) \sim 1/z$; then take the limit by letting $z \to 0$.

$i\lambda_n$ in place of $q = (z/a)^{1/2}$, then simplifying the result by means of Eq. (7–326), we have from Eq. (7–329)

$$\left(\frac{dQ}{dz}\right)_{z=-a\lambda_n^2} = \frac{\pi}{2}[R_oY_0(\lambda_n R_o)J_0(\lambda_n R_i) - R_oJ_0(\lambda_n R_o)Y_0(\lambda_n R_i)$$
$$- R_iY_1(\lambda_n R_o)J_1(\lambda_n R_i) + R_iJ_1(\lambda_n R_o)Y_1(\lambda_n R_i)]\frac{1}{2ia\lambda_n}.$$

(7–330)

Similarly,

$$P(-a\lambda_n^2) = i\frac{\pi}{2}e^{-a\lambda_n^2 t}[Y_1(\lambda_n R_o)J_0(\lambda_n r) - J_1(\lambda_n R_o)Y_0(\lambda_n r)]. \quad (7\text{–}331)$$

Finally, the use of Eqs. (7–330) and (7–331) in Eq. (7–305), and the result together with Eq. (7–328) in Eq. (7–327) gives

$$\frac{\theta(r, t)}{\theta_\infty} = 1 - 2\sum_{n=1}^{\infty} \frac{e^{-a\lambda_n^2 t}[Y_1(\lambda_n R_o)J_0(\lambda_n r) - J_1(\lambda_n R_o)Y_0(\lambda_n r)]}{(\lambda_n R_o)\Delta_0(R_o, R_i) - (\lambda_n R_i)\Delta_1(R_o, R_i)}, \quad (7\text{–}332)$$

where

$$\Delta_0(R_o, R_i) = Y_0(\lambda_n R_o)J_0(\lambda_n R_i) - J_0(\lambda_n R_o)Y_0(\lambda_n R_i),$$
$$\Delta_1(R_o, R_i) = Y_1(\lambda_n R_o)J_1(\lambda_n R_i) - J_1(\lambda_n R_o)Y_1(\lambda_n R_i).$$

Since the well-known transformation $\theta(r, t) = \psi(r, t)/r$ [see Eq. (5–60)] reduces the problems of one-dimensional spherical geometry to those of cartesian geometry, examples illustrating spherical geometry are not considered here.†

In the next two examples we demonstrate the use of the inversion theorem in two-dimensional geometry.

Example 7–20. Let us find the solution of Example 5–11 by means of the inversion theorem.

The formulation of the problem, taken from Example 5–11, is

$$\frac{\partial \theta}{\partial t} = a\left(\frac{\partial^2 \theta}{\partial \xi^2} + \frac{\partial^2 \theta}{\partial y^2}\right), \quad \theta(\xi, y, 0) = 0, \quad \theta(0, y, t) = 0, \quad \theta(L, y, t) = \theta_0,$$

$$\frac{\partial \theta(\xi, 0, t)}{\partial y} = 0, \quad \theta(\xi, l, t) = 0. \quad (5\text{–}71)$$

The transformation of Eq. (5–71) gives

$$\frac{\partial^2 \bar{\theta}}{\partial \xi^2} + \frac{\partial^2 \bar{\theta}}{\partial y^2} - q^2\bar{\theta} = 0, \quad \bar{\theta}(0, y, p) = 0, \quad \bar{\theta}(L, y, p) = \frac{\theta_0}{p},$$

$$\frac{\partial \bar{\theta}(\xi, 0, p)}{\partial y} = 0, \quad \bar{\theta}(\xi, l, p) = 0. \quad (7\text{–}333)$$

---

† See, however, Problems 7–28, 7–29, and 7–30.

Note that the use of Laplace transforms reduces the formulation of a multi-dimensional problem, a partial differential equation, to another partial differential equation involving one less independent variable, rather than to an ordinary differential equation as in one-dimensional problems. The solution of this transformed formulation requires further considerations. Equation (7–333), for example, may be solved by the separation of variables in a manner similar to steady problems of two-dimensional cartesian geometry.† The steps of the solution of Eq. (7–333) by the separation of variables, being trivial at this point, are not given. The result is

$$\frac{\bar{\theta}(\xi, y, p)}{\theta_0} = 2 \sum_{n=0}^{\infty} \frac{(-1)^n \sinh (\lambda_n^2 + q^2)^{1/2} \xi}{(\lambda_n l) \sinh (\lambda_n^2 + q^2)^{1/2} L} \cos \lambda_n y, \qquad (7\text{–}334)$$

where $\lambda_n l = (2n + 1)\pi/2$, $n = 0, 1, 2, \ldots$ Then from the inversion theorem we have

$$\frac{\theta(\xi, y, t)}{\theta_0} = 2 \sum_{n=0}^{\infty} \frac{(-1)^n}{(\lambda_n l)} \left[\frac{1}{2\pi i} \int_{\gamma-i\infty}^{\gamma+i\infty} \frac{e^{tz} \sinh (\lambda_n^2 + z/a)^{1/2} \xi}{z \sinh (\lambda_n^2 + z/a)^{1/2} L} \, dz \right] \cos \lambda_n y. \qquad (7\text{–}335)$$

The only singularities of the integrand of Eq. (7–335) are the simple poles corresponding to $z = 0$ and to

$$(\lambda_n^2 + z/a)^{1/2} = i\mu_m \quad \text{or} \quad z = -a(\lambda_n^2 + \mu_m^2),$$

where $\mu_m L = m\pi$, $m = 1, 2, 3, \ldots$ It follows then from Contour I that

$$\frac{\theta(\xi, y, t)}{\theta_0} = 2 \sum_{n=0}^{\infty} \frac{(-1)^n}{(\lambda_n l)} \left\{ \text{Res}\,(0) + \sum_{m=1}^{\infty} \text{Res}\,[-a(\lambda_n^2 + \mu_m^2)] \right\} \cos \lambda_n y. \qquad (7\text{–}336)$$

The evaluation of residues is straightforward. The result is

$$\text{Res}\,(0) = \frac{\sinh \lambda_n \xi}{\sinh \lambda_n L}, \qquad (7\text{–}337)$$

$$\text{Res}\,[-a(\lambda_n^2 + \mu_m^2)] = 2 \frac{(-1)^m (\mu_m L) e^{-a(\lambda_n^2 + \mu_m^2)t}}{(\lambda_n^2 + \mu_m^2) L^2} \sin \mu_m \xi. \qquad (7\text{–}338)$$

---

† This, of course, implies the existence of a more general method of solution, the successive use of two or more solution methods which are separately known to exist. A full discussion on this matter requires further background on the transform calculus and other solution techniques, and hence is beyond the scope of the text. See, however, Problem 7–14, involving the successive use of Laplace transforms.

The use of Eqs. (7–337) and (7–338) in Eq. (7–336) gives finally

$$\frac{\theta(\xi, y, t)}{\theta_0} = 2 \sum_{n=0}^{\infty} \frac{(-1)^n}{(\lambda_n l)} \left(\frac{\sinh \lambda_n \xi}{\sinh \lambda_n L}\right) \cos \lambda_n y$$

$$+ 4 \sum_{n=0}^{\infty} \sum_{m=1}^{\infty} (-1)^{n+m} \left(\frac{\mu_m L}{\lambda_n l}\right) \frac{e^{-a(\lambda_n^2 + \mu_m^2)t}}{(\lambda_n^2 + \mu_m^2) L^2} \sin \mu_m \xi \cos \lambda_n y, \quad (7\text{–}339)$$

which is identical to the result of Example 5–11, the sum of Eqs. (4–123) and (5–80).

We now have obtained the solution of Example 5–11 both by the separation of variables and by the successive use of Laplace transforms and separation of variables. The importance of the latter procedure will become evident in the next example, for which the former method is not suitable (why?).

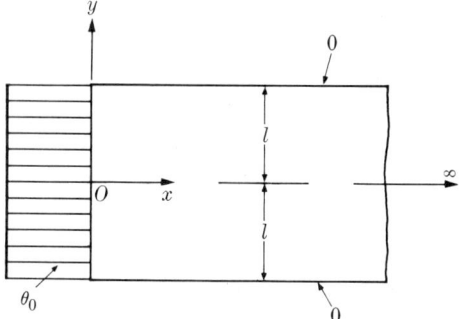

FIG. 7–25

Example 7–21. Let us reconsider the previous example, assuming now that the geometry of the problem extends to infinity in the x-direction.† We wish to find the unsteady temperature of the fin.

The formulation of the problem in terms of the coordinate system shown in Fig. 7–25 is

$$\frac{\partial \theta}{\partial t} = a \left(\frac{\partial^2 \theta}{\partial x^2} + \frac{\partial^2 \theta}{\partial y^2}\right), \quad \theta(x, y, 0) = 0, \quad \theta(0, y, t) = \theta_0, \quad \theta(\infty, y, t) = 0,$$

$$\frac{\partial \theta(x, 0, t)}{\partial y} = 0, \quad \theta(x, l, t) = 0. \quad (7\text{–}340)$$

The transform of Eq. (7–340) is readily found to be

$$\frac{\partial^2 \bar{\theta}}{\partial x^2} + \frac{\partial^2 \bar{\theta}}{\partial y^2} - q^2 \bar{\theta} = 0, \quad \bar{\theta}(0, y, p) = \frac{\theta_0}{p}, \quad \bar{\theta}(\infty, y, p) = 0,$$

$$\frac{\partial \bar{\theta}(x, 0, p)}{\partial y} = 0, \quad \bar{\theta}(x, l, p) = 0. \quad (7\text{–}341)$$

---

† As $t \to \infty$ the problem approaches Example 4–2.

Proceeding as in Example 7–20, and applying the separation of variables to Eq. (7–341), we get

$$\frac{\bar{\theta}(x, y, p)}{\theta_0} = 2 \sum_{n=0}^{\infty} \frac{(-1)^n}{(\lambda_n l)} \frac{e^{-(\lambda_n^2 + q^2)^{1/2} x}}{p} \cos \lambda_n y, \qquad (7\text{–}342)$$

where $\lambda_n l = (2n + 1)\pi/2$, $n = 0, 1, 2, \ldots$ The inversion theorem is not needed for the inversion of Eq. (7–342). In fact, TP(28) appropriately combined with TP(IV) and TP(VI) gives

$$\frac{\theta(x, y, t)}{\theta_0} = \frac{2}{\pi^{1/2}} \sum_{n=0}^{\infty} \frac{(-1)^n}{(\lambda_n l)} \left[\int_0^t \left(\frac{x^2}{4a\tau}\right)^{1/2} e^{-(a\lambda_n^2 \tau + x^2/4a\tau)} \frac{d\tau}{\tau} \right] \cos \lambda_n y. \qquad (7\text{–}343)$$

This example completes our study of the exact solution of unsteady problems by Laplace transforms. The next section is devoted to approximate solutions that are valid for either small or large values of time and that are obtained by considering the appropriate series expansions of the transformed solution. The methods to be developed are general, and may be applied to suitable space variables as well; cases of the latter however, occur less frequently than expansions in time.

7–10. Solutions Valid for Small or Large Values of Time

For small values of time the series solutions of unsteady transient problems,† obtained in Chapter 5 by the separation of variables and in Section 7–9 by the inversion theorem, are usually slowly convergent and not suitable for numerical computations. Thus, although these solutions can conveniently be used for large values of time, it becomes necessary to supplement them by other solutions which converge rapidly for small values of time. Furthermore, as illustrated by Example 5–6, the method of separation of variables is not suitable to unsteady periodic problems, and the method of complex temperature given in Chapter 6 is useful only for steady periodic solutions. In this section we employ Laplace transforms to find solutions valid for small or large values of time, and transient periodic solutions. Actually, transient periodic solutions could have been obtained in Section 7–9 by the inversion theorem; the reason for leaving these solutions to this section is that their steady part is best explained in terms of solutions valid for large values of time.

Thus the present section is devoted to a systematic study of solutions convenient for small or large values of time. Despite their convenience, the basic concept behind these solutions is surprisingly simple; it amounts to an appropriate series expansion of the transformed solution in terms of the Laplace

---

† Refer to Fig. 5–1 for the classification of unsteady problems.

transform parameter p. To explain the point further, let us first consider the definition of Laplace transforms,

$$L\{f(t)\} = \bar{f}(p) = \int_0^\infty e^{-pt} f(t)\, dt, \qquad (7\text{-}4)$$

and note the fact that p is, generally, a finite parameter which may be restricted, depending on the particular function $f(t)$ under consideration. Next, by interpreting Eq. (7-4) as an ordinary integral having a finite integrand, we conclude that the product pt must be finite. Therefore

$$\text{small } t \rightarrow \text{large } p,$$
$$\text{large } t \rightarrow \text{small } p.$$

Consequently, the expansion of the transformed solution of a problem into a convergent series of ascending powers of $1/p$ or p, and subsequent term-by-term inversion gives a solution useful for small or large values of time, respectively. Whenever convenient, the same concept may of course be applied to generalized expansions in terms of $1/f(p)$ or of $g(p)$, where $f(p)$ and $g(p)$ both are increasing functions of p, to be determined according to the particular problem in question. These expansions are demonstrated below by a number of examples.

1. *Solutions useful for small values of time.* Let us start with the lumped formulation of a problem. Assume that the transformed solution $\bar{T}(p)$ of such a formulation can be expanded into a power series in the form

$$\bar{T}(p) = \sum_{n=1}^{\infty} \frac{a_n}{p^n}. \qquad (7\text{-}344)$$

The direct use of the inversion theorem or TP(3) readily gives the inversion of Eq. (7-344) as

$$T(t) = \sum_{n=1}^{\infty} \frac{a_n t^{n-1}}{(n-1)!}. \qquad (7\text{-}345)$$

It is clear that Eq. (7-345) converges rapidly for small values of time. Since the transformed solution of lumped problems always has the form of a rational function of p, it can always be expanded into a convergent series like Eq. (7-344), leading to a solution in the form of Eq. (7-345).

Example 7-22. Let us find a solution of the lumped formulation of Example 2-2 useful for small values of time.

According to Example 2-2 the lumped formulation of the problem is

$$\rho c L \frac{dT}{dt} = -h(T - T_\infty) + u''' L, \qquad (2\text{-}149)$$

$$T(0) = T_\infty. \qquad (2\text{-}147)$$

Simplifying Eqs. (2–147) and (2–149) in terms of $\theta = T - T_\infty$, then taking their transforms, we obtain

$$\bar{\theta}(p) = \frac{n}{p(p+m)}, \qquad (7\text{–}346)$$

where $m = h/\rho c L$ and $n = u'''/\rho c$. Equation (7–346), first rearranged, then expanded in powers of $1/p$, gives

$$\bar{\theta}(p) = \frac{n}{p^2(1 + m/p)} = n\left(\frac{1}{p^2} - \frac{m}{p^3} + \frac{m^2}{p^4} - \cdots\right). \qquad (7\text{–}347)$$

Finally, term-by-term inversion of Eq. (7–347) yields a rapidly converging solution for small values of time in the form

$$\frac{T(t) - T_\infty}{u'''L/h} = mt - \frac{1}{2!}(mt)^2 + \frac{1}{3!}(mt)^3 - \cdots \qquad (7\text{–}348)$$

Note that Eq. (7–348) is the Maclaurin expansion of the complete solution given by Eq. (2–150).

For distributed problems over a finite region, the transform solutions contain hyperbolic or cylindrical (Bessel) functions of $(p/a)^{1/2}$ or more complicated arguments. These solutions can more conveniently be expanded in series of negative exponentials rather than in powers of $1/p$. Then term-by-term inversion gives solutions valid for small values of time in terms of exponentials and error functions rather than in powers of t.

For cartesian geometry the foregoing method usually yields one or the other of two types of solutions, depending on the boundary conditions of the problem. (i) For simple boundary conditions, such as prescribed temperature or prescribed heat flux, all terms of the expansion have simple coefficients, leading to solutions which are useful for all values of time and not merely for small values of time. (ii) For more complicated boundary conditions, such as heat transfer to the ambient by convection, the coefficients of the expansion are progressively more and more complicated functions of negative exponentials, so that only the inversion of the first few terms can be obtained. The resulting solutions are therefore useful only for relatively small values of time. For other geometries even the simplest boundary conditions yield series of progressively more complicated terms. Solutions thus obtained are again valid for only relatively small values of time.

Example 7–23. Let us find a solution of Example 7–10 useful for small values of time.

The solution of the transformed formulation may be taken from Example 7–10. Thus

$$\frac{\bar{\theta}(x, p)}{\theta_0} = \frac{1}{p} - \frac{\cosh qx}{p \cosh qL}. \qquad (7\text{–}248)$$

Expressing the hyperbolic functions in terms of exponentials, we obtain from Eq. (7-248)

$$\frac{\bar{\theta}(x, p)}{\theta_0} = \frac{1}{p} - \frac{e^{qx} + e^{-qx}}{p(e^{qL} + e^{-qL})}. \tag{7-349}$$

Further rearrangement of Eq. (7-349) by factoring the denominator of the second term on the right by e^{qL} gives

$$\frac{\bar{\theta}(x, p)}{\theta_0} = \frac{1}{p} - \frac{e^{qx} + e^{-qx}}{pe^{qL}(1 + e^{-2qL})}. \tag{7-350}$$

Noting that

$$\frac{1}{1 + e^{-2qL}} = \sum_{n=0}^{\infty} (-1)^n e^{-2qnL}, \tag{7-351}$$

we may write Eq. (7-350) in the form

$$\frac{\bar{\theta}(x, p)}{\theta_0} = \frac{1}{p} - \sum_{n=0}^{\infty} \frac{(-1)^n}{p} \{e^{-[(2n+1)L-x]q} + e^{-[(2n+1)L+x]q}\}. \tag{7-352}$$

Then the successive use of TP(30) readily gives the inversion of Eq. (7-352) as

$$\frac{T(x, t) - T_\infty}{T_0 - T_\infty} = 1 - \sum_{n=0}^{\infty} (-1)^n \left\{ \operatorname{erfc}\left[\frac{(2n+1) - x/L}{2(at/L^2)^{1/2}}\right] \right.$$

$$\left. - \operatorname{erfc}\left[\frac{(2n+1) + x/L}{2(at/L^2)^{1/2}}\right] \right\}. \tag{7-353}$$

Equation (7-353) is convergent for all values of time. When the dimensionless number, the so-called Fourier modulus $\mathrm{Fo} = at/L^2$, is less than 1 only the first two or three terms of Eq. (7-353) are needed for four-place accuracy; the other solution of the problem, given by Eq. (5-10), would require the consideration of hundreds of terms for the same order of accuracy.† On the other hand, for large values of Fo Eq. (7-353) is slowly convergent, yet the first few terms of Eq. (5-10) provide a sufficiently accurate answer. Thus for numerical computations Eqs. (5-10) and (7-353) complement each other, one being suitable to small values of time and the other to large values of time.

Example 7-24. Let us find a solution of Example 7-11 valid for small values of time.

According to Example 7-11 the solution of the transformed formulation is

$$\frac{\bar{\theta}(x, p)}{\theta_0} = \frac{1}{p} - \frac{\cosh qx}{p[\cosh qL + (k/h)q \sinh qL]}. \tag{7-257}$$

---

† Equation (5-10) is also convergent for all values of time.

Equation (7–257) may be expressed in terms of the exponential functions to give

$$\frac{\bar{\theta}(x, p)}{\theta_0} = \frac{1}{p} - \frac{e^{qx} + e^{-qx}}{p[(e^{qL} + e^{-qL}) + (k/h)q(e^{qL} - e^{-qL})]}. \quad (7\text{–}354)$$

Now, factoring the bracketed portion by e^{qL} and rearranging, we obtain

$$\frac{\bar{\theta}(x, p)}{\theta_0} = \frac{1}{p} - \frac{e^{-(L-x)q} + e^{-(L+x)q}}{p[(1 + qk/h) + (1 - qk/h)e^{-2qL}]}. \quad (7\text{–}355)$$

Finally, factoring the bracketed terms of Eq. (7–355) by $(1 + qk/h)$, we get

$$\frac{\bar{\theta}(x, p)}{\theta_0} = \frac{1}{p} - \frac{e^{-(L-x)q} + e^{-(L+x)q}}{p(1 + qk/h)[1 + (1 - qk/h)e^{-2qL}/(1 + qk/h)]}, \quad (7\text{–}356)$$

and noting that

$$\frac{1}{1 + (1 - qk/h)e^{-2qL}/(1 + qk/h)} = 1 - \left(\frac{1 - qk/h}{1 + qk/h}\right)e^{-2qL} + \cdots,$$

we may further rearrange Eq. (7–356) to give

$$\frac{\bar{\theta}(x, p)}{\theta_0} = \frac{1}{p} - \frac{h/k}{p(h/k + q)}[e^{-(L-x)q} + e^{-(L+x)q}]$$

$$+ \frac{(h/k)(h/k - q)}{p(h/k + q)^2}[e^{-(3L-x)q} + e^{-(3L+x)q}] + \cdots \quad (7\text{–}357)$$

Employing TP(1) and TP(37) we get the inversion of the first two terms of Eq. (7–357) in the form

$$\frac{T(x, t) - T_\infty}{T_0 - T_\infty} = 1 - \left\{\text{erfc}\left[\frac{1 - x/L}{2(at/L^2)^{1/2}}\right] + \text{erfc}\left[\frac{1 + x/L}{2(at/L^2)^{1/2}}\right]\right\}$$

$$+ e^{(hL/k)(1-x/L)+(hL/k)^2(at/L^2)} \text{erfc}\left[\left(\frac{hL}{k}\right)\left(\frac{at}{L^2}\right)^{1/2} + \frac{1 - x/L}{2(at/L^2)^{1/2}}\right]$$

$$+ e^{(hL/k)(1+x/L)+(hL/k)^2(at/L^2)} \text{erfc}\left[\left(\frac{hL}{k}\right)\left(\frac{at}{L^2}\right)^{1/2} + \frac{1 + x/L}{2(at/L^2)^{1/2}}\right] + \cdots$$

$$(7\text{–}358)$$

Because of the boundary condition given by Eq. (7–256), the terms of Eq. (7–357) become progressively more complicated algebraically. Consequently, we have considered the inversion of only the first two terms of this equation, and hence Eq. (7–358) is valid for relatively small values of time only.

The next example, taken from cylindrical geometry, demonstrates that even the simplest boundary conditions employed with this geometry result in series having progressively more complex terms.

Example 7-25. Let us find a solution of Example 7-16 useful for small values of time.

The solution of the transformed formulation may be taken from Example 7-16. Thus

$$\frac{\bar{\theta}(r, p)}{\theta_0} = \frac{1}{p} - \frac{I_0(qr)}{pI_0(qR)}. \tag{7-301}$$

Remembering that small values of time correspond to large p (or q), we employ the asymptotic expansions of $I_0(qr)$ and $I_0(qR)$ [see Eq. (3-156)], and obtain

$$\frac{\bar{\theta}(r, p)}{\theta_0} = \frac{1}{p} - \frac{[e^{qr}/(2\pi qr)^{1/2}](1 + 1/8qr + 9/128q^2r^2 + \cdots)}{p[e^{qR}/(2\pi qR)^{1/2}](1 + 1/8qR + 9/128q^2R^2 + \cdots)}. \tag{7-359}$$

Dividing the numerator by the denominator, we may rearrange Eq. (7-359) in the form

$$\frac{\bar{\theta}(r, p)}{\theta_0} = \frac{e^{-q(R-r)}}{(r/R)^{1/2}p}\left(1 + \frac{R-r}{8Rrq} + \frac{9R^2 - 2Rr - 7r^2}{128R^2r^2q^2} + \cdots\right). \tag{7-360}$$

The use of TP(30), TP(31), and TP(32) in Eq. (7-360) gives

$$\frac{T(r, t) - T_\infty}{T_0 - T_\infty} = \frac{1}{(r/R)^{1/2}} \operatorname{erfc}\left[\frac{1 - r/R}{2(at/R^2)^{1/2}}\right]$$

$$+ \frac{(1 - r/R)(at/R^2)^{1/2}}{4(r/R)^{3/2}} i \operatorname{erfc}\left[\frac{1 - r/R}{2(at/R^2)^{1/2}}\right]$$

$$+ \frac{9 - 2(r/R) - 7(r/R)^2}{32(r/R)^{5/2}}\left(\frac{at}{R^2}\right)^{1/2} i^2 \operatorname{erfc}\left[\frac{1 - r/R}{2(at/R^2)^{1/2}}\right] + \cdots, \tag{7-361}$$

where $i \operatorname{erfc} x = \int_x^\infty \operatorname{erfc} \xi \, d\xi$ and $i^2 \operatorname{erfc} x = \int_x^\infty i \operatorname{erfc} \xi \, d\xi$. Since we have used the asymptotic expansion of the Bessel functions in Eq. (7-359), the validity of Eq. (7-361) for small values of time must be so restricted that r/R is not too small, and qr always remains large (see Problem 7-38).

Again, the problems of spherical geometry are not considered because they may be reduced to those of cartesian geometry by the transformation $\theta(r, t) = \psi(r, t)/r$.

Since the solutions obtained by the separation of variables or by the inversion theorem converge rapidly for large values of time, and those obtained by the expansion of the solution of the transformed formulation in powers of $1/p$ converge rapidly for small values of time, the simultaneous use of both types of solution would be convenient and sufficient for numerical investigation of our problems. It remains to be determined, however, what happens when we expand the solution of a transformed formulation in powers of p rather than in powers of $1/p$. We know, of course, that the new solution would be valid for *small $p \to$ large t*. What we are interested in is the availability of this solution; this will be discussed next.

2. *Solutions useful for large values of time.* First, three illustrative examples are worked out; later the procedure is generalized.

Let us reconsider Example 7–22, and the solution of its transformed formulation

$$\bar{\theta}(p) = \frac{n}{p(p+m)}. \qquad (7\text{–}346)$$

Equation (7–346) can be readily expanded in powers of p in the form

$$\bar{\theta}(p) = \frac{n}{mp(1+p/m)} = \frac{n}{m}\left(\frac{1}{p} - \frac{1}{m} + \frac{p}{m^2} - \frac{p^2}{m^3} + \cdots\right). \qquad (7\text{–}362)$$

Noting that all terms of the series of Eq. (7–362) except the first one are analytic, we get immediately from the inversion theorem

$$T(t) - T_\infty = n/m = u'''L/h,$$

which is the steady solution of the problem. Hence in place of an asymptotic solution, which would give the behavior of the present problem for large values of time, we obtained the asymptote, namely, the steady solution itself.

Second, let us reconsider Example 7–10, whose transformed solution is

$$\frac{\bar{\theta}(x,p)}{\theta_0} = \frac{1}{p} - \frac{\cosh qx}{p \cosh qL}. \qquad (7\text{–}248)$$

Now, expanding the numerator and the denominator of the second term on the right-hand side of Eq. (7–248) in powers of q (or p), we have

$$\frac{\bar{\theta}(x,p)}{\theta_0} = \frac{1}{p} - \frac{1 + q^2 x^2/2! + q^4 x^4/4! + \cdots}{p(1 + q^2 L^2/2! + q^4 L^4/4! + \cdots)}. \qquad (7\text{–}363)$$

Equation (7–363) may be rearranged in the usual manner to give

$$\frac{\bar{\theta}(x,p)}{\theta_0} = \frac{1}{2a}(L^2 - x^2) - \frac{p}{4a^2}(L^2 - x^2)[L^2 - \tfrac{1}{6}(L^2 + x^2)] + \cdots \qquad (7\text{–}364)$$

Since all terms of Eq. (7–364) are analytic, the inversion theorem yields readily $[T(x,t) - T_\infty]/(T_0 - T_\infty) = 0$, which again is the steady solution of the problem.

In fact, term-by-term inversion of the expansion of the transformed solution in powers of p gives the asymptote† rather than the asymptotic behavior for

---

† Some problems do not have a steady solution. In this case the asymptote implies the form of the solution as $t \to \infty$. See Example 7–13.

all lumped problems, and for all distributed problems that are solvable by Contour I of the inversion theorem. Remember that Contour I applies to problems whose transforms do not possess any branch points, that is, to problems whose geometry is finite in their orthogonal directions; these problems are also solvable by the separation of variables. However, for distributed problems whose transformed solutions do possess branch points, that is, for problems whose solutions may be obtained by Contour II of the inversion theorem,† the expansion of the transformed solution in powers of p and term-by-term inversion of the result gives the asymptotic behavior, rather than the asymptote. Let us illustrate this point by the following example.

Example 7–26. Let us find a solution of Example 7–3 valid for large values of time.

The solution of the transformed formulation of the problem may be taken from Example 7–3,

$$\frac{\bar{\theta}(x, p)}{\theta_0} = \frac{1}{p} - \frac{1}{p} e^{-qx}. \tag{7-51}$$

Expanding the second term on the right-hand side of Eq. (7–51) in powers of q, we obtain

$$\frac{\bar{\theta}(x, p)}{\theta_0} = \frac{x}{a^{1/2} p^{1/2}} - \frac{x^2}{a 2!} + \frac{p^{1/2} x^3}{a^{3/2} 3!} - \frac{p x^4}{a^2 4!} + \frac{p^{3/2} x^5}{a^{5/2} 5!} - \cdots \tag{7-365}$$

Eliminating its analytic terms, we may readily find the inversion of the remainder of Eq. (7–365) by TP(4), TP(8), and TP(9). The result is

$$\frac{T(x, t) - T_\infty}{T_0 - T_\infty} = \frac{1}{\pi^{1/2}} \left[\left(\frac{x^2}{at}\right)^{1/2} - \frac{1}{2 \cdot 3!} \left(\frac{x^2}{a}\right)^{3/2} + \frac{3}{4 \cdot 5!} \left(\frac{x^2}{at}\right)^{5/2} - \cdots \right]. \tag{7-366}$$

Thus, by contrast with the first two examples, we obtained the asymptotic behavior rather than the asymptote of our problem for large values of time.

We may now generalize the results of the foregoing three problems. Let the solution of the transformed formulation of a problem, say $T(\mathbf{r}, t)$, be $\bar{T}(\mathbf{r}, p)$, $\mathbf{r}$ being the position vector. If $\bar{T}(\mathbf{r}, p)$ has a number of singularities, an asymptotic expansion of $T(\mathbf{r}, t)$ can be deduced from the behavior of $\bar{T}(\mathbf{r}, p)$ near its singularity with largest real part.‡ Suppose this singularity to be at $p = \gamma_0$,

---

† It should be remembered that these problems cannot be solved by the separation of variables and the Fourier series (or the functions orthogonal over a finite interval) of Chapters 4 and 5. Only a few simple cases may be solved by these methods, and the procedure is somewhat elaborate even for simple problems. See, for example, Example 7–8.

‡ Note that relative to the possible roots on the negative half (most probably on the negative real axis) of the complex plane, $p = 0$ is the singularity with largest real part involved in the series expansions of the transforms of the last three examples.

where $\gamma_0 < \gamma$.† Then assume that $\overline{T}(\mathbf{r}, p)$ can be expanded near $p = \gamma_0$ in the form

$$\overline{T}(\mathbf{r}, p) = \sum_{n=0}^{\infty} a_n(\mathbf{r})(p - \gamma_0)^{n-1} + (p - \gamma_0)^{\beta-1} \sum_{n=0}^{\infty} b_n(\mathbf{r})(p - \gamma_0)^n, \qquad (7\text{-}367)$$

where $0 < \beta < 1$. All terms of the first series of Eq. (7-367) except the first term are analytic and do not contribute to the solution. The inversion of the first term of the first series gives the asymptote of the problem. This asymptote plus the inversion of the second series and its rearrangement by the well-known property of the Gamma functions $\Gamma(\nu)\Gamma(1 - \nu) = \pi/\sin \pi\nu$, where ν may be a noninteger, give the asymptotic behavior of $T(\mathbf{r}, t)$ for large values of time in the form

$$T(\mathbf{r}, t) = e^{\gamma_0 t}\left[a_0(\mathbf{r}) + \frac{\sin \pi\beta}{\pi} \sum_{n=0}^{\infty} (-1)^n b_n(\mathbf{r})\Gamma(n + \beta)t^{-(n+\beta)}\right]. \qquad (7\text{-}368)$$

If there are several singularities with the same real part, there must be a term of the type found in Eq. (7-368) corresponding to each of them.

The most frequent special case of Eq. (7-367) corresponds to $\gamma_0 = 0$, for which Eq. (7-367) becomes

$$\overline{T}(\mathbf{r}, p) = \sum_{n=0}^{\infty} a_n(\mathbf{r})p^{n-1} + p^{\beta-1} \sum_{n=0}^{\infty} b_n(\mathbf{r})p^n, \qquad (7\text{-}369)$$

and then Eq. (7-368) yields

$$T(\mathbf{r}, t) = a_0(\mathbf{r}) + \frac{\sin \pi\beta}{\pi} \sum_{n=0}^{\infty} (-1)^n b_n(\mathbf{r})\Gamma(n + \beta)t^{-(n+\beta)}. \qquad (7\text{-}370)$$

Furthermore, if $\beta = \frac{1}{2}$, Eq. (7-369) is reduced to

$$\overline{T}(\mathbf{r}, p) = \sum_{n=0}^{\infty} a_n(\mathbf{r})p^{n-1} + \sum_{n=0}^{\infty} b_n(\mathbf{r})p^{n-1/2}, \qquad (7\text{-}371)$$

and Eq. (7-370) gives

$$T(\mathbf{r}, t) = a_0(\mathbf{r}) + \frac{1}{\pi^{1/2}} \sum_{n=0}^{\infty} (-1)^n b_n(\mathbf{r})(n - \tfrac{1}{2})(n - \tfrac{3}{2}) \cdots \tfrac{3}{2} \cdot \tfrac{1}{2} t^{-(n+1/2)}. \qquad (7\text{-}372)$$

Since the second series of Eq. (7-369) is zero for the first two examples, only the asymptotes of these examples are obtained rather than their asymptotic

---

† See Fig. 7-20 for the definition of γ.

behaviors. On the other hand, Eq. (7–365) of the third example may be rearranged in the form of Eq. (7–371), which implies the existence of a solution giving the asymptotic behavior of the problem.

3. *Periodic solutions.* In Chapter 5 our attempt to solve problems subject to periodic disturbances by the separation of variables led us to a somewhat involved procedure. Despite the complexity of the procedure, we were able to obtain only the steady part of the periodic solution of a simple problem such as Example 5–6. Then in Chapter 6 we learned a convenient procedure for getting the steady solution of periodic problems, namely, the method of complex temperature; however, the complete solution of periodic problems has not yet been treated. Here we discuss the complete solution of these problems by the inversion theorem, and interpret their steady (periodic) part by means of their asymptotic behavior. Since no new concept is needed, these problems may best be explained by a couple of illustrative examples. First, however, let us note that a term (residue) contributed to the solution of a problem by a *simple* pole may be interpreted in physical terms by considering the location of this pole on the complex plane. The residue evaluated at a pole by Eq. (7–229) may depend on time only through the exponential term e^{tz} of the inversion theorem; therefore, interpreting e^{tz} by the location of a pole, we may draw the following conclusions when a problem under consideration has a simple pole: (a) at the origin, the contribution of this pole to the solution of the problem is a steady term; (b) on the negative real axis, the contribution is an exponentially decaying term; (c) on the imaginary axis, the contribution is a periodic term with constant amplitude; (d) on the negative half containing parts both on the imaginary and the negative real axes, the contribution is a periodic term with decaying amplitude.

Example 7–27. Let us find the complete periodic solution of the problem of Example 5–1 by the use of the inversion theorem.

The formulation of the problem is

$$\frac{d\theta}{dt} + m\theta = n \cos \omega t, \qquad (5\text{–}1)$$

$$\theta(0) = 0. \qquad (5\text{–}2)$$

The solution of the transformed formulation may readily be obtained in the form

$$\bar{\theta}(p) = \frac{np}{(p+m)(p^2+\omega^2)}. \qquad (7\text{–}373)$$

Inserting Eq. (7–373) into the inversion theorem, we find that the solution is

$$\frac{\theta(t)}{n} = \frac{1}{2\pi i} \int_{\gamma-i\infty}^{\gamma+i\infty} \frac{z e^{tz} \, dz}{(z+m)(z^2+\omega^2)}. \qquad (7\text{–}374)$$

The integrand of Eq. (7–374) has simple poles at $z = -m$ and at $z = \pm i\omega$. Thus from Contour I we have

$$\frac{\theta(t)}{n} = \text{Res}\,(-m) + \text{Res}\,(\pm i\omega). \tag{7-375}$$

The residue at $z = -m$ is

$$\text{Res}\,(-m) = -\frac{me^{-mt}}{(m^2 + \omega^2)},$$

and the residues at $z = \pm i\omega$ are

$$\text{Res}\,(i\omega) = \frac{e^{i\omega t}}{2(m + i\omega)}, \qquad \text{Res}\,(-i\omega) = \frac{e^{-i\omega t}}{2(m - i\omega)}.$$

Using these results in Eq. (7–375), we obtain the previously found solution

$$\frac{T(t) - T_\infty}{n/m} = \frac{m}{(m^2 + \omega^2)^{1/2}} \cos(\omega t - \alpha) - \frac{m^2 e^{-mt}}{(m^2 + \omega^2)}, \quad \alpha = \tan^{-1}(\omega/m). \tag{5-3}$$

As we have learned, the asymptotic behavior of Eq. (5–3) may be deduced from the behavior of Eq. (7–373) near its singularities with the largest real part, that is, near the two poles on the imaginary axis, $z = \pm i\omega$ (Fig. 7–26). Thus the first term on the right-hand side of Eq. (5–3), corresponding to the sum of the residues at $z = \pm i\omega$ only, is periodic with a constant amplitude, that is, it constitutes the steady part of the complete periodic solution.

Finally, let us consider the complete periodic solution of Example 5–6, which has been avoided so far.

Example 7–28. Let us find the complete periodic solution of Example 5–6 by the inversion theorem.

The formulation of the problem may be taken from Example 5–6. Thus

$$\frac{\partial \theta}{\partial t} = a \frac{\partial^2 \theta}{\partial x^2}, \tag{5-35}$$

$$\theta(x, 0) = 0, \tag{5-36}$$

$$\theta(0, t) = \theta_0 \cos \omega t, \tag{5-37}$$

$$\theta(\infty, t) = 0. \tag{5-38}$$

The solution of the transformed formulation is

$$\frac{\bar{\theta}(x, p)}{\theta_0} = \frac{pe^{-qx}}{(p^2 + \omega^2)}, \tag{7-376}$$

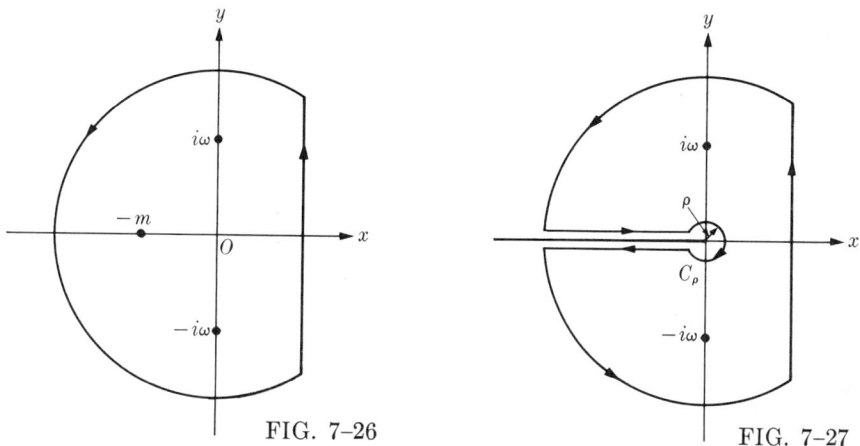

FIG. 7-26   FIG. 7-27

and the inversion theorem gives

$$\frac{\theta(x, t)}{\theta_0} = \frac{1}{2\pi i} \int_{\gamma-i\infty}^{\gamma+i\infty} \frac{z e^{tz-(z/a)^{1/2}x}}{(z^2 + \omega^2)} \, dz. \quad (7\text{-}377)$$

The integrand of Eq. (7-377) has a branch point at $z = 0$ and two simple poles at $z = \pm i\omega$ (Fig. 7-27). Thus from Contour II and Eq. (7-236) we have

$$\frac{\theta(x, t)}{\theta_0} = \text{Res} \, (\pm i\omega) + \lim_{\rho \to 0} \int_{C_\rho} \frac{z e^{tz-(z/a)^{1/2}x}}{(z^2 + \omega^2)} \, dz$$

$$- \frac{1}{2\pi i} \int_0^\infty \frac{\rho e^{-t\rho} [e^{i(\rho/a)^{1/2}x} - e^{-i(\rho/a)^{1/2}x}]}{(\rho^2 + \omega^2)} \, d\rho. \quad (7\text{-}378)$$

The residues at $z = \pm i\omega$ are

$$\text{Res} \, (i\omega) = \tfrac{1}{2} e^{i\omega t - (1+i)(\omega/2a)^{1/2}x}, \qquad \text{Res} \, (-i\omega) = \tfrac{1}{2} e^{-i\omega t - (1-i)(\omega/2a)^{1/2}x}.$$

The sum of these residues,

$$\text{Res} \, (i\omega) + \text{Res} \, (-i\omega) = e^{-(\omega/2a)^{1/2}x} \cos [\omega t - (\omega/2a)^{1/2}x], \quad (5\text{-}44)$$

is the steady part of the solution. Since its integrand approaches zero as ρ (or z) $\to 0$, the integral around the small circle C_ρ vanishes.

Employing these results in Eq. (7-378), and expressing the last integral of this equation in terms of a circular function, we obtain finally

$$\frac{\theta(x, t)}{\theta_0} = e^{-(\omega/2a)^{1/2}x} \cos [\omega t - (\omega/2a)^{1/2}x] - \frac{1}{\pi} \int_0^\infty \frac{\rho e^{-t\rho} \sin (\rho/a)^{1/2}x}{(\rho^2 + \omega^2)} \, d\rho. \quad (7\text{-}379)$$

The integral involved in Eq. (7-379) cannot be integrated explicitly; a numerical integration procedure may be employed for its parametric study.

References

1. H. S. Carslaw and J. C. Jaeger, *Conduction of Heat in Solids*. Oxford: Clarendon Press, 1959.
2. H. S. Carslaw and J. C. Jaeger, *Operational Methods in Applied Mathematics*. Oxford: Clarendon Press, 1948.
3. N. W. McLachlan, *Complex Variable Theory and Transform Calculus*. Cambridge, England: Cambridge University Press, 1953.
4. N. W. McLachlan, *Modern Operational Calculus*. New York: Macmillan, 1948.
5. R. V. Churchill, *Operational Mathematics*. New York: McGraw-Hill, 1958.
6. R. V. Churchill, *Complex Variables and Applications*. New York: McGraw-Hill, 1960.
7. W. Kaplan, *Operational Methods for Linear Systems*. Reading, Mass.: Addison-Wesley, 1962.
8. F. B. Hildebrand, *Advanced Calculus for Engineers*. Englewood Cliffs: Prentice-Hall, 1948.
9. E. J. Scott, *Transform Calculus*. New York: Harper, 1955.
10. G. A. Campbell and R. M. Foster, *Fourier Integrals for Practical Applications*. Princeton: Van Nostrand, 1948.
11. A. Erdélyi, *Tables of Integral Transforms* I. New York: McGraw-Hill, 1954.
12. M. Jakob, *Heat Transfer* II. New York: Wiley, 1957.
13. J. W. Rizika, "Thermal Lags in Flowing Incompressible Fluid Systems Containing Heat Capacitors." *Trans. ASME*, **78,** 1407 (1956).
14. J. A. Clark, V. S. Arpaci, and K. M. Treadwell, "Dynamic Response of Heat Exchangers Having Internal Heat Sources I." *Trans. ASME*, **80,** 612 (1958).
15. V. S. Arpaci and J. A. Clark, "Dynamic Response of Heat Exchangers Having Internal Heat Sources II." *Trans. ASME*, **80,** 625 (1958).
16. V. S. Arpaci and J. A. Clark, "Dynamic Response of Heat Exchangers Having Internal Heat Sources III." *Trans. ASME, C, Journal of Heat Transfer*, **81,** 253 (1959).
17. W. J. Yang, J. A. Clark, and V. S. Arpaci, "Dynamic Response of Heat Exchangers Having Internal Heat Sources IV." *Trans. ASME, C, Journal of Heat Transfer*, **83,** 321 (1961).
18. V. S. Arpaci, J. A. Clark, and W. O. Winer, "Dynamic Response of Fluid and Wall Temperatures During Pressurized Discharge of a Liquid from a Container." *Advances in Cryogenic Engineering*, Plenum, **6,** 310 (1961).
19. V. S. Arpaci and J. A. Clark, "Dynamic Response of Fluid and Wall Temperatures During Pressurized Discharge for Simultaneous Time-Dependent Inlet Gas Temperature, Ambient Temperature, and/or Ambient Heat Flux." *Advances in Cryogenic Engineering*, Plenum, **7,** 419 (1962).
20. V. S. Arpaci, "Transient Conduction in Coaxial Cylinders with Relative Motion and Heat Generation." *Trans. ASME, E, Journal of Applied Mechanics*, **27,** 623 (1960).

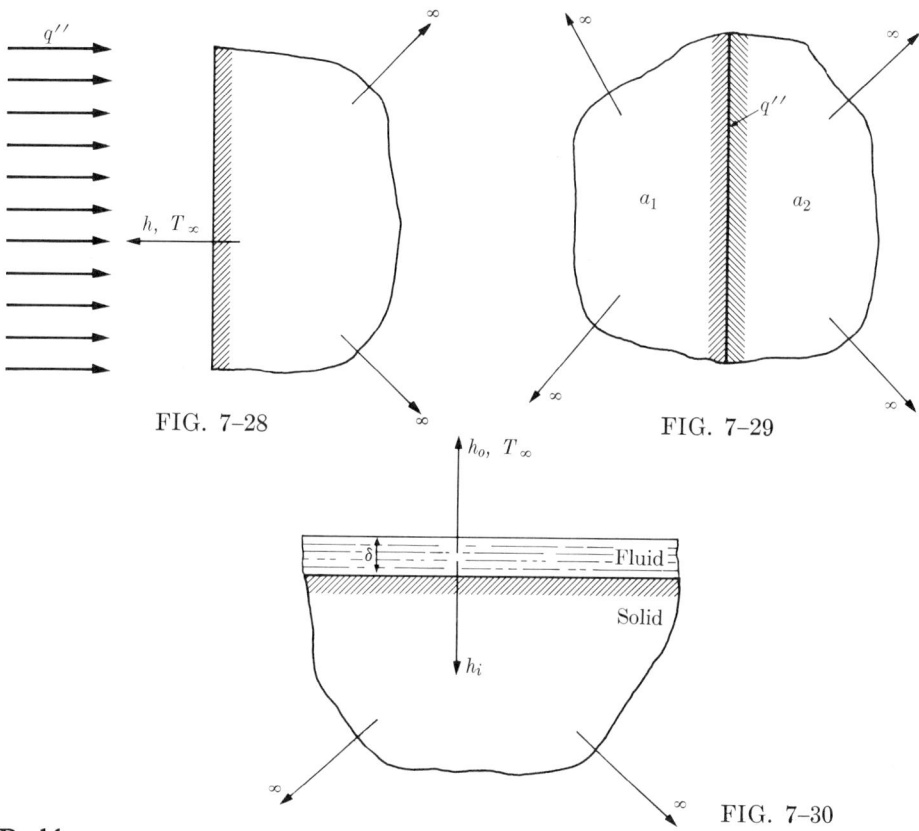

FIG. 7-28

FIG. 7-29

FIG. 7-30

Problems

7-1. A semi-infinite solid is suddenly subjected to the radiant heat flux q'' acting at a distance (Fig. 7-28). Initially the solid is at the ambient temperature. The heat transfer coefficient between the solid and the ambient is h. Find the unsteady temperature in the system.

7-2. Find the unsteady temperature in the problem of Example 3-9 corresponding to a sudden change in the base temperature from 0 to θ_0.

7-3. An infinite medium is composed of two semi-infinite solids having different diffusivities, say a_1 and a_2 (Fig. 7-29). The initial temperature of the system is uniform; then the interface suddenly assumes the heat flux q'', which may result from an electrically heated plate of negligible thickness. Find the unsteady temperature in the solids.

7-4. A fluid at temperature T_0 is suddenly poured over a semi-infinite solid at temperature T_∞. Assume that the fluid instantaneously becomes a thin layer of thickness δ covering the entire surface of the solid (Fig. 7-30). The heat transfer coefficient between the fluid and the ambient is h_o. Find the unsteady temperature in the system, assuming that the heat transfer coefficient h_i between the fluid and the solid is (a) infinite, (b) finite.

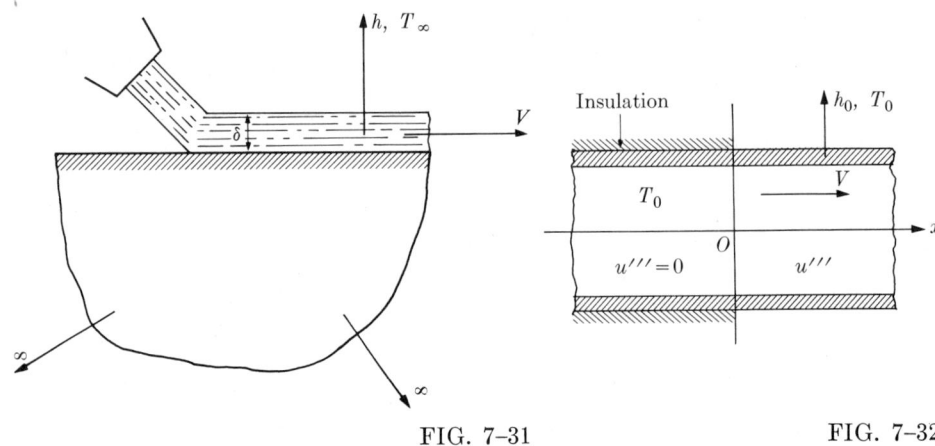

FIG. 7-31   FIG. 7-32

7-5. A thin layer of fluid having the thickness δ, velocity V, and temperature T_0 attaches to a semi-infinite solid at a point, and thereafter flows over the solid (Fig. 7-31). The initial temperature of the solid is T_∞. Let h denote the heat transfer coefficient between the fluid and the ambient, and assume that the heat transfer coefficient between the liquid and the solid is large. Conduction is negligible in the direction of flow, both in the fluid and in the solid. Find the unsteady temperature in the system.

7-6. A nuclear fluid at temperature T_0 flows steadily through an infinitely long tube (Fig. 7-32). The upstream half of the tube is insulated, while the downstream half may transfer heat with a heat transfer coefficient h_0 to an ambient at temperature T_0. Assume now that constant nuclear energy u''' is uniformly generated in the downstream half of the fluid. Axial conduction is negligible both in the fluid and in the tube walls. Find the unsteady temperatures in the system, on the basis of a radially lumped and axially distributed analysis.

7-7. Re-solve Example 7-7 in terms of an integral formulation and appropriately selected approximate temperature profiles.

7-8. Re-solve the insulated case of Example 7-5, assuming that the inlet temperature T_0 of the liquid is changing as shown in Fig. 7-33, rather than being changed suddenly.

7-9. A cross-flow heat exchanger may be simulated by three parallel plates of negligible thicknesses. The upper and lower surfaces of the exchanger are insulated. Assume that the fluids flow steadily with the velocities U_1 and U_2 (Fig. 7-34), that the inlet temperatures of the fluids, T_{1_0} and T_{2_0}, are uniform, and that axial conduction is negligible. Find the *steady* temperature in the system on the basis of a transversely lumped but otherwise differential analysis.

7-10. The unsteady thermal behavior of closed containers during discharge of the container fluid by a pressurant gas finds important application in current problems of space technology. The following analytical model is suggested for the investigation of this problem (Fig. 7-35). (i) The top and the bottom of the container are insulated. (ii) Axial conduction in the system is negligible. (iii) The physical properties are assumed to be constant. (iv) The time dependence of the pressurant gas velocity is

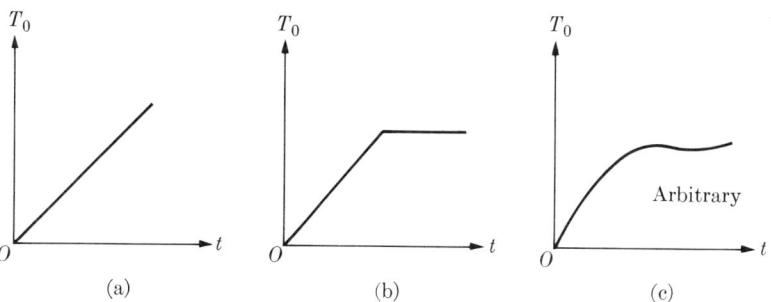

FIG. 7-33

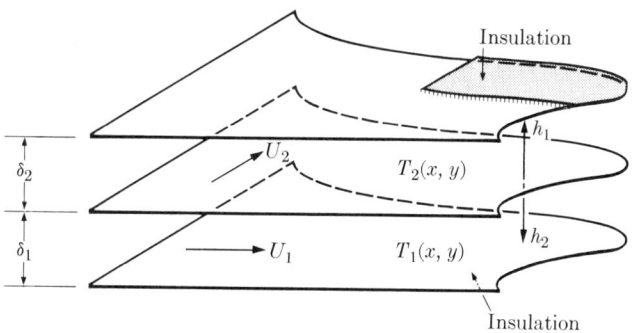

FIG. 7-34

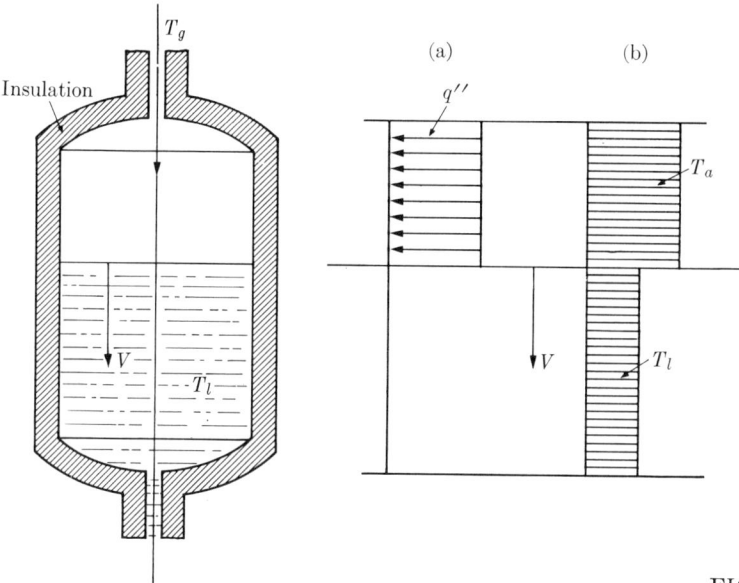

FIG. 7-35

negligible compared to that of its temperature. (v) The heat transfer between the fluid and the ambient is small compared to that between the pressurant gas and the ambient.† (vi) The periphery of the container is subjected to a uniform heat flux q'', or transfers heat to an ambient at temperature T_a. Since the heat transfer below the gas-fluid interface is negligible according to (v), q'' and T_a may be assumed to be moving with the interface velocity as shown in Fig. 7–35. (vii) Initially the container is filled with a cryogenic fluid at temperature T_l. Then the fluid is drained out of the container with constant velocity V by a pressurized gas having the inlet temperature T_g. Find the unsteady temperatures in the system on the basis of a radially lumped and axially distributed analysis.

7–11. Two fluids flow steadily in a parallel-flow heat exchanger (Fig. 7–36). The inlet temperature and the velocity of the fluids are T_i, θ_i and U, V, respectively. The tube walls which separate the fluids have negligible heat capacity.‡ The outer surface of the heat exchanger is insulated. (a) Find the steady temperatures in the system on the basis of a radially lumped and axially distributed analysis. (b) Assume that the inlet temperature of one fluid is suddenly changed to another temperature, say from T_i to T_i'. Find the unsteady temperatures in the system.

7–12. Re-solve Problem 7–11 for (i) an infinitely long, (ii) a finite counterflow heat exchanger.

7–13. Re-solve Example 3–10 by the use of Laplace transforms. The following procedure is suggested. Starting from the formulation of the problem,

$$d^2\theta/dx^2 - m^2\theta = 0, \qquad (3\text{–}163)$$

$$d\theta(0)/dx = 0, \qquad (3\text{–}170)$$

$$\theta(L) = \theta_0, \qquad (3\text{–}171)$$

integrate Eq. (3–163) twice over the interval $(0, x)$ to get

$$\theta(x) = \theta(0) + m^2 \int_0^x \int_0^x \theta(\xi)\, d\xi\, d\xi, \qquad (7\text{–}380)$$

where ξ is the dummy variable; using Leibnitz's integral formula,§ show from the integral

$$I = \int_0^x (x - \xi)\theta(\xi)\, d\xi$$

that

$$\int_0^x \int_0^x \theta(\xi)\, d\xi\, d\xi = \int_0^x (x - \xi)\theta(\xi)\, d\xi; \qquad (7\text{–}381)$$

then rearrange Eq. (7–380) by considering Eq. (7–381); take the Laplace transforms of the result in x; eliminate $\theta(0)$ from the solution by using Eq. (3–171).

7–14. Find the *unsteady* behavior of the system of Problem 7–9 corresponding to a sudden change in the inlet temperature of one fluid. [*Hint:* Use Laplace trans-

---

† See Reference 18.
‡ See Problem 5–5 for the formulation of the problem corresponding to the case of counterflow and including the heat capacity of the tube walls.
§ See Problem 2–8.

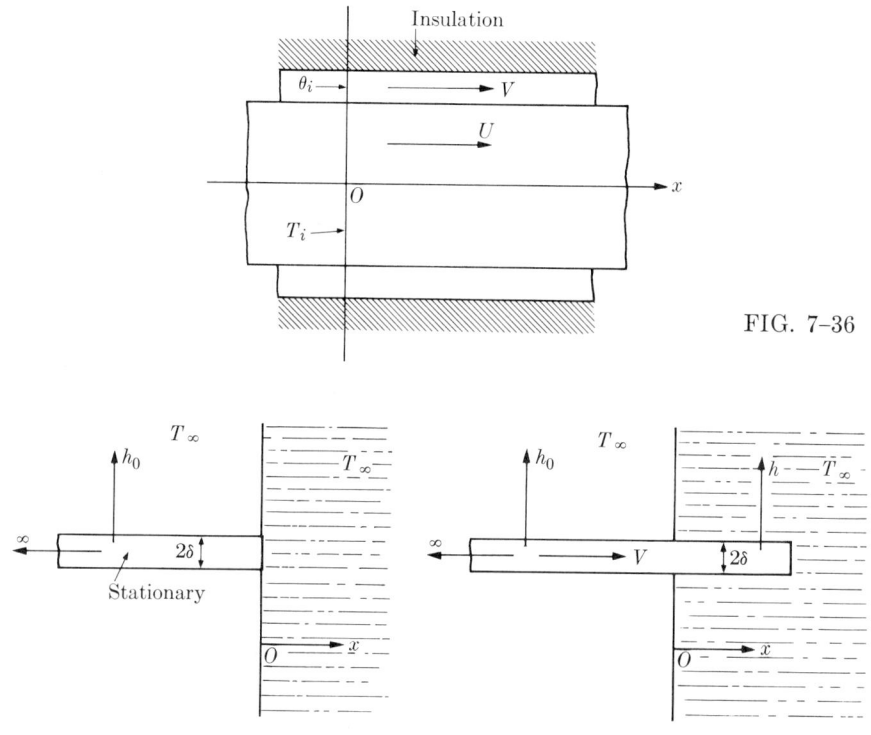

FIG. 7-36

FIG. 7-37

forms successively in the time and the space in which the inlet temperature of the fluid is kept unchanged.]

7-15. Consider a semi-infinite ($x < 0$) metal sheet of thickness 2δ. The *instantaneous* initial temperature of the sheet is T_0; then the sheet suddenly assumes a uniform velocity V and moves into the second half ($x > 0$) of an infinite domain (Fig. 7-37). The ambient temperature of the first and second half-domains is the same, say T_∞. The second half-domain is a liquid metal bath. The momentum (velocity) of the moving sheet penetrates to a small depth in the liquid metal. This depth may be neglected for the intended investigation, and the entire second half-domain may be assumed stationary. *Axial conduction is negligible.* Find the unsteady temperature in the second half-domain corresponding to (i) an infinite heat transfer coefficient h, (ii) a finite heat transfer coefficient h.

7-16. Re-solve Problem 5-9.

7-17. Re-solve Problem 5-10.

7-18. Re-solve Problem 5-11.

7-19. Re-solve Problem 5-12.

7-20. The outside walls of a house are assumed to be made of composite walls consisting of a flat plate of thickness L and a semi-infinite body (Fig. 7-38). On a cloudy summer day the inside and the outside of the house may be assumed to have

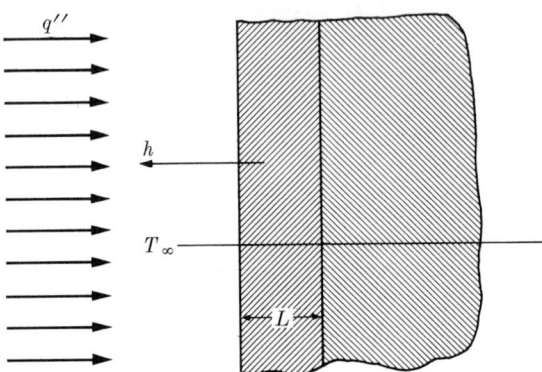

FIG. 7-38

the same temperature, T_∞. During the day the sun suddenly starts to shine and imposes the uniform heat flux q'' on the walls for a period of time. Find the unsteady temperature in the outside walls.

7-21. Re-solve Problem 5-15.

7-22. Re-solve Problem 5-16.

7-23. Re-solve Problem 5-17.

7-24. Re-solve Example 7-19, assuming that the tube is not insulated and that it transfers heat with a heat transfer coefficient h_0 to the ambient at temperature T_∞.

7-25. Re-solve Problem 5-18.

7-26. The process of drilling rocks may be simulated by an infinite solid having a cylindrical hole that contains an infinitely long drill (Fig. 7-39). Thus the axial motion of the drill is eliminated from the formulation of the problem. The initial temperature of the system is uniform, say T_∞; then the drill starts drawing a constant power from the power supply. It may be assumed that the total power drawn by the drill is uniformly transferred to the solid. Find the unsteady temperature in the solid.

7-27. Re-solve the preceding problem, assuming that the drill is a solid rod and that only a part of the total power drawn from the power supply is (uniformly) transferred to the solid.

7-28. Re-solve Problem 5-23.

7-29. Re-solve Problem 5-24.

7-30. Assume that the radiant energy generated in a spherical hole in an infinite solid may be considered as a sudden heat flux imposed on the surface of the hole (Fig. 7-40). The initial temperature of the solid is uniform, say T_∞. Find the unsteady temperature in the solid.

7-31. Re-solve Problem 5-27.

7-32. Re-solve Example 7-21 by means of the Laplace transforms successively considered in time and in the space (direction) in which the geometry of the problem extends to infinity. [*Hint:* First establish the property

$$L\left\{\frac{d^2 f(t)}{dt^2}\right\} = p^2 \bar{f}(p) - pf(0+) - \frac{df(0+)}{dt}$$

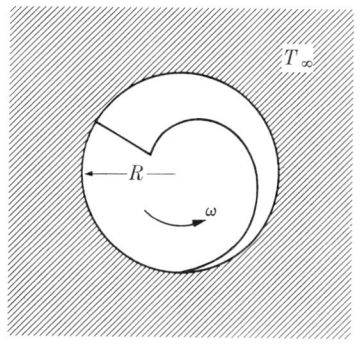
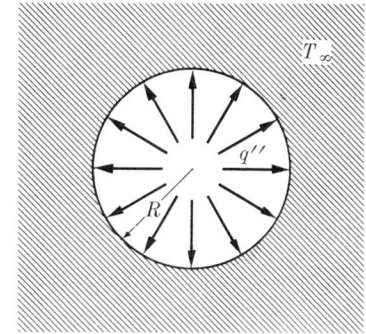

FIG. 7-39   FIG. 7-40

in a manner analogous to the establishment of TP(II). Recall that t does not necessarily imply time.]

7-33. Solve the unsteady problem associated with Example 4-21, assuming that the initial temperature of the rod is zero and that the base temperature is suddenly changed to θ_0.

7-34. Re-solve Example 5-15.

7-35. Are there any solutions of Problem 7-1 through Problem 7-5 useful for small or large values of time? First state your answers for each problem. Next consider at least one of these problems and, employing the method of solution suitable to small or large values of time, back up your answers.

7-36. Repeat Problem 7-35 for Problems 7-6 through 7-12 except Problems 7-7 and 7-9.

7-37. Repeat Problem 7-35 for Problems 7-15 through 7-20.

7-38. Find a solution of Example 7-25 for small values of time and small values of r/R. [*Hint:* Use the small-argument expansion of the numerator of the transformed solution.]

7-39. Repeat Problem 7-35 for Problems 7-21 through 7-27.

7-40. Repeat Problem 7-35 for Problems 7-28 through 7-34.

7-41. Find solutions of Problem 4-25 useful for small or large values of time corresponding to a sudden change in the internal energy generation in the rod from 0 to u'''.[†]

7-42. Find the transient periodic solution of Problem 6-6.

7-43. Find the transient periodic solution of Example 6-4.

7-44. Find the transient periodic solution of Problem 6-7.

7-45. Find the transient periodic solution of Problem 6-9.

7-46. Find the transient periodic solution of Example 6-3.

---

† See Reference 20.

PART III | FURTHER METHODS OF FORMULATION AND SOLUTION

CHAPTER 8

VARIATIONAL FORMULATION.
SOLUTION BY APPROXIMATE PROFILES

In Chapter 2 we classified the formulations of problems as *lumped* and *distributed* (integral, differential, variational, and difference). So far we have discussed the lumped, integral, and differential formulations, and the solution methods suitable to each. Since both the variational and difference formulations can be obtained from the differential formulation by means of mathematics, we have postponed the study of these formulations until the necessary mathematical foundation could be laid. This chapter is devoted to variational formulation and solution, and Chapters 9 and 10 to difference formulation and solution.

Here we first introduce the concept of variational calculus and briefly develop its properties; then we shall apply the results to some problems. Also, we shall discuss at length the selection of profiles, which until now has been loosely treated.

8–1. Basic Problem of Variational Calculus

An important problem of differential calculus is to find the extreme (stationary) values of a function. Given, for example, a function of one independent variable, $y = y(x)$, the necessary condition for the existence of extreme values of this function is $dy/dx = 0$. The sufficient condition for the function to be minimum or maximum is $d^2y/dx^2 > 0$ or $d^2y/dx^2 < 0$, respectively, at the point where $dy/dx = 0$. Variational calculus owes its origin to a similar but more complicated problem, that of finding a function $y(x)$ such that a definite integral whose integrand is a function of this function shall assume an extreme value. The necessary condition for the existence of an extreme value of the integral is the subject matter of this chapter. (The sufficient condition for the extreme value to correspond to a minimum or a maximum may be obvious for a given physical problem, but it is rather difficult to prove and will not be considered here.)

As an introduction, we present one of the classical examples of variational calculus, a problem which readily permits physical interpretation of the extremum being sought; let us find the equation of the shortest plane curve joining two given points (Fig. 8–1). The differential length of a curve $y = y(x)$ connecting these points is

$$ds = (dx^2 + dy^2)^{1/2},$$

which can be rearranged as

$$ds = (1 + y'^2)^{1/2}.$$

The total length of this curve,

$$S = \int_a^b (1 + y'^2)^{1/2}\, dx, \quad (8\text{--}1)$$

is now required to be a minimum.

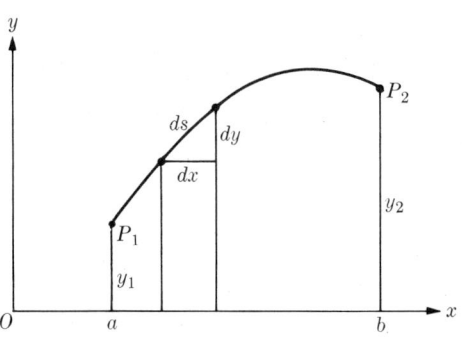

FIG. 8–1

Before proceeding with the solution of this problem, let us consider the general problem of finding the extreme values of the integral

$$I = \int_a^b F(x, y, y')\, dx, \quad (8\text{--}2)$$

where $F(x, y, y')$ is a function known from the statement of the problem, while y is the unknown to be determined such that the integral becomes stationary. A function defined by a definite integral whose integrand itself depends on functions is said to be a *functional*. The integrand of this integral may also be called a functional when the independent variable x is considered fixed but the dependent variable y is varied. The integral of Eq. (8–2) takes on different values along different paths connecting the fixed points P_1, P_2. We wish to find the particular curve $y(x)$ which makes the integral, for example, a minimum.

If it is assumed that $y(x)$ actually minimizes this integral, any function in the neighborhood of $y(x)$ can be represented by the form $y(x) + \epsilon\eta(x)$, where $\eta(x)$ is a continuously differentiable function that vanishes at the ends of the interval, and ϵ is a parameter (Fig. 8–2). Then, in terms of these functions, the integral of Eq. (8–2) can be written in the form

$$I(\epsilon) = \int_a^b F(x, y + \epsilon\eta, y' + \epsilon\eta')\, dx, \quad (8\text{--}3)$$

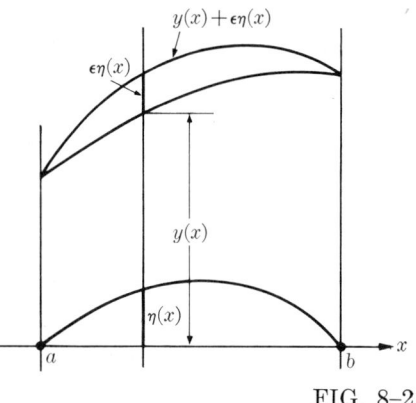

FIG. 8–2

which, according to the assumption about $y(x)$, assumes its minimum value when $\epsilon = 0$.

Since in the (I, ϵ)-plane $I(\epsilon)$ has a minimum for $\epsilon = 0$, according to differential calculus $I(\epsilon)$ satisfies the necessary condition for extrema at this point, namely

$$dI(\epsilon)/d\epsilon = 0 \quad \text{when} \quad \epsilon = 0. \quad (8\text{--}4)$$

Now, defining

$$F(\epsilon) = F(x, y + \epsilon\eta, y' + \epsilon\eta'), \quad (8\text{--}5)$$

and *assuming* that the order of integration and differentiation can be interchanged, we combine Eqs. (8–3), (8–4), and (8–5) to give

$$\frac{dI(\epsilon)}{d\epsilon} = \int_a^b \frac{dF(\epsilon)}{d\epsilon} dx. \tag{8-6}$$

Since for a fixed value of x the derivative of Eq. (8–5) with respect to ϵ is

$$\frac{dF(\epsilon)}{d\epsilon} = \frac{\partial F(\epsilon)}{\partial(y + \epsilon\eta)} \eta + \frac{\partial F(\epsilon)}{\partial(y' + \epsilon\eta')} \eta', \tag{8-7}$$

noting that $F(\epsilon)|_{\epsilon=0} = F$ and using Eqs. (8–3), (8–4), (8–5), (8–6), and (8–7), we may rearrange Eq. (8–4) in the form

$$\left.\frac{dI(\epsilon)}{d\epsilon}\right|_{\epsilon=0} = \int_a^b \left(\frac{\partial F}{\partial y}\eta + \frac{\partial F}{\partial y'}\eta'\right) dx = 0.$$

Integrating the second term of this integral by parts, we obtain

$$\int_a^b \left[\frac{\partial F}{\partial y} - \frac{d}{dx}\left(\frac{\partial F}{\partial y'}\right)\right] \eta\, dx + \frac{\partial F}{\partial y'}\eta \bigg|_a^b = 0. \tag{8-8}$$

Since, by definition, η vanishes at the ends of the interval, the second term of Eq. (8–8) is zero. Hence

$$\int_a^b \left[\frac{\partial F}{\partial y} - \frac{d}{dx}\left(\frac{\partial F}{\partial y'}\right)\right] \eta\, dx = 0. \tag{8-9}$$

The integral of Eq. (8–9) vanishes for any $\eta(x)$ when

$$\frac{\partial F}{\partial y} - \frac{d}{dx}\left(\frac{\partial F}{\partial y'}\right) = 0. \tag{8-10}$$

Equation (8–10) is the *Euler equation* associated with the variational problem given by Eq. (8–2). Thus *the condition necessary for $y(x)$ to minimize (or maximize) the integral of Eq. (8–2) is that $F(x, y, y')$ must satisfy the corresponding Euler equation*. Note that the partial derivatives $\partial F/\partial y$ and $\partial F/\partial y'$ of Eq. (8–10) are formed by assuming y and y' as independent variables; this is necessary since the variation is on these, not on the independent variable x.

Let us now return to our specific problem, the plane curve which gives the shortest distance between two points. Using Eq. (8–1) and noting that $F = (1 + y'^2)^{1/2}$, we find that the Euler equation of the problem is

$$\frac{d}{dx}\left[\frac{y'}{(1 + y'^2)^{1/2}}\right] = 0,$$

which integrates to
$$y'/(1+y'^2)^{1/2} = C \quad \text{or} \quad y' = C_1.$$
The next integration gives
$$y = C_1 x + C_2,$$
where C, C_1, and C_2 are constants. Thus we are led to the well-known fact that the shortest plane curve joining two points is a straight line.

The general problem discussed above may be extended by assuming that $y(x)$ is fixed at one end, say at $x = a$, and is arbitrary at $x = b$. Now, defining $\eta(x)$ such that it is continuously differentiable yet is zero at $x = a$ only, we may write Eq. (8–8) in the form

$$\int_a^b \left[\frac{\partial F}{\partial y} - \frac{d}{dx}\left(\frac{\partial F}{\partial y'}\right)\right] \eta \, dx + \frac{\partial F}{\partial y'} \eta \bigg|_{x=b} = 0. \tag{8–11}$$

One way to satisfy Eq. (8–11) is to let each of its two terms separately be equal to zero. The first term then gives the previous result, the Euler equation, while the second term, since $\eta(b)$ is arbitrary, introduces the additional condition for the new problem,

$$\frac{\partial F}{\partial y'}\bigg|_{x=b} = 0, \tag{8–12}$$

as a constraint on the Euler equation. Clearly, a further extension would involve $y(x)$ arbitrary at both ends.

8–2. Meaning and Rules of Variational Calculus

The differential of a function is a first-order approximation to the change in the value of this function along the particular curve it describes, while the variation of the same function is a first-order approximation to the change in this function from the curve it describes to another curve by virtue of a contained parameter (Fig. 8–3). The change $\epsilon\eta(x)$ in $y(x)$ is called the *variation of y* and is denoted by

$$\delta y = \epsilon \eta. \tag{8–13}$$

Now, corresponding to the variation of y, the change in the functional $F(x, y, y')$ for a fixed x is

$$\Delta F = F(x, y + \epsilon\eta, y' + \epsilon\eta') - F(x, y, y'). \tag{8–14}$$

Assuming y and y' as independent variables, and expanding the right-hand side of Eq. (8–14) into a two-dimensional Taylor series around (y, y'), we obtain

$$\Delta F = \frac{\partial F}{\partial y} \epsilon\eta + \frac{\partial F}{\partial y'} \epsilon\eta' + O(\epsilon^2). \tag{8–15}$$

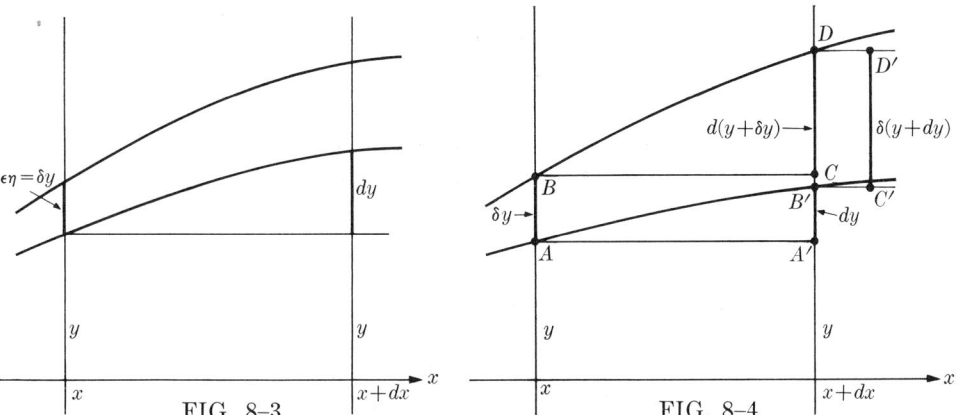

FIG. 8-3   FIG. 8-4

In a manner analogous to the definition of the differential of a function, the first two terms of the right-hand member of Eq. (8–15) are defined to be the variation of the functional $F(x, y, y')$ which, with the use of Eq. (8–13), may be written in the form

$$\delta F = \frac{\partial F}{\partial y} \delta y + \frac{\partial F}{\partial y'} \delta y'. \tag{8-16}$$

Noting that $\delta x = 0$, and thus adding the identically zero term $(\partial F/\partial x)\, \delta x$ to Eq. (8–16), we clearly see the analogy between the definitions of the differential of a function and the variation of a functional.

From the definition of variation, it may be shown that the rules for the variation of sums, products, ratios, powers, and so on, are completely analogous to the corresponding rules of differentiation.

The commutative property, that *the differential of the variation of a function $y(x)$ is identical to the variation of the differential of the same function*, is one of the most important properties of the variational calculus. This statement may readily be proved by means of Fig. 8–4 as follows:

$$AB + CD = A'B' + C'D'$$
$$\delta y + d(y + \delta y) = dy + \delta(y + dy)$$
$$d(\delta y) = \delta(dy). \tag{8-17}$$

Finally, summing up our general discussion of one-dimensional variational problems, we shall show that *the necessary condition for the existence of an extremum of the functional*

$$I = \int_a^b F(x, y, y')\, dx \tag{8-18}$$

is the vanishing of its variation,

$$\delta I = \delta \int_a^b F(x, y, y')\, dx = 0 \tag{8-19}$$

provided

$$\frac{\partial F}{\partial y'} \delta y \bigg|_a^b = 0 \qquad (8\text{-}20)$$

is satisfied.*

Since variation and integration, like variation and differentiation, are commutative, we may now write Eq. (8–19) in the form

$$\delta I = \int_a^b \delta F(x, y, y')\, dx = 0$$

or, using Eq. (8–16),

$$\delta I = \int_a^b \left(\frac{\partial F}{\partial y} \delta y + \frac{\partial F}{\partial y'} \delta y' \right) dx = 0. \qquad (8\text{-}21)$$

Next, rearranging $\delta y'$ by means of Eq. (8–17), we obtain

$$\delta y' = \delta \left(\frac{dy}{dx} \right) = \frac{\delta(dy)}{dx}$$

$$= \frac{d(\delta y)}{dx} = \frac{d}{dx}(\delta y),$$

and integrating the second term of Eq. (8–21) by parts gives

$$\delta I = \int_a^b \left[\frac{\partial F}{\partial y} - \frac{d}{dx}\left(\frac{\partial F}{\partial y'} \right) \right] \delta y\, dx + \frac{\partial F}{\partial y'} \delta y \bigg|_a^b = 0.$$

This equation is proportional to Eq. (8–8), ϵ being the proportionality constant, and with Eq. (8–20) reduces to

$$\delta I = \int_a^b \left[\frac{\partial F}{\partial y} - \frac{d}{dx}\left(\frac{\partial F}{\partial y'} \right) \right] \delta y\, dx = 0, \qquad (8\text{-}22)$$

or to the Euler equation-given by Eq. (8–10), the condition necessary for I to be stationary.

8–3. Steady One-Dimensional Problems

On the basis of the preceding discussion, we may obtain the variational formulation of steady one-dimensional conduction problems. First, however, let us note a basic difference between the variational formulation of geometric and of

---

* Equation (8–20) is often referred to as the *natural boundary condition* of the variational problem.

physical problems. *The statement of geometric problems* (for example, the distance between two points, the surface of revolution passing through two points, etc.) *is the functional of their variational formulations.* The use of the corresponding Euler's equation gives then the extreme values of this functional (the shortest distance between two points, the smallest surface of revolution passing through two points, etc.). By contrast, *the statement of physical problems cannot explicitly be related to a functional to be extremized on the basis of a physical reasoning.* Reconsider the general problem discussed in connection with Eqs. (8–2) through (8–10). Conceptually speaking, what we have learned by this problem has been *the existence of a differential equation as the condition necessary for an appropriate integral to be stationary.* Since physical problems are, in general, expressible in terms of differential equations, we may obtain the variational formulation of these problems by following the steps of the variational formulation of geometric problems in the reverse order. Thus we start by considering the differential formulation of a physical problem to be the Euler equation associated with its desired variational formulation. The point will further be clarified by the following example.

Example 8–1. Let us find the variational formulation of Example (3–10). The differential formulation of the problem is

$$\frac{d^2\theta}{dx^2} - m^2\theta = 0, \tag{3-163}$$

$$\frac{d\theta(0)}{dx} = 0, \tag{3-170}$$

$$\theta(L) = \theta_0. \tag{3-171}$$

Now, taking the governing differential equation of this problem, Eq. (3–163), as the Euler equation of the desired variational problem, we have from Eq. (8–22)

$$\delta I = \int_0^L \left(\frac{d^2\theta}{dx^2} - m^2\theta\right) \delta\theta \, dx = 0. \tag{8-23}$$

Integrating the first term of Eq. (8–23) by parts gives

$$\left.\frac{d\theta}{dx} \delta\theta\right|_0^L - \int_0^L \left(\frac{d\theta}{dx}\right) \frac{d}{dx} (\delta\theta) \, dx - m^2 \int_0^L \theta \, \delta\theta \, dx = 0. \tag{8-24}$$

Using the commutative property

$$\frac{d}{dx}(\delta\theta) = \delta\left(\frac{d\theta}{dx}\right)$$

for the middle term, and noting that

$$\left(\frac{d\theta}{dx}\right)\delta\left(\frac{d\theta}{dx}\right) = \tfrac{1}{2}\delta\left(\frac{d\theta}{dx}\right)^2, \qquad \theta\,\delta\theta = \tfrac{1}{2}\delta\theta^2,$$

we may rearrange Eq. (8–24) as follows:

$$\left(\frac{d\theta}{dx}\right)\delta\theta\bigg|_0^L - \tfrac{1}{2}\delta\int_0^L\left[\left(\frac{d\theta}{dx}\right)^2 + m^2\theta^2\right]dx = 0. \qquad (8\text{–}25)$$

The boundary conditions $\theta = \text{const}$ and $(d\theta/dx) = 0$ for which the first term of Eq. (8–25) vanishes are the natural boundary conditions of the problem. Thus the variational formulation of the problem is

$$\delta\int_0^L\left[\left(\frac{d\theta}{dx}\right)^2 + m^2\theta^2\right]dx = 0, \qquad (8\text{–}26)$$

and the corresponding functional is

$$I = \int_0^L\left[\left(\frac{d\theta}{dx}\right)^2 + m^2\theta^2\right]dx. \qquad (8\text{–}27)$$

Let us now reconsider the problem of Example 8–1, and keep the end insulated but replace the base temperature with the specified heat flux

$$+k\frac{d\theta(L)}{dx} = q''. \qquad (8\text{–}28)$$

The variational formulation of Example 8–1 applies equally to the present case before the use of boundary conditions. Introducing Eqs. (3–170) and (8–28) into Eq. (8–25), we find that the variational form of the new problem is

$$\frac{q''}{k}\delta\theta(L) - \tfrac{1}{2}\delta\int_0^L\left[\left(\frac{d\theta}{dx}\right)^2 + m^2\theta^2\right]dx = 0$$

or

$$\delta\left\{\frac{q''}{k}\theta(L) - \tfrac{1}{2}\int_0^L\left[\left(\frac{d\theta}{dx}\right)^2 + m^2\theta^2\right]dx\right\} = 0. \qquad (8\text{–}29)$$

Thus, *unlike the differential formulation, which involves the governing differential equation and the boundary conditions separately, the variational formulation includes the boundary conditions in the governing variational equation. Therefore, a problem may yield different variational formulations, depending on the boundary conditions.*

We may extend the foregoing cartesian problem to other geometries by considering a general form of linear second-order differential equations:*

$$\frac{d}{dx}\left[p(x)\frac{dy}{dx}\right] - q(x)y = f(x). \tag{8-30}$$

Let us find the variational form of Eq. (8–30) over the interval (a, b). Taking Eq. (8–30) as the Euler equation of the desired variational problem, we have

$$\int_a^b \left[\frac{d}{dx}\left(p\frac{dy}{dx}\right) - qy - f\right]\delta y\, dx = 0.$$

Integrating the first term of this equation by parts gives

$$\left(p\frac{dy}{dx}\right)\delta y\Big|_a^b - \int_a^b \left[\left(p\frac{dy}{dx}\right)\frac{d}{dx}(\delta y) + qy\,\delta y + f\,\delta y\right] dx = 0,$$

which may be rearranged in the form

$$\left(p\frac{dy}{dx}\right)\delta y\Big|_a^b - \delta\int_a^b \left[\tfrac{1}{2}p\left(\frac{dy}{dx}\right)^2 + \tfrac{1}{2}qy^2 + fy\right] dx = 0.$$

If the boundary conditions are natural, the first term is zero, and the variational form of Eq. (8–30) is

$$\delta\int_a^b \left[\tfrac{1}{2}p\left(\frac{dy}{dx}\right)^2 + \tfrac{1}{2}qy^2 + fy\right] dx = 0. \tag{8-31}$$

If the boundary conditions are of the general type

$$\frac{dy(a)}{dx} + B_1 y(a) = 0, \qquad \frac{dy(b)}{dx} + B_2 y(b) = 0,$$

---

* A linear second-order differential equation written in the form

$$\frac{d^2y}{dx^2} + f_1(x)\frac{dy}{dx} + f_2(x)y = f_3(x)$$

may be reduced to Eq. (8–30) by the transformation:

$$p(x) = \exp\left[\int f_1(x)\,dx\right], \qquad q(x) = -f_2(x)p(x), \qquad \text{and} \qquad f(x) = f_3(x)p(x).$$

In Chapter 4 we considered a similar but homogeneous equation, Eq. (4–16), in connection with characteristic-value problems.

the variational form becomes

$$\delta\left\{-\tfrac{1}{2}B_2 p(b)y^2(b) + \tfrac{1}{2}B_1 p(a)y^2(a) - \int_a^b \left[\tfrac{1}{2}p\left(\frac{dy}{dx}\right)^2 + \tfrac{1}{2}qy^2 + fy\right]dx\right\} = 0. \tag{8-32}$$

Other forms corresponding to other boundary conditions may be obtained in a similar manner.

Although cartesian geometry requires no rearrangement, a linear second-order differential equation written in terms of cylindrical, spherical, or other coordinates cannot be transformed into a variational form unless it is first expressed in the form of Eq. (8–30). [Take, for example, cylindrical geometry and try the variational form of $r^2(d^2\theta/dr^2) + r(d\theta/dr)$ or $d^2\theta/dr^2 + (1/r)(d\theta/dr)$.]

Now that we have learned how to obtain the variational form of steady one-dimensional problems, we shall make use of this form to obtain approximate solutions to these problems by a procedure analogous to that used with the integral formulation, as follows. It is possible to construct a number of geometrical profiles which satisfy the prescribed boundary conditions of a given problem; however, among these there is only one physically possible profile which, by satisfying the appropriate Euler equation, makes the functional of the corresponding variational problem stationary. It is often difficult to guess which profile this is. Instead, we try approximate profiles for which the functional has a value approximately equal to its stationary value. As we learned in Chapter 2, in the selection of approximate profiles there are two procedures, the so-called *Ritz* and *Kantorovich* methods. We shall now examine in detail the Ritz method. The Kantorovich method, which is suitable for use with multi-dimensional and unsteady problems, will be discussed in Sections 8–8 through 8–11.

8–4. Ritz Method

This method is based on the selection of a convergent sequence of functions

$$y(x) = \sum_{n=0}^{N} a_n \varphi_n(x), \tag{8-33}$$

depending on N parameters, where $\varphi_n(x)$ for all values of n satisfies the boundary conditions; otherwise the choice of the functions $\varphi_n(x)$ is to a large extent arbitrary. In physical problems the general behavior of the desired solution is usually known and, depending on the nature of the problem, special functions such as polynomials or circular, hyperbolic, cylindrical, or spherical functions may conveniently be employed.

Inserting Eq. (8–33) into Eq. (8–2), we obtain the functional in terms of N parameters, $I(a_0, a_1, a_2, \ldots, a_N)$. The stationary values of this functional may be obtained in the usual manner. The result is N simultaneous algebraic equations

$$\partial I/\partial a_n = 0, \quad n = 0, 1, 2, \ldots, N, \tag{8-34}$$

sufficient for the evaluation of the parameters of Eq. (8-33). As may be expected, the approximate solution given by Eq. (8-33) becomes closer to the exact solution when the number N of the approximations is increased; the convergence of this solution largely depends on the choice of functions $\varphi_n(x)$.

A more elaborate procedure consists in a sequence of approximations

$$y_0(x) = a_0\varphi_0(x),$$
$$y_1(x) = a_0\varphi_0(x) + a_1\varphi_1(x),$$
$$\vdots$$
$$y_N(x) = a_0\varphi_0(x) + a_1\varphi_1(x) + \cdots + a_N\varphi_N(x),$$

to be evaluated separately. With this procedure, we can check by successive comparison the accuracy attained at each order of approximation. Note that for the same coefficient different values are obtained, depending on the order of approximation.

Example 8-2. Let us obtain an approximate solution to the problem of Example (8-1) by means of the Ritz method.

In terms of polynomials, for example, the series

$$\theta(\xi)/\theta_0 = 1 - (1 - \xi^2)(a_0 + a_1\xi^2 + a_2\xi^4 + \cdots) \tag{8-35}$$

or

$$\theta(\xi)/\theta_0 = 1 - (1 - \xi^2)a_0 - (1 - \xi^2)^2 a_1 - (1 - \xi^2)^3 a_2 - \cdots, \tag{8-36}$$

both of which satisfy the boundary conditions of the problem, may be used conveniently. Here $\xi = x/L$ is the dimensionless distance and θ_0 the base temperature. The first of these series has already been used in Example 3-10 for an approximate solution to the same problem by the integral method; since this series involves less algebra than that of Eq. (8-36) in the evaluation of the unknown parameters a_n, it will also be used here. Thus we shall be able to compare the results of the integral method with those obtained by the variational technique.

Here we shall limit ourselves to the first- and second-order approximations, but we shall carry out both of them so that we may compare the accuracy attained at each order of approximation. Although Eq. (8-26) is the only variational form of the problem, Eq. (8-23), being identical to Eq. (8-26) and involving less algebra in the evaluation of the unknown parameters, is more convenient to use than Eq. (8-26).

The first approximation of Eq. (8-35),

$$\theta(\xi)/\theta_0 = 1 - (1 - \xi^2)a_0, \tag{3-195}$$

introduced into the dimensionless form of Eq. (8–26),

$$\int_0^1 \left(\frac{d^2\theta}{d\xi^2} - \mu^2\theta\right) \delta\theta \, d\xi = 0, \tag{8-37}$$

yields

$$-\int_0^1 \{2a_0 - \mu^2[1 - (1 - \xi^2)a_0]\}(1 - \xi^2) \, \delta a_0 \, d\xi = 0.$$

This integral may readily be evaluated by the successive use of Eq. (8–129).* The result is

$$[2(1 + 2\mu^2/5)a_0 - \mu^2] \, \delta a_0 = 0.$$

Since δa_0 is arbitrary, the coefficient in brackets must vanish, giving

$$a_0 = \frac{\mu^2/2}{1 + 2\mu^2/5}.$$

Thus a first-order approximation of the temperature distribution is

$$\frac{\theta(\xi)}{\theta_0} = 1 - \frac{\mu^2/2}{1 + 2\mu^2/5}(1 - \xi^2). \tag{8-38}$$

The second approximation of Eq. (8–35),

$$\theta(\xi)/\theta_0 = 1 - (1 - \xi^2)(a_0 + a_1\xi^2), \tag{3-215}$$

introduced into Eq. (8–37) yields

$$-\int_0^1 \{2a_0 - 2a_1 + 12a_1\xi^2 - \mu^2[1 - (1 - \xi^2)(a_0 + a_1\xi^2)]\} \\ \times (1 - \xi^2)(\delta a_0 + \xi^2 \, \delta a_1) \, d\xi = 0.$$

This integral may be evaluated by the repeated use of Eqs. (8–129) and (8–131). The result is

$$\tfrac{4}{3}[(1 + \tfrac{2}{5}\mu^2)a_0 + \tfrac{1}{5}(1 + \tfrac{2}{7}\mu^2)a_1 - \tfrac{1}{2}\mu^2] \, \delta a_0 \\ + \tfrac{4}{15}[(1 + \tfrac{2}{7}\mu^2)a_0 + \tfrac{1}{7}(11 + \tfrac{2}{3}\mu^2)a_1 - \tfrac{1}{2}\mu^2] \, \delta a_1 = 0.$$

Since δa_0 and δa_1 are arbitrary, their coefficients must vanish, giving the simultaneous algebraic equations

$$(1 + \tfrac{2}{5}\mu^2)a_0 + \tfrac{1}{5}(1 + \tfrac{2}{7}\mu^2)a_1 = \tfrac{1}{2}\mu^2,$$
$$(1 + \tfrac{2}{7}\mu^2)a_0 + \tfrac{1}{7}(11 + \tfrac{2}{3}\mu^2)a_1 = \tfrac{1}{2}\mu^2.$$

---

* See Section 8–12.

Solving these equations for a_0 and a_1 yields

$$a_0 = \frac{\mu^2(3 + \mu^2/12)}{6 + 8\mu^2/3 + 2\mu^4/21}, \qquad a_1 = \frac{\mu^4/4}{6 + 8\mu^2/3 + 2\mu^4/21}.$$

Thus a second-order approximation of the temperature distribution is

$$\frac{\theta(\xi)}{\theta_0} = 1 - \frac{\mu^2(1 - \xi^2)}{6 + 8\mu^2/3 + 2\mu^4/21}\left[\left(3 + \frac{\mu^2}{12}\right) + \frac{\mu^2\xi^2}{4}\right]. \tag{8-39}$$

Let us now compare the first- and second-order approximations obtained by the variational calculus and the integral method with the exact solution of the problem. As the basis of comparison, we shall use the total heat loss from the fin, since this is more important and meaningful for the purpose than the tip temperature. Although the heat transfer by conduction from the base of the fin is equal to the heat loss by convection from the peripheral surface of the fin, an approximate profile results in different values for these. For obvious reasons, we choose to use the peripheral heat transfer, whose dimensionless form is

$$\frac{q}{kA\theta_0/L} = \mu^2 \int_0^1 \theta \, d\xi. \tag{8-40}$$

Inserting Eqs. (8-38) and (8-39) into Eq. (8-40) gives the first- and second-order approximations of the heat loss from the fin,

$$\frac{q}{kA\theta_0/L} = \frac{\mu^2(1 + \mu^2/15)}{(1 + 2\mu^2/5)}, \qquad \frac{q}{kA\theta_0/L} = \frac{\mu^2(3 + \mu^2/3 + \mu^4/315)}{(3 + 4\mu^2/3 + \mu^4/21)}. \tag{8-41}$$

In Table 8-1, we compare the exact heat loss from the fin with the results of the integral technique (see Table 3-4) and those of the variational calculus. This table clearly shows the improved accuracy of results obtained by use of the variational calculus as compared with those obtained by the integral method when the same profile is employed for both cases.

In the light of the foregoing problem, let us emphasize the basic difference between the variational calculus and the integral method. Suppose that we wish to find the nth approximation of a problem by either of the two methods. The variational calculus results directly in n simultaneous algebraic equations that yield the values of the n parameters, weighed to give the closest fit to the exact solution. On the other hand, as a result of integration of the governing differential equation, the integral method yields only one algebraic relation among the unknown parameters. The remaining $n - 1$ conditions are, to some extent, arbitrarily determined (recall Section 3-11). Therefore, the accuracy of the approximation obtained by the integral method depends on individual skill and experience. In general, this accuracy does not exceed that of the approximation obtained by the variational calculus.

448 VARIATIONAL FORMULATION AND SOLUTION [8–4]

TABLE 8–1

			μ	$\tfrac{1}{2}$	1	2	4
		Exact		0.2311	0.7616	1.9281	3.9973
Approximate	First	Integral Error, %		0.2308 0.130	0.7500 1.52	1.7143 11.1	2.5263 36.8
		Variational Error, %		0.2311 0.000	0.7619 0.041	1.9487 1.07	4.4685 11.8
	Second	Integral I Error, %		0.2311 0.000	0.7616 0.000	1.9269 0.062	3.9223 1.88
		Integral II Error, %		0.2311 0.000	0.7614 0.026	1.9130 0.783	3.6791 7.96
		Variational Error, %		0.2311 0.000	0.7616 0.000	1.9281 0.000	4.0066 0.233

Example 8–3. Reconsider Problem 3–30 (Fig. 8–5). Let us find the steady temperature distribution in the system. The case of constant wear readily accepts an exact solution; hence we focus our attention on the more involved case of constant pressure.

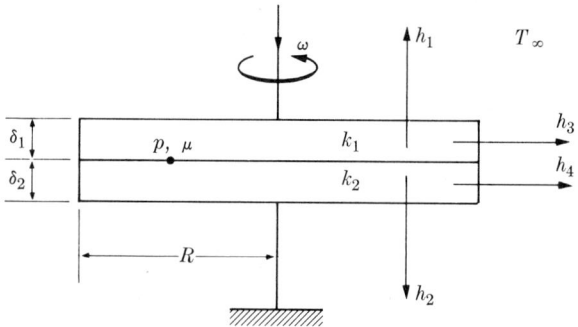

FIG. 8–5

Under steady conditions, with the axial temperature distribution neglected, the differential formulation of the problem is

$$\frac{d}{d\rho}\left(\rho\frac{d\theta}{d\rho}\right) - \mu^2\rho\theta + \nu\rho^2 = 0, \qquad (8\text{–}42)$$

where

$$\theta = T - T_\infty, \quad \rho = \frac{r}{R}, \quad \mu^2 = \left(\frac{h_1 + h_2}{k_1\delta_1 + k_2\delta_2}\right)R^2, \quad \nu = \frac{\mu p\omega R^3}{(k_1\delta_1 + k_2\delta_2)}.$$

The variational form of Eq. (8–42) is identical to

$$\int_0^1 \left[\frac{d}{d\rho}\left(\rho\frac{d\theta}{d\rho}\right) - \mu^2\rho\theta + \nu\rho^2\right]\delta\theta\, d\rho = 0. \qquad (8\text{–}43)$$

Let us neglect the heat transfer from the peripheral surface of the disks. Then a first approximation of the temperature distribution which satisfies the end conditions may be written in the form

$$\theta(\rho) = a_0(1 + 3\rho^2 - 2\rho^3). \tag{8-44}$$

Introducing Eq. (8-44) into Eq. (8-43), then integrating and rearranging the result gives

$$a_0 = \frac{(21/52)\nu}{21/52 + \mu^2}.$$

Thus a first-order approximation of the temperature distribution is

$$\frac{\theta(\rho)}{\nu} = \frac{21/52}{21/52 + \mu^2}(1 + 3\rho^2 - 2\rho^3). \tag{8-45}$$

Similarly, a second-order approximation of the temperature may be constructed in the form

$$\theta(\rho) = a_0 + (3\rho^2 - 2\rho^3)a_1. \tag{8-46}$$

Insertion of Eq. (8-46) into Eq. (8-43) and subsequent integration yields the simultaneous algebraic equations

$$\mu^2 a_0 + \tfrac{7}{10}\mu^2 a_1 = \tfrac{2}{3}\nu,$$
$$\tfrac{7}{20}\mu^2 a_0 + \tfrac{2}{7}(\mu^2 + \tfrac{21}{10})a_1 = \tfrac{4}{15}\nu.$$

Solving these equations for a_0 and a_1 gives

$$a_0 = \frac{16\nu(\mu^2 + 105)}{171\mu^2(\mu^2 + 840/57)}, \qquad a_1 = \frac{140\nu}{171(\mu^2 + 840/57)}.$$

Thus a second-order approximation of the temperature distribution is

$$\frac{\theta(\rho)}{\nu} = \frac{16(\mu^2 + 105)}{171\mu^2(\mu^2 + 840/57)} + \frac{140}{171(\mu^2 + 840/57)}(3\rho^2 - 2\rho^3). \tag{8-47}$$

Let us compare, for example, the centerline temperature of the first and second approximations to see how well the two approximations agree. Introducing $\rho = 0$ into Eqs. (8-45) and (8-47), and establishing the ratio of centerline temperatures yields

$$\frac{\theta_1(0)}{\theta_2(0)} = \frac{3591\mu^2(\mu^2 + 840/57)}{832(\mu^2 + 105)(\mu^2 + 21/52)}.$$

This ratio is given in Table 8-2 as a function of μ. Inspection of the table reveals a large and variable discrepancy between the two approximations which is quite contrary to the results of Example 8-2. For problems with no specified temperature at the boundaries, a first approximation such as Eq. (8-44), intro-

TABLE 8-2

μ	$\frac{1}{2}$	1	2	4
$\theta_1(0)/\theta_2(0)$	0.2350	.0.4564	0.6739	1.069

ducing a constraint between the centerline and the peripheral temperatures, gives a rather poor result and should not be used. This point will be discussed at greater length in the next section. To illustrate the accuracy of the approximate solutions of the foregoing problem, the reader should carry out the third approximation and compare it with the second approximation given by Eq. (8–47) (see Problem 8–4).

8–5. Steady One-Dimensional Ritz Profiles

As may be expected, profiles subject to homogeneous boundary conditions are, in general, easier to construct than profiles subject to nonhomogeneous boundary conditions. Thus, using the principle of superposition, we first reduce the problem of selecting the latter type to that of selecting the former type. The procedure is best explained in terms of frequently encountered boundary conditions as follows:

(a) *The function is specified at the boundaries,*

$$y(0) = Y_2, \qquad y(L) = Y_1.$$

Let us separate the desired profile into three profiles as shown in Fig. 8–6. Thus we have

$$y(x) = y_1(x) + y_2(x) + y_3(x).$$

Each of the first two of these profiles satisfies one nonhomogeneous boundary condition of the problem. The selection of $y_1(x)$ and $y_2(x)$ as straight lines is for convenience only. It follows then that

$$y(x) = \left(\frac{x}{L}\right) Y_1 + \left(\frac{L-x}{L}\right) Y_2 + y_3(x), \qquad (8\text{–}48)$$

where the last profile satisfies the corresponding homogeneous boundary conditions,

$$y_3(0) = 0, \qquad y_3(L) = 0.$$

The use of polynomials or circular functions in the construction of $y_3(x)$ readily gives

$$y_3(x) = x(L - x)(a_0 + a_1 x + a_2 x^2 + \cdots) \qquad (8\text{–}49)$$

or

$$y_3(x) = a_0 \sin(\pi x/L) + a_1 \sin(2\pi x/L) + a_2 \sin(3\pi x/L) + \cdots \qquad (8\text{–}50)$$

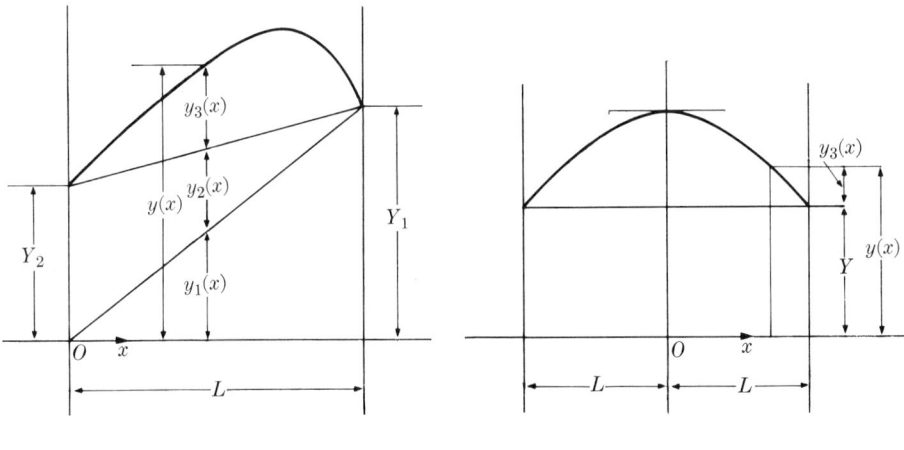

FIG. 8-6 FIG. 8-7

As a special case, the same profiles may be geometrically and thermally symmetric (Fig. 8-7). In such a case, given the fact that $Y_1 = Y_2 = Y$, it follows that $y_1(x) = y_2(x) = Y$. Thus

$$y(x) = Y + y_3(x), \tag{8-51}$$

where

$$y_3(x) = (L^2 - x^2)(a_0 + a_1 x^2 + a_2 x^4 + \cdots) \tag{8-52}$$

or

$$y_3(x) = a_0 \cos(\pi x/2L) + a_1 \cos(3\pi x/2L) + a_2 \cos(5\pi x/2L) + \cdots \tag{8-53}$$

(b) *The gradient of the function is specified at the boundaries,*

$$y'(0) = m_2, \qquad y'(L) = m_1.$$

Again let the required profile be composed of three profiles,

$$y(x) = y_1(x) + y_2(x) + y_3(x),$$

such that each of the first two satisfies one nonhomogeneous boundary condition and the homogeneous form of the other boundary condition, as shown in Fig. 8-8(a). The simplest physically possible profiles which satisfy these boundary conditions are parabolas. Thus we have

$$y(x) = \frac{x^2}{2L} m_1 - \frac{(L^2 - x^2)}{2L} m_2 + y_3(x), \tag{8-54}$$

where the third profile, $y_3(x)$, satisfies the corresponding homogeneous boundary conditions

$$y_3'(0) = 0, \qquad y_3'(L) = 0.$$

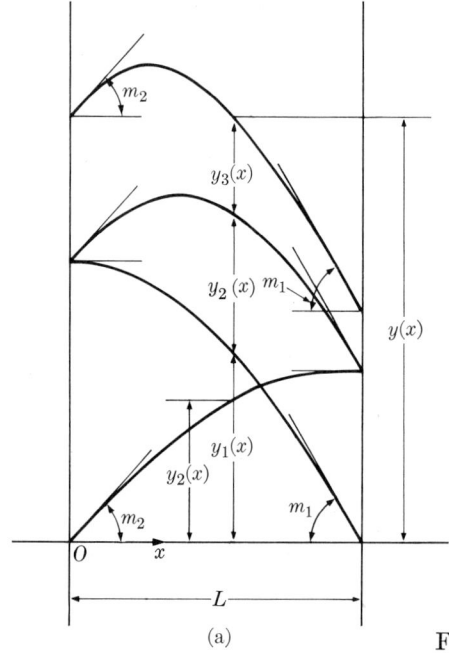
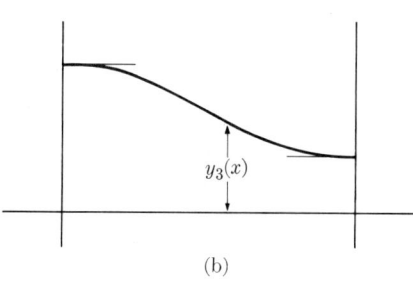

FIG. 8-8

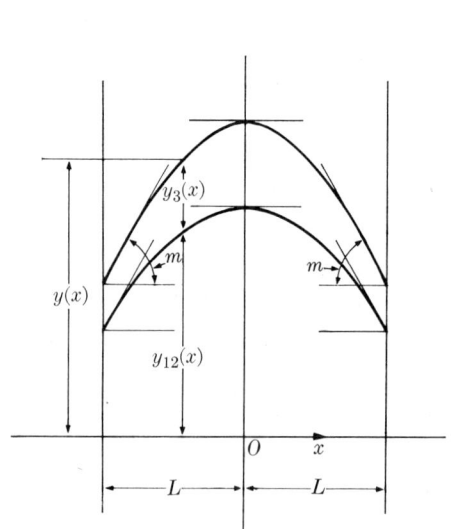

FIG. 8-9

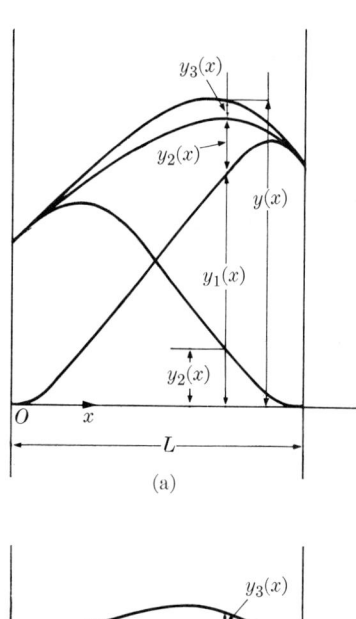

FIG. 8-10

It follows from Fig. 8–8(b) that in terms of polynomials

$$y_3(x) = a_0 + a_1(3Lx^2 - 2x^3) + a_2(L - x)^2 x^2 + a_3(L - x)^2 x^3 \\ + a_4(L - x)^2 x^4 + \cdots \tag{8-55}$$

or

$$y_3(x) = a_0 + a_1(3Lx^2 - 2x^3) + a_2(4Lx^3 - 3x^4) \\ + a_3(5Lx^4 - 4x^5) + \cdots \tag{8-56}$$

The use of circular functions results in

$$y_3(x) = a_0 + a_1 \cos(\pi x/L) + a_2 \cos(2\pi x/L) + a_3 \cos(3\pi x/L) + \cdots \tag{8-57}$$

The geometrically and thermally symmetric case of the same profiles is shown in Fig. 8–9. Since in this case $m_1 = m_2 = m$, the profiles $y_1(x)$ and $y_2(x)$ of the general case may now be replaced by a single profile, say $y_{12}(x)$. Thus

$$y(x) = y_{12}(x) + y_3(x).$$

Assuming that $y_{12}(x)$ is subject to the nonhomogeneous boundary condition of the problem, and satisfying this boundary condition again by a parabola, we have

$$y(x) = -\frac{(L^2 - x^2)}{2L} m + y_3(x), \tag{8-58}$$

where $y_3(x)$, having the general form of Fig. 8–8(b), may be represented by the profile of Eq. (8–56) or Eq. (8–57).

(c) *A linear combination of the function and its derivative is prescribed at the boundaries,*

$$y'(0) + B_2 y(0) = n_2, \qquad y'(L) + B_1 y(L) = n_1.$$

As before, we let

$$y(x) = y_1(x) + y_2(x) + y_3(x),$$

where $y_1(x)$ satisfies the second boundary condition and the homogeneous boundary condition corresponding to the first boundary condition, and $y_2(x)$ satisfies the first boundary condition and the homogeneous boundary condition corresponding to the second boundary condition (Fig. 8–10a). Note that the homogeneous boundary conditions are conveniently satisfied by letting the function and its derivative, rather than their sum, be separately equal to zero. Since $y_1(x)$ and $y_2(x)$ now have saddle points, the simplest physically meaningful profiles corresponding to these functions must necessarily be constructed in terms of third-order polynomials. Thus we have

$$y_1(x) = x^2(x + C_1) \qquad \text{and} \qquad y_2(x) = (L - x)^2(x + C_2),$$

where C_1 and C_2 are to be evaluated from the second and the first boundary conditions, respectively. The result is

$$y_1(x) = x^2 \left(x - L - \frac{L - n_1/L}{2 + B_1 L} \right) \tag{8-59}$$

and

$$y_2(x) = (L - x)^2 \left(x + \frac{L - n_2/L}{2 - B_2 L} \right). \tag{8-60}$$

The general form of $y_3(x)$ is shown in Fig. 8–10(b). It readily follows, if we use polynomials, that

$$y_3(x) = a_0(L - x)^2 x^2 + a_1(L - x)^2 x^3 + a_2(L - x)^2 x^4 + \cdots, \tag{8-61}$$

or if we use circular functions, that

$$y_3(x) = a_0(1 - \cos \pi x/L) + a_1(1 - \cos 2\pi x/L) + a_2(1 - \cos 3\pi x/L) + \cdots \tag{8-62}$$

The geometrically and thermally symmetric case of these profiles is shown in Fig. 8–11(a). As in the preceding symmetric cases, we may write

$$y(x) = y_{12}(x) + y_3(x),$$

where $y_{12}(x)$ may be constructed in terms of the parabola

$$y_{12}(x) = C - x^2,$$

to be subjected to the boundary condition

$$y'(L) + By(L) = n.$$

Thus we have

$$y_{12}(x) = L^2 - x^2 + \frac{n + 2L}{B}. \tag{8-63}$$

The general form of $y_3(x)$ is shown in Fig. 8–11(b). Using polynomials, we obtain

$$y_3(x) = a_0 + a_1(L - x)^2 x^2 + a_2(L - x)^2 x^3 + a_3(L - x)^2 x^4 + \cdots, \tag{8-64}$$

or using circular functions, we get

$$y_3(x) = a_0(1 + \cos \pi x/L) + a_1(1 + \cos 3\pi x/L) + a_2(1 + \cos 5\pi x/L) + \cdots \tag{8-65}$$

Having thus completed our investigation of steady one-dimensional problems we now proceed to the variational formulation of steady two-dimensional problems.

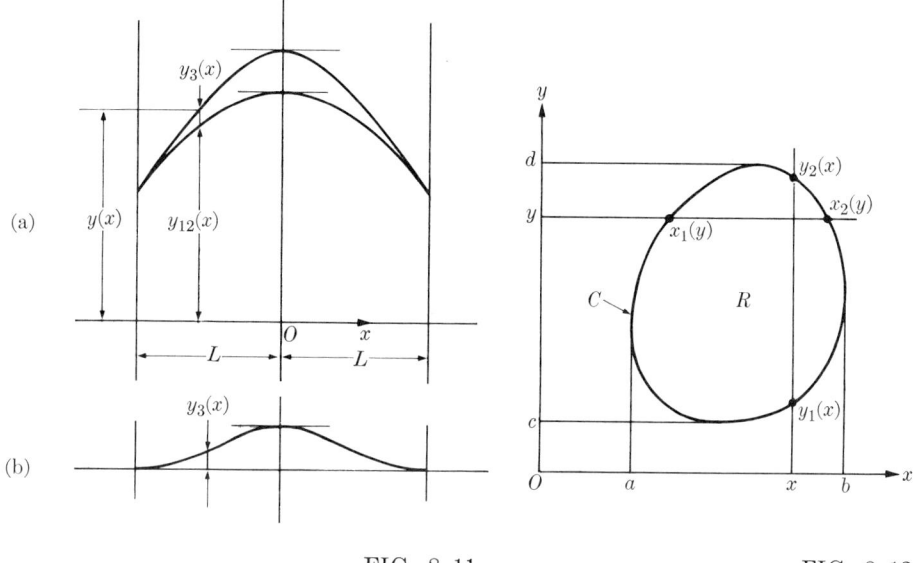

FIG. 8–11   FIG. 8–12

8–6. Steady Two-Dimensional Problems

We begin our discussion by considering a two-dimensional (cartesian) functional

$$I = \iint_R F(x, y, T, T_x, T_y) \, dx \, dy, \qquad (8\text{–}66)$$

where x and y are independent variables, $T(x, y)$ is the temperature function to be determined, and T_x, T_y are the partial derivatives with respect to x and y; the functional F and the integration domain R are known from the statement of the problem.

It follows from the rules of the variational calculus (Section 8–2) that the necessary condition, $\delta I = 0$, for the functional given by Eq. (8–66) to be stationary is

$$\delta I = \iint_R \left(\frac{\partial F}{\partial T} \delta T + \frac{\partial F}{\partial T_x} \delta T_x + \frac{\partial F}{\partial T_y} \delta T_y \right) dx \, dy = 0. \qquad (8\text{–}67)$$

The Euler equation of this functional will now be obtained by expressing the last two terms of the integrand in terms of the variation δT. First we consider, for example, the last term of the integrand, which can be rearranged by using Fig. 8–12 as follows:

$$\iint_R \frac{\partial F}{\partial T_y} \delta T_y \, dx \, dy = \int_a^b \left(\int_{y_1(x)}^{y_2(x)} \frac{\partial F}{\partial T_y} \delta T_y \, dy \right) dx. \qquad (8\text{–}68)$$

Since
$$\delta T_y = \frac{\partial}{\partial y}(\delta T),$$

Eq. (8-68) integrated by parts in the y-direction gives

$$\iint_R \frac{\partial F}{\partial T_y} \delta T_y \, dx \, dy = \int_a^b \frac{\partial F}{\partial T_y} \delta T \Big|_{y_1(x)}^{y_2(x)} dx - \iint_R \frac{\partial}{\partial y}\left(\frac{\partial F}{\partial T_y}\right) \delta T \, dx \, dy. \quad (8\text{-}69)$$

Similarly, the second term in the integrand of Eq. (8-67) results in

$$\iint_R \frac{\partial F}{\partial T_x} \delta T_x \, dx \, dy = \int_c^d \frac{\partial F}{\partial T_x} \delta T \Big|_{x_1(y)}^{x_2(y)} dy - \iint_R \frac{\partial}{\partial x}\left(\frac{\partial F}{\partial T_x}\right) \delta T \, dx \, dy. \quad (8\text{-}70)$$

Using Fig. 8-13, we see that the first right-hand term of Eqs. (8-69) and (8-70) becomes

$$\int_a^b \frac{\partial F}{\partial T_y} \delta T \Big|_{y_1(x)}^{y_2(x)} dx + \int_c^d \frac{\partial F}{\partial T_x} \delta T \Big|_{x_1(y)}^{x_2(y)} dy = \oint_C \left(\frac{\partial F}{\partial T_x} \cos \phi + \frac{\partial F}{\partial T_y} \sin \phi\right) \delta T \, ds. \quad (8\text{-}71)$$

Combining Eqs. (8-67), (8-69), (8-70), and (8-71) yields the variational formulation

$$\delta I = \iint_R \left[\frac{\partial F}{\partial T} - \frac{\partial}{\partial x}\left(\frac{\partial F}{\partial T_x}\right) - \frac{\partial}{\partial y}\left(\frac{\partial F}{\partial T_y}\right)\right] \delta T \, dx \, dy$$
$$+ \oint_C \left(\frac{\partial F}{\partial T_x} \cos \phi + \frac{\partial F}{\partial T_y} \sin \phi\right) \delta T \, ds = 0. \quad (8\text{-}72)$$

The natural boundary conditions of the problem, according to Eq. (8-72), satisfy

$$\oint_C \left(\frac{\partial F}{\partial T_x} \cos \phi + \frac{\partial F}{\partial T_y} \sin \phi\right) \delta T \, ds = 0. \quad (8\text{-}73)$$

One way of satisfying Eq. (8-73) is to select T such that it is continuously differentiable in R, and δT vanishes on C. This may be accomplished, for example, by specifying T on C. The variational formulation, subject to the natural boundary conditions, then becomes

$$\delta I = \iint_R \left[\frac{\partial F}{\partial T} - \frac{\partial}{\partial x}\left(\frac{\partial F}{\partial T_x}\right) - \frac{\partial}{\partial y}\left(\frac{\partial F}{\partial T_y}\right)\right] \delta T \, dx \, dy = 0, \quad (8\text{-}74)$$

which implies that

$$\frac{\partial F}{\partial T} - \frac{\partial}{\partial x}\left(\frac{\partial F}{\partial T_x}\right) - \frac{\partial}{\partial y}\left(\frac{\partial F}{\partial T_y}\right) = 0. \quad (8\text{-}75)$$

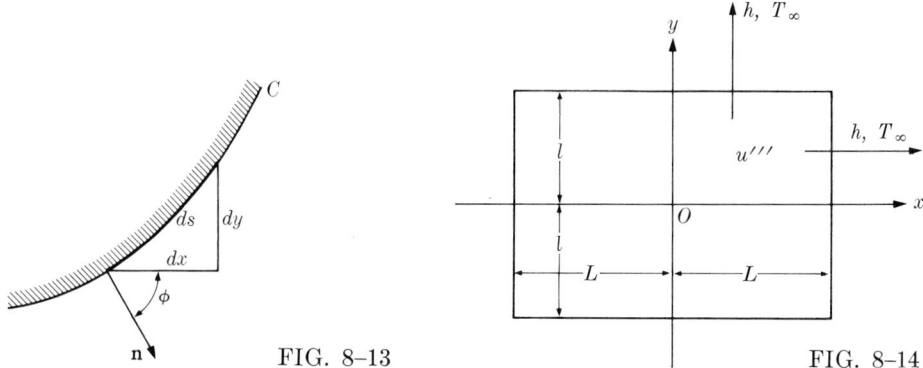

FIG. 8–13   FIG. 8–14

Equation (8–75) is the Euler equation associated with the functional given by Eq. (8–66). Thus, in a manner analogous to the one-dimensional case, $T(x, y)$, subject to the natural boundary conditions, makes the integral of Eq. (8–66) stationary, if it satisfies the corresponding Euler equation.

Example 8–4. Reconsider Example 2–1 (Fig. 8–14). Assume the case of a large heat transfer coefficient. Let us find the steady variational formulation of the problem.

The governing differential equation is

$$\frac{\partial^2 \theta}{\partial x^2} + \frac{\partial^2 \theta}{\partial y^2} + \frac{u'''}{k} = 0,$$

subject to the boundary conditions

$$\theta(\pm L, y) = \theta(x, \pm l) = 0,$$

where θ is measured above the ambient temperature T_∞.

The general procedure outlined in Eqs. (8–66) through (8–75) is now followed in the reverse order; taking the differential equation of the problem as the Euler equation of the variational formulation, we have from Eq. (8–74)

$$\delta I = \int_{-L}^{L} \int_{-l}^{l} \left(\frac{\partial^2 \theta}{\partial x^2} + \frac{\partial^2 \theta}{\partial y^2} + \frac{u'''}{k} \right) \delta\theta \, dx \, dy = 0. \tag{8–76}$$

The first term in the integrand of Eq. (8–76) integrated by parts in the x-direction results in

$$\int_{-L}^{L} \int_{-l}^{l} \frac{\partial^2 \theta}{\partial x^2} \delta\theta \, dx \, dy = \int_{-l}^{l} \left(\int_{-L}^{L} \frac{\partial^2 \theta}{\partial x^2} \delta\theta \, dx \right) dy$$

$$= \int_{-l}^{l} \frac{\partial \theta}{\partial x} \delta\theta \bigg|_{-L}^{L} dy - \int_{-l}^{l} \int_{-L}^{L} \delta \left[\frac{1}{2} \left(\frac{\partial \theta}{\partial x} \right)^2 \right] dx \, dy. \tag{8–77}$$

Similarly, the second term integrated by parts in the y-direction yields

$$\int_{-L}^{L}\int_{-l}^{l}\frac{\partial^2 \theta}{\partial y^2}\delta\theta\,dx\,dy = \int_{-L}^{L}\frac{\partial\theta}{\partial y}\delta\theta\bigg|_{-l}^{l}\,dx - \int_{-L}^{L}\int_{-l}^{l}\delta\left[\frac{1}{2}\left(\frac{\partial\theta}{\partial y}\right)^2\right]dx\,dy. \tag{8-78}$$

Furthermore, note that

$$\int_{-l}^{l}\frac{\partial\theta}{\partial x}\delta\theta\bigg|_{-L}^{L}\,dy + \int_{-L}^{L}\frac{\partial\theta}{\partial y}\delta\theta\bigg|_{-l}^{l}\,dx = \oint_{C}\frac{\partial\theta}{\partial n}\delta\theta\,ds, \tag{8-79}$$

where n is the outward normal to the periphery of the rectangle, ds the differential arc-length on this periphery, and C the boundaries of the rectangle (see Fig. 8–13). Finally, combining Eqs. (8–77), (8–78), (8–79), and introducing the result into Eq. (8–76) gives the variational form

$$\delta\iint_{R}\left\{-\frac{1}{2}\left[\left(\frac{\partial\theta}{\partial x}\right)^2 + \left(\frac{\partial\theta}{\partial y}\right)^2\right] + \frac{u'''}{k}\theta\right\}dx\,dy + \int_{C}\frac{\partial\theta}{\partial n}\delta\theta\,ds = 0. \tag{8-80}$$

Inspection of the line integral of Eq. (8–80) reveals that, as with steady one-dimensional problems, either a prescribed surface temperature or an insulated boundary will make this integral vanish; hence both are natural boundary conditions by definition. Since the temperature of the boundaries is prescribed in the present case, the variational form of the problem reduces to

$$\delta\iint_{R}\left\{-\frac{1}{2}\left[\left(\frac{\partial\theta}{\partial x}\right)^2 + \left(\frac{\partial\theta}{\partial y}\right)^2\right] + \frac{u'''}{k}\theta\right\}dx\,dy = 0. \tag{8-81}$$

8–7. Steady Two-Dimensional Ritz Method

The Ritz method of steady one-dimensional problems may readily be extended to steady two-dimensional problems. We select an approximate solution $\theta(x, y)$ which depends on N parameters and has the form of a convergent sequence of functions

$$\theta(x, y) = \sum_{n=0}^{N} a_n \varphi_n(x, y). \tag{8-82}$$

Here $\varphi_n(x, y)$ for all values of n satisfies the boundary conditions. Furthermore, if $\varphi_n(x, y)$ is assumed to be a product consisting of a function of x alone and a function of y alone,

$$\varphi_n(x, y) = X_n(x) Y_n(y),$$

then $X_n(x)$ and $Y_n(y)$ may be constructed such that $X_n(x)$ satisfies only the boundary conditions in the x-direction and $Y_n(y)$ only the boundary conditions in the y-direction. Thus Eq. (8–82) becomes

$$\theta(x, y) = \sum_{n=0}^{N} a_n X_n(x) Y_n(y). \tag{8-83}$$

Example 8–5. Using the Ritz method, let us find an approximate solution of Example 8–4.

In terms of polynomials, for example, the series

$$\theta(x, y) = (L^2 - x^2)(l^2 - y^2)(a_0 + a_1 x^2 + a_2 y^2 + \cdots) \tag{8-84}$$

or

$$\theta(x, y) = a_0(L^2 - x^2)(l^2 - y^2) + a_1(L^2 - x^2)^2(l^2 - y^2)^2 + \cdots, \tag{8-85}$$

both of which satisfy the boundary conditions of the problem, may be used conveniently. Because of the thermal and geometric symmetry of the problem with respect to the selected reference frame, odd powers of x and y, being identically zero, are excluded from the series of Eqs. (8–84) and (8–85). The first of these series has already been used in Examples 2–1 and 4–11 in connection with the integral formulation and its solution. The same series, permitting a comparison between the results of the integral and variational techniques, will also be used here.

Again, the first two approximations are carried out. The first approximation of Eqs. (8–84),

$$\theta(x, y) = (L^2 - x^2)(l^2 - y^2)a_0, \tag{2-134}$$

inserted into Eq. (8–76) yields

$$4 \int_0^L \int_0^l [-2(l^2 - y^2)a_0 - 2(L^2 - x^2)a_0 + u'''/k] \\ \times (L^2 - x^2)(l^2 - y^2) \, \delta a_0 \, dx \, dy = 0.$$

The result of this integral gives

$$a_0 = \frac{5}{8} \frac{u'''/k}{(l^2 + L^2)}.$$

Thus the first-order Ritz profile based on the variational formulation is

$$\frac{\theta(x, y)}{u''' l^2/k} = \frac{5}{8} \frac{[1 - (x/L)^2][1 - (y/l)^2]}{1 + (l/L)^2}. \tag{8-86}$$

Similarly, the second approximation of Eq. (8–84),

$$\theta(x, y) = (L^2 - x^2)(l^2 - y^2)(a_0 + a_1 x^2), \tag{4-135}$$

introduced into Eq. (8–76) gives

$$4\int_0^L \int_0^l [-2(l^2 - y^2)(a_0 - a_1 L^2 + 6a_1 x^2) - 2(L^2 - x^2)(a_0 + a_1 x^2) + u'''/k]$$
$$\times (L^2 - x^2)(l^2 - y^2)(\delta a_0 + x^2 \, \delta a_1) \, dx \, dy = 0.$$

This integral yields the simultaneous algebraic equations

$$(l^2 + L^2)a_0 + \left(\frac{l^2}{5} + \frac{L^2}{7}\right) L^2 a_1 = \frac{5}{8} \frac{u'''}{k},$$

$$\left(\frac{l^2}{5} + \frac{L^2}{7}\right) a_0 + \frac{1}{7}\left(\frac{11}{5} l^2 + \frac{L^2}{3}\right) L^2 a_1 = \frac{1}{8} \frac{u'''}{k}.$$

Solving these equations for a_0 and a_1 gives

$$a_0 = \frac{(u'''/16k)(L^2/3 + 24l^2/5)}{(L^4/21 + 8L^2 l^2/15 + 12l^4/25)}, \qquad a_1 = \frac{(u'''/16k)}{(L^4/21 + 8L^2 l^2/15 + 12l^4/25)}.$$

Hence the second-order Ritz profile based on the variational formulation is

$$\frac{\theta(x, y)}{u'''l^2/k} = \frac{1}{16}\left[1 - \left(\frac{x}{L}\right)^2\right]\left[1 - \left(\frac{y}{l}\right)^2\right]\left[\frac{1/3 + (24/5)(l/L)^2 + (x/L)^2}{1/21 + (8/15)(l/L)^2 + (12/25)(l/L)^4}\right].$$

(8–87)

Before comparing the foregoing approximate solutions with the solution by the integral method and the exact solution, let us consider the Kantorovich method, to which we have so far given no attention.

8–8. Kantorovich Method

Up to now, we have constructed two-dimensional Ritz profiles in the form of Eq. (8–83) by specifying completely the functions $X_n(x)$ and $Y_n(y)$ in the x- and y-directions. However, in many physical problems, the general behavior of the unknown temperature $\theta(x, y)$ in one direction, say the y-direction, may be predicted more accurately than that in the other direction. For such cases, the Ritz profile may be generalized by specifying completely the functions $Y_n(y)$, but leaving the functions $X_n(x)$ unspecified, to be determined by the variational calculus. The use of this profile in variational formulations leads to a set of ordinary differential equations in terms of $X_n(x)$, to be considered with proper boundary conditions in the x-direction. This procedure constitutes the basis of the *Kantorovich method*.

Example 8–6. Using the Kantorovich method, let us find an approximate solution of Example 8–4.

If we approximate the temperature dependence in the y-direction by, for example, a parabola, the function

$$\theta(x, y) = (l^2 - y^2)X(x), \qquad (2\text{-}137)$$

which satisfies the boundary conditions of the problem in the y-direction, may be used as a first approximation. Since a function over a short interval can be approximated closely by a specified function, when $L > l$ Eq. (2-137) is suitable. On the other hand, when $l > L$ the function

$$\theta(x, y) = (L^2 - x^2)Y(y)$$

should be used instead. Similarly, since a function over a long interval can be approximated more closely by higher-degree polynomials, when $L > l$ the function

$$\theta(x, y) = (L^2 - x^2)(l^2 - y^2)(a_0 + a_1 x^2) \qquad (4\text{-}135)$$

should be used for the second approximation of the Ritz profile; but when $l > L$ the function

$$\theta(x, y) = (L^2 - x^2)(l^2 - y^2)(a_0 + a_1 y^2)$$

is more suitable. The preceding remarks point once again to the fact that the accuracy of an approximate procedure rests heavily on the individual's understanding of the physics of the problem and on his skill in the choice of profiles.

Returning now to the solution of the problem by the Kantorovich method, let us insert Eq. (2-137) into Eq. (8-76). This gives

$$4\int_0^L \int_0^l [(l^2 - y^2)X'' - 2X + u'''/k](l^2 - y^2) \, \delta X \, dx \, dy = 0.$$

Integrating this integral in the y-direction, we obtain

$$\left[\tfrac{8}{3}l^3 \int_0^L (\tfrac{4}{5}l^2 X'' - 2X + u'''/k) \, dx\right] \delta X = 0. \qquad (8\text{-}88)$$

Since δX is arbitrary, the integral of Eq. (8-88) must vanish, which implies that the integrand be identically zero. Thus the unspecified function $X(x)$ is found to satisfy the differential equation

$$X'' - \frac{5}{2l^2} X = -\frac{5u'''}{4kl^2}, \qquad (8\text{-}89)$$

subject to the boundary conditions

$$X'(0) = 0, \qquad X(L) = 0. \qquad (8\text{-}90)$$

The solution of Eq. (8–89) which satisfies Eq. (8–90) is

$$X(x) = \frac{u'''}{2k}\left(1 - \frac{\cosh\sqrt{\tfrac{5}{2}}(x/l)}{\cosh\sqrt{\tfrac{5}{2}}(L/l)}\right).$$

Thus a first-order Kantorovich profile obtained by the use of the variational form of the problem is

$$\frac{\theta(x,y)}{u'''l^2/k} = \frac{1}{2}\left[1 - \left(\frac{y}{l}\right)^2\right]\left(1 - \frac{\cosh\sqrt{\tfrac{5}{2}}(x/l)}{\cosh\sqrt{\tfrac{5}{2}}(L/l)}\right). \tag{8–91}$$

Similarly, the second-order Kantorovich profile from Example 4–11,

$$\theta(x,y) = (l^2 - y^2)[X_1(x) + y^2 X_2(x)], \tag{4–141}$$

introduced into Eq. (8–76) gives

$$4\int_0^L \int_0^l [(l^2 - y^2)(X_1'' + y^2 X_2'') - 2X_1 + 2X_2(l^2 - 6y^2) + u'''/k]$$
$$\times (l^2 - y^2)(\delta X_1 + y^2\, \delta X_2)\, dx\, dy = 0.$$

Integrating this integral in the y-direction, we obtain

$$\tfrac{8}{3}l^3\left[\int_0^L \left(\tfrac{4}{5}l^2 X_1'' + \tfrac{4}{35}l^4 X_2'' - 2X_1 - \tfrac{2}{5}l^2 X_2 + \frac{u'''}{k}\right)dx\right]\delta X_1$$
$$+ \tfrac{8}{15}l^5\left[\int_0^L \left(\tfrac{4}{7}l^2 X_1'' + \tfrac{4}{21}l^4 X_2'' - 2X_1 - \tfrac{22}{7}l^2 X_2 + \frac{u'''}{k}\right)dx\right]\delta X_2 = 0. \tag{8–92}$$

Since δX_1 and δX_2 are arbitrary, the integrals of Eq. (8–92) must vanish, which implies that the integrands be identically zero. Then $X_1(x)$ and $X_2(x)$ satisfy the simultaneous differential equations

$$\tfrac{4}{5}l^2 X_1'' + \tfrac{4}{35}l^4 X_2'' - 2X_1 - \tfrac{2}{5}l^2 X_2 + \frac{u'''}{k} = 0,$$
$$\tfrac{4}{7}l^2 X_1'' + \tfrac{4}{21}l^4 X_2'' - 2X_1 - \tfrac{22}{7}l^2 X_2 + \frac{u'''}{k} = 0. \tag{8–93}$$

Again, by introduction of the linear operator $D \equiv d/dx$ Eq. (8–93) may be written in the equivalent operational form

$$2(\tfrac{2}{5}l^2 D^2 - 1)X_1 + \tfrac{2}{5}l^2(\tfrac{2}{7}l^2 D^2 - 1)X_2 = -\frac{u'''}{k},$$
$$2(\tfrac{2}{7}l^2 D^2 - 1)X_1 + \tfrac{2}{7}l^2(\tfrac{2}{3}l^2 D^2 - 11)X_2 = -\frac{u'''}{k}.$$

Using the theory of determinants, or by direct elimination, we obtain the following differential equations for X_1 and X_2:

$$(l^4 D^4 - 28 l^2 D^2 + 63) X_1 = (63/2) u'''/k,$$
$$(l^4 D^4 - 28 l^2 D^2 + 63) X_2 = 0. \qquad (8\text{-}94)$$

The homogeneous parts of X_1 and X_2 have the same characteristic equation

$$l^4 r^4 - 28 l^2 r^2 + 63 = 0,$$

yielding four real roots, $\pm \alpha/l$, $\pm \beta/l$, where $\alpha = [14 - (133)^{1/2}]^{1/2}$ and $\beta = [14 + (133)^{1/2}]^{1/2}$. Thus four independent solutions, conveniently expressed in terms of hyperbolic functions, are $\cosh \alpha(x/l)$, $\sinh \alpha(x/l)$, $\cosh \beta(x/l)$, $\sinh \beta(x/l)$. The thermal and geometric symmetry of the problem with respect to the selected reference frame eliminates $\sinh \alpha(x/l)$ and $\sinh \beta(x/l)$ from the solution. Furthermore, noting the particular solutions $X_1 = u'''/2k$, $X_2 = 0$, we may write the general solutions of X_1 and X_2 in the form

$$X_1 = u'''/2k + A \cosh \alpha(x/l) + B \cosh \beta(x/l),$$
$$X_2 = C \cosh \alpha(x/l) + D \cosh \beta(x/l). \qquad (8\text{-}95)$$

To determine the relationship which must exist among A, B, C, and D, Eq. (8-95) is inserted into either one of the differential equations given by Eq. (8-93). Using the first equation, for example, and requiring that the result to be identically zero, we get

$$(\tfrac{2}{5}\alpha^2 - 1)A + \tfrac{1}{5}l^2(\tfrac{2}{7}\alpha^2 - 1)C = 0, \qquad (\tfrac{2}{5}\beta^2 - 1)B + \tfrac{1}{5}l^2(\tfrac{2}{7}\beta^2 - 1)D = 0.$$

Expressing, say C in terms of A, and D in terms of B, and introducing the result into the second equation of Eq. (8-95), we have

$$X_1 = \frac{u'''}{2k} + A \cosh \alpha \frac{x}{l} + B \cosh \beta \frac{x}{l},$$

$$X_2 = -\frac{5}{l^2}\left(\frac{2\alpha^2/5 - 1}{2\alpha^2/7 - 1}\right) A \cosh \alpha \frac{x}{l} - \frac{5}{l^2}\left(\frac{2\beta^2/5 - 1}{2\beta^2/7 - 1}\right) B \cosh \beta \frac{x}{l}. \qquad (8\text{-}96)$$

Since the temperature symmetry in the x-direction has already been used, the remaining condition, $\theta(L, y) = 0$, is now satisfied by requiring that

$$X_1(L) = 0, \qquad X_2(L) = 0.$$

Using this condition, from Eq. (8-96) we obtain

$$A \cosh \alpha \frac{L}{l} + B \cosh \beta \frac{L}{l} = -\frac{u'''}{2k},$$

$$\left(\frac{2\alpha^2/5 - 1}{2\alpha^2/7 - 1}\right) A \cosh \alpha \frac{L}{l} + \left(\frac{2\beta^2/5 - 1}{2\beta^2/7 - 1}\right) B \cosh \beta \frac{L}{l} = 0.$$

Solving these equations for A and B results in

$$A = -\frac{35}{4}\left(\frac{q'''}{2k}\right)\frac{(2\beta^2/5 - 1)(2\alpha^2/7 - 1)}{(\alpha^2 - \beta^2)\cosh\alpha\,(L/l)},$$

$$B = \frac{35}{4}\left(\frac{q'''}{2k}\right)\frac{(2\alpha^2/5 - 1)(2\beta^2/7 - 1)}{(\alpha^2 - \beta^2)\cosh\beta\,(L/l)}.$$

(8-97)

Finally, combining Eqs. (4-141), (8-96), and (8-97), we find that a second-order Kantorovich profile based on the variational formulation of the problem is

$$\frac{\theta(x,y)}{u'''l^2/k} = \frac{1}{2}\left[1 - \left(\frac{y}{l}\right)^2\right]$$

$$\times \left\{1 + \frac{35}{4}\left(\frac{2\beta^2/5 - 1}{\beta^2 - \alpha^2}\right)[(2\alpha^2/7 - 1) - 5(2\alpha^2/5 - 1)(y/l)^2]\right.$$

$$\times \frac{\cosh\alpha(x/l)}{\cosh\alpha(L/l)} - \frac{35}{4}\left(\frac{2\alpha^2/5 - 1}{\beta^2 - \alpha^2}\right)$$

$$\left.\times [(2\beta^2/7 - 1) - 5(2\beta^2/5 - 1)(y/l)^2]\frac{\cosh\beta(x/l)}{\cosh\beta(L/l)}\right\}. \quad (8\text{-}98)$$

A comparison can now be made between the exact solution of the problem and the approximate Ritz and Kantorovich profiles obtained by the integral and variational techniques. To simplify the algebra of this comparison, only the first approximations will be taken into account, and the space dependence will be eliminated by considering a single location only. For this location let us take the origin of the reference frame, noting that the maximum error will occur at this point (why?). The exact and first-order approximate values of the origin temperature are obtained by inserting $x = 0$, $y = 0$ into Eqs. (4-133), (2-136), (2-143), (8-86), and (8-91). The comparison is summarized in Table 8-3 for three values of L/l.

Tables 8-1 and 8-3, and the experience gained in the integral and variational techniques from the previous sections, lead us to the following conclusions:

(a) In terms of a given profile, the accuracy attained by the variational calculus is better than that of the integral method.

(b) Kantorovich profiles employed with the variational or integral techniques give more accurate results than Ritz profiles, because only part of any Kantorovich profile is chosen *a priori*, the remainder being determined in accordance with the physics of the problem.

(c) It can be shown, using Example 8-6 and Table 8-3, for example, that the accuracy of a Ritz profile decreases with increasing L/l for two-dimensional problems. Clearly, if l is assumed constant but L increasing, the actual temperature distribution becomes difficult to approximate closely by a selected simple profile. Thus the limiting one-dimensional problem which corresponds to $L/l = \infty$ cannot be obtained from the two-dimensional Ritz profile by simply letting $L/l \to \infty$ in this profile. By contrast, the Kantorovich profile converges rapidly to the actual distribution with increasing L/l, because the differential equation obtained by the Kantorovich technique accepts a particular solution as $L/l \to \infty$. Furthermore, if the y-direction of the two-dimensional Kantorovich profile is properly chosen, which was the case in Example 8–6, this profile approaches the exact one-dimensional solution as $L/l \to \infty$.

The reason the Kantorovich method yields a relatively poor result for $L/l = 1$, compared to the results obtained by the same method for $L/l > 1$, is that for $L/l = 1$ the actual temperature distribution is symmetric, hence cannot well be approximated by Kantorovich profiles, which are by nature asymmetric.

TABLE 8–3

	L/l	1	2	4
	Exact	0.29469	0.45687	0.49807
First Approximations	Integral-Ritz	0.37500	0.60000	0.70584
	Error, %	27.3	31.3	41.7
	Variational-Ritz	0.31250	0.50000	0.58824
	Error, %	6.05	9.44	18.1
	Integral-Kantorovich	0.32844	0.48723	0.49902
	Error, %	11.5	6.64	0.190
	Variational-Kantorovich	0.30259	0.45774	0.49821
	Error, %	2.68	0.189	0.0269

(d) Circular or other special functions may be used as well as polynomials in constructing approximate profiles. Although it is difficult to comment on the relative accuracy of these profiles, experience indicates that greater accuracy is obtained using polynomials as compared with circular functions (see Problems 8–1, 8–6). The reason for this is not quite clear. However, the following argument given in terms of Example 8–2 may be helpful. If only the first approximation is desired, polynomials may be more natural than circular functions, since the exact solution of a large number of simple conduction problems is found in terms of polynomials rather than of circular functions. Furthermore, because of the oscillating character of the higher circular harmonics (Fig. 8–15), the convergence of the circular functions is slower than that of polynomials. Nevertheless, circular profiles have two advantages: First, if

First approximation

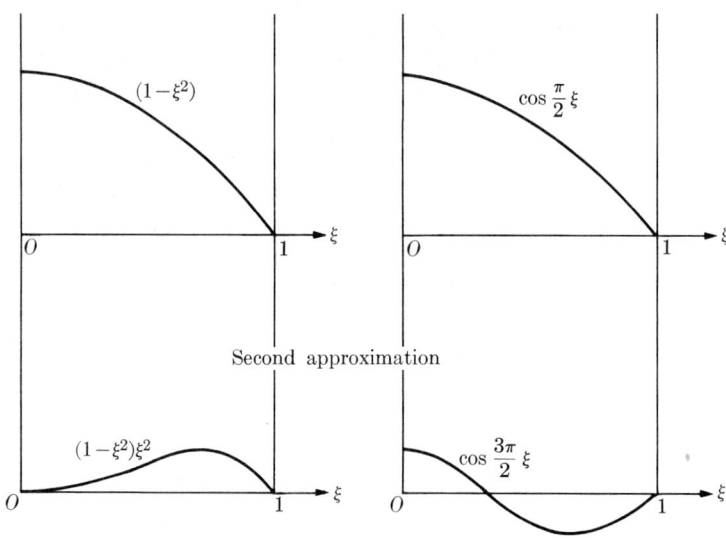

FIG. 8-15

these are selected such that they constitute a set of orthogonal functions, the evaluation of unknown parameters involves less algebra, because the products of different harmonics vanish; second, the values of unknown parameters do not depend on the order of approximation, since they are the Fourier coefficients in the expansion of the exact solution (see Problem 8–1).

8–9. Kantorovich Method Extended

The Kantorovich method is conveniently applied to a general class of steady two-dimensional problems characterized by the internal energy generation $u'''(x, y)$ in a region of the x-y-plane that is bounded by the curves $y = \pm f(x)$ and the straight lines $x = a$, $x = b$ on which the temperature remains constant, say zero (Fig. 8–16).

Using the basic concept of the Kantorovich method, we may construct a first approximation of the temperature distribution by leaving the x-direction to be determined by an unknown function $X(x)$ and satisfying the y-direction by the equations of the boundaries $y = \pm f(x)$. Thus we have

$$\theta(x, y) = (f^2 - y^2) X(x). \tag{8-99}$$

Introducing Eq. (8–99) into Eq. (8–76) gives

$$\int_a^b \int_{-f}^f [(f^2 - y^2)X'' + 4ff'X' + 2(f'^2 + ff'' - 1)X + u'''/k] \\ \times (f^2 - y^2) \, \delta X \, dx \, dy = 0,$$

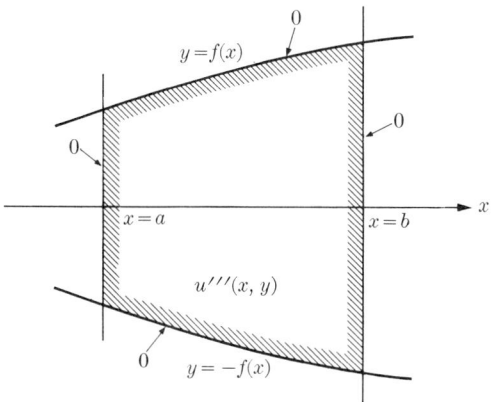

FIG. 8-16

and integrating this equation in the y-direction, we obtain the equation to be satisfied by X as follows:

$$\tfrac{2}{5}f^2 X'' + 2ff'X' + (f'^2 + ff'' - 1)X = 3\Phi/8kf^3, \qquad (8\text{-}100)$$

where

$$\Phi(x) = -\int_{-f}^{f} u'''(x, y)(f^2 - y^2)\, dy.$$

In particular, when $u'''(x, y)$ is constant Eq. (8-100) reduces to

$$\tfrac{2}{5}f^2 X'' + 2ff'X' + (f'^2 + ff'' - 1)X = -u'''/2k. \qquad (8\text{-}101)$$

Three special cases of Eq. (8-101) are considered below:

(a) $f(x) = l = \text{const}$ results in

$$X'' - \frac{5}{2l^2} X = -\frac{5u'''}{4kl^2}, \qquad (8\text{-}89)$$

which is identical to Eq. (8-89) of Example 8-6, as expected.

(b) $f(x) = mx$ yields the Euler (Cauchy) equation in the form

$$x^2 X'' + 5xX' + \frac{5(m^2 - 1)}{2m^2} X = -\frac{5u'''}{4km^2}, \qquad (8\text{-}102)$$

whose solution is given in Section 3-3 (see Table 3-1).

(c) $f(x) = mx^n$, $n \neq 0$, $n \neq 1$ yields an equation reducible to a form of the Bessel equation. The procedure, being rather involved, is not given here.

From Chapter 2 until now we have solved a number of problems by the integral and variational techniques, as well as by the methods of exact solution. Thus we have been able to compare the accuracy of different approximate methods. We have also become aware of the *convenience* of using approximate methods for problems with *complicated exact solutions*,* and of the *importance* of using them *for those with no exact solutions*. We now show how approximate methods are easily applied to problems whose exact solution is not available.

Example 8–7. The uniform internal energy u''' is generated in a triangular bar (Fig. 8–17). The surfaces of the bar are kept at zero temperature. We wish to find the steady temperature of the bar.

According to the selected reference frame, $f(x) = x$ and $m = 1$. Equation (8–102) gives then

$$x^2 X'' + 5xX' = -5u'''/4k. \qquad (8\text{–}103)$$

The general solution of Eq. (8–103) is

$$X(x) = A + \frac{B}{x^4} - \frac{5u'''}{16k} \ln x. \qquad (8\text{–}104)$$

Since

$$\lim_{x \to 0} x^2 \ln x \to 0,$$

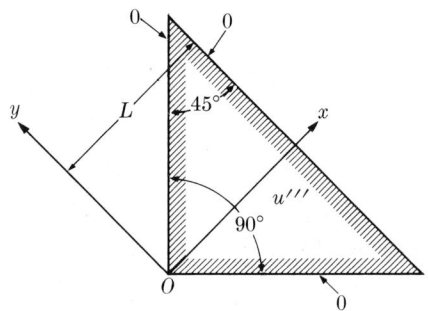

FIG. 8–17

the condition of finite temperature at the origin, $\theta(0, 0) \to$ finite, considered with the combination of Eqs. (8–99) and (8–104) yields $B = 0$. Then the other boundary condition, $X(L) = 0$, gives

$$A = (5u'''/16k) \ln L.$$

Introducing the values of A and B into Eq. (8–104) and the result into Eq. (8–99) yields the first-order Kantorovich profile

$$\frac{\theta(x, y)}{u''' L^2/k} = \frac{5}{16}\left[\left(\frac{x}{L}\right)^2 - \left(\frac{y}{L}\right)^2\right] \ln\left(\frac{L}{x}\right). \qquad (8\text{–}105)$$

The same problem may also be solved by the Ritz method. Thus we assume a profile in the form

$$\theta(x, y) = (x^2 - y^2)(L - x)a_0, \qquad (8\text{–}106)$$

which satisfies all the boundaries; introducing this profile into Eq. (8–76) yields

$$\int_0^L \int_{-x}^x (-4a_0 x + u'''/k)(x^2 - y^2)(L - x)\, \delta a_0\, dx\, dy = 0. \qquad (8\text{–}107)$$

Integrating Eq. (8–107) with respect to x and y gives

$$a_0 = 6u'''/16kL.$$

---

* Recall the exact solution of Problem 3–30 for the case of constant pressure.

Introducing the value of a_0 into Eq. (8–106) results in the first-order Ritz profile

$$\frac{\theta(x, y)}{u'''L^2/k} = \frac{6}{16}\left[\left(\frac{x}{L}\right)^2 - \left(\frac{y}{L}\right)^2\right]\left(1 - \frac{x}{L}\right). \qquad (8\text{–}108)$$

Note that if we expand $\ln (L/x)$ into a Taylor series around $x/L = 1$ as

$$\ln (L/x) = -\ln (x/L) = (1 - x/L) + \tfrac{1}{2}(1 - x/L)^2 - \cdots,$$

then Eq. (8–105) may be rearranged in the form

$$\frac{\theta(x, y)}{u'''L^2/k} = \frac{5}{16}\left[\left(\frac{x}{L}\right)^2 - \left(\frac{y}{L}\right)^2\right]\left[\left(1 - \frac{x}{L}\right) + \frac{1}{2}\left(1 - \frac{x}{L}\right)^2 - \cdots\right]. \qquad (8\text{–}109)$$

Thus the Ritz profile given by Eq. (8–108) corresponds, except for a constant, to the first term in the power series expansion of the Kantorovich profile, Eq. (8–109). A similar relationship may be shown to exist between the Ritz and Kantorovich profiles of other problems.

8–10. Construction of Steady Two-Dimensional Profiles

Let us now consider the general rules useful in constructing steady two-dimensional profiles:

(a) *The Ritz profile of problems involving zero temperature on the boundaries.* This case may be reduced to that of finding the equation of the boundaries with respect to the selected reference frame. For example, a suitable Ritz profile for the steady two-dimensional problem given by Fig. 8–18 may readily be written in the form

$$\theta(x, y) = xy(1 - x/L - y/b)(1 + x/a - y/l)(a_0 + a_1x + b_1y + \cdots),$$

where the first four factors, when separately equated to zero, give the respective equations of the boundaries.

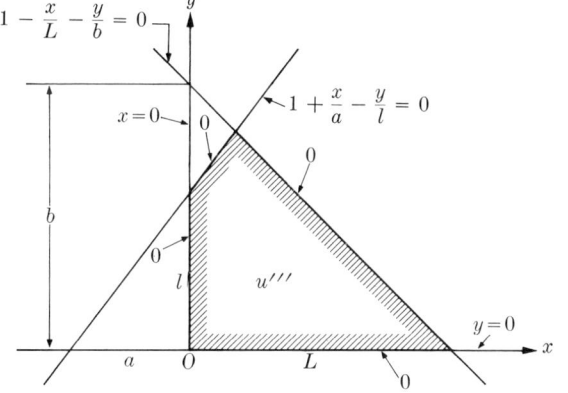

FIG. 8–18

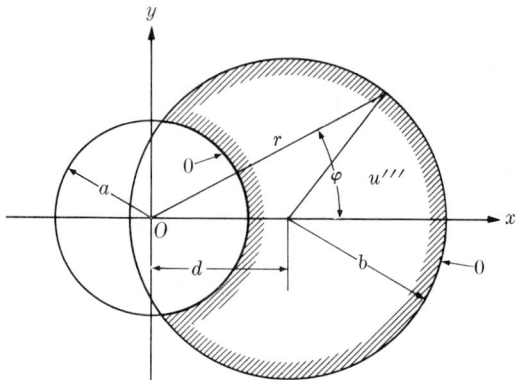

FIG. 8–19

Similarly, the Ritz profile of the steady two-dimensional problem of Fig. 8–19 may be constructed as

$$\theta(x, y) = (a^2 - x^2 - y^2)[b^2 - (x - d)^2 - y^2](a_0 + a_1 x + b_1 y^2 + \cdots),$$

where the first two factors are obtained from the equations of the circles. The same problem may also be expressed in terms of polar coordinates in the form

$$\theta(r, \varphi) = (a - r)(b^2 - d^2 - r^2 + 2rd \cos \varphi)(a_0 + a_1 r \cos \varphi + \cdots).$$

In general, the Ritz procedure is powerful for handling problems involving irregular geometry with boundaries kept at a constant temperature, say zero; these problems are not suitable to exact analytical methods. However, when the boundary conditions are other than those of constant temperature, the Ritz profile often becomes difficult to contruct. The Kantorovich method is also inconvenient for these problems.

(b) *The Ritz and Kantorovich profiles of problems involving regular geometry, and nonzero temperature on the boundaries.* By means of the principle of superposition, the Ritz profile of these problems may be expressed in terms of problems having zero temperature on the boundaries plus one or more suitable terms satisfying the nonzero temperatures on the boundaries.

Consider, for example, that the temperature of one boundary of a steady two-dimensional problem is specified, say by a parabola, while the remaining boundaries are kept at zero temperature (Fig. 8–20). Assuming $\theta(x, y)$ as the temperature distribution to be found, a new function $\psi(x, y)$ is defined such that the difference $\theta(x, y) - \psi(x, y)$ satisfies the boundary conditions of the problem. The simplest possible profile for this difference is

$$\theta(x, y) - \psi(x, y) = (x/L)[1 - (y/l)^2]T_0.$$

Here $\psi(x, y)$, satisfying the homogeneous boundary conditions

$$\psi(x, \pm l) = \psi(0, y) = \psi(L, y) = 0,$$

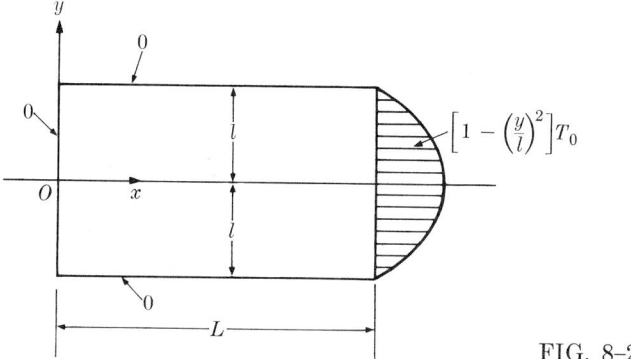

FIG. 8-20

can readily be determined. The Ritz profile of $\psi(x, y)$ is

$$\psi(x, y) = x(L - x)(l^2 - y^2)(a_0 + a_1 x + b_1 y^2 + \cdots).$$

Unlike the Ritz profile, the Kantorovich profile requires no superposition, and can be directly expressed in the form

$$\theta(x, y) = T_0[1 - (y/l)^2][X_0(x) + y^2 X_1(x) + \cdots],$$

subject to boundary conditions

$$X_0(0) = 0, \quad X_0(L) = 1,$$
$$X_1(0) = 0, \quad X_1(L) = 0,$$
$$\vdots \qquad\qquad \vdots$$

On the other hand, if the nonhomogeneous temperature is constant (Fig. 8-21), the boundaries of the problem cannot be represented adequately by a finite number of approximations. Here, expressing the uniform temperature by an orthogonal (or nonorthogonal) set and leaving the x-direction to be determined, we may conveniently employ the Kantorovich method. For

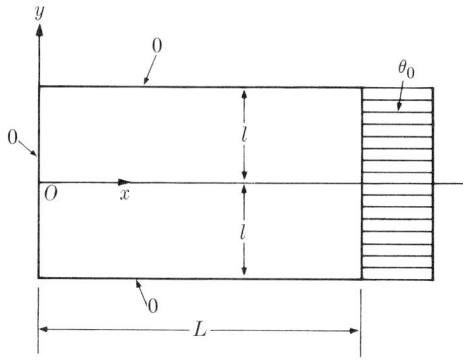

FIG. 8-21

example, using the orthogonal set which is the result of the characteristic-value problem

$$Y'' + \lambda^2 Y = 0, \qquad Y'(0) = 0, \qquad Y(l) = 0,$$

we may write the Kantorovich profile in the form

$$\frac{\theta(x,y)}{\theta_0} = 2 \sum_{n=0}^{\infty} \frac{(-1)^n}{(\lambda_n l)} X_n(x) \cos \lambda_n y, \qquad \lambda_n l = (2n+1)\frac{\pi}{2}, \qquad n = 0, 1, 2, \ldots,$$

with the appropriate boundary conditions

$$X_n(0) = 0, \qquad X_n(L) = 1;$$

this leads to the *exact* solution of the problem.

The Ritz procedure, requiring the superposition of two profiles, say,

$$\frac{\theta(x,y)}{\theta_0} = \psi(x,y) + 2 \sum_{n=0}^{\infty} \frac{(-1)^n}{(\lambda_n l)} \left(\frac{x}{L}\right)^{n+1} \cos \lambda_n y, \qquad \lambda_n l = (2n+1)\frac{\pi}{2},$$

$$n = 0, 1, 2, \ldots,$$

is tedious. Here $\psi(x,y)$ vanishes on the boundaries of the problem.

(c) *The Ritz and Kantorovich profiles of problems involving regular geometry and heat transfer to the ambient.* In terms of a product solution the problem is reduced to the selection of one-dimensional problems in each direction of the problem. For example, a Ritz profile for Example 2–1 with a moderate h (see Fig. 8–14) may readily be written in the form

$$\theta(x,y) = \left[\left(1 + \frac{2}{hL/k}\right)L^2 - x^2\right]\left[\left(1 + \frac{2}{hl/k}\right)l^2 - y^2\right]a_0 + (L-x)^2(l-y)^2$$

$$\times [a_1 + a_2 x + a_3 x^2 + \cdots + b_2 y + b_3 y^2 + \cdots$$

$$+ c_3 xy + c_4 x^2 y + c_5 xy^2 + \cdots], \qquad x \geq 0, \; y \geq 0,$$

and a Kantorovich profile may be written in the form

$$\theta(x,y) = \left[\left(1 + \frac{2}{hl/k}\right)l^2 - y^2\right]X_0(x)$$

$$+ (l-y)^2[X_1(x) + yX_2(x) + y^2 X_3(x) + \cdots], \qquad y \geq 0,$$

together with the boundary conditions

$$X_0'(0) = 0, \qquad X_0'(L) + \frac{h}{k}X_0(L) = 0,$$

$$X_1'(0) = 0, \qquad X_1'(L) = X_1(L) = 0,$$

$$\vdots \qquad\qquad \vdots$$

Profiles for irregular geometry with specified temperature or heat flux, or heat transfer to the ambient, are often difficult to construct, and need further investigation.

The procedure for steady two-dimensional cartesian problems may readily be extended to steady three-dimensional problems. Similarly, steady one-, two-, and three-dimensional problems of cylindrical and spherical geometry require no additional basic knowledge.

8–11. Unsteady Problems

As with steady problems, by taking the governing differential equation of an unsteady problem as the Euler equation of the desired variational formulation we readily obtain

$$\int_{t_1}^{t_2} \int_\mathcal{V} \left(\nabla^2 \theta - \frac{1}{a} \frac{\partial \theta}{\partial t} \right) \delta\theta \, d\mathcal{V} \, dt = 0, \tag{8-110}$$

where $d\mathcal{V}$ is the differential volume element. Although the proof is too involved for us to consider here, it is possible to establish the equality of Eq. (8–110) to the variational formulation, $\delta I = 0$. Thus Eq. (8–110) will be used in this section as an equivalent form of the variational formulation of unsteady problems.

The difficulty in determining the timewise variation of the temperature makes the Ritz procedure unsuitable to unsteady problems. However, if we leave the time direction undetermined, the Kantorovich method may conveniently be used. We assume a profile,*

$$\theta(\mathcal{V}, t) = \theta[\mathcal{V}, \tau_0(t), \tau_1(t), \ldots, \tau_n(t)] \tag{3-211}$$

which is specified in space, but depends on the unknown functions $\tau_0(t)$, $\tau_1(t), \ldots, \tau_n(t)$ in time. Inserting Eq. (3–211) into Eq. (8–110) gives

$$\int_{t_1}^{t_2} \int_\mathcal{V} \left(\nabla^2 \theta - \frac{1}{a} \frac{\partial \theta}{\partial t} \right) \left(\frac{\partial \theta}{\partial \tau_0} \delta\tau_0 + \frac{\partial \theta}{\partial \tau_1} \delta\tau_1 + \cdots + \frac{\partial \theta}{\partial \tau_n} \delta\tau_n \right) d\mathcal{V} \, dt = 0. \tag{8-111}$$

Since Eq. (8–111) is true for an arbitrary time interval (t_1, t_2), the integrand relative to time integration must vanish everywhere in this interval.† Thus we have

$$\int_\mathcal{V} \left(\nabla^2 \theta - \frac{1}{a} \frac{\partial \theta}{\partial t} \right) \left(\frac{\partial \theta}{\partial \tau_0} \delta\tau_0 + \frac{\partial \theta}{\partial \tau_1} \delta\tau_1 + \cdots + \frac{\partial \theta}{\partial \tau_n} \delta\tau_n \right) d\mathcal{V} = 0. \tag{8-112}$$

---

* See Section 3–10.
† This reasoning is identical to that following Eq. (2–138).

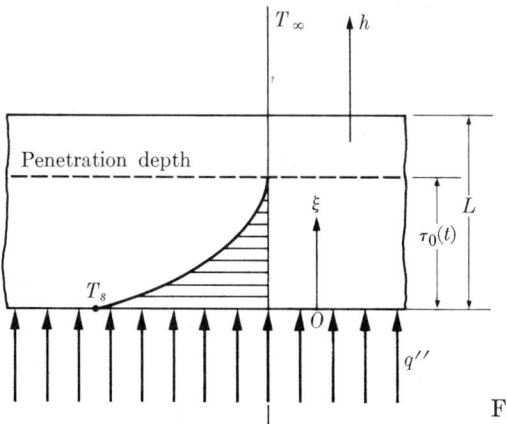

FIG. 8–22

Furthermore, since the variations $\delta\tau_0, \delta\tau_1, \ldots, \delta\tau_n$ are arbitrary, their coefficients must vanish in Eq. (8–112), giving the equations

$$\int_{\mathcal{V}} \left(\nabla^2 \theta - \frac{1}{a} \frac{\partial \theta}{\partial t} \right) \frac{\partial \theta}{\partial \tau_i} d\mathcal{V} = 0, \quad i = 0, 1, 2, \ldots, n. \quad (8\text{–}113)$$

After space integration, Eq. (8–113) yields $(n + 1)$ simultaneous ordinary differential equations in terms of the unknown functions $\tau_0(t), \tau_1(t), \ldots, \tau_n(t)$.

Example 8–8. Let us find a first-order approximate solution for the problem of Example 2–3.

Here Eq. (8–113) reduces to

$$\int_0^L \left(\frac{\partial^2 \theta}{\partial x^2} - \frac{1}{a} \frac{\partial \theta}{\partial t} \right) \frac{\partial \theta}{\partial \tau_0} dx = 0 \quad (8\text{–}114)$$

with the assumption that the time dependence is adequately described by a single function τ_0. Again the time domain is conveniently divided into two regions.* In the first domain the penetration depth $\tau_0(t)$ is less than the thickness of the plate. The corresponding profile, in ξ measured from the bottom of the plate (Fig. 8–22), may be written in the form

$$\theta(\xi, t) = \frac{q''}{2k} \tau_0 \left(1 - \frac{\xi}{\tau_0} \right)^2. \quad (8\text{–}115)$$

[Why can we not use the moving coordinate axis of Eq. (2–3) for the present

---

* See the integral formulation of Example 2–3.

problem?] Introducing Eq. (8–115) into Eq. (8–114) and rearranging gives

$$\int_0^{\tau_0} \left\{ \frac{1}{\tau_0} - \frac{1}{2a} \left[\left(1 - \frac{\xi}{\tau_0}\right)^2 + 2\left(1 - \frac{\xi}{\tau_0}\right)\frac{\xi}{\tau_0} \right] \frac{d\tau_0}{dt} \right\}$$

$$\times \left[\left(1 - \frac{\xi}{\tau_0}\right)^2 + 2\left(1 - \frac{\xi}{\tau_0}\right)\frac{\xi}{\tau_0} \right] d\xi = 0.$$

The integration results in

$$d\tau_0^2 = 5a\, dt.$$

The solution of this equation which satisfies the initial condition $\tau_0(0) = 0$ is

$$\tau_0(t) = (5at)^{1/2}. \tag{8–116}$$

Replacing τ_0 by L in Eq. (8–116), we find that the penetration time of the heat flux to the upper surface is*

$$t_0 = L^2/5a. \tag{8–117}$$

The profile constructed from Fig. 2–42 for the second time domain of Example 2–3,

$$\theta(x, t) = \left(\frac{q''}{h} - \tau_1\right)\frac{hx^2}{2kL} + \left(1 + \frac{hx}{k}\right)\tau_1, \tag{2–193}$$

may also be used in the present problem. Thus, introducing Eq. (2–193) into Eq. (8–114), we obtain

$$\int_0^L \left[\frac{1}{kL}(q'' - h\tau_1) - \frac{1}{a}\left(1 + \frac{hx}{k} - \frac{hx^2}{2kL}\right)\frac{d\tau_1}{dt} \right]\left(1 + \frac{hx}{k} - \frac{hx^2}{2kL}\right) dx = 0.$$

The integration gives the differential equation

$$\left[1 + \frac{2}{3}\left(\frac{hL}{k}\right) + \frac{2}{15}\left(\frac{hL}{k}\right)^2\right]\frac{d\tau_1}{dt} + \frac{h}{\rho cL}\left(1 + \frac{hL}{3k}\right)\left(\tau_1 - \frac{q''}{h}\right) = 0. \tag{8–118}$$

The solution of Eq. (8–118), subject to $\tau_1(t_0) = 0$, is

$$\tau_1(t) = \frac{q''}{h}\left\{1 - \exp\left[-\frac{(1 + hL/3k)h(t - t_0)/\rho cL}{1 + (2/3)(hL/k) + (2/15)(hL/k)^2}\right]\right\}.$$

Finally, the complete solution of the problem expressed in terms of ξ is found

---

* Compare Eqs. (8–116) and (8–117) with Eqs. (2–184) and (2–186), respectively.

to be

$$\theta(\xi, t) = \left(\frac{q''}{2k}\right)(5at)^{1/2}\left[1 - \frac{\xi}{(5at)^{1/2}}\right]^2, \quad \text{when} \quad 0 \le t \le 0, \quad (8\text{--}119)$$

$$\frac{\theta(x,t)}{q''L/k} = \frac{1}{2}\left(1 - \frac{\xi}{L}\right)^2 + \left\{\frac{k}{hL} + \frac{1}{2}\left[1 - \left(\frac{\xi}{L}\right)^2\right]\right\}$$

$$\times \left\{1 - \exp\left[-\frac{(1 + hL/3k)h(t - t_0)/\rho cL}{1 + (2/3)(hL/k) + (2/15)(hL/k)^2}\right]\right\}, \quad \text{when} \quad t \ge t_0. \quad (8\text{--}120)$$

Example 8–9. Let us find the solution of the unsteady problem associated with Example 8–3. Initially the system is at the ambient temperature. Then the upper disk suddenly assumes the angular velocity ω.

The dimensionless form of Eq. (8–114) suitable to our problem is

$$\int_0^1 \left[\frac{\partial}{\partial \rho}\left(\rho \frac{\partial \theta}{\partial \rho}\right) - \mu^2 \rho \theta + \nu \rho^2 - \frac{\partial \theta}{\partial t}\right]\frac{\partial \theta}{\partial \tau_i} d\rho = 0, \quad i = 0, 1, 2, \ldots \quad (8\text{--}121)$$

where $t = at^*/R^2$, t^* being the actual time; the rest of the notation is the same as that of Example 8–3.

Assuming an instantaneous temperature distribution identical to the steady temperature given by Eq. (8–46),† and replacing the parameters a_0, a_1 with the time-dependent functions $a_0(t)$, $a_1(t)$, we may write the unsteady profile of the problem in the form

$$\theta(\rho, \tau) = a_0(t) + (3\rho^2 - 2\rho^3)a_1(t). \quad (8\text{--}122)$$

Introducing Eq. (8–122) into Eq. (8–121) yields

$$\int_0^1 \Bigg[6(2\rho - 3\rho^2)a_1 - \mu^2 \rho[a_0 + (3\rho^2 - 2\rho^3)a_1]$$

$$+ \nu \rho^2 - \rho\left[\frac{da_0}{dt} + (3\rho^2 - 2\rho^3)\frac{da_1}{dt}\right]\Bigg] d\rho = 0,$$

$$\int_0^1 \Bigg[6(2\rho - 3\rho^2)a_1 - \mu^2 \rho[a_0 + (3\rho^2 - 2\rho^3)a_1]$$

$$+ \nu \rho^2 - \rho\left[\frac{da_0}{dt} + (3\rho^2 - 2\rho^3)\frac{da_1}{dt}\right]\Bigg](3\rho^2 - 2\rho^3) d\rho = 0.$$

---

† See also Eqs. (8–55) and (8–56).

These integrals give the simultaneous differential equations

$$\frac{da_0}{dt} + \frac{7}{10}\frac{da_1}{dt} + \mu^2 a_0 + \frac{7}{10}\mu^2 a_1 = \frac{2}{3}\nu, \tag{8-123}$$

$$\frac{7}{20}\frac{da_0}{dt} + \frac{2}{7}\frac{da_1}{dt} + \frac{7}{20}\mu^2 a_0 + \left(\frac{2}{7}\mu^2 + \frac{3}{5}\right)a_0 = \frac{4}{15}\nu. \tag{8-124}$$

By means of the linear operator $D \equiv d/dt$, Eqs. (8-123) and (8-124) may be written in the operational forms

$$(D + \mu^2)a_0 + \tfrac{7}{10}(D + \mu^2)a_1 = \tfrac{2}{3}\nu,$$
$$\tfrac{7}{20}(D + \mu^2)a_0 + \tfrac{2}{7}(D + \mu^2 + \tfrac{21}{10})a_1 = \tfrac{4}{15}\nu.$$

Solving these equations for a_0 and a_1, replacing D by d/dt, and solving the resulting differential equations gives

$$a_0(t) = Ae^{-\mu^2 t} + Be^{-(\mu^2 + 840/57)t} + \frac{16\nu}{171\mu^2}\left(\frac{\mu^2 + 105}{\mu^2 + 840/57}\right), \tag{8-125}$$

$$a_1(t) = Ce^{-\mu^2 t} + De^{-(\mu^2 + 840/57)t} + \frac{140\nu}{171(\mu^2 + 840/57)}. \tag{8-126}$$

To determine the relationship which exists among the constants A, B, C, and D, Eqs. (8-125) and (8-126) are inserted, for example, into Eq. (8-124). Thus we have

$$A = 0, \quad B = -\tfrac{10}{7}D.$$

Introducing A and B into Eqs. (8-125) and (8-126), and using the initial conditions $a_0(0) = 0$, $a_1(0) = 0$ results in

$$C = -\frac{84\nu(9/5 + 14/\mu^2)}{171(\mu^2 + 840/57)}, \quad D = \left(\frac{7}{10}\right)\frac{16\nu(\mu^2 + 105)}{171\mu^2(\mu^2 + 840/57)}.$$

Thus inserting the values of A, B, C, and D into Eqs. (8-125) and (8-126), and the result into Eq. (8-122) yields the approximate Kantorovich profile

$$\frac{\theta(\rho, \tau)}{\nu} = \frac{16(\mu^2 + 105)}{171\mu^2(\mu^2 + 840/57)}[1 - e^{-(\mu^2 + 840/57)t}] + \frac{140(3\rho^2 - 2\rho^3)}{171(\mu^2 + 840/57)}$$

$$\times \left[1 - \frac{3}{5}\left(\frac{9}{5} + \frac{14}{\mu^2}\right)e^{-\mu^2 t} + \frac{2}{5}\left(\frac{1}{5} + \frac{21}{\mu^2}\right)e^{-(\mu^2 + 840/57)t}\right]. \tag{8-127}$$

For large values of time the foregoing Kantorovich profile approaches the steady Ritz profile given by Eq. (8-47), as expected.

8-12. Some Definite Integrals

Here, for convenience, we list the results of some definite integrals appearing frequently in problems of variational calculus:

$$\int_0^l x^m(l-x)^n\,dx = \frac{m!n!}{(m+n+1)!} l^{(m+n+1)}, \tag{8-128}$$

$$\int_0^l (l^2 - x^2)^m\,dx = \frac{(2m)!!}{(2m+1)!!} l^{(2m+1)}, \tag{8-129}$$

$$\int_0^l x^{2m+1}(l^2 - x^2)^n\,dx = \frac{(2m)!!(2n)!!}{(2m+2n+2)!!} l^{(2m+2n+2)}, \tag{8-130}$$

$$\int_0^l x^{2m+2}(l^2 - x^2)^n\,dx = \frac{(2m+1)!!(2n)!!}{(2m+2n+3)!!} l^{(2m+2n+3)}, \tag{8-131}$$

$$\int_0^{\pi/2} \sin^m\phi\,d\phi = \int_0^{\pi/2} \cos^m\phi\,d\phi = \begin{cases} \dfrac{(m-1)!!}{m!!}, & m \text{ odd,} \\[2mm] \dfrac{(m-1)!!}{m!!}\dfrac{\pi}{2}, & m \text{ even,} \end{cases} \tag{8-132}$$

$$\int_0^{\pi/2} \sin^m\phi\cos^n\phi\,d\phi = \begin{cases} \dfrac{(m+n-1)!!}{(m+n)!!}, & m \text{ or } n \text{ odd,} \\[2mm] \dfrac{(m+n-1)!!}{(m+n)!!}\dfrac{\pi}{2}, & m \text{ and } n \text{ even,} \end{cases} \tag{8-133}$$

where, by definition,

$$0! = 1, \qquad m! = m(m-1)(m-2)\cdots,$$
$$0!! = 1, \qquad m!! = m(m-2)(m-4)\cdots$$

For integrals other than those listed above, the reader is referred to integral tables.

References

1. L. V. KANTOROVICH and V. I. KRYLOV, *Approximate Methods of Higher Analysis*. New York–Gröningen (Holland): Interscience-Noordhoff, 1958.

2. R. COURANT and D. HILBERT, *Methods of Mathematical Physics I*. New York: Interscience, 1953.

3. P. M. MORSE and H. FESHBACH, *Methods of Theoretical Physics I*. New York: McGraw-Hill, 1953.

4. I. S. SOKOLNIKOFF, *Mathematical Theory of Elasticity*. New York: McGraw-Hill, 1956.

5. F. B. HILDEBRAND, *Methods of Applied Mathematics.* Englewood Cliffs: Prentice-Hall, 1952.

6. R. WEINSTOCK, *Calculus of Variations.* New York: McGraw-Hill, 1952.

Problems

8–1. Reconsider Example 8–2. Find the first- and second-order approximate solutions in terms of orthogonal circular functions by using (a) Eq. (8–23), (b) Eq. (8–26). (c) Compare the complexity of the algebra involved in both cases. (d) Show that the value of the unknown parameters is independent of the order of approximation. (e) Find the higher-order approximations by inspecting the first two approximations. (f) Show that part (e) is equivalent to the expansion of the exact solution $\cosh \mu \xi / \cosh \mu$ into a Fourier cosine series. (g) Compare the accuracy of the results obtained using circular profiles with that of results obtained using polynomials, given in Example 8–2.

8–2. In a small steam boiler, the rims connecting the furnace to the body are made up of thin rods (Fig. 8–23). (a) Show that the steady temperature problem associated with the rims can be made identical to a variational problem. (b) Construct approximate temperature profiles in terms of polynomials. (c) Find the temperature of the rims by using the polynomials of part (b).

8–3. Condensing steam at temperature T_0 flows through a thick-walled pipe (Fig. 8–24). The inner and outer radii of the pipe are R_i, R_o, respectively. The heat transfer coefficient between the outer surface of the pipe and the ambient is h, and the ambient temperature is T_∞.* Under steady conditions, (a) find the functional asso-

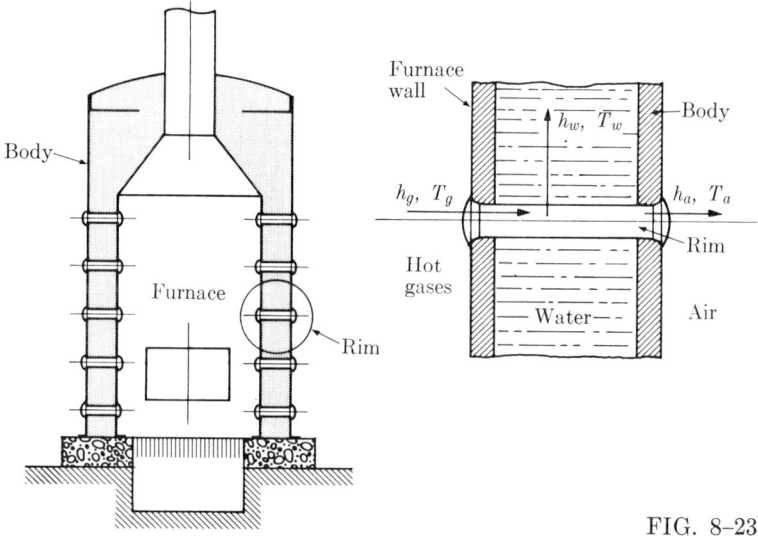

FIG. 8–23

---

* This problem is the steady case of Problem 7–24.

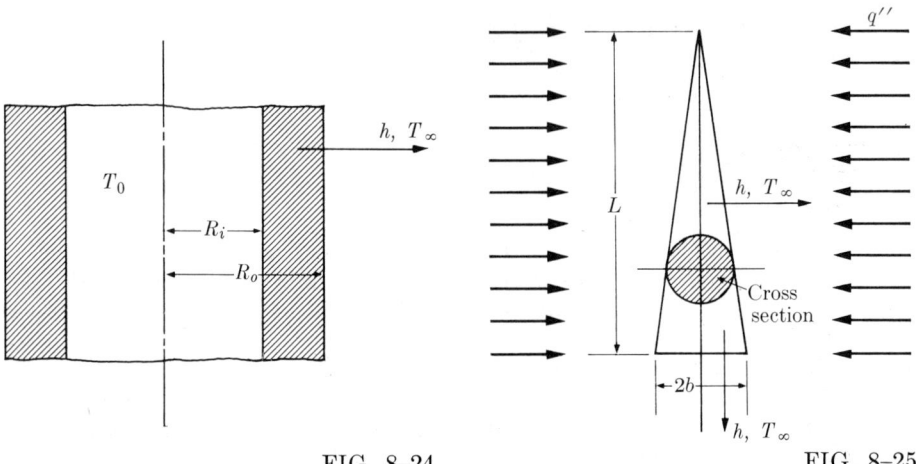

FIG. 8-24   FIG. 8-25

ciated with the temperature of the pipe walls; (b) construct approximate temperature profiles in terms of polynomials; (c) obtain the temperature of the pipe walls by using the polynomials of part (b).

8-4. Reconsider Example 8-3. (a) Find the third-order approximation of the temperature of the disks in terms of polynomials. (b) Compare the centerline temperature of the third-order approximation to that of the second-order approximation as a function of μ.

8-5. A flying object of solid conical shape receives the peripheral radiant heat q'' (Fig. 8-25). The ambient temperature is T_∞, and the heat transfer coefficient is h. The bottom radius and the height of the cone are b and L, respectively, such that $b/L \ll 1$. (a) Find the functional associated with the steady temperature of the object. (b) Obtain an approximate temperature distribution in terms of polynomials.

8-6. Reconsider Example 8-4. (a) Find the first two approximations of the steady heater temperature for a large heat transfer coefficient by using the Ritz and Kantorovich profiles expressed in terms of circular functions. (b) Compare the accuracy of these circular profiles with that of profiles based on polynomials, using the first approximations only. (See Examples 8-5 and 8-6.)

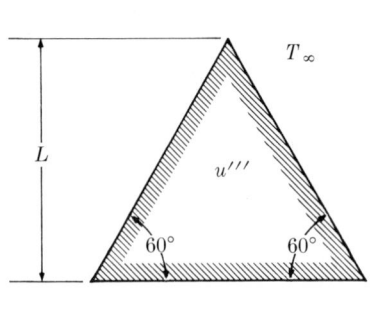

FIG. 8-26

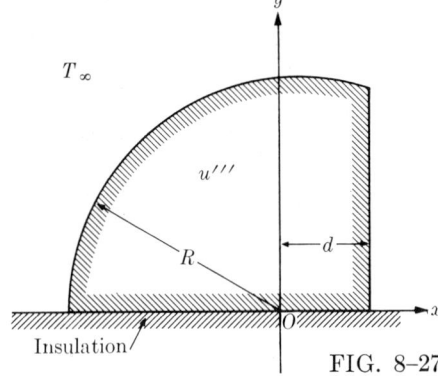

FIG. 8-27

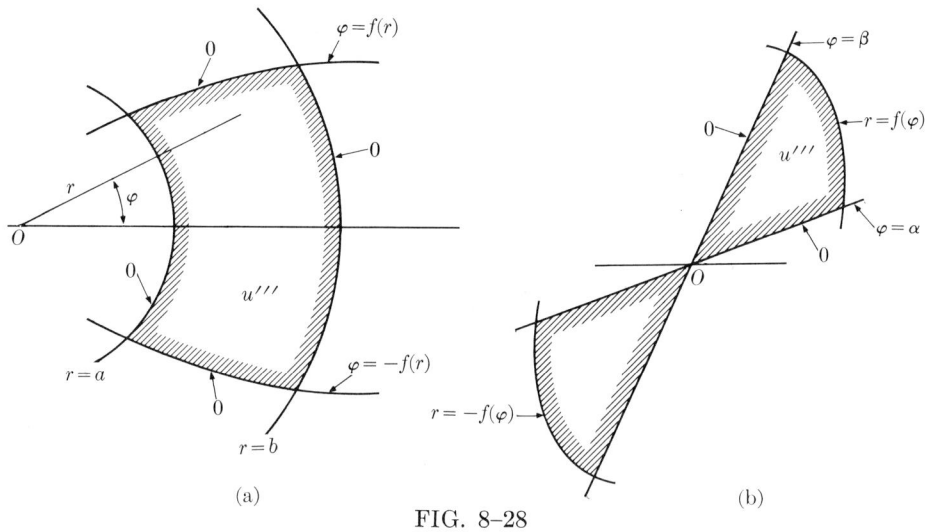

(a) (b)

FIG. 8–28

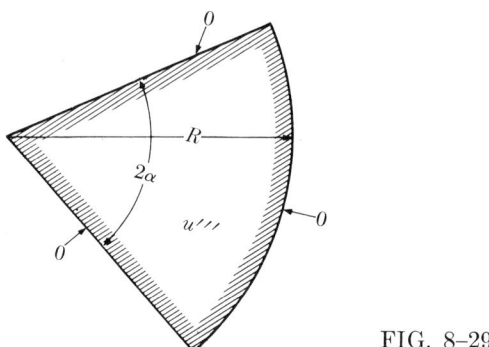

FIG. 8–29

8–7. The internal energy u''' is generated uniformly in a solid triangular rod (Fig. 8–26). The heat transfer coefficient is large. Calculate the steady temperature of this rod, using Ritz and Kantorovich profiles expressed in terms of polynomials.

8–8. The internal energy u''' is generated uniformly in a rod whose cross section is shown in Fig. 8–27. One surface of the rod is insulated while the others transfer heat to the ambient with a large heat transfer coefficient. Calculate the steady temperature of this rod, using (a) cartesian Ritz and Kantorovich profiles expressed in terms of polynomials, (b) polar Ritz profiles.

8–9. Generalize the procedure given in Section 8–9 to polar geometry, as follows. (a) Using Fig. 8–28, find the formulas equivalent to those of Eqs. (8–100) and (8–101). (b) Apply the result of Fig. 8–28 to the problem given by Fig. 8–29.

8–10. Write a steady Kantorovich profile for the problems shown in Fig. 8–30.

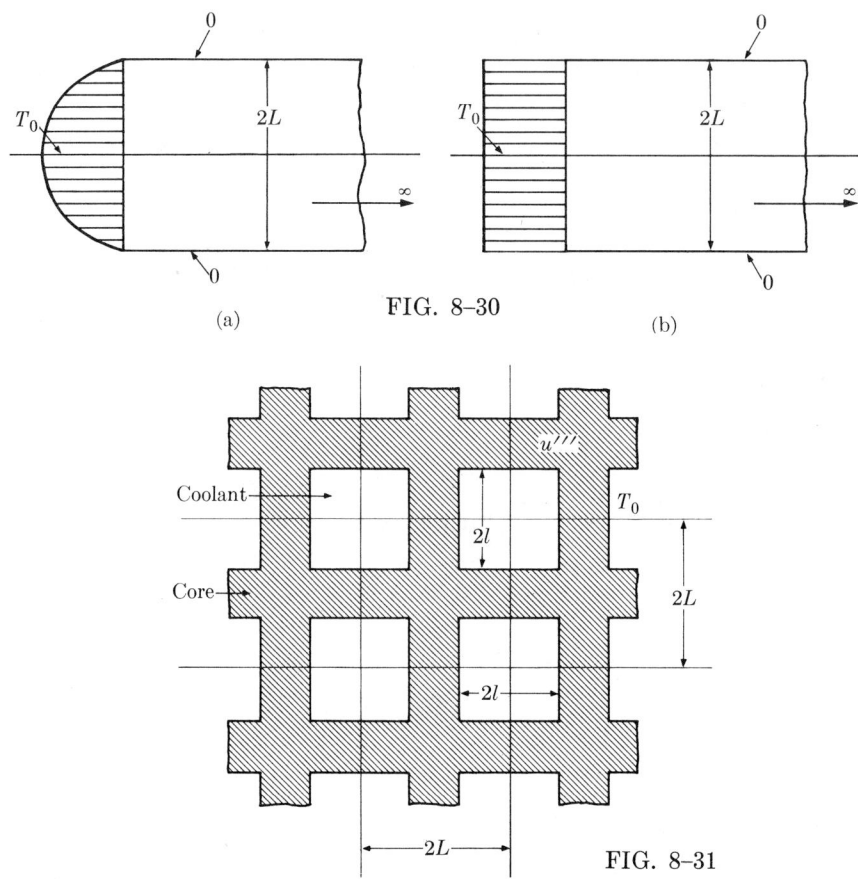

FIG. 8–30

FIG. 8–31

8–11. The internal energy u''' generated uniformly in the core of a nuclear reactor is removed by a coolant flowing at temperature T_0 through the core (Fig. 8–31). Find the first two approximations of the core temperature under steady conditions, assuming (a) a large heat transfer coefficient, (b) a moderate heat transfer coefficient h between the core and the coolant.

8–12. Reconsider Example 2–2. Find the first two approximations of the temperature of the fuel plates, assuming (a) a large heat transfer coefficient, (b) a moderate heat transfer coefficient h between the plates and the coolant.

8–13. Re-solve Problem 5–11.   8–14. Re-solve Problem 5–12.
8–15. Re-solve Example 5–7.   8–16. Re-solve Problem 5–18.
8–17. Re-solve Problem 5–22.   8–18. Re-solve Problem 5–24.

8–19. Find the first- and second-order profiles of the *unsteady* problem associated with Example 2–1.

8–20. Find the first-order profile of the *unsteady* problem associated with Problem 8–7.

CHAPTER 9

DIFFERENCE FORMULATION.
NUMERICAL AND GRAPHICAL SOLUTIONS

Except for a few trivial cases, so far we have considered only single-domain problems of regular geometry. When the boundaries are irregular, that is, not parallel to coordinates of a system, or when we are dealing with a multidomain problem, there is no exact solution available by the known analytical methods. We learned in previous chapters that many problems of this type are solvable by the approximate integral and variational techniques; however, even these techniques are not flexible enough to handle all irregular geometry and/or multidomain problems. The difference method is the only procedure suitable particularly to these problems. Hence we devote this chapter to the difference formulation and its solution by numerical and graphical means. The analog solution of the same formulation will be the subject matter of Chapter 10. Since the difference formulation of steady problems requires solution techniques different from those used for unsteady problems, we shall investigate steady and unsteady problems separately.

9-1. Difference Formulation of Steady Problems

First we shall consider two-dimensional cartesian geometry; then we shall extend the results to three-dimensional cartesian geometry, and to cylindrical and spherical geometries. We illustrate two-dimensional cartesian geometry in terms of Example 2-1 (Fig. 9-1).

Let us follow again the five steps of the formulation (see Section 2-9) and rearrange them for the difference formulation.

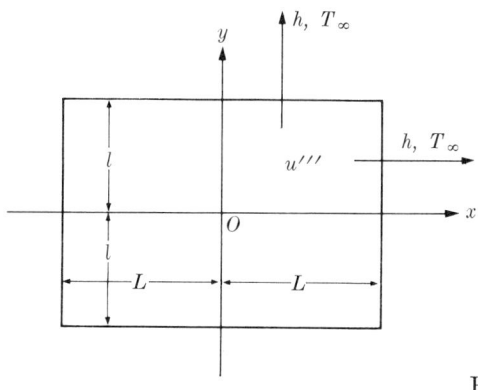

FIG. 9-1

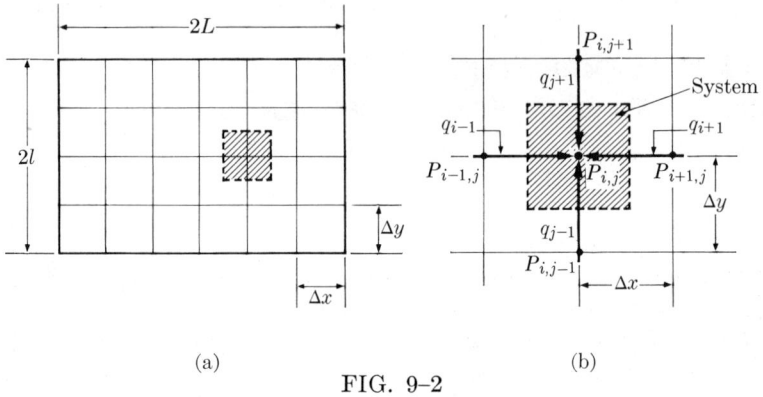

FIG. 9-2

(i) *Define a system* by subdividing the electric heater into a number of squares of small but finite size, forming the so-called *network* (Fig. 9-2a). Here a square network is chosen for convenience only. The geometry or other factors may indicate the use of rectangular, triangular, or hexagonal networks instead (see Section 9-4). For example, suppose that the geometry of the electric heater of Fig. 9-2 did not permit a square network of desired size. Then the use of a rectangular network or a suitable combination of square and rectangular networks would become necessary. The intersections of networks are called the *nodal points* P, and are identified by two subscripts, say i and j, to indicate the row and the column of the point, respectively. Nodal points are classified according to their location as *inner* and *boundary* nodal points. The system for a typical inner nodal point is shown in Fig. 9-2(b). The systems for boundary nodal points are considered in step (v) of the formulation.

(ii) *State the general law* in terms of the system defined by the preceding step. The first law of thermodynamics applied to the system of Fig. 9-2(b) gives

$$0 = q_{i-1}(\Delta y \cdot 1) + q_{j-1}(\Delta x \cdot 1)$$
$$+ q_{i+1}(\Delta y \cdot 1) + q_{j+1}(\Delta x \cdot 1) + u'''(\Delta x\, \Delta y \cdot 1), \qquad (9\text{-}1)$$

where 1 denotes the unit thickness. The directions of heat fluxes are arbitrarily selected toward the nodal point for algebraic convenience. Thus all signs appearing in Eq. (9-1) are made positive.

(iii) *State the particular law* in terms of the preceding steps. Relating q_{i-1}, q_{j-1}, q_{i+1}, and q_{j+1} to the temperature by Fourier's law of conduction we have

$$q_{i-1} = k\frac{T_{i-1,j} - T_{i,j}}{\Delta x}, \qquad q_{j-1} = k\frac{T_{i,j-1} - T_{i,j}}{\Delta y},$$
$$q_{i+1} = k\frac{T_{i+1,j} - T_{i,j}}{\Delta x}, \qquad q_{j+1} = k\frac{T_{i,j+1} - T_{i,j}}{\Delta y}, \qquad (9\text{-}2)$$

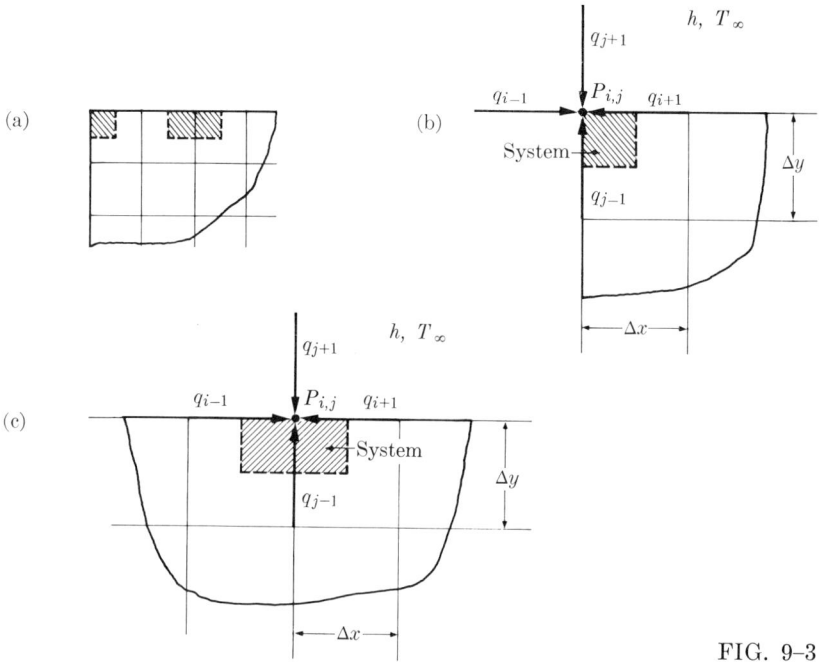

FIG. 9-3

where T denotes the temperature, with appropriate subscripts corresponding to the nodal points.

(iv) *Obtain the governing difference equation* by inserting Eq. (9-2) into Eq. (9-1). Noting that $\Delta x = \Delta y$ (say Δl), and assuming the material of the heater to be homogeneous and isotropic, we find that

$$T_{i-1,j} + T_{i,j-1} + T_{i+1,j} + T_{i,j+1} - 4T_{i,j} + u'''(\Delta l)^2/k = 0. \qquad (9\text{-}3)$$

Eq. (9-3) applies equally to *all* inner nodal points of Fig. 9-2(a).

(v) *Write the boundary conditions* in terms of typical boundary nodal points and their associated systems, shown in Fig. 9-3(a) and elaborated in Fig. 9-3(b, c). Note that the difference form of boundary conditions, unlike their differential form, is related to a volume. Thus the statement of this form of boundary conditions requires that the first four steps of formulation be applied to the boundary systems shown in Fig. 9-3(b, c).

The first law of thermodynamics applied to the system of Fig. 9-3(b) gives

$$0 = q_{i-1}\left(\frac{\Delta y}{2} \cdot 1\right) + q_{j-1}\left(\frac{\Delta x}{2} \cdot 1\right) + q_{i+1}\left(\frac{\Delta y}{2} \cdot 1\right)$$
$$+ q_{j+1}\left(\frac{\Delta x}{2} \cdot 1\right) + u'''\left(\frac{\Delta x}{2} \frac{\Delta y}{2} \cdot 1\right). \qquad (9\text{-}4)$$

Relating q_{i-1} and q_{j+1} to the temperature by the definition of h,

$$q_{i-1} = q_{j+1} = h(T_\infty - T_{i,j}),$$

and q_{j-1} and q_{i+1} to the temperature by Eq. (9-2), and noting that

$$\Delta x = \Delta y = \Delta l,$$

we have

$$\frac{1}{2}\left(T_{i,j-1} + T_{i+1,j}\right) - \left(1 + \frac{h\,\Delta l}{k}\right) T_{i,j} + \frac{h\,\Delta l}{k} T_\infty + \frac{u'''(\Delta l)^2}{4k} = 0. \quad (9\text{-}5)$$

Equation (9-5) applies to the corners of the heater. Following the foregoing procedure in terms of Fig. 9-3(c) we find that the difference equation for the remaining boundaries is

$$\frac{1}{2}\left(T_{i-1,j} + T_{i+1,j}\right) + T_{i,j-1} - \left(2 + \frac{h\,\Delta l}{k}\right) T_{i,j} + \frac{h\,\Delta l}{k} T_\infty + \frac{u'''(\Delta l)^2}{2k} = 0. \quad (9\text{-}6)$$

Thus the formulation of the problem is reduced to a set of simultaneous algebraic equations composed of three typical equations given by Eqs. (9-3), (9-5), and (9-6). (How many equations and unknowns?) The numerical solution of these algebraic equations will be discussed in Section 9-7.

9-2. Relation Between Difference and Differential Formulations

In Section 2-9 we learned that problems may be formulated according to the physical or the mathematical approach. Throughout the text, we have preferred to use the physical approach, and in the preceding section we obtained the difference formulation of steady problems by that same approach. Now, using the mathematical approach, we shall show the relation between the difference and the differential formulations. For this purpose, consider again the differential formulation of Example 2-1. From Chapter 2, we have

$$\frac{\partial^2 T}{\partial x^2} + \frac{\partial^2 T}{\partial y^2} + \frac{u'''}{k} = 0. \quad (2\text{-}128)$$

Since Eq. (2-128) applies locally, it may be written for an inner nodal point $P_{i,j}$ as follows:

$$\left(\frac{\partial^2 T}{\partial x^2} + \frac{\partial^2 T}{\partial y^2} + \frac{u'''}{k}\right)_{P_{i,j}} = 0. \quad (9\text{-}7)$$

Now, replacing the partial derivatives by the equivalent difference forms yields

$$\left(\frac{\partial^2 T}{\partial x^2}\right)_{P_{i,j}} = \left(\frac{\Delta_x^2 T}{\Delta x^2}\right)_{P_{i,j}} = \left.\frac{\Delta_x(\Delta_x T/\Delta x)}{\Delta x}\right|_{P_{i,j}}, \quad (9\text{-}8)$$

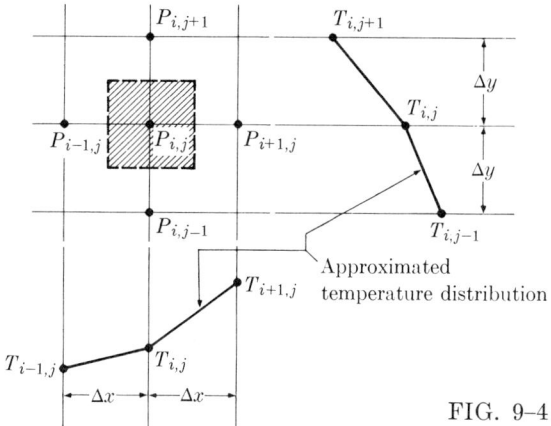

FIG. 9-4

where Δ_x denotes the difference in the direction of x. The numerator of the last expression in Eq. (9-8) is simply the difference in $\Delta_x T/\Delta x$ at point $P_{i,j}$ in the direction of x. It follows from Fig. 9-4 that

$$\left(\frac{\partial^2 T}{\partial x^2}\right)_{P_{i,j}} = \frac{1}{\Delta x}\left(\frac{T_{i+1,j} - T_{i,j}}{\Delta x} - \frac{T_{i,j} - T_{i-1,j}}{\Delta x}\right)$$

$$= \frac{1}{(\Delta x)^2}(T_{i+1,j} + T_{i-1,j} - 2T_{i,j}). \qquad (9\text{-}9)$$

Similarly,

$$\left(\frac{\partial^2 T}{\partial y^2}\right)_{P_{i,j}} = \frac{1}{(\Delta y)^2}(T_{i,j+1} + T_{i,j-1} - 2T_{i,j}). \qquad (9\text{-}10)$$

Introducing Eqs. (9-9) and (9-10) into Eq. (9-7), and again assuming a square network so that $\Delta x = \Delta y = \Delta l$, we obtain Eq. (9-3), as expected. [Obtain the difference form of the boundary conditions from their differential form. Is the result identical to Eqs. (9-5) and (9-6)? Why?]

9-3. Error in Difference Formulation

We now investigate the error involved in the difference formulation of steady problems, again in terms of Example 2-1. Recalling the expression of the one-dimensional Taylor series in the neighborhood of a point, say ζ_0,

$$T(\zeta) = T(\zeta_0) + (\zeta - \zeta_0)\left(\frac{dT}{d\zeta}\right)_{\zeta=\zeta_0} + \frac{1}{2!}(\zeta - \zeta_0)^2\left(\frac{d^2 T}{d\zeta^2}\right)_{\zeta=\zeta_0}$$
$$+ \frac{1}{3!}(\zeta - \zeta_0)^3\left(\frac{d^3 T}{d\zeta^3}\right)_{\zeta=\zeta_0} + \frac{1}{4!}(\zeta - \zeta_0)^4\left(\frac{d^4 T}{d\zeta^4}\right)_{\zeta=\zeta_0} + \cdots,$$

we may express $T_{i-1,j}$, $T_{i+1,j}$, $T_{i,j-1}$, and $T_{i,j+1}$ in terms of $T_{i,j}$ as follows:

$$T_{i-1,j} = T_{i,j} - \Delta x \left(\frac{\partial T}{\partial x}\right)_{P_{i,j}} + \frac{1}{2!}(\Delta x)^2 \left(\frac{\partial^2 T}{\partial x^2}\right)_{P_{i,j}}$$
$$- \frac{1}{3!}(\Delta x)^3 \left(\frac{\partial^3 T}{\partial x^3}\right)_{P_{i,j}} + \frac{1}{4!}(\Delta x)^4 \left(\frac{\partial^4 T}{\partial x^4}\right)_{P_{i,j}} - \cdots,$$

$$T_{i+1,j} = T_{i,j} + \Delta x \left(\frac{\partial T}{\partial x}\right)_{P_{i,j}} + \frac{1}{2!}(\Delta x)^2 \left(\frac{\partial^2 T}{\partial x^2}\right)_{P_{i,j}}$$
$$+ \frac{1}{3!}(\Delta x)^3 \left(\frac{\partial^3 T}{\partial x^3}\right)_{P_{i,j}} + \frac{1}{4!}(\Delta x)^4 \left(\frac{\partial^4 T}{\partial x^4}\right)_{P_{i,j}} + \cdots,$$

$$T_{i,j-1} = T_{i,j} - \Delta y \left(\frac{\partial T}{\partial y}\right)_{P_{i,j}} + \frac{1}{2!}(\Delta y)^2 \left(\frac{\partial^2 T}{\partial y^2}\right)_{P_{i,j}}$$
$$- \frac{1}{3!}(\Delta y)^3 \left(\frac{\partial^3 T}{\partial y^3}\right)_{P_{i,j}} + \frac{1}{4!}(\Delta y)^4 \left(\frac{\partial^4 T}{\partial y^4}\right)_{P_{i,j}} - \cdots,$$

$$T_{i,j+1} = T_{i,j} + \Delta y \left(\frac{\partial T}{\partial y}\right)_{P_{i,j}} + \frac{1}{2!}(\Delta y)^2 \left(\frac{\partial^2 T}{\partial y^2}\right)_{P_{i,j}}$$
$$+ \frac{1}{3!}(\Delta y)^3 \left(\frac{\partial^3 T}{\partial y^3}\right)_{P_{i,j}} + \frac{1}{4!}(\Delta y)^4 \left(\frac{\partial^4 T}{\partial y^4}\right)_{P_{i,j}} + \cdots$$

Adding these expressions together, we have for a square network

$$T_{i-1,j} + T_{i,j-1} + T_{i+1,j} + T_{i,j+1} - 4T_{i,j}$$
$$= (\Delta l)^2 \left(\frac{\partial^2 T}{\partial x^2} + \frac{\partial^2 T}{\partial y^2}\right)_{P_{i,j}} + O[(\Delta l)^4], \qquad (9\text{--}11)$$

where all terms containing fourth or higher powers of Δl are included in $O[(\Delta l)^4]$. Combining Eq. (9–7) and (9–11), we obtain

$$T_{i-1,j} + T_{i,j-1} + T_{i+1,j} + T_{i,j+1} - 4T_{i,j} + u'''(\Delta l)^2/k = O[(\Delta l)^4]. \qquad (9\text{--}12)$$

Neglecting $O[(\Delta l)^4]$, which becomes smaller as $\Delta l \to 0$, we obtain the difference formulation given by Eq. (9–3). This formulation is therefore an approximation to the differential formulation only when $(\Delta l)^4$ is small. $O[(\Delta l)^4]$ is called the *truncation error* of Eq. (9–3). Clearly, this error is minimized by selecting as small a network as is practicable. On the other hand, since the smaller the network, the larger the number of nodal points and associated difference equations, from the viewpoint of the algebraic simplicity of the solution we would prefer to select as coarse a network as possible. The best network size is therefore a compromise, and there is no rule other than insight and experience for choosing the optimum size in a given situation. Until much practice has been gained, the only way to judge the errors introduced by too coarse a network is to check the change in temperature with successively finer networks.

9–4. Finer, Graded, Triangular, and Hexagonal Networks

To determine the accuracy of a given difference formulation, it is necessary to formulate the problem for at least two different sizes of network. Successively *finer networks* may be chosen, often each to have squares one-quarter as big as those of the previous network.

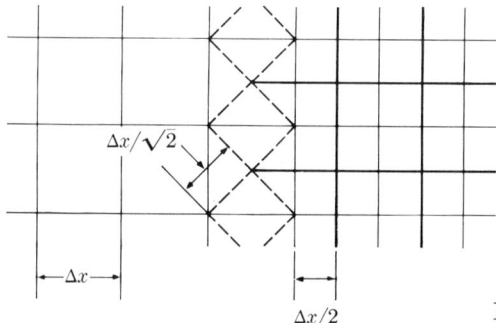

FIG. 9–5

In the difference formulation of some problems a fine network may be needed only in some part of the integration domain. For example, we may wish to have a greater accuracy in one part of the solution; or for the same accuracy we may be forced to a smaller network in one part of the field than in the remainder because the temperature may be varying more rapidly in that part. It is possible, of course, to extend the fine network to the entire domain, but this would entail unnecessary expenditure of time and effort in the solution. Therefore, it is convenient to know how to use networks whose size differs in different parts of the field. A network which exhibits this feature is said to be a *graded network*. We may illustrate the point on a two-dimensional network by halving one part of the network while keeping the remainder unchanged (Fig. 9–5). (Note that an intermediate network of side $\Delta x/\sqrt{2}$ is required to join the coarse and fine networks and that the intermediate network is oriented at 45° to the other two networks.)

It is geometrically possible to cover an integration domain by a uniform network made of squares, or *equilateral triangles* (Fig. 9–6a), or *equiangular hexagons* (Fig. 9–6b). So far we have dealt with square networks. On rare occasions the triangular or hexagonal networks are also used. For example, if

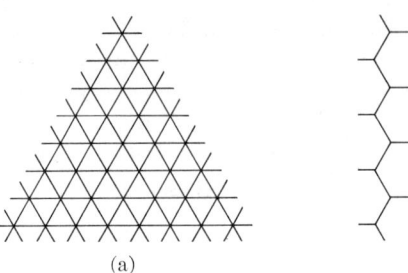

FIG. 9-6

the shape of the boundary is such that a triangular or hexagonal network exactly fits to the field, it is more convenient as well as more accurate to employ these networks.*

9–5. Cylindrical and Spherical Geometries

We now consider the difference formulation of cylindrical problems. Since no additional concept is needed for spherical problems, we leave these to the reader as an exercise. Let us illustrate the cylindrical case in terms of the electric heater of Problem 4–20 (Fig. 9–7). In accordance with the physical approach, we take the appropriate systems as indicated on one-quarter of the heater in Fig. 9–8.

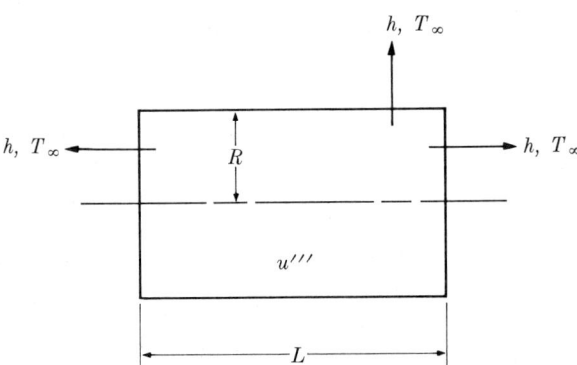

FIG. 9-7

For the system corresponding to a typical inner nodal point the first law of thermodynamics gives

$$0 = q_{i-1} 2\pi \left(R_m - \frac{\Delta l}{2} \right) \Delta l + q_{j-1} 2\pi R_m \, \Delta l$$
$$+ q_{i+1} 2\pi \left(R_m + \frac{\Delta l}{2} \right) \Delta l + q_{j+1} 2\pi R_m \, \Delta l + u''' 2\pi R_m (\Delta l)^2. \qquad (9\text{–}13)$$

---

* For further discussion of triangular and hexagonal networks see, for example, Reference 3, Chapter 10 and Reference 11, Chapter III.

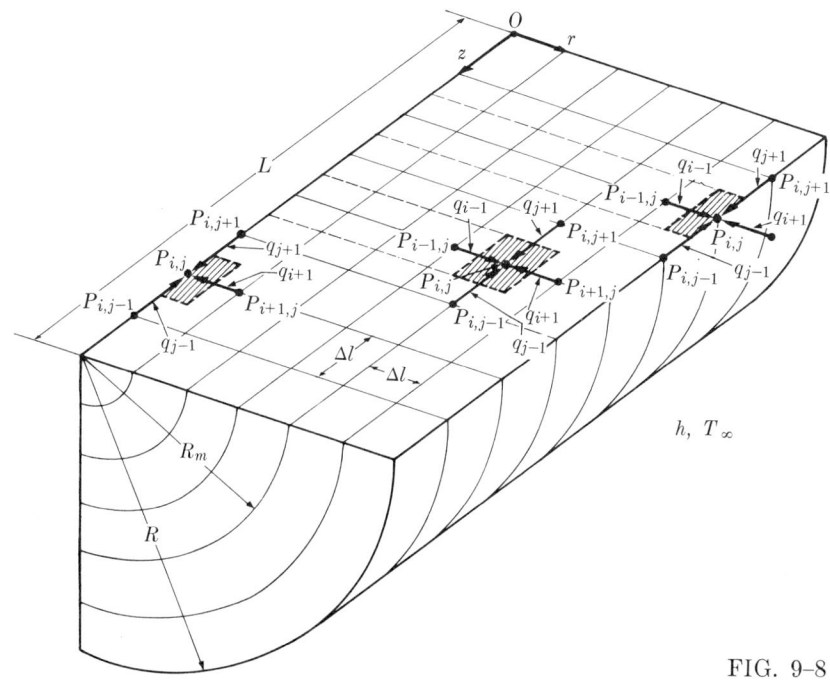

FIG. 9-8

We may relate the heat fluxes of Eq. (9-13) to the temperatures by Fourier's law as stated by Eq. (9-2), provided Eq. (9-2) is interpreted in terms of polar coordinates and a common difference length Δl is used; thus we obtain the difference equation

$$T_{i,j-1} + T_{i,j+1} + \left(1 - \frac{\Delta l}{2R_m}\right) T_{i-1,j} + \left(1 + \frac{\Delta l}{2R_m}\right) T_{i+1,j}$$

$$- 4T_{i,j} + \frac{u'''(\Delta l)^2}{k} = 0. \quad (9\text{-}14)$$

For the system corresponding to a typical centerline nodal point the first law of thermodynamics gives

$$0 = q_{j-1} \frac{\pi(\Delta l)^2}{4} + q_{i+1} \pi \Delta l \, \Delta l + q_{j+1} \frac{\pi(\Delta l)^2}{4} + u''' \frac{\pi(\Delta l)^2}{4} \Delta l. \quad (9\text{-}15)$$

Expressing the heat fluxes of Eq. (9-15) by means of the appropriate form of Fourier's law, we have

$$T_{i+1,j} + \tfrac{1}{4}(T_{i,j-1} + T_{i,j+1}) - \tfrac{3}{2} T_{i,j} + \frac{u'''(\Delta l)^2}{4k} = 0. \quad (9\text{-}16)$$

Similarly, for the typical boundary nodal point indicated in Fig. 9–8 we get

$$\left(1 - \frac{\Delta l}{2R}\right) T_{i-1,j} + \tfrac{1}{2}(T_{i,j-1} - T_{i,j+1}) - \left(2 + \frac{h\,\Delta l}{k} - \frac{\Delta l}{2R}\right) T_{i,j}$$
$$+ \frac{h\,\Delta l}{k} T_\infty + \frac{u'''(\Delta l)^2}{2k} = 0. \qquad (9\text{–}17)$$

(How many types of difference equation does the foregoing problem have?)

The difference equations given by Eqs. (9–14) and (9–16) may also be obtained by the mathematical approach, that is, by considering the term-by-term difference approximation of the differential equation

$$\frac{\partial^2 T}{\partial r^2} + \frac{1}{r}\frac{\partial T}{\partial r} + \frac{\partial^2 T}{\partial z^2} + \frac{u'''}{k} = 0. \qquad (9\text{–}18)$$

However, it should be noted that the nodal points on the axis $r = 0$ are exceptional; there the term $(1/r)(\partial T/\partial r)$ in Eq. (9–18) is indeterminate and must be replaced by its limit as $r \to 0$. It follows from L'Hôpital's rule that

$$\lim_{r \to 0} \frac{1}{r}\frac{\partial T}{\partial r} = \lim_{r \to 0} \frac{\partial T/\partial r}{r} \to \frac{\partial^2 T}{\partial r^2}.$$

Because of the inherent nature of cylindrical problems, the use of a logarithmic scale in the radial direction improves the difference formulation and hence the accuracy of the solution of these problems, but at the expense of increased labor. When the boundaries of a cylindrical problem are arcs of concentric cylinders, it is possible to bring these boundaries into a more convenient (cartesian) shape by conformal mapping.*

9–6. Irregular Boundaries

Let us reconsider Example 2–1 as discussed in Section 9–1, but assume that the boundaries of the electric heater are now irregular and therefore not suitable to simple coordinate systems (Fig. 9–9). In practice, such irregular geometry often occurs—for example, in turbine blades. This type of problem could be solved by approximating the boundaries in terms of, say, a cartesian coordinate system as shown in Fig. 9–9(b). However, the approximation thus introduced may not always be desirable. It is therefore of vital importance that we be able to apply the techniques of difference formulation to irregular geometry. This may, in fact, be done, although only at the expense of the introduction of modified difference equations. Figure 9–9(a) and its elaboration in Fig. 9–9(c) show the typical system suitable to irregular geometry. The first law of thermo-

---

* See, for example, Reference 7, p. 317.

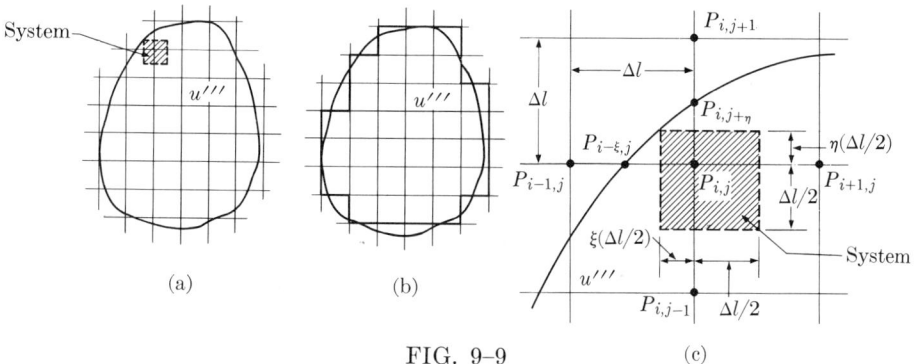

FIG. 9-9

dynamics applied to this system and rearranged by Fourier's law yields

$$\frac{2}{1+\xi}T_{i+1,j} + \frac{2}{1+\eta}T_{i,j-1} + \frac{2}{\xi(1+\xi)}T_{i-\xi,j}$$
$$+ \frac{2}{\eta(1+\eta)}T_{i,j+\eta} - 2\left(\frac{1}{\eta}+\frac{1}{\xi}\right)T_{i,j} + \frac{u'''(\Delta l)^2}{k} = 0. \quad (9\text{-}19)$$

Equation (9-19) reduces, as of course it should, to Eq. (9-3) as $\xi \to 1$ and $\eta \to 1$. By a procedure analogous to that used in Section 9-3, we may also obtain Eq. (9-19) using the appropriate Taylor series expansion of Eq. (9-7). It is possible to show by this procedure that the terms neglected in Eq. (9-19) are of $O[(\Delta l)^3]$. Therefore, for a given size of network, there is more approximation introduced in a boundary difference equation when the boundary is irregular than when it is regular.

9-7. Solution of Steady Problems. Relaxation Method

In the preceding sections we have learned how to reduce the formulation of steady problems to a set of algebraic equations by means of the difference method. We know from mathematics, in principle at least, how to solve these equations by elimination or by determinants. However, when the number of equations increases, say to 10, 20, or even more, the labor required in the use of these procedures becomes excessive, and we are forced to employ other techniques, such as the use of *high-speed computers* and the *relaxation method*.

Computers may be classified as *analog* and *digital* computers. Analog computers include devices which replace the physical variables by electric potentials; their over-all accuracy is limited to the range 0.1–5%. In digital computation all numbers are quantized; there is no clear bound on the accuracy attainable. Despite the astonishingly rapid development of computers in recent years, common-language digital computers are not yet available. Therefore, there is still no place for digital computer solutions in a text of this nature. The analog solution of unsteady problems will be discussed in the next chapter.

Here we consider the solution of algebraic sets by the relaxation method only. However, the reader is urged to solve suitable problems of this chapter by means of available digital computers.

The basic rule of relaxation and some of its widely used modifications may best be explained by employing them to solve a simple pair of algebraic equations,

$$3x - y = 65, \qquad -x + 2y = 40. \tag{9-20}$$

We first rearrange these equations with all their terms on one side, then replace the zeros on the other side by the *residuals* R_1 and R_2 as follows:

$$R_1 = -3x + y + 65, \qquad R_2 = x - 2y + 40. \tag{9-21}$$

The basic rule of relaxation is to start with a pair of x and y, and while introducing orderly changes in the values of x and y, to reduce the currently larger residual to zero, or as nearly zero as desired. The starting values for x and y are, in general, arbitrary for algebraic equations, so they may conveniently be taken equal to zero. However, for problems of physical significance, a thorough understanding of the problem and past experience with similar problems may make it possible to select starting values different from zero, thus shortening the labor of solution to a large extent.

An *operation table*, which shows the effect on the residuals of a positive unit increment in the variables x and y, is convenient in the reduction of residuals. For the problem under consideration, the operation table is

	ΔR_1	ΔR_2
$\Delta x = 1$	-3	1
$\Delta y = 1$	1	-2

$$\tag{9-22}$$

where Δ denotes an incremental change in a quantity. Note that the change Δx or Δy affects both residuals.

Starting with the value of zero for both x and y, we may systematically carry out the algebra in a *relaxation table* as shown below.

x	y	R_1	R_2
0	0	65	40
22		-1	62
	31	30	0
10		0	10
	5	5	0
2		-1	2
	1	0	0
34	37	0	0

$$\tag{9-23}$$

[9-7] SOLUTION OF STEADY PROBLEMS. RELAXATION METHOD 495

Thus, following the basic rule of relaxation, first we change the value of x by $\Delta x = 22$ in order to reduce the larger residual (almost) to zero. This operation decreases x by -66 and increases y by 22, as may readily be seen from the operation table. The same operation reduces R_1 to -1 and increases R_2 to 62 in the relaxation table. Second, we increase the value of y by $\Delta y = 31$, which reduces R_2 to zero and increases R_1 to 30 in the relaxation table, and so on. Note the convergence of the process by observing the continual decrease in the sum of the residuals as the process continues. Since in the last line of the table the residuals are reduced to zero, the final values $x = 34$ and $y = 37$ actually are the exact solution of Eq. (9–20). This may easily be verified by inserting these values into one equation of Eq. (9–20).

In the preceding relaxation table, first we worked with integer numbers and obtained a solution correct to the nearest integers. If necessary, the calculations may be refined to the first decimal place by means of decimal increments, and so on.

One important advantage of the relaxation method is that if errors are made in the process of solving a problem, we do not need to go back and check each step of the solution procedure. The effect of errors may quickly be removed starting from the erroneous result. For example, in the pair of algebraic equations under consideration, suppose that the value 65 of R_1 in the relaxation table (9–23) is accidentally taken to be 55, and the table is completed accordingly, as shown on the left below.

x	y	R_1	R_2		x	y	R_1	R_2
0	0	(55)	40		30	35	10	0
18		1	58	Error	3		1	3
	29	30	0			2	3	-1
10		0	10		1		0	0
	5	5	0		34	37	0	0
2		-1	2					
	1	0	0					
30	35	10	0					

The error becomes evident from the fact that R_1 in the last line, obtained by inserting $x = 30$ and $y = 35$ into the first equation of Eq. (9–21), does not agree with R_1 in the preceding line, obtained by the relaxation procedure. We do not need to know the location of the error—we simply carry on the table until the residuals are actually reduced to zero, as shown on the right above. This time, inserting the values 34 and 37 into Eq. (9–21), we observe the agreement between residuals. No relaxation procedure may be assumed complete without this final check.

The basic rule of relaxation, the continual and systematic reduction of residuals to zero, may readily be applied to the solution of more complex problems. However, by three widely used modifications of this rule—*overrelaxation*, *block relaxation*, and *group relaxation*—and with a little attention, considerable time and effort can be saved. We next consider these modifications.

It is clear even before numerical computations that the use of the basic rule in the first step of relaxation reduces R_1 but increases R_2; in the second step R_2 is reduced but R_1 is increased, and so on. Now, instead of reducing R_1 in the first step, it is, of course, possible to deliberately make it negative such that the second step yields a smaller value for R_1 than that obtained by the basic rule. This procedure is known as *overrelaxation*. To what extent R_1 should be made negative is a matter of experience. However, assuming the convergence of residuals as the relaxation procedure continues, we may take an increment of about half as much as that which reduces R_1 to zero. Thus, instead of the increment $\Delta x = 22$ obtained by the basic rule, we deliberately overrelax x about 50% and take $\Delta x = 33$. The remainder of the relaxation table is completed by the basic rule, as shown below.

Overrelaxation

x	y	R_1	R_2
0	0	65	40
33		−34	73
	37	3	−1
1		0	0
34	37	0	0

A second and even more important modification of the basic rule is known as *block relaxation*. This procedure consists of using a unit block operation; that is, we apply increments to all unknowns simultaneously. For our illustrative example the unit block operation is $\Delta x = \Delta y = 1$; the corresponding change in residuals may be calculated directly from Eq. (9–21) or by adding together the changes produced by the basic unit operations in the operation table (9–22). Thus we have the operation table shown below.

Unit block operation

	ΔR_1	ΔR_2
$\Delta x = 1$	−3	1
$\Delta y = 1$	1	−2
$\Delta x = \Delta y = 1$	−2	−1

We now start a new relaxation table, reducing the initial value of both residuals simultaneously by a block relaxation. Note that the sum of the initial residuals is $105 (= 65 + 40)$, and the unit block operation affects a change in that sum of $-3 (= -2 - 1)$. First considering the block operation $\Delta x = \Delta y = 35(105/3)$, next completing the table by basic operations, we obtain the following table.

[9-7] SOLUTION OF STEADY PROBLEMS. RELAXATION METHOD

Block relaxation

	x	y	R_1	R_2
	0	0	65	40
	35	35	−5	5
		2	−3	1
	−1		0	0
	34	37	0	0

The third modification of the basic rule is *group relaxation*, which consists of using different increments on some unknowns. In general, in a group relaxation our objective is to cause certain residuals to be of the same order but different in sign. These residuals may then be rapidly reduced by basic operations. The extension of our operation table (9–22) by a suitable group operation is as follows.

Group operation

	ΔR_1	ΔR_2
$\Delta x = 1$	−3	1
$\Delta y = 1$	1	−2
$\Delta x = 3, \Delta y = 4$	−5	−5

In terms of this group operation we begin a relaxation table with a group relaxation and obtain the table shown below.

Group relaxation

	x	y	R_1	R_2
	0	0	65	40
	30	40	15	−10
	5		0	−5
		−3	−3	1
	−1		0	0
	34	37	0	0

Because of the triviality of our illustrative example, we had to employ overrelaxation, block relaxation, and group relaxation procedures at the beginning of our relaxation tables. In actual cases these procedures should be adopted at whatever point they are appropriate.

Here we terminate our study of the relaxation method. However, the reader is urged to consider, within the spirit of the method, the possibility of further modifications of the basic rule of relaxation in conjunction with special situations. Now, starting with a trivial example, we illustrate the use of the relaxation method by applying it to the solution of three examples of physical significance.

Example 9-1. Consider a flat plate of thickness $2L = 4$ in. Assume that internal energy is generated in this plate at the rate of

$$u''' = 14{,}400 \text{ Btu/ft}^3\cdot\text{hr.}$$

The thermal conductivity of the plate is $k = 10$ Btu/ft·hr·°F. The ambient temperatures are 50°F and 100°F. The heat transfer coefficients are large. We wish to find the temperature distribution in the plate.

Let us assume a somewhat crude network, and divide the plate into four equal parts of thickness Δx. The temperatures of the nodal points are denoted by T_0, T_1, T_2, T_3, and T_4. Since the surface temperatures are given, the only temperatures to be determined are T_1, T_2, and T_3.

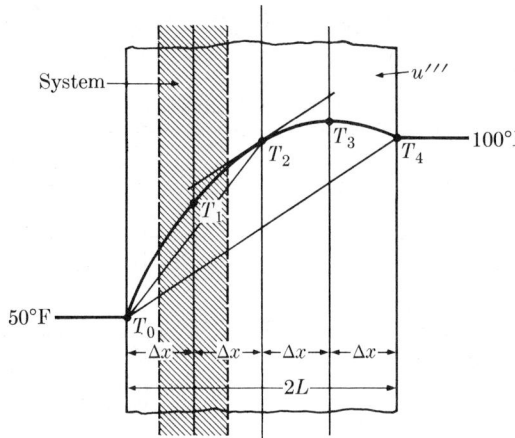

FIG. 9-10

We apply the first law of thermodynamics to the typical system shown in Fig. 9-10; the result related to the temperatures by Fourier's law gives

$$0 = T_0 + T_2 - 2T_1 + u'''(\Delta x)^2/k.$$

The difference equations corresponding to nodal points 2 and 3 are the same, provided the temperature subscripts are changed accordingly. Noting that

$$\frac{u'''(\Delta x)^2}{k} = \frac{14{,}400(1/12)^2}{10} = 10°\text{F}, \tag{9-24}$$

and introducing the residuals R_1, R_2, and R_3, we have

$$\begin{aligned} R_1 &= T_2 - 2T_1 + 60, \\ R_2 &= T_1 + T_3 - 2T_2 + 10, \\ R_3 &= T_2 - 2T_3 + 110. \end{aligned} \tag{9-25}$$

[9-7] SOLUTION OF STEADY PROBLEMS. RELAXATION METHOD

We then have the following operation table.

		ΔR_1	ΔR_2	ΔR_3
	$\Delta T_1 = 1$	-2	1	0
	$\Delta T_2 = 1$	1	-2	1
	$\Delta T_3 = 1$	0	1	-2
Block	$\Delta T_1 = \Delta T_2 = \Delta T_3 = 1$	-1	0	-1

In the absence of energy generation the temperature distribution would be linear, and we would have $T_1 = 62.5°F$, $T_2 = 75°F$, and $T_3 = 87.5°F$. The energy generation should increase these temperatures by about 10°F according to Eq. (9-24). However, if we add this amount to T_2, then T_1 and T_3 must be augmented by an amount between 5°F and 10°F.* Assuming, for example, 7.5°F as the temperature to be added to these temperatures, we may start our relaxation table with

T_1	R_1	T_2	R_2	T_3	R_3
70	5	85	5	95	5

where the residuals have been evaluated from Eq. (9-25). The equality of these residuals suggests the search for a group operation affecting the residuals equally. In fact, after a few trials, we find the appropriate extension of the operation table.

		ΔR_1	ΔR_2	ΔR_3
Group I	$\Delta T_1 = \Delta T_3 = 1, \ \Delta T_2 = 2$	0	-2	0
Group II	$\Delta T_1 = \Delta T_3 = 3, \ \Delta T_2 = 4$	-2	-2	-2
= Group I + 2 Block				

Now the relaxation table may be easily completed as shown below.

T_1	R_1	T_2	R_2	T_3	R_3
70	5	85	5	95	5
7.5	0	10	0	7.5	0
77.5	0	95	0	102.5	0

---

* See the temperature distribution sketched in Fig. 9-10, and note the straight line connecting T_0 to T_2.

Despite the crude network used, the foregoing temperatures correspond to the *exact* values of these temperatures (why?).

Example 9–2. Two surfaces of a square rod are kept at 100°F, while the other two surfaces are kept at 0°F. We wish to find the temperature distribution in the rod.

Let us assume the network shown in Fig. 9–11. Clearly, a finer network would be necessary to represent the temperature distribution; here, however, our purpose is to demonstrate the features of the relaxation method rather than to obtain an accurate solution. Because of the thermal symmetry of the problem, we need consider only the temperature of the four locations indicated in the figure. The corresponding difference equations, modified by the residuals, are

$$R_1 = 2T_2 + 2T_3 - 4T_1,$$
$$R_2 = T_1 + 2T_4 + 100 - 4T_2,$$
$$R_3 = T_1 + 2T_4 - 4T_3,$$
$$R_4 = T_2 + T_3 + 100 - 4T_4.$$

Thus we have the operation table shown below.

	ΔR_1	ΔR_2	ΔR_3	ΔR_4
$\Delta T_1 = 1$	-4	1	1	0
$\Delta T_2 = 1$	2	-4	0	1
$\Delta T_3 = 1$	2	0	-4	1
$\Delta T_4 = 1$	0	2	2	-4

Now we may start the relaxation table by guessing the temperature of locations 1, 2, 3, and 4. The exact temperature of location 1 is $T_1 = 50°F$ (why?). Then a linear interpolation gives $T_2 = 75°F$ and $T_3 = 25°F$; the corresponding residuals are as follows.

T_1	R_1	T_2	R_2	T_3	R_3	T_4	R_4
50	0	75	-50	25	50	50	0

The values of these residuals suggest the use of a group operation which inversely and equally affects R_2 and R_3 without influencing R_1 and R_4. Inspection of the part of the operation table already established reveals the desired grouping as shown below.

		ΔR_1	ΔR_2	ΔR_3	ΔR_4
Group	$\Delta T_2 = -1$, $\Delta T_3 = 1$	0	4	-4	0

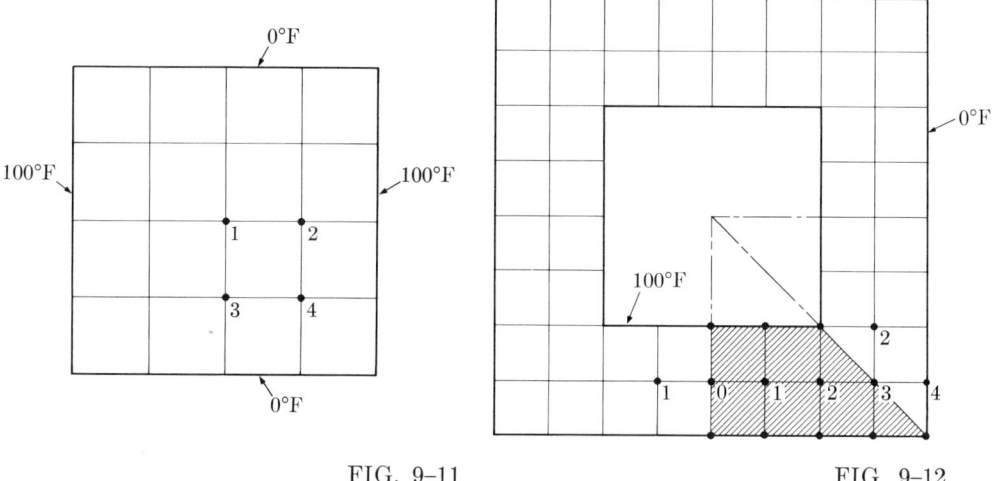

FIG. 9-11   FIG. 9-12

Then we readily complete the relaxation table in the following form.

T_1	R_1	T_2	R_2	T_3	R_3	T_4	R_4
50	0	75 −12.5	−50 0	25 12.5	50 0	50	0
50	0	62.5	0	37.5	0	50	0

(Is it possible to guess these temperatures without going through a relaxation procedure?)

Note that neither Example 9–2 nor the trivial Example 9–1 required the use of the relaxation method, for both may be solved by exact and approximate methods with no difficulty. Next, however, we consider a problem whose exact solution is not available, and whose approximate solution by integral or variational means represents considerable difficulty.

Example 9–3. The inner and outer surfaces of a chimney are kept at 100°F and 0°F, respectively. The ratio between the outer and inner sides is 2 to 1. We wish to find the temperature distribution in the chimney.

Again we are interested in the relaxation method itself rather than an accurate solution of the problem.* Therefore, we assume the coarse network shown in Fig. 9–12. Because of the thermal symmetry, we need consider only one-eighth of the chimney, as shaded in Fig. 9–12.

---

* Approximate solutions based on finer networks may be found in the literature. See, for example, Reference 6, Chapter 18.

The appropriate difference equations, modified by the residuals, are

$$R_0 = 2T_1 + 100 - 4T_0,$$
$$R_1 = T_0 + T_2 + 100 - 4T_1,$$
$$R_2 = T_1 + T_3 + 100 - 4T_2,$$
$$R_3 = 2T_2 - 4T_3.$$

Then we have the following operation table.

	ΔR_0	ΔR_1	ΔR_2	ΔR_3
$\Delta T_0 = 1$	-4	1	0	0
$\Delta T_1 = 1$	2	-4	1	0
$\Delta T_2 = 1$	0	1	-4	2
$\Delta T_3 = 1$	0	0	1	-4

Next we start guessing the temperature of the nodal points. Because of the thickness of the chimney, the temperature is approximately one-dimensional in the neighborhood of the nodal point 0. Thus we have $T_0 = 50°F$ and, by assuming a parabolic distribution between the nodal points 0 and 4, $T_1 = 47°F$, $T_2 = 38°F$, and $T_3 = 22°F$. The complete relaxation table is shown below.

	T_0	R_0	T_1	R_1	T_2	R_2	T_3	R_3	Steps
	50	-6	47	0	38	17	22	-12	Start
				4	4	1		-4	1
Group I	-1	0	1	0	1	-3	-1	2	2
				-1	-1	1		0	3
		-0.6	-0.3	0.2		0.7			4
				0.4	0.2	-0.1		0.4	5
	-0.2	0.2		0.2					6
						0	0.1	0	7
Group II	0.1	0	0.1	-0.1		0.1			8
	48.9	0	47.8	-0.1	42.2	0.1	21.1	0	

Note that we start with the basic rule. Then the resulting residuals suggest the following group operation.

	ΔR_1	ΔR_2	ΔR_3	ΔR_4
Group I $\Delta T_0 = \Delta T_3 = -1$, $\Delta T_1 = \Delta T_2 = 1$	6	-4	-4	6

After the fourth step calculations are refined to the first decimal place; again we employ the basic rule except in the eighth step, which implies the group operation shown below.

		ΔR_1	ΔR_2	ΔR_3	ΔR_4
Group II	$\Delta T_0 = \Delta T_1 = 1$	-2	-3	1	0

Inspection of the residuals of the third and eighth steps reveals that these differ by a decimal point only. This property allows us to carry the solution of difference equations to an unlimited number of decimal places. For example, the results to be obtained by carrying the calculations to the next decimal place would give $T_0 = 48.89$, $T_1 = 47.78$, $T_2 = 42.22$, and $T_3 = 21.11$, and the residuals $R_0 = 0$, $R_1 = -0.01$, $R_2 = 0.01$, and $R_3 = 0$. However, it should be remarked that the foregoing improvement in accuracy has no effect on the approximation involved in the difference formulation of the problem.

9–8. Difference Formulation of Unsteady Problems. Stability

As with steady problems, we introduce the difference formulation of unsteady problems in terms of an illustrative example taken from cartesian geometry. Thus we consider a flat electric heater plate of negligible thickness between two plates of thickness L, density ρ, specific heat c, and thermal conductivity k (Fig. 9–13). The energy generated in the heater is transferred through the plates to the ambient. The heat transfer coefficient is large, and the temperatures are measured relative to the ambient temperature.* Assume now that the heater is turned off. We wish to find the difference formulation of the subsequent unsteady problem.

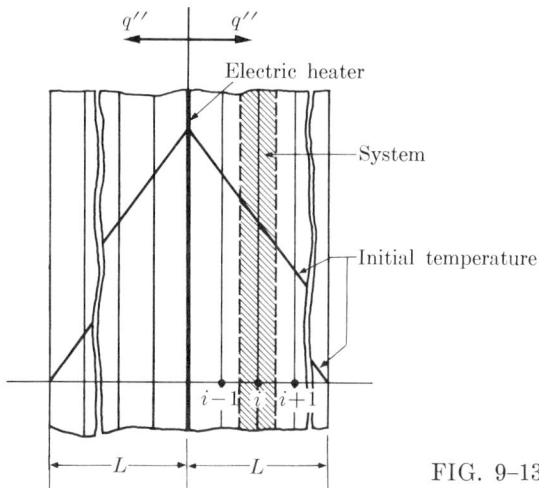

FIG. 9–13

---

* The initial steady problem is a special case of that associated with Fig. 2–15.

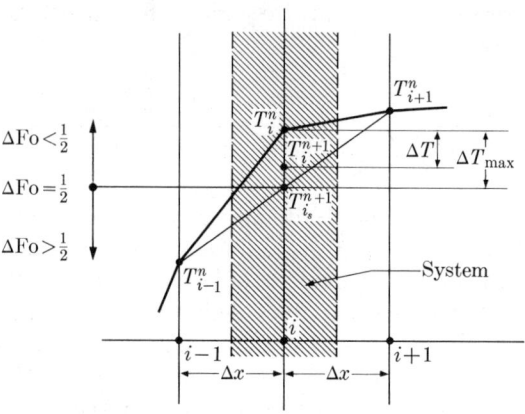

FIG. 9–14

Applying the first law of thermodynamics to the system shown and elaborated in Figs. 9–13 and 9–14, respectively, and relating the result to the temperatures by means of Fourier's law, we obtain

$$\rho c(\Delta x)\frac{T_i^{n+1} - T_i^n}{\Delta t} = k\frac{T_{i-1}^n - T_i^n}{\Delta x} + k\frac{T_{i+1}^n - T_i^n}{\Delta x}, \quad (9\text{–}26)$$

and subsequently

$$T_i^{n+1} - T_i^n = \Delta\text{Fo}\,(T_{i+1}^n - 2T_i^n + T_{i-1}^n), \quad (9\text{–}27)$$

where, as before, the subscripts denote locations; the newly introduced superscripts denote time intervals, and $\Delta\text{Fo} = a(\Delta t)/(\Delta x)^2$ is Fourier's modulus based on the difference values of time and space.

If we know the temperatures T_{i+1}^n, T_i^n, and T_{i-1}^n, as well as ΔFo, the temperature change at location i from T_i^n to T_i^{n+1} during the time interval

$$(n+1)\,\Delta t - n\,\Delta t$$

may be found from Eq. (9–27). Clearly, selecting some values for the intervals Δt and Δx, and getting the value of a (for the plate material) from a table of properties, we can evaluate ΔFo. Postponing comment on the solution of unsteady problems to Section 9–9, let us assume for the time being that we have carried out the calculations for two slightly different values of ΔFo, say $\frac{5}{11}$ and $\frac{5}{9}$. This corresponds to slightly different values for the time intervals if the plate is divided into the same equal subintervals, say 10, for both cases. The results are compared with the exact solution in Figs. 9–15 and 9–16. Inspection of these figures reveals that although the results corresponding to the value $\Delta\text{Fo} = \frac{5}{11}$ are quite satisfactory, those corresponding to $\Delta\text{Fo} = \frac{5}{9}$ are influenced by some sort of accumulated and amplified error. The phenomenon occurring in the latter case is called *instability*. The error thus introduced has nothing to do with the truncation error discussed in Section 9–2 or the *round-off*

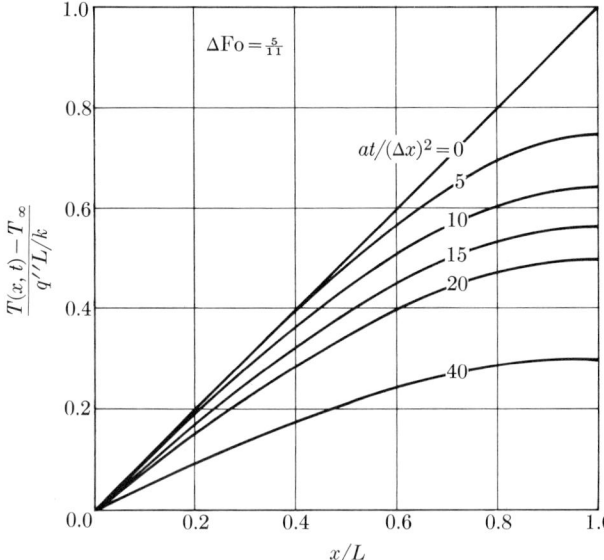

FIG. 9-15

error resulting from the use of rounded numbers, both of which are negligibly small in the present example; it is, rather, a property of difference equations, and it increases with successively smaller values of Δx unless Δt is also reduced properly. We therefore learn the important fact that *the difference formulation of unsteady problems is never complete without the statement of the stability criteria for the difference equations.* Formal but somewhat lengthy mathematical proofs on stability may be found in the literature.* Here we suggest a rather simple physical argument based on the approximation involved in the problems of field theory: *the spacewise variation of volume terms is negligible as compared with their timewise variation; the timewise variation of surface terms is negligible as compared with their spacewise variation.*† Actually, we have always used this approximation in the formulation of our problems without emphasizing it. For example, in the development of Eq. (9–26) we neglected the spacewise variation in the internal energy, which is a volume term, and the timewise variation in the heat fluxes by conduction, which are surface terms. Therefore, in the time interval $(n+1)\Delta t - n\Delta t$, the temperature T_i^n changes to T_i^{n+1} while the temperatures T_{i-1}^n and T_{i+1}^n remain constant in order of magnitude. The limit of T_i^{n+1} is its steady value $T_{i_s}^{n+1}$, which may be obtained by assuming that T_i^{n+1} reaches this value in the interval $(n+1)\Delta t - n\Delta t$. Since the steady one-dimensional temperature distribution in a plate of constant thermal conductivity is a straight line, $T_{i_s}^{n+1}$ is at the intersection of the isothermal plane of location i with a straight line connecting T_{i-1}^n and T_{i+1}^n. Correspond-

---

* See, for example, Reference 4.
† Recall Problem 5–7.

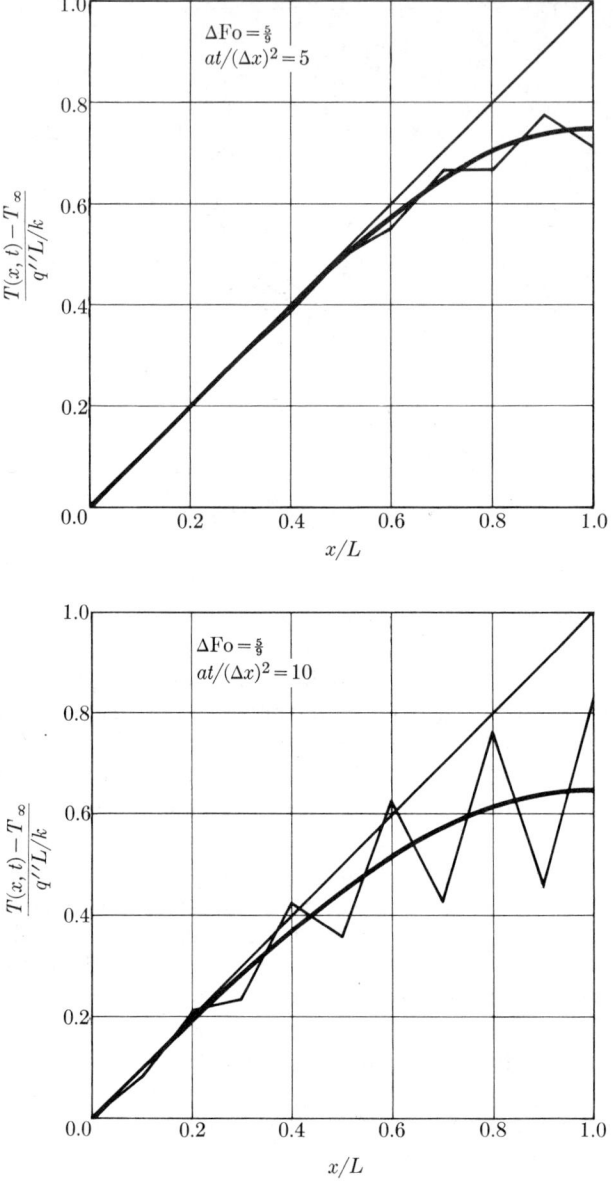

FIG. 9-16

ingly, the limiting steady form of Eq. (9-26) or Eq. (9-27) is

$$0 = T_{i+1}^n - 2T_{i_s}^{n+1} + T_{i-1}^n. \qquad (9\text{-}28)$$

Now, subtraction of Eq. (9-28) from (Eq. 9-27) and subsequent rearrangement

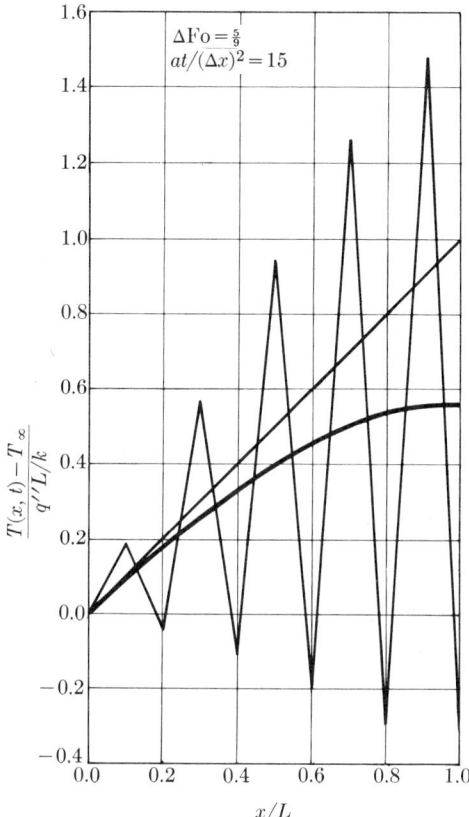

FIG. 9–16 (cont.)

in terms of the nomenclature of Fig. 9–14 gives

$$\Delta T = 2(\Delta \text{Fo}) \Delta T_{\max}.$$

Furthermore, noting from the same figure that $\Delta T \leq \Delta T_{\max}$, we find that the stability criterion of Eq. (9–27) is

$$\Delta \text{Fo} \leq \tfrac{1}{2}. \tag{9-29}$$

Thus the reason behind the selection of the values $\tfrac{5}{11}$ and $\tfrac{5}{9}$ for ΔFo in Figs. 9–15 and 9–16 becomes clear. It is remarkable that the case $\Delta \text{Fo} > \tfrac{1}{2}$, which causes instability in the difference equations, also violates the physics of the problem by giving a temperature that exceeds its steady value. For the limiting case $\Delta \text{Fo} = \tfrac{1}{2}$, Eq. (9–27) reduces to Eq. (9–28) or, equivalently, to

$$T_{i_s}^{n+1} = \frac{T_{i+1}^n + T_{i-1}^n}{2}. \tag{9-30}$$

Equation (9–30) readily admits a graphical interpretation as follows. The temperature change $T_{i_s}^{n+1}$ at location i in the time interval $(n+1)\Delta t - n\Delta t$

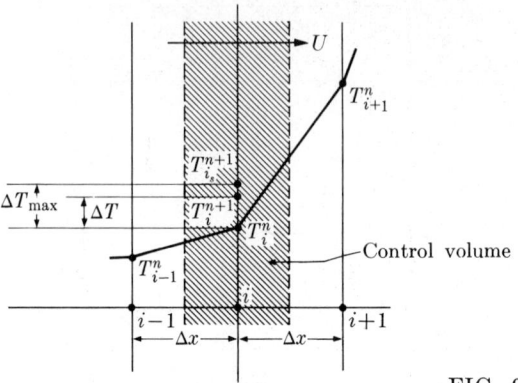

FIG. 9–17

is the arithmetic mean of the neighboring temperatures T_{i+1}^n and T_{i-1}^n taken from the preceding interval; thus the intersection of the straight line connecting T_{i+1}^n to T_{i-1}^n with the isothermal plane of location i gives $T_{i_s}^{n+1}$ (Fig. 9–14). This is the basis of the *Binder-Schmidt graphical method*, which will be discussed in Section 9–9.

We continue our discussion by considering a moving one-dimensional continuum (Fig. 9–17).* The first law of thermodynamics and Fourier's law successively applied to the control volume of Fig. 9–17 give

$$\rho c(\Delta x)\frac{T_i^{n+1} - T_i^n}{\Delta t} = \rho c U(T_{i-1}^n - T_i^n) + k\frac{T_{i-1}^n - T_i^n}{\Delta x} + k\frac{T_{i+1}^n - T_i^n}{\Delta x},$$

which may be rearranged in the form

$$T_i^{n+1} - T_i^n = \Delta\text{Fo}\,[T_{i+1}^n - (2 + \Delta\text{Pé})\,T_i^n + (1 + \Delta\text{Pé})T_{i-1}^n], \qquad (9\text{--}31)$$

where $\Delta\text{Pé} = U(\Delta x)/a$ is the Péclèt modulus based on the difference value of the space variable. Proceeding as in the stationary case, we find that the limiting steady value of Eq. (9–31) based on the temperatures T_{i+1}^n, $T_{i_s}^{n+1}$, and T_{i-1}^n is

$$0 = \Delta\text{Fo}\,[T_{i+1}^n - (2 + \Delta\text{Pé})\,T_{i_s}^{n+1} + (1 + \Delta\text{Pé})\,T_{i-1}^n]. \qquad (9\text{--}32)$$

Subtracting Eq. (9–32) from Eq. (9–31), and noting that $\Delta T \leq \Delta T_{\max}$, we obtain the stability criterion for Eq. (9–31) in the form

$$\Delta\text{Fo} \leq 1/(2 + \Delta\text{Pé}). \qquad (9\text{--}33)$$

Strictly speaking, the graphical interpretation of Eq. (9–32), which actually constitutes the graphical solution of the equation, belongs in Section 9–9.

---

* Recall the irrelevance of the temperature distribution with regard to the formulation.

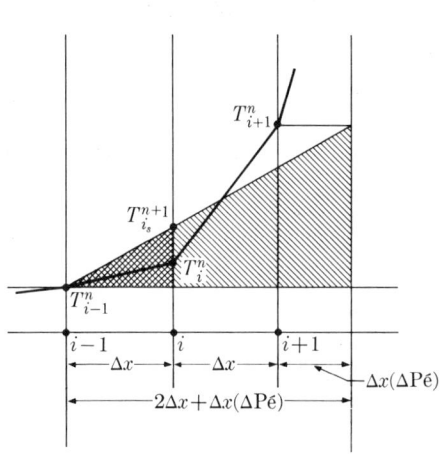
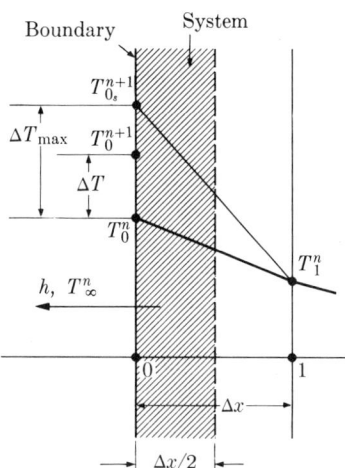

FIG. 9–18   FIG. 9–19

However, because of its simplicity this interpretation is given here. Rearranging Eq. (9-32), and dividing the result by Δx, we have

$$\frac{T_{i_s}^{n+1} - T_{i-1}^n}{\Delta x} = \frac{T_{i+1}^n - T_{i-1}^n}{2\,\Delta x + \Delta x(\Delta\text{Pé})}. \qquad (9\text{-}34)$$

The similarity of the triangles shown in Fig. 9–18 readily yields Eq. (9-34). Note that the temperature $T_{i_s}^{n+1}$ can no longer be obtained by connecting the neighboring temperatures with a straight line, because of the terms in Eq. (9-32) related to enthalpy flow. (What happens to Fig. 9–18 as $U \to 0$?)

For two particular problems, so far we have established difference equations applicable to a typical inner nodal point. Now, we turn to boundary nodal points and assume, for example, heat transfer by convection through the boundary of a stationary one-dimensional problem. The first law of thermodynamics and Fourier's law successively applied to the system shown in Fig. 9–19 yield

$$\rho c \left(\frac{\Delta x}{2}\right) \frac{T_0^{n+1} - T_0^n}{\Delta t} = h(T_\infty^n - T_0^n) + k\frac{T_1^n - T_0^n}{\Delta x}, \qquad (9\text{-}35)$$

where T_∞^n denotes the ambient temperature in the time interval n. Rearranging Eq. (9-35), we obtain

$$T_0^{n+1} - T_0^n = 2\Delta\text{Fo}\,[T_1^n - (1 + \Delta\text{Bi})\,T_0^n + (\Delta\text{Bi})\,T_\infty^n], \qquad (9\text{-}36)$$

where $\Delta\text{Bi} = h\,\Delta x/k$ denotes the Biot modulus based on the difference value of the space variable. For the limiting steady case, we also have

$$0 = T_1^n - (1 + \Delta\text{Bi})\,T_{0_s}^{n+1} + (\Delta\text{Bi})\,T_\infty^n. \qquad (9\text{-}37)$$

In the usual manner, subtracting Eq. (9–37) from Eq. (9–36) and noting again that $\Delta T \leq \Delta T_{max}$, we obtain the stability criterion for Eq. (9–36) as

$$\Delta \text{Fo} \leq 1/2(1 + \Delta \text{Bi}). \tag{9-38}$$

Similarly, Eq. (9–37) readily accepts a geometric interpretation when rearranged in the form

$$\frac{T_\infty^n - T_{0_s}^{n+1}}{\Delta x / \Delta \text{Bi}} = \frac{T_{0_s}^{n+1} - T_1^n}{\Delta x}. \tag{9-39}$$

Now, if we extend the boundary of the problem by the distance $\Delta x / \Delta \text{Bi}$ and assume that the extended boundary has the temperature T_∞^n in the time interval n, the intersection of the straight line connecting T_∞^n to T_1^n with the original boundary gives $T_{0_s}^{n+1}$ (Fig. 9–20). Clearly, the similarity of the triangles indicated in Fig. 9–20 satisfies Eq. (9–39).

In preceding chapters we have indicated the difficulty of solving multi-domain problems. Only in Chapter 7 were we able to consider even two-domain problems.* The difference formulation and its solution by numerical or graphical means, however, present no difficulty and hence are indispensable for these problems. Consider, for example, a composite wall made of two parallel flat plates. The difference equation given by Eq. (9–27) is directly applicable to the inner nodal points of both plates provided $\Delta \text{Fo} \leq \frac{1}{2}$. Excluding for now the graphical case corresponding to $\Delta \text{Fo} = \frac{1}{2}$, let us assume the values

$$\Delta \text{Fo}_1 = m_1 \quad \text{and} \quad \Delta \text{Fo}_2 = m_2, \tag{9-40}$$

where $m_1, m_2 < \frac{1}{2}$. Since the time interval to be used must be the same for both plates, we obtain from Eq. (9–40) the relation between the space intervals of the plates in the form

$$\frac{\Delta x_1}{\Delta x_2} = \left(\frac{a_1}{a_2} \frac{m_2}{m_1} \right)^{1/2}. \tag{9-41}$$

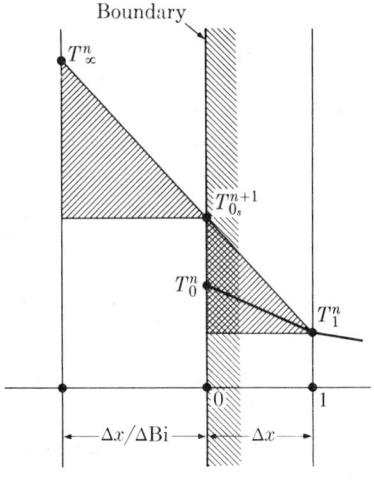

FIG. 9–20

Therefore, we need to select the interval size of only one plate; Eq. (9–41) gives the interval size of the other.

Unlike the difference equation for the inner nodal points, the difference equation for the interface is not available from preceding sections. However, the derivation of this equation is straightforward. The result in terms of the

---

* See Example 7–15.

system shown in Fig. 9–21 is

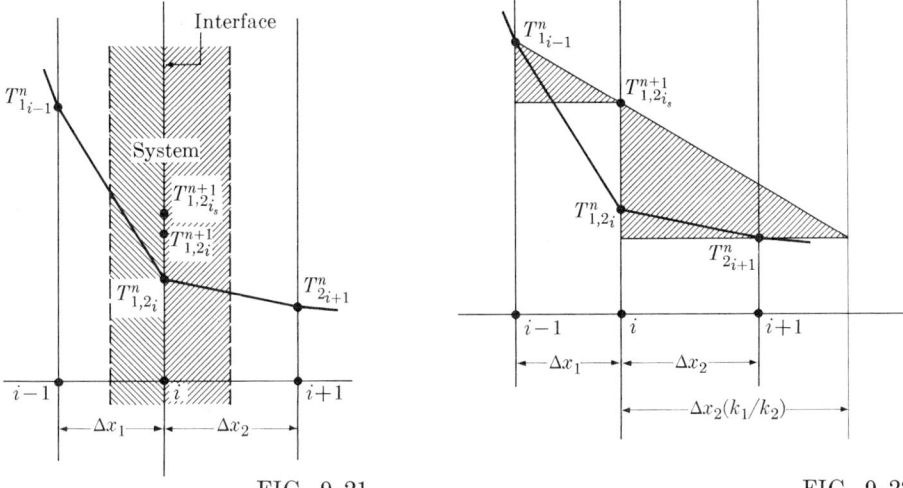

FIG. 9–21   FIG. 9–22

$$T^{n+1}_{1,2_i} - T^n_{1,2_i} = \frac{(k_2/\Delta x_2)T^n_{2_{i+1}} - (k_1/\Delta x_1 + k_2/\Delta x_2)T^n_{1,2_i} + (k_1/\Delta x_1)T^n_{1_{i-1}}}{(\rho_1 c_1 \Delta x_1 + \rho_2 c_2 \Delta x_2)/2 \, \Delta t}; \quad (9\text{–}42)$$

the steady solution is

$$0 = (k_2/\Delta x_2)T^n_{2_{i+1}} - (k_1/\Delta x_1 + k_2/\Delta x_2)T^{n+1}_{1,2_{i,s}} + (k_1/\Delta x_1)T^n_{1_{i-1}}, \quad (9\text{–}43)$$

and the stability criterion for Eq. (9–42) is

$$\left(\frac{k_1/\Delta x_1 + k_2/\Delta x_2}{\rho_1 c_1 \Delta x_1 + \rho_2 c_2 \Delta x_2}\right) \Delta t \leq \frac{1}{2}. \quad (9\text{–}44)$$

[Note the special case of Eq. (9–44) for $k_1 = k_2$ and $\rho_1 c_1 = \rho_2 c_2$.] The geometric interpretation of Eq. (9–43) may be obtained from its rearrangement in the form

$$\frac{T^n_{1_{i-1}} - T^{n+1}_{1,2_{i,s}}}{\Delta x_1} = \frac{T^{n+1}_{1,2_{i,s}} - T^n_{2_{i+1}}}{\Delta x_2(k_1/k_2)}, \quad (9\text{–}45)$$

which is readily satisfied by the similarity of the triangles shown in Fig. 9–22. However, it is easier to consider the graphical procedure in terms of the intervals $\Delta \xi_1 = \Delta x_1/k_1$ and $\Delta \xi_2 = \Delta x_2/k_2$, thereby eliminating the use of Fig. 9–22. In this case, introducing the variable transformation $\xi = x/k$, we find the differential formulation of the problem to be

$$\frac{\partial T}{\partial t} = \frac{1}{\rho c k} \frac{\partial^2 T}{\partial \xi^2}. \quad (9\text{–}46)$$

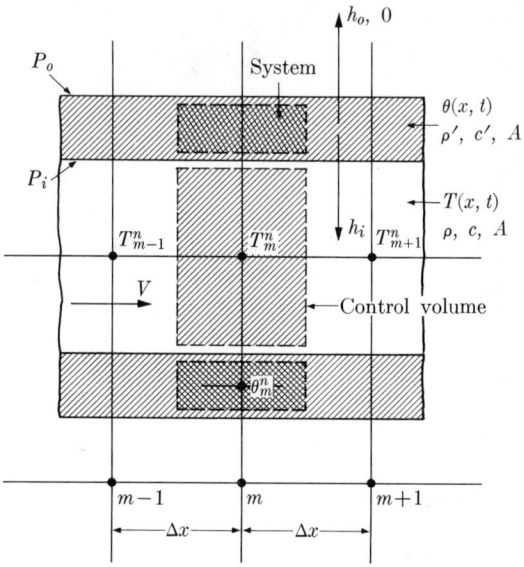

FIG. 9-23

Thus $1/\rho ck$ replaces a, and the stability criterion for the difference equation corresponding to Eq. (9-46) is

$$\frac{\Delta t}{\rho ck(\Delta \xi)^2} \leq \frac{1}{2}. \tag{9-47}$$

Considering the equality in Eq. (9-47) for both plates, and equating the time intervals, we obtain as the relation between the new space intervals,

$$\frac{\Delta \xi_1}{\Delta \xi_2} = \left(\frac{\rho_2 c_2 k_2}{\rho_1 c_1 k_1}\right)^{1/2}. \tag{9-48}$$

The extension of the difference formulation to unsteady problems of other geometries and the geometric interpretation of its limiting steady case require no additional knowledge. Therefore any elaboration on these points seems unnecessary.

Next we illustrate by an example how physical reasoning can readily be used to establish the stability criteria of complicated problems.

Example 9-4. Let us find the difference equations and their stability criteria for the typical inner nodal point associated with Example 6-3.

The first law of thermodynamics and the definition of the heat transfer coefficient applied successively to the control volume for the fluid and the system for the tube walls shown in Fig. 9-23 give for the fluid

$$\rho c A (\Delta x) \frac{T_m^{n+1} - T_m^n}{\Delta t} = \rho c A V (T_{m-1}^n - T_m^n) + h_i P_i (\Delta x)(\theta_m^n - T_m^n),$$

and for the tube walls

$$\rho'c'A'(\Delta x)\frac{\theta_m^{n+1}-\theta_m^n}{\Delta t} = h_iP_i(\Delta x)(T_m^n - \theta_m^n) + h_oP_o(\Delta x)(0 - \theta_m^n).$$

The rearrangement of these equations yields

$$T_m^{n+1} - T_m^n = \frac{V(\Delta t)}{\Delta x}T_{m-1}^n - \left(\frac{V}{\Delta x} + \frac{h_iP_i}{\rho cA}\right)(\Delta t)T_m^n + \frac{h_iP_i(\Delta t)}{\rho cA}\theta_m^n, \quad (9\text{-}49)$$

$$\theta_m^{n+1} - \theta_m^n = \frac{h_iP_i(\Delta t)}{\rho'c'A'}T_m^n - \left(\frac{h_iP_i + h_oP_o}{\rho'c'A'}\right)(\Delta t)\theta_m^n. \quad (9\text{-}50)$$

The corresponding steady equations are

$$0 = \frac{V(\Delta t)}{\Delta x}T_{m-1}^n - \left(\frac{V}{\Delta x} + \frac{h_iP_i}{\rho cA}\right)(\Delta t)T_{m_s}^{n+1} + \frac{h_iP_i(\Delta t)}{\rho cA}\theta_m^n, \quad (9\text{-}51)$$

$$0 = \frac{h_iP_i(\Delta t)}{\rho'c'A'}T_m^n - \left(\frac{h_iP_i + h_oP_o}{\rho'c'A'}\right)(\Delta t)\theta_{m_s}^{n+1}. \quad (9\text{-}52)$$

Subtracting Eq. (9–51) from Eq. (9–49), and Eq. (9–52) from Eq. (9–50), we obtain

$$T_m^{n+1} - T_m^n = \left(\frac{V}{\Delta x} + \frac{h_iP_i}{\rho cA}\right)(\Delta t)(T_{m_s}^{n+1} - T_m^n), \quad (9\text{-}53)$$

$$\theta_m^{n+1} - \theta_m^n = \left(\frac{h_iP_i + h_oP_o}{\rho'c'A'}\right)(\Delta t)(\theta_{m_s}^{n+1} - \theta_m^n). \quad (9\text{-}54)$$

Thus the stability criterion for Eq. (9–49) is

$$\left(\frac{V}{\Delta x} + \frac{h_iP_i}{\rho cA}\right)\Delta t \leq 1, \quad (9\text{-}55)$$

and that for Eq. (9–50) is

$$\left(\frac{h_iP_i + h_oP_o}{\rho'c'A'}\right)\Delta t \leq 1. \quad (9\text{-}56)$$

Needless to say, the common time interval to be used for the fluid and tube-wall difference equations must satisfy both Eq. (9–55) and Eq. (9–56).

So far, we have considered the difference formulation and its criteria of stability for a number of unsteady problems. The use of the formulations thus obtained becomes somewhat cumbersome when, in the interests of accuracy, a rather small Δx is selected; the Δt allowed by the stability criterion is then so small that an enormous amount of calculations may be required. We now

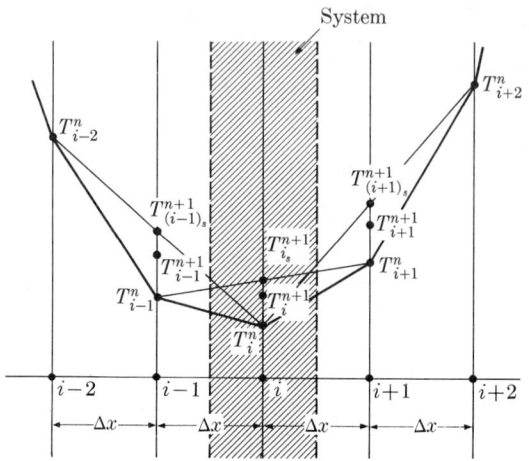

FIG. 9-24

intend to eliminate this difficulty by giving different forms to difference equations. Let us take the case of one-dimensional conduction in stationary problems for which we obtained the difference equation given by Eq. (9–27). Consider the system shown in Fig. 9–24 and formulate the problem this time in terms of *backward* rather than *forward* differences in time; that is, decrease the time from $(n + 1)\,\Delta t$ to $n\,\Delta t$ rather than following the usual procedure of increasing it from $n\,\Delta t$ to $(n + 1)\,\Delta t$. Thus we obtain

$$T_i^{n+1} - T_i^n = \Delta\text{Fo}\,(T_{i+1}^{n+1} - 2T_i^{n+1} + T_{i-1}^{n+1}). \tag{9-57}$$

Subtracting from Eq. (9–57) its initial steady value given by Eq. (9–28) yields

$$T_i^{n+1} - T_i^n = \Delta\text{Fo}\,[(T_{i+1}^{n+1} - T_{i+1}^n) + 2(T_{i_s}^{n+1} - T_i^{n+1}) + (T_{i-1}^{n+1} - T_{i-1}^n)]. \tag{9-58}$$

Since all differences appearing in Eq. (9–58) are positive according to Fig. 9–24,* this equation is satisfied under all circumstances. Therefore, it is *unconditionally stable*. We obtain this stability, however, at the cost of a new algebraic complexity. Recall Eq. (9–27), in which all temperatures except T_i^{n+1} are known, and the latter is obtained by solution of the equation. By contrast, in Eq. (9–57) only T_i^n is known, and the application of this equation to the nodal points yields a set of simultaneous algebraic equations which must then be solved by the relaxation method discussed in Section 9–7 or by computer. Equations (9–27) and (9–57) are referred to as *explicit* and *implicit* difference equations, respectively. Although there exists only one explicit equation

---

* The only other possibility is a downward concave temperature distribution for which all differences are negative, and Eq. (9–58) is again unconditionally satisfied.

corresponding to each nodal point of a problem, the implicit equation may be written in a number of forms. For example, we may express the right-hand side of Eq. (9–57) in the form

$$T_i^{n+1} - T_i^n = \Delta\text{Fo}\,[\eta(T_{i+1}^{n+1} - 2T_i^{n+1} + T_{i-1}^{n+1}) \\ + (1 - \eta)(T_{i+1}^n - 2T_i^n + T_{i-1}^n)], \tag{9-59}$$

where $0 \leq \eta \leq 1$. The reader is referred to Reference 4* for further discussion on this point.

The extension of the difference formulation to multidimensional problems is trivial, and will not be considered here.

9–9. Solution of Unsteady Problems. Step-by-Step Numerical Solution. Binder-Schmidt Graphical Method

Because of the inherent nature of the difference formulation of unsteady problems, we could not avoid discussing the numerical and graphical solutions of these problems in the process of their formulation. Here, after some recapitulation, we shall indicate a number of salient points of these solutions.

Unsteady problems are initial-value problems. The solution of their explicit difference formulation is trivial; it amounts to the evaluation of the temperature change at each location and in each time interval in terms of the temperature of the same location and the neighboring temperatures taken from the preceding time interval; the procedure is carried out numerically and step by step to following intervals. The solution of the implicit difference formulation reduces at each location and in each time interval to an algebraic equation involving the temperature of the same location and the neighboring temperatures expressed in terms of the following time interval; thus we obtain a set of algebraic equations to be solved by iteration, either by the relaxation method or by computer.

It is important to note that although steady problems are in general boundary-value problems, those having motion and negligible conduction in a given direction are conceptually initial-value problems. There have been some attempts to convert initial-value problems to boundary-value problems by means of a transformation which doubles the order of each derivative involved in the former. However, these methods have remained somewhat limited to simple cases, and are not considered here.†

Let us now examine by means of an example the Binder-Schmidt graphical procedure, which we mentioned briefly in the preceding section.

---

* See pages 16 and 93.
† See, for example, Reference 3, Chapter 15, page 225.

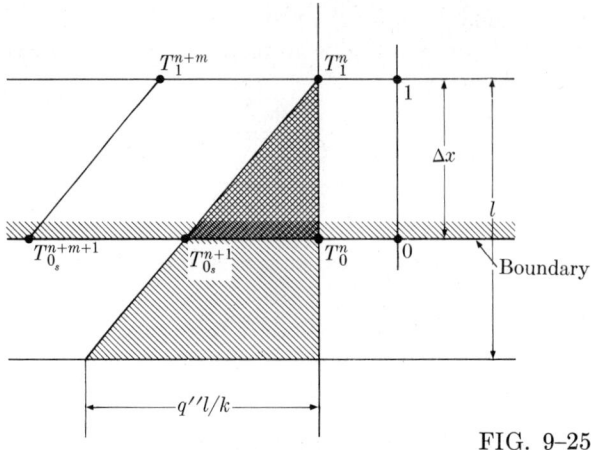

FIG. 9–25

Example 9–5. Reconsider Example 2–3, assuming the following numerical values: $L = 1$ in., $\rho = 500$ lbm/ft^3, $c = 0.1$ Btu/lbm·°F, $k = 10$ Btu/ft·hr·°F, $h = 100$ Btu/ft^2·hr·°F, and $q'' = 4800$ Btu/ft^2·hr. We wish to find the temperature rise on the upper surface two minutes after the application of heat flux to the bottom of the plate.

For purposes of illustration, let us consider a coarse network, with the plate divided into four equal layers. The temperature of the inner nodal points is obtained by taking graphically the arithmetic mean of the neighboring temperatures. As Fig. 9–20 shows, the upper boundary is taken care of by extending this boundary by an amount of $\Delta x/\Delta\mathrm{Bi}$. Since $\Delta x = L/4 = \frac{1}{4}$ in. and $\Delta\mathrm{Bi} = h\,\Delta x/k = 100 \times (\frac{1}{4}/12)/10 = 1/4.8$, the extension is $\Delta x/\Delta\mathrm{Bi} = 4.8\,\Delta x$. However, so far we have developed no graphical procedure suitable to the lower boundary, which must be considered first. Since we are interested in the graphical solution only, under steady conditions we have in terms of Fig. 9–25

$$0 = k\frac{T_1^n - T_{0_s}^{n+1}}{\Delta x} + q''. \tag{9-60}$$

This may be rearranged in the form

$$\frac{T_{0_s}^{n+1} - T_1^n}{\Delta x} = \frac{q''l/k}{l}, \tag{9-61}$$

where l, an arbitrary length introduced for convenience in the graphical procedure, may be taken equal to Δx. Once the temperature gradient given by Eq. (9–61) is established by the similarity of the triangles shown in Fig. 9–25, in any time interval, say $n + m$, we need only draw a straight line from T_1^{n+m} parallel to this gradient; the intersection of the straight line with the boundary gives $T_{0_s}^{n+m+1}$.

[9-9] SOLUTION OF UNSTEADY PROBLEMS 517

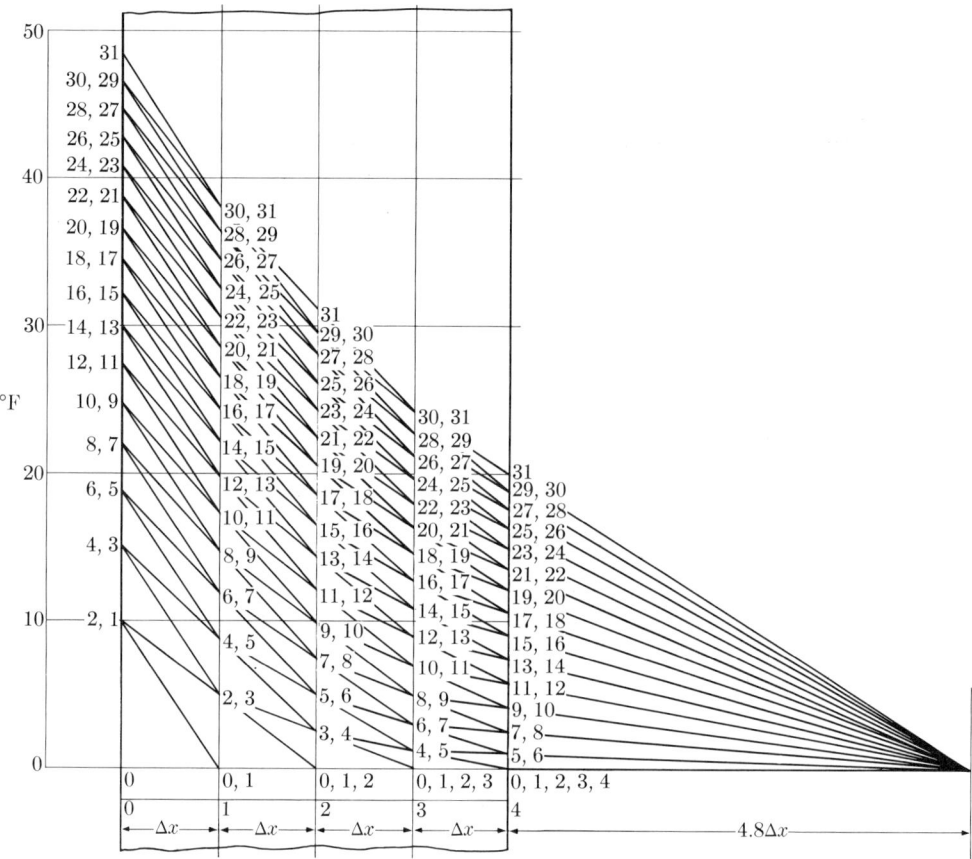

FIG. 9-26

Having completed the graphical construction suitable to the inner and to both boundary nodal points, we now proceed to the solution as indicated in Fig. 9-26. By marking the time intervals on the nodal points we clearly show the instantaneous temperature distribution associated with each interval. The temperature scale on Fig. 9-26 is obtained by rearranging Eq. (9-61) in the form

$$T^1_{0_s} - T^0_1 = \frac{q''(\Delta x)}{k} = \frac{4800 \times 1/4}{12 \times 10} = 10°F.$$

The number n of the time intervals is found from the specified total time $t = n(\Delta t)$; since

$$a = k/\rho c = 10/(500 \times 0.1) = 0.2$$

and

$$\Delta Fo = \tfrac{1}{2} = a(\Delta t)/(\Delta x)^2$$

yield

$$\Delta t = (\Delta x)^2/2a = (\tfrac{1}{4}/12)^2 \times 3600/(2 \times 0.2) \cong 3.9 \text{ sec},$$

we have $n = 2 \times 60/3.9 \cong 31$ intervals. Figure 9–26 gives 20°F for the temperature of the upper surface at the end of 31 time intervals. (Compare the penetration time to be obtained from Fig. 9–26 with the times found in Examples 2–3 and 8–8.)

References

1. H. W. Emmons, "The Numerical Solution of Heat Conduction Problems." *Trans. ASME*, **65**, 607 (1943).
2. H. W. Emmons, "The Numerical Solution of Partial Differential Equations." *Quart. App. Math.*, **2**, 173 (1944).
3. D. N. de G. Allen, *Relaxation Methods*. New York: McGraw-Hill, 1954.
4. R. D. Richtmyer, *Difference Methods for Initial-Value Problems*. New York: Interscience, 1957.
5. G. M. Dusinberre, *Heat Transfer Calculations by Finite Differences*. Scranton: International, 1961.
6. M. Jakob, *Heat Transfer I*. New York: Wiley, 1949.
7. F. B. Hildebrand, *Methods of Applied Mathematics*. Englewood Cliffs: Prentice-Hall, 1952.
8. S. H. Crandall, *Engineering Analysis*. New York: McGraw-Hill, 1956.
9. R. V. Southwell, *Relaxation Methods in Engineering Science*. Oxford: Clarendon Press, 1940.
10. R. V. Southwell, *Relaxation Methods in Theoretical Physics*. Oxford: Clarendon Press, 1946.
11. L. V. Kantorovich and V. I. Krylov, *Approximate Methods of Higher Analysis*. New York–Gröningen: Interscience-Noordhoff, 1958.
12. F. B. Hildebrand, *Introduction to Numerical Analysis*. New York: McGraw-Hill, 1956.
13. J. Todd, *A Survey of Numerical Analysis*. New York: McGraw-Hill, 1962.
14. G. E. Forsythe and W. R. Wasow, *Finite-Difference Methods for Partial Differential Equations*. New York: Wiley, 1960.

Problems

In the following problems, if a relaxation or step-by-step solution is to be carried out by hand or by desk calculator, a gross answer is satisfactory, and numerical values may be assigned to dimensions and properties given in symbols; if a digital computer is to be used, an accurate answer is required, depending on a parametric study in terms of the dimensionless numbers involved.

PROBLEMS

9-1. A two-dimensional extended surface made of two different materials is shown in Fig. 9-27. Find the steady temperature of the extended surface.

9-2. A Prony brake is simulated by the two-dimensional model shown in Fig. 9-28. Find the steady temperature of the brake.

9-3. An approximation of a turbine blade is shown in Fig. 9-29. Assume that the heat transfer coefficient linearly decreases from the upstream value h to the downstream value $h/2$. Find the steady temperature of the blade.

9-4. A solid cylinder rotates while pressed on another solid cylinder (Fig. 9-30). Find the steady temperature of the system.*

9-5. The hydrodynamic friction generated in the bearings of a shaft may be assumed to be a constant heat flux acting on the shaft (Fig 9-31). Find the steady temperature of the shaft.

9-6. Find the stability criterion of the explicit difference equation corresponding to the differential equation

$$\frac{\partial T}{\partial t} + u\frac{\partial T}{\partial x} + v\frac{\partial T}{\partial y} + w\frac{\partial T}{\partial z} = a\left(\frac{\partial^2 T}{\partial x^2} + \frac{\partial^2 T}{\partial y^2} + \frac{\partial^2 T}{\partial z^2}\right).$$

9-7. Find the stability criterion of the explicit difference equations corresponding to the boundary conditions

$$\pm k\left(\frac{\partial T}{\partial n}\right)_\sigma + q'' = h(T_\sigma - T_\infty)$$

and

$$\pm k_1\left(\frac{\partial T_1}{\partial n}\right)_\sigma + \mu pV = \pm k_2\left(\frac{\partial T_2}{\partial n}\right)_\sigma.$$

9-8. Reconsider Problem 5-5. Assume that the length of the heat exchanger is L. (a) Obtain the typical difference equations and the associated stability criteria. (b) Find the unsteady temperature of the heat exchanger by numerical means.

9-9. A torch moving with constant velocity V applies the constant heat flux q'' to a small circular area of a flat plate (Fig. 9-32). Find the unsteady temperature of the plate by numerical means. [*Hint:* Analyze the problem by assuming that the torch is stationary and the plate is moving.]

9-10. Consider a fully developed laminar flow in a tube (Fig. 9-33). While the upstream half of the tube is insulated, the downstream half is suddenly subjected to (i) a constant heat flux q'', (ii) a periodic heat flux $q_0''(1 + \epsilon \sin \omega t)$. The axial conduction is appreciable both in the fluid and tube-walls. (a) Obtain the typical difference equations and the associated stability criteria. (b) Show the effect of axial conduction, tube material, and unsteadiness on the distribution of the steady heat transfer coefficient.

9-11. Consider a flat plate that is initially at a uniform temperature (Fig. 9-34). Assume that the constant internal energy u''' begins to be generated in the plate. (a) Modify the graphical procedure by including the effect of energy generation.

---

* This problem corresponds to part (f) of Problem 4-29.

520   DIFFERENCE FORMULATION

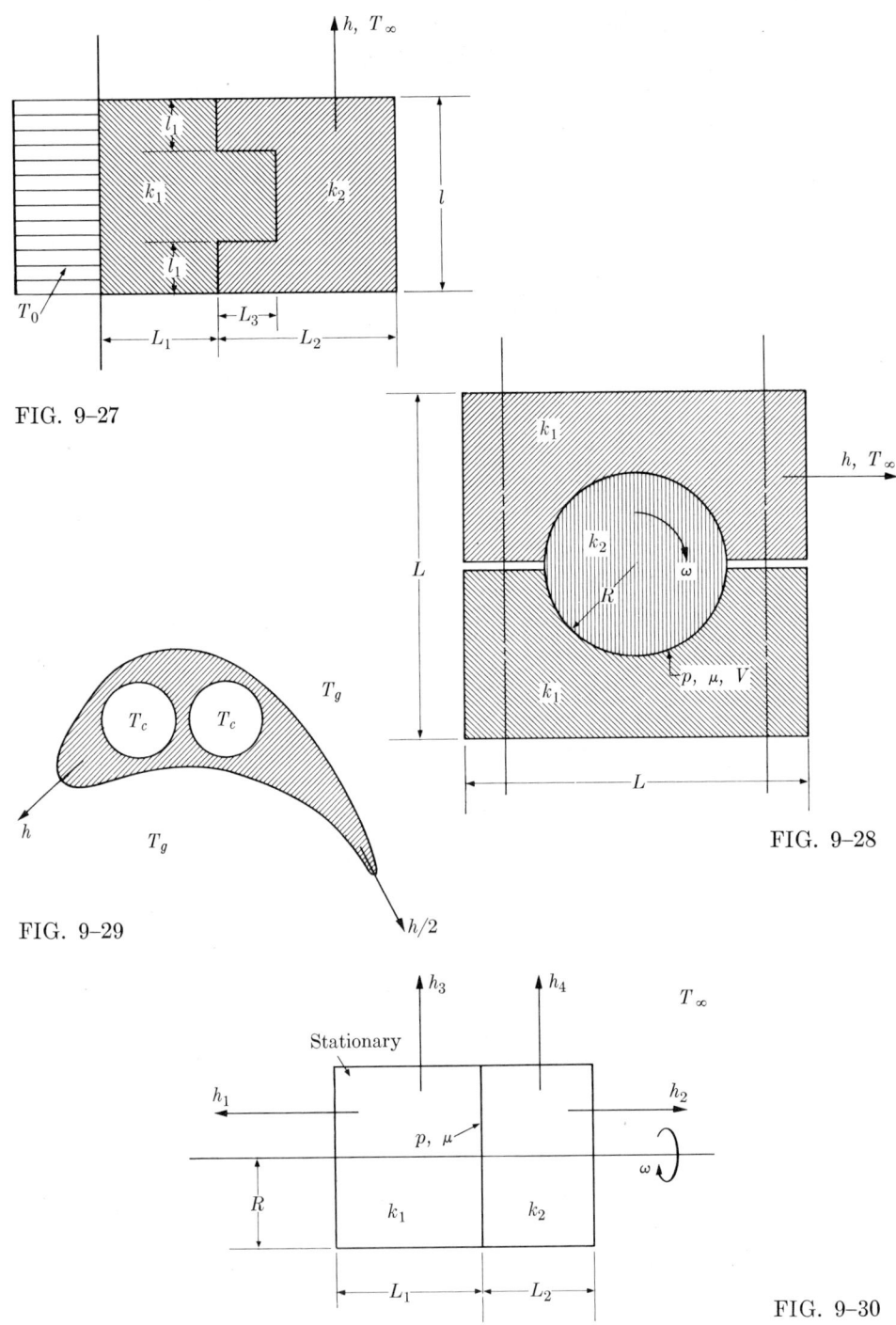

FIG. 9-27

FIG. 9-29

FIG. 9-28

FIG. 9-30

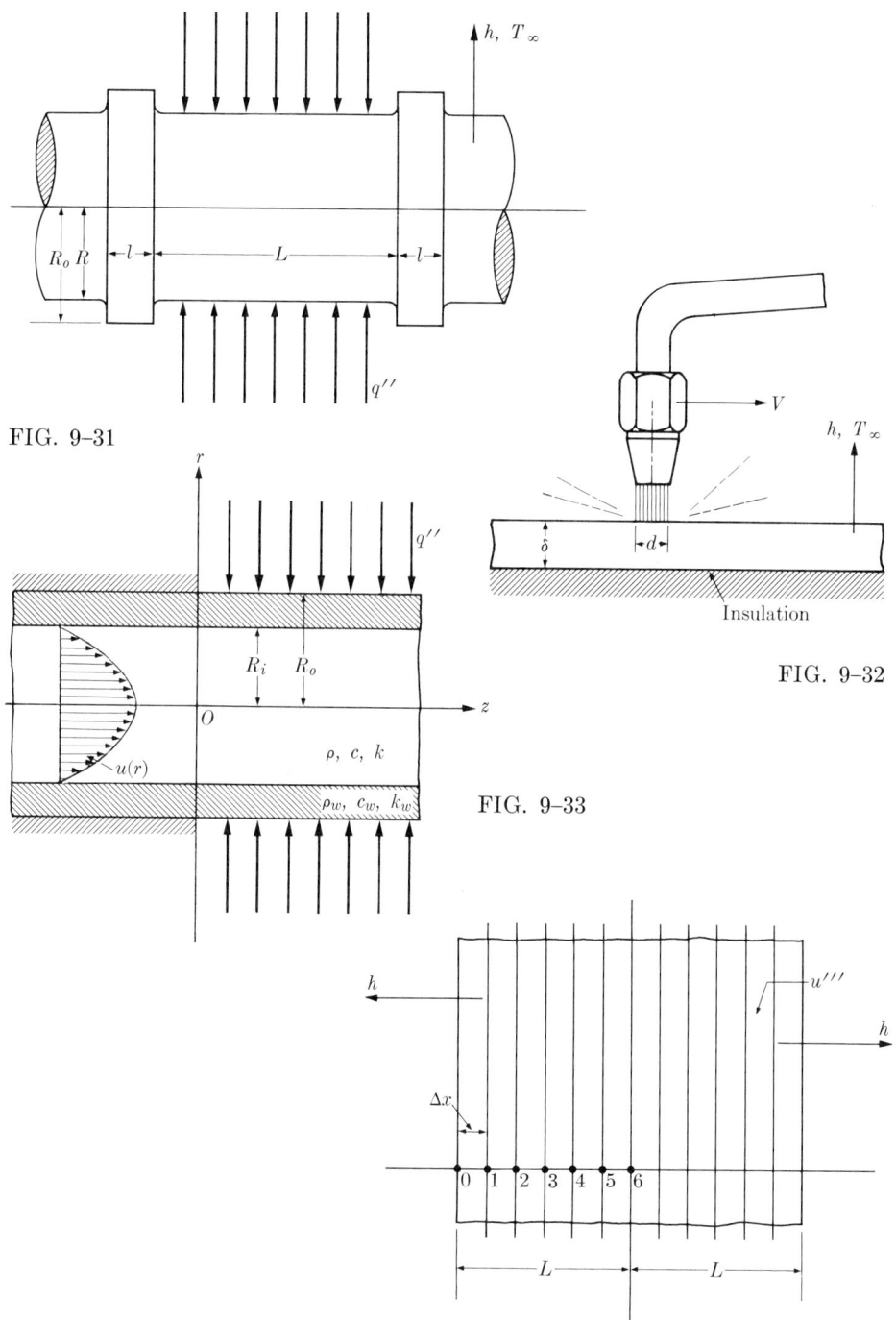

FIG. 9-31

FIG. 9-32

FIG. 9-33

FIG. 9-34

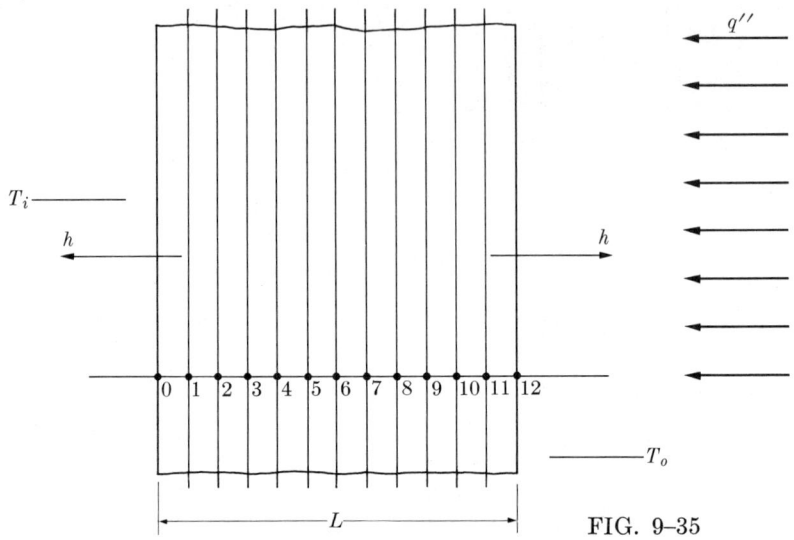

FIG. 9-35

(b) Using part (a), find the temperature distribution in the plate after 10 min have elapsed. Data:

$$L = 6 \text{ in.}, \quad \rho = 480 \text{ lbm/ft}^3, \quad c = 0.1 \text{ Btu/lbm·°F},$$
$$k = 10 \text{ Btu/ft·hr·°F}, \quad h = 100 \text{ Btu/ft}^2\text{·hr·°F}, \quad u''' = 14{,}400 \text{ Btu/ft}^3\text{·hr}.$$

9-12. Develop graphical procedures suitable to cylindric and spherical geometry.

9-13. Consider the brick walls of a house. The inside and outside heat transfer coefficients are the same. The inside temperature T_i is greater than the outside temperature T_o. There is a steady heat loss from the house to the atmosphere. Now let us assume that we start collecting solar radiation q'' on the outer surface of the walls (Fig. 9-35). Using the graphical procedure, determine the time required to stop the heat loss from the house. Data:

$$L = 1 \text{ ft}, \quad \rho = 100 \text{ lbm/ft}^3, \quad c = 0.2 \text{ Btu/lbm·°F},$$
$$k = 0.4 \text{ Btu/ft·hr·°F}, \quad h = 2 \text{ Btu/ft}^2\text{·hr·°F},$$
$$q'' = 440 \text{ Btu/ft}^2\text{·hr}, \quad T_i = 75°\text{F}, \quad T_o = -30°\text{F}.$$

9-14. Reconsider the distributed case of Problem 6-2. Using the graphical procedure, carry out the temperature variation of the plate for the first twenty intervals of time. Data:

$$\delta = \tfrac{1}{2} \text{ in.}, \quad \Delta x = \delta/12, \quad \rho = 500 \text{ lbm/ft}^3,$$
$$c = 0.1 \text{ Btu/lbm·°F}, \quad k = 10 \text{ Btu/ft·hr·°F},$$
$$h = 2000 \text{ Btu/ft}^2\text{·hr·°F}, \quad T_o = 100°\text{F}, \quad \omega = 5 \text{ min}^{-1}.$$

9-15. Reconsider Problem 2-9. Assume that the heat transfer coefficient and all properties are constant except the thermal conductivity of the insulation, which varies in the form

$$k_i = k_\infty(1 + \beta \Delta T),$$

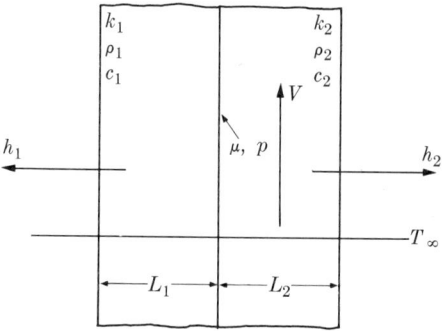

FIG. 9-36

where ΔT is measured relative to the ambient temperature. Using the graphical procedure, carry out the temperature variation of the system for the first twenty intervals of time. Data:

$$\delta = \tfrac{1}{16} \text{ in.}, \quad L = \tfrac{1}{2} \text{ in.}, \quad \Delta x = L/8,$$
$$k_\infty = 0.1 \text{ Btu/ft·hr·°F}, \quad \beta = 0.01,$$
$$k_h = 200 \text{ Btu/ft·hr·°F}, \quad a_i = 0.01 \text{ ft}^2/\text{hr},$$
$$a_h = 4 \text{ ft}^2/\text{hr}, \quad h = 2 \text{ Btu/ft}^2\text{·hr·°F},$$
$$u''' = 148{,}000 \text{ Btu/ft}^3\text{·hr}.$$

9-16. Simulate a dry friction brake by two flat plates (Fig. 9-36). The initial temperature of the brake is equal to that of the ambient, T_∞. Then one of the plates suddenly assumes a constant velocity V. The coefficient of dry friction and the pressure between the plates are μ and p, respectively. Using the graphical procedure, carry out the temperature variation of the brake for the first twenty intervals of time. Data:

$$L_1 = L_2 = \tfrac{1}{2} \text{ in.}, \quad \mu p V = 144{,}000 \text{ Btu/ft}^2\text{·hr},$$
$$\rho_1 = \rho_2 = 500 \text{ lbm/ft}^3, \quad c_1 = c_2 = 0.1 \text{ Btu/lbm·°F},$$
$$k_1 = 20 \text{ Btu/ft·hr·°F}, \quad k_2 = 30 \text{ Btu/ft·hr·°F},$$
$$h_1 = 400 \text{ Btu/ft}^2\text{·hr·°F}, \quad h_2 = 600 \text{ Btu/ft}^2\text{·hr·°F}.$$

CHAPTER 10

DIFFERENTIAL-DIFFERENCE FORMULATION. ANALOG SOLUTION

The final method of solution to be considered in this text is analogy. A number of lumped and distributed models for conduction problems are available, based on mechanical, hydrodynamic, and electrical systems. Networks of electrical resistors, capacitors, and sometimes inductors are the most important simulators of lumped systems; on rare occasions, mechanical systems comprised of masses, springs, and dashpots are also used for this purpose. Electrolytic tanks, conductive paper, stretched membranes, soap film, fluid mappers, and polarized light are some of the distributed models occasionally used. Here, because of its versatility, we shall make use of the electrical analogy.

10–1. Analogy Between Conduction and Electricity

Let us first formulate the diffusion of electricity through a resistive and capacitive solid.† We illustrate the point in terms of one-dimensional cartesian geometry. Following the steps of formulation given in Section 2–9, we select the system shown in Fig. 10–1. Applying the general law of electricity, *the conservation of electric charge q* (coulomb), to this system yields‡

$$\partial \rho_e / \partial t = - \partial j_x / \partial x, \qquad (10\text{–}1)$$

where $\rho_e(\text{coulomb}/L^3) = q/A(\Delta x)$ is the charge density and $j_x(\text{ampere}/L^2) = i/A$ is the electric current density which, by definition, is the charge flux (diffusion of charge per unit time and area); $i(\text{ampere})$ is the electric current or diffusion of charge per unit time,

$$i = dq/dt. \qquad (10\text{–}2)$$

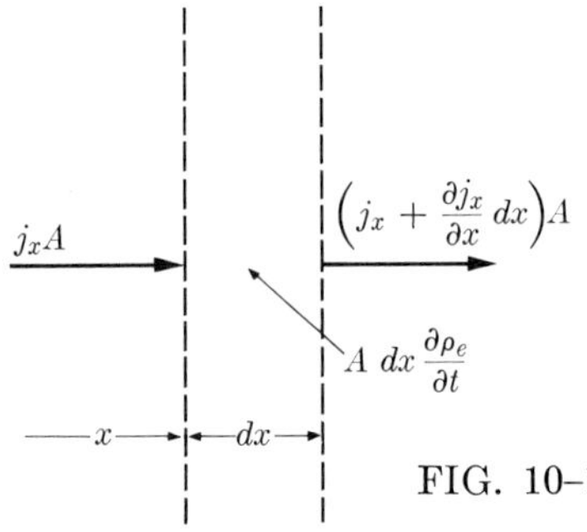

FIG. 10–1

---

† A practical example of such a solid would be thin layers of conducting and dielectric materials composed parallel in the direction of current.
‡ The general differential form of Eq. (10–1) is

$$\partial \rho_e / \partial t + \nabla \cdot \mathbf{j} = 0,$$

and the corresponding integral formulation is

$$\int_\mathcal{V} (\partial \rho_e / \partial t) \, d\mathcal{V} + \int_\sigma \mathbf{j} \cdot \mathbf{n} \, d\sigma = 0.$$

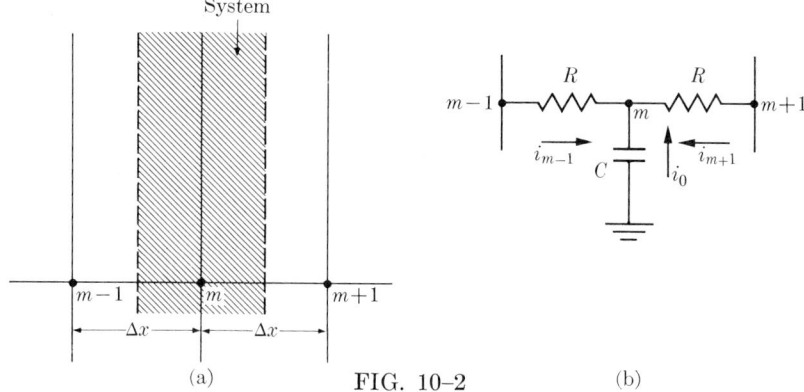

FIG. 10–2

Let us express Eq. (10–1) in terms of the electric potential E(volt) by means of the particular laws†

$$\rho_e = C^*E, \quad \text{Coulomb's law}, \tag{10-3}$$

$$j_x = -\frac{1}{R^*}\frac{\partial E}{\partial x}, \quad \text{Ohm's law}, \tag{10-4}$$

where C^*(farad/L^3) is the capacity per unit volume and R^*(ohm/L) the resistivity per unit length of the solid. Thus we obtain as the governing equation for the one-dimensional diffusion of electricity‡

$$R^*C^*\frac{\partial E}{\partial t} = \frac{\partial^2 E}{\partial x^2}, \tag{10-5}$$

which is analogous to that for the one-dimensional conduction of heat

$$\frac{1}{a}\frac{\partial T}{\partial t} = \frac{\partial^2 T}{\partial x^2}. \tag{10-6}$$

Therefore, the solution of Eq. (10–6) may be found by solving Eq. (10–5), that is, by measuring the diffusion of electricity in a resistive-capacitive solid. Furthermore, expressing the space derivative of Eq. (10–5) in difference form (Fig. 10–2a), and subsequently rearranging we have

$$R^*C^*(\Delta x)^2 \frac{dE_m}{dt} = E_{m+1} + E_{m-1} - 2E_m, \tag{10-7}$$

which is the *timewise differential, spacewise difference formulation* of the problem.

The important fact is that a formulation of this type can be simulated by a lumped circuit element consisting of one capacitor and two resistors as shown

---

† Discuss the meaning and equivalents of Eqs. (10–1), (10–3), and (10–4) in conduction.
‡ This equation becomes the one-dimensional electromagnetic wave equation when the solid is further assumed to have inductance.

in Fig. 10–2(b). The governing equation of this circuit is identical to Eq. (10–7), provided R and C are the total resistance and capacitance of the system, respectively, so that

$$R(\text{ohm}) = R^*(\Delta x)/A \quad \text{and} \quad C(\text{farad}) = C^*A(\Delta x). \quad (10\text{–}8)$$

To show this fact we apply the steps of formulation to Fig. 10–2(b). The system of the distributed case now reduces to nodal point m, and the general law given by Eq. (10–1) becomes *Kirchhoff's current law* applied to this point. It follows then that

$$i_{m-1} + i_0 + i_{m+1} = 0. \quad (10\text{–}9)$$

Furthermore, for convenience the particular laws may now be written in terms of charge and current rather than in terms of their densities. Thus Coulomb's and Ohm's laws, the former rearranged by Eq. (10–2) and the latter expressed in terms of the potential difference, become

$$i = C\frac{dE}{dt}, \quad (10\text{–}10)$$

and

$$i = \frac{\Delta E}{R}, \quad (10\text{–}11)$$

respectively. Interpreting Eqs. (10–10) and (10–11) in terms of Fig. 10–2(b), and inserting the result into Eq. (10–9) gives the governing equation

$$RC\frac{dE_m}{dt} = E_{m+1} + E_{m-1} - 2E_m, \quad (10\text{–}12)$$

which is identical to Eq. (10–7), as may readily be shown by means of Eq. (10–8). Thus we prove that the circuit element given by Fig. 10–2(b) is the approximate *analog solution* of the one-dimensional heat conduction governed by Eq. (10–6). Furthermore, since the differential-difference formulation that corresponds to Eq. (10–6) is

$$\frac{(\Delta x)^2}{a}\frac{dT_m}{dt} = T_{m+1} + T_{m-1} - 2T_m, \quad (10\text{–}13)$$

the comparison between Eqs. (10–12) and (10–13) yields that if temperatures are made proportional to potentials, then $(\Delta x)^2/a$ becomes proportional to RC.

Inspection of Eq. (10–12) reveals that the analog solution of two-dimensional conduction is governed by the equation

$$RC\frac{dE_{m,n}}{dt} = E_{m-1,n} + E_{m,n-1} + E_{m+1,n} + E_{m,n+1} - 4E_{m,n}, \quad (10\text{–}14)$$

which is satisfied by the circuit element shown in Fig. 10–3. The extension of the procedure to three-dimensional cartesian geometry, moving solids, boundary conditions, and other geometry presents no difficulty. This is left to the reader as an exercise (see Problems 10-1 and 10-2).

From Eqs. (10-12) and (10-14) we may generalize the fact that the approximate analog solution of conduction problems requires a circuit capable of performing multiplication by a constant, addition, integration, and differentiation. Electric circuits that can accomplish these operations are called *passive* circuits if they include, like those in Figs. 10-2(b) and 10-3, only fixed resistors and capacitors, and possibly inductors and transformers. They are called *active* circuits if, like electronic amplifiers, they involve additional elements drawing energy from an external source. Let us now examine the operations mentioned above as performed by passive circuit elements.

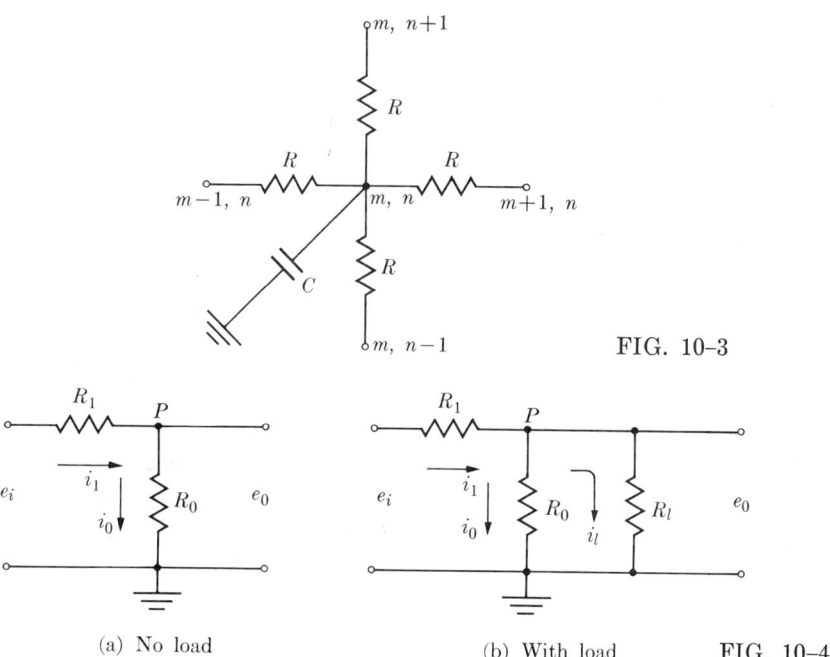

FIG. 10-3

(a) No load (b) With load FIG. 10-4

10-2. Passive Circuit Elements

We start with multiplication, which can be accomplished by the resistive circuit element shown in Fig. 10-4(a). Kirchhoff's current law applied to nodal point P yields

$$i_1 - i_0 = 0.$$

Expressing the currents in terms of potentials by the use of Ohm's law, we obtain

$$\frac{e_i - e_0}{R_1} - \frac{e_0}{R_0} = 0,$$

which may be rearranged in the form

$$e_0 = \left(\frac{R_0}{R_0 + R_1}\right) e_i. \qquad (10\text{--}15)$$

Thus the circuit element of Fig. 10–4(a) multiplies the potential e_i with the positive constant $R_0/(R_0 + R_1)$. However, when this element is used in a circuit its output is, in general, *loaded*. Let us now see how Eq. (10–15) is affected by the resistance R_l equivalent to the output load (Fig. 10–4b). Kirchhoff's current law applied to nodal point P now becomes

$$i_1 - i_0 - i_l = 0,$$

and relating the currents to potentials by Ohm's law we obtain

$$\frac{e_i - e_0}{R_1} - \frac{e_0}{R_0} - \frac{e_0}{R_l} = 0,$$

which may be rearranged to give

$$e_0 = \left(\frac{R_0}{R_0 + R_1 + R_0 R_1/R_l}\right) e_i. \qquad (10\text{--}16)$$

Note that Eq. (10–16) approaches Eq. (10–15) as $R_l \to \infty$ only; in any other circumstances the output potential is affected by the loading. Since we are interested in multiplying the input potential by a predetermined positive constant, the deviation of the output potential from Eq. (10–15) is called the *error* of the circuit element due to loading.

Next we consider integration, which may be approximately achieved by the circuit shown in Fig. 10–5. Again we write Kirchhoff's current law for nodal point P,

$$i_1 - i_0 = 0.$$

Relating i_0 and i_1 to the potentials by Ohm's and Coulomb's laws, respectively, gives

$$\frac{e_i - e_0}{R} + C\frac{de_0}{dt} = 0.$$

Now, if we assume that $e_i \gg e_0$, it follows that

$$e_0 = \frac{1}{RC}\int_0^t e_i \, d\tau, \qquad (10\text{--}17)$$

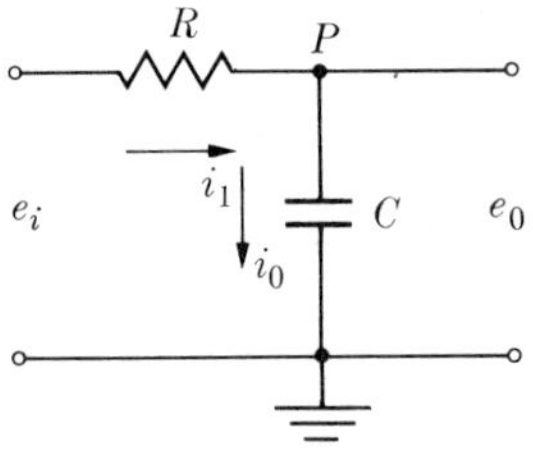

FIG. 10–5

which is the integral of the input potential. Note that the circuit element for integration is *approximate* even in the absence of any load because of the assumption $e_i \gg e_0$. The effect of loading may be determined by a procedure similar to that followed for the preceding circuit element. This is left to the reader (see Problem 10–4).

It is clear from the foregoing discussion that all passive circuit elements suffer from loading errors. In general, these elements operate correctly only if the impedance at the output of the circuit element is very high. Since active circuit elements such as electronic amplifiers have input impedances often in excess of several megohms, the coupling of a passive circuit element with an amplifier eliminates the loading error on the element. Of course, the use of amplifiers increases both the first cost and the operating cost of the analog circuit (hereafter to be referred to as the *analog computer*); however, this expense may be well justified by the increased accuracy of the solution. For most general-purpose analog computers, the high-gain DC amplifier discussed in the next section is employed almost exclusively. In general, AC analog computers are less expensive to build and require simpler auxiliaries; however, the phase shifts in AC circuits tend to make this type of computer less accurate than DC computers.

10–3. Active Circuit Elements. High-Gain DC Amplifiers

The most commonly employed active elements are high-gain DC amplifiers, function generators, and function multiplier-dividers. We shall consider only amplifiers.

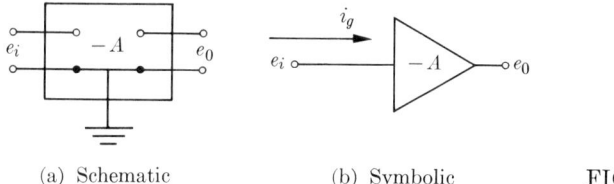

(a) Schematic (b) Symbolic   FIG. 10–6

The design of amplifiers and the function of its components are beyond the scope of the text. Interested readers may refer to the references at the end of this chapter. Here we picture the amplifier as simply a black box (Fig. 10–6), and discuss only those amplifier characteristics conceptually important for analog computer solutions. Designating the input and output potentials relative to the ground by e_i and e_0, respectively, we first define the gain A of the amplifier by the ratio†

$$A = -e_0/e_i.$$

The basic requirement of the amplifier is that it have as high a gain as possible. The practical range of this gain is 10^4–10^8. Assuming that for the purpose of recording, the output potential is to be kept between ± 100 volts, we obtain

---

† The sign of this ratio will be discussed in connection with Fig. 10–7.

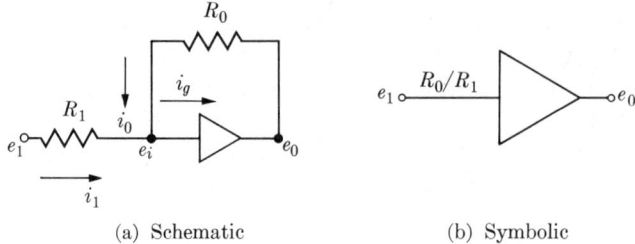

(a) Schematic (b) Symbolic   FIG. 10-7

a maximum possible value of 0.01 volts for the input potential. Since this potential is usually connected to a vacuum-tube grid which has an extremely high impedance, the input current is very small. Hence the basic and associated properties required from the amplifier as an analog component are: (i) a very high gain A, (ii) approximately zero input potential e_i, (iii) negligibly small input current i_g. Of course, there are other important properties required from the amplifier, such as linear output over a wide range, flat frequency response, low noise level, etc. These, however, are means to further improvement in the accuracy of the solution, rather than basic requirements for the use of the amplifier.

Let us now consider a circuit consisting of a high-gain DC amplifier, an input resistor R_1, and a feedback resistor R_0 (Fig. 10-7). A potential, say e_1, is applied to the input of the amplifier through the resistor R_1. Since $i_g = 0$ according to property (iii), Kirchhoff's law applied to the input of the amplifier gives

$$i_1 + i_0 = 0.$$

Then, relating currents to potentials by Ohm's law, and noting that $e_i = 0$ because of property (ii), we obtain

$$e_1/R_1 + e_0/R_0 = 0,$$

which may be rearranged in the form

$$e_0 = -\frac{R_0}{R_1} e_1. \tag{10-18}$$

Thus we have shown that the circuit element given by Fig. 10-7 is a *multiplier*. Clearly, the same circuit can be used as a *sign changer* by taking the ratio R_0/R_1 equal to unity.

Note that in analog computer applications, the DC amplifier always includes a feedback circuit as exemplified by Fig. 10-7. It is for this reason that the amplifier gain A has to be negative. For, if it were positive, an increase at the output potential would increase the input potential which, in turn, would further increase the output potential, causing instability. The sign of amplifier gain is made negative by using an odd number of amplification stages. This is

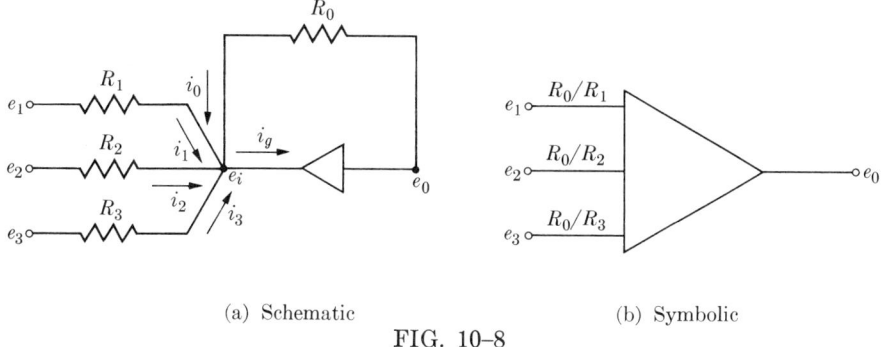

(a) Schematic (b) Symbolic

FIG. 10-8

a problem of electronics, however, and is not our concern. Note, moreover, that this sign has nothing to do with that of Eq. (10–18).

Let us now reconsider the circuit element given by Fig. 10–7 and replace the input resistor by three parallel resistors as shown in Fig. 10–8. Again, Kirchhoff's law applied to the input of the amplifier in the light of property (iii), and rearranged by Ohm's law and property (ii) gives

$$e_1/R_1 + e_2/R_2 + e_3/R_3 + e_0/R_0 = 0,$$

which yields

$$e_0 = -\left(\frac{R_0}{R_1} e_1 + \frac{R_0}{R_2} e_2 + \frac{R_0}{R_3} e_3\right). \tag{10-19}$$

We learn from Eq. (10–19) that the circuit element of Fig. 10–8 can be used as an *adder*, as clearly seen from the special case $R_0/R_1 = R_0/R_2 = R_0/R_3 = 1$.

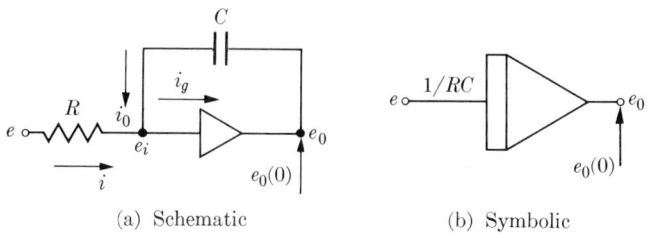

(a) Schematic (b) Symbolic

FIG. 10-9

We now return once more to the circuit element given by Fig. 10–7, and this time replace the feedback element by a capacitor C as indicated in Fig. 10–9. Kirchhoff's law applied to the input of the amplifier and rearranged by Ohm's and Coulomb's laws then gives

$$\frac{e}{R} + C\frac{de_0}{dt} = 0$$

which, by integration, may be rearranged in the form

$$e_0 = -\frac{1}{RC}\int_0^t e\,d\tau + e_0(0), \tag{10-20}$$

where $e_0(0)$ is the initial condition, the potential to which the capacitor is charged at $t = 0$. Equation (10–20) indicates that the circuit element of Fig. 10–9 can be employed as an *integrator*.

Clearly, a DC amplifier may also be used as a *differentiator*. (Draw the corresponding circuit.) However, whenever possible this operation should be avoided because of noise.

Thus we have seen the use of the DC amplifier as a multiplier, a sign changer, an adder, and an integrator. The two other important active circuit elements, function generators and function multiplier-dividers, require a somewhat lengthy and special treatment, and will not be considered here. The interested reader may consult the references cited at the end of this chapter.

We now proceed to the use of DC amplifiers in the solution of conduction problems. We begin by considering the active element corresponding to the passive element shown in Fig. 10–2 and governed by Eq. (10–12). If this equation is rearranged in the form

$$\frac{E_{m+1}}{R} + \frac{E_{m-1}}{R} + \frac{-E_m}{R/2} + C\frac{d(-E_m)}{dt} = 0, \tag{10-21}$$

then the desired active circuit element is one that involves a DC amplifier whose input nodal point satisfies Kirchhoff's current law corresponding to Eq. (10–21) (Fig. 10–10). The extension of the foregoing procedure to two-dimensional geometry follows immediately. Thus we obtain the active element (Fig. 10–11)

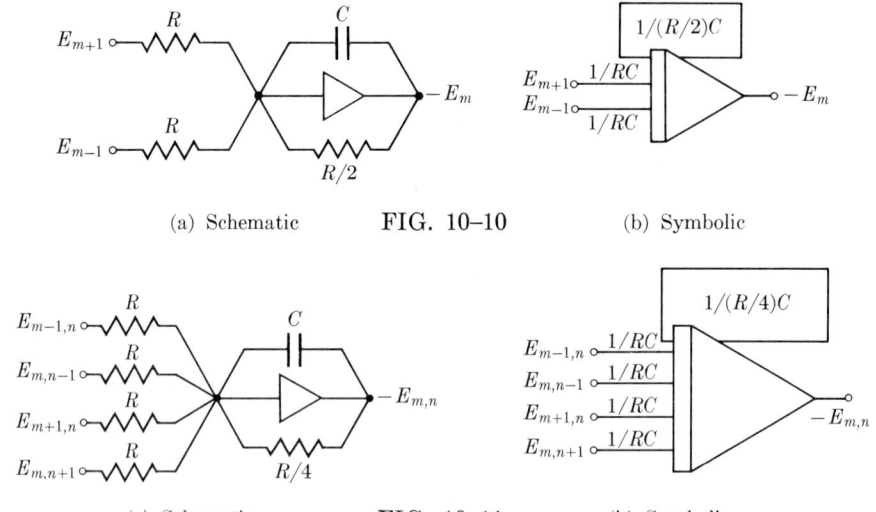

(a) Schematic   FIG. 10–10   (b) Symbolic

(a) Schematic   FIG. 10–11   (b) Symbolic

corresponding to the passive element given in Fig. 10–3 and governed by Eq. (10–14). Further extension to three-dimensional cartesian geometry, moving solids, boundary conditions, and other geometries presents no difficulty. This is left to the reader as an exercise.

10–4. Examples

For further experience in analog solutions we next consider a number of lumped and distributed problems. Since active circuits involve negligible error as compared with passive circuits, they are more commonly used in practice; hence we shall discuss the solution of these problems in terms of active circuits only.

Example 10–1. One surface of a flat plate is suddenly subjected to constant heat flux q''; the other surface is insulated (Fig. 10–12). The thickness of the plate is δ, and the initial temperature T_0. We wish to find the analog solution of the problem.

Assuming that the temperature of the plate may be lumped, we find the formulation of the problem to be

$$dT/dt - m = 0, \quad T(0) = T_0, \quad (10\text{–}22)$$

where $m = q''/\rho c \delta$. Equation (10–22) can readily be simulated by

$$C\frac{dE}{dt} + \frac{(-E_1)}{R} = 0, \quad E(0) = E_0. \quad (10\text{–}23)$$

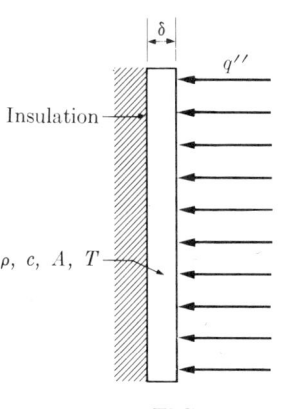

FIG. 10–12

The desired analog solution is obtained from an active circuit that includes a DC amplifier whose input nodal point satisfies Kirchhoff's current law corresponding to Eq. (10–23) (Fig. 10–13). If E is made proportional to T, consequently E_0 to T_0, the values of E_1, R, and C must be so adjusted that E_1/RC is proportional to m in the same sense. The circuit operates as follows: two switches, both possibly activated by the same relay coil, are employed to apply

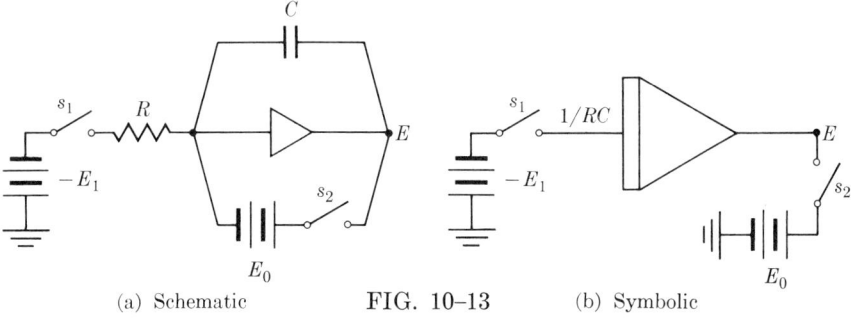

(a) Schematic   FIG. 10–13   (b) Symbolic

the initial condition and driving potential; initially S_1 is open and S_2 is closed; suddenly S_1 is closed and S_2 is opened.

Example 10–2. A square plate of thickness δ is pressed on a solid surface by force P (Fig. 10–14). The plate suddenly assumes constant velocity V on the surface. The coefficient of dry friction between the plate and the solid is μ. The conductivity of the solid is much smaller than that of the plate. Initially the plate is at ambient temperature T_∞. The heat transfer coefficient is h. We wish to find the analog solution of the problem.

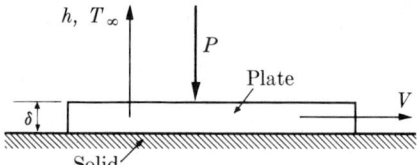

FIG. 10–14

Assuming that the plate temperature may be lumped, and neglecting the heat transfer to the solid, we find that the formulation of the problem in terms of $\theta = T - T_\infty$ is

$$d\theta/dt + m\theta = n, \qquad \theta(0) = 0, \tag{10-24}$$

where $m = h/\rho c \delta$ and $n = \mu PV/\rho c \delta A$. The corresponding analog formulation is

$$C\frac{dE}{dt} + \frac{E}{R} + \frac{(-E_0)}{R_0} = 0. \tag{10-25}$$

An active circuit which satisfies Eq. (10–25) at the inlet nodal point of the DC amplifier provides the required analog solution (Fig. 10–15). The values of R_0, R, and C are selected by considering the proportionality of m to $1/RC$ and that of n to $E_0/R_0 C$.

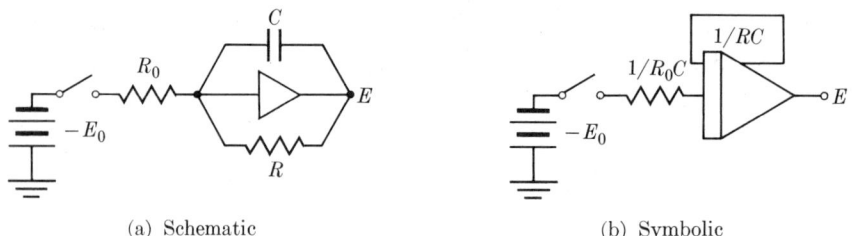

(a) Schematic (b) Symbolic

FIG. 10–15

Example 10–3. Let us find the analog solution of Problem 5–3.

The solution of this problem by means of Laplace transforms was discussed in Example 7–2. Referring to that example, and after some rearrangements, we write the formulation of the problem in the form†

$$\frac{d\psi}{dt} + b_1\psi - b_1\phi = 0, \qquad (10\text{-}26)$$

$$\frac{d\phi}{dt} + (b_2 + b_3)\phi - b_2\psi = 0, \qquad (10\text{-}27)$$

$$\psi(0) = \phi(0) = \psi_0. \qquad (10\text{-}28)$$

Simulating ψ by E and ϕ by e, we find that the corresponding analog equations are

$$C_1 \frac{dE}{dt} + \frac{E}{R_1} + \frac{(-e)}{R_1} = 0, \qquad (10\text{-}29)$$

$$C_2 \frac{de}{dt} + \frac{e}{R_3} + \frac{(-E)}{R_2} = 0, \qquad (10\text{-}30)$$

$$E(0) = e(0) = E_0. \qquad (10\text{-}31)$$

Although the first two terms of Eqs. (10–29) and (10–30) are easily satisfied at the input of two DC amplifiers, the third term of each equation requires the reverse value of the potential associated with the first two terms of the other equation. Thus the use of another amplifier as a sign changer at the output of each of the first two amplifiers becomes necessary. The analog solution of the problem is shown in Fig. 10–16. However, by simply changing the sign of Eq. (10–29) or Eq. (10–30), we can eliminate the amplifiers employed as sign changers; multiplying one of these equations, say the latter, by -1 gives

$$C_2 \frac{d(-e)}{dt} + \frac{(-e)}{R_3} + \frac{E}{R_2} = 0. \qquad (10\text{-}32)$$

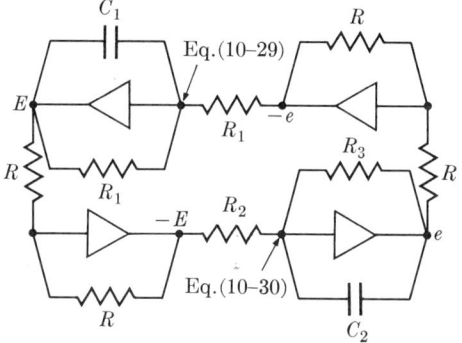

FIG. 10–16

† See Eqs. (7–29), (7–30), and (7–31).

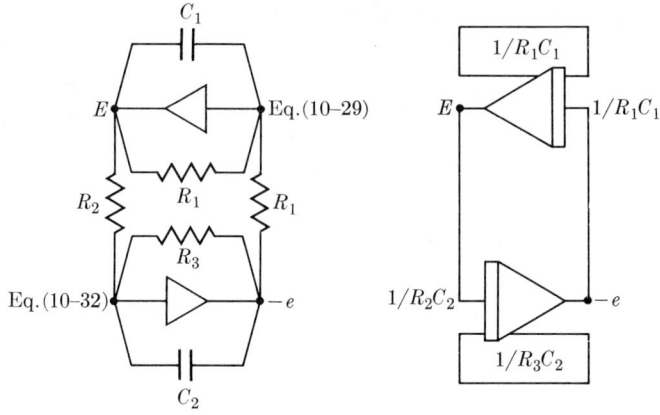

(a) Schematic (b) Symbolic FIG. 10–17

Now we need only E and $-e$. In fact, the circuit indicated in Fig. 10–17 satisfies Eqs. (10–29) and (10–32). Note that both capacitors must initially be loaded according to Eq. (10–31). Also, the values of R_1, R_2, R_3, C_1, and C_2 must be so selected that b_1 is proportional to $1/R_1C_1$, $b_2 + b_3$ to $1/R_3C_2$, and b_2 to $1/R_2C_2$.

Example 10–4. Let us find the analog solution of Example 9–5.

Since again our purpose is demonstration rather than an accurate solution, let us use the coarse subdivision considered in Example 9–5. For the typical inner nodal point we may use the circuit element in Fig. 10–10; however, we need to develop the circuit elements corresponding to boundary nodal points.

In terms of Fig. 9–25, the differential-difference formulation of the boundary associated with the imposed heat flux is

$$\frac{(\Delta x)^2}{2a} \frac{dT_0}{dt} = T_1 - T_0 + \frac{q''(\Delta x)}{k}. \tag{10–33}$$

Since $(\Delta x)^2/a$ has been made proportional to RC in connection with our discussion of the inner nodal points,† the simulation of Eq. (10–33) gives

$$\frac{E_1}{R/2} + \frac{E''}{R/2} + \frac{(-E_0)}{R/2} + C\frac{d(-E_0)}{dt} = 0, \tag{10–34}$$

where E'' is proportional to $q''(\Delta x)/k$. The circuit element that satisfies Eq. (10–34) is given in Fig. 10–18.

In terms of Fig. 9–20 and its rearranged nomenclature, the differential-difference formulation of the boundary that transfers heat to the ambient is

$$\frac{(\Delta x)^2}{2a} \frac{dT_4}{dt} = T_3 - T_4 + \Delta \text{Bi}\,(T_\infty - T_4). \tag{10–35}$$

---

† Recall the comparison between Eqs. (10–12) and (10–13).

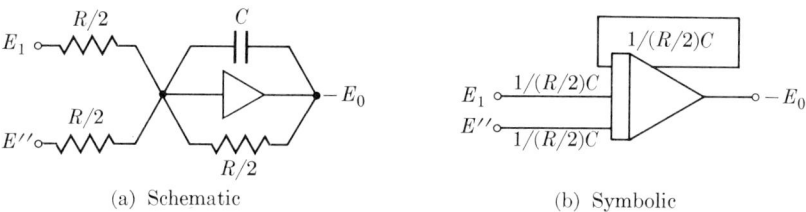

(a) Schematic (b) Symbolic

FIG. 10–18

The simulation of Eq. (10–35) yields

$$\frac{E_3}{R/2} + \frac{E_\infty}{R/2\Delta\mathrm{Bi}} + \frac{(-E_4)}{R/2(1+\Delta\mathrm{Bi})} + C\frac{d(-E_4)}{dt} = 0, \qquad (10\text{--}36)$$

where E_∞ is proportional to T_∞. The circuit element indicated in Fig. 10–19 satisfies Eq. (10–36).

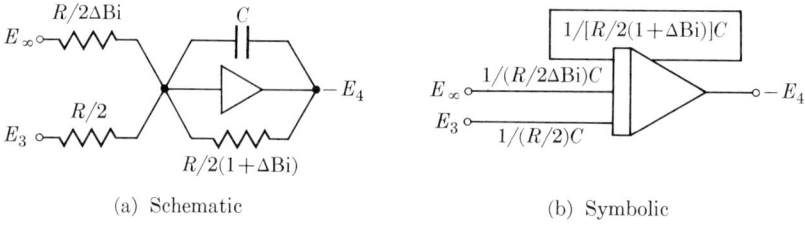

(a) Schematic (b) Symbolic

FIG. 10–19

If the numerical value of $\Delta\mathrm{Bi}$ requires a fraction of an available resistance, any necessary adjustment can be made by means of a potentiometer (Fig. 10–20).

When the appropriate circuit elements are assigned to nodal points, the existence of both plus and minus potentials requires the use of additional amplifiers as sign changers. However, these amplifiers can be eliminated as in the preceding example, by simply multiplying the potentials of alternate circuit elements by -1. The analog solution thus obtained is shown in Fig. 10–21.

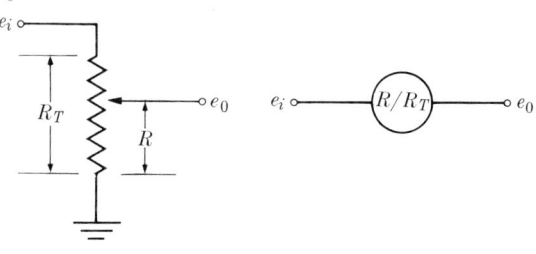

(a) Schematic (b) Symbolic FIG. 10–20

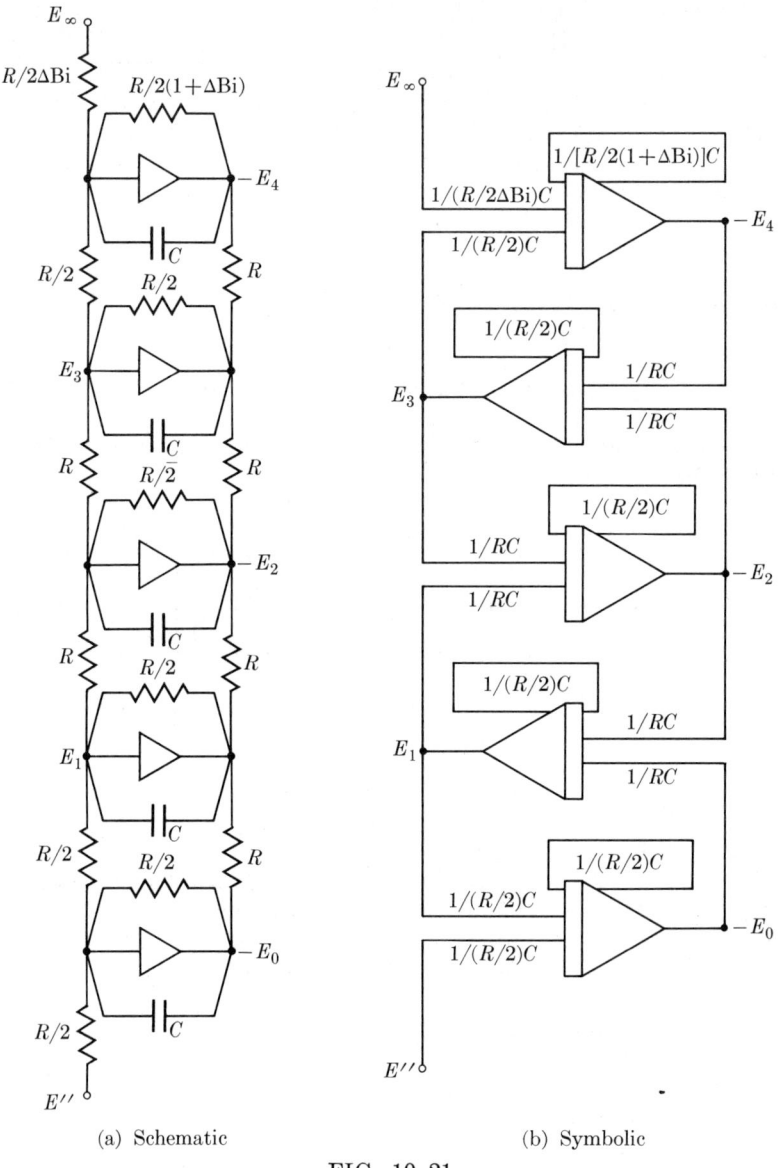

FIG. 10-21

The circuit operates as follows: initially the input and output potentials have the same value E_∞, which may conveniently be assumed zero; then the input potential is suddenly changed to E''.

Example 10-5. Let us find the circuit element associated with Example 9-4. In terms of Fig. 9-23 the differential-difference formulation that corresponds to

[10-4] EXAMPLES

the difference formulation given in Example 9-4 is

$$\rho c A (\Delta x) \frac{dT_m}{dt} = \rho c A V (T_{m-1} - T_m) + h_i P_i (\Delta x)(\theta_m - T_m),$$

$$\rho' c' A' (\Delta x) \frac{d\theta_m}{dt} = h_i P_i (\Delta x)(T_m - \theta_m) + h_o P_o (\Delta x)(0 - \theta_m),$$

which may be rearranged in the form

$$\frac{dT_m}{dt} = \frac{V}{\Delta x} T_{m-1} + \left(\frac{V}{\Delta x} + \frac{h_i P_i}{\rho c A} \right)(-T_m) + \frac{h_i P_i}{\rho c A} \theta_m, \quad (10\text{-}37)$$

$$\frac{d\theta_m}{dt} = \frac{h_i P_i}{\rho' c' A'} T_m + \left(\frac{h_i P_i + h_o P_o}{\rho' c' A'} \right)(-\theta_m) + \frac{h_o P_o}{\rho' c' A'} 0. \quad (10\text{-}38)$$

The zero ambient temperature is intentionally left in Eq. (10-38) to determine the ground connection of the corresponding analog circuit element. Simulating Eqs. (10-37) and (10-38), and reversing the signs of the second equation thus obtained, we get

$$\frac{E_{m-1}}{R_1} + \frac{e_m}{R_2} + \frac{(-E_m)}{R_3} + C_1 \frac{d(-E_m)}{dt} = 0, \quad (10\text{-}39)$$

$$\frac{(-E_m)}{R_4} + \frac{(-0)}{R_5} + \frac{e_m}{R_6} + C_2 \frac{de_m}{dt} = 0, \quad (10\text{-}40)$$

where the potentials E and e are used for the temperatures T and θ, respectively. Equations (10-39) and (10-40) are satisfied by the circuit element shown in Fig. 10-22. An appropriate number of such circuit elements should now be connected in series to cover the entire length of the tube. This, however, presents no problem and is left to the reader.

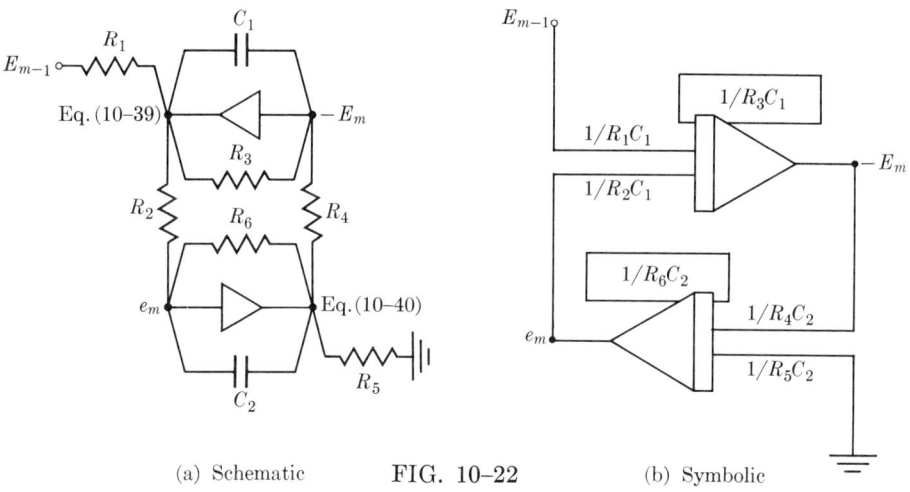

(a) Schematic FIG. 10-22 (b) Symbolic

10–5. Miscellaneous

In this chapter we have given a short introduction to the analog solution of conduction problems. In conclusion, we discuss briefly a number of important points related to mechanics of computer operation or to advanced topics.

Units. The units commonly employed in practice are the megohm (10^6 ohms) for resistors and the microfarad (10^{-6} farad) for capacitors. The product RC is then expressed in ohm $\times$ farad = sec.

Amplitude and time scale factors. Variables of conduction and related computer problems have limitations on their possible magnitudes and rates of change with time; hence in practice the corresponding variables of the two systems must be related by amplitude and time scale factors. More specifically, an amplitude scale factor is needed because the computer variable, the output potential of a DC amplifier, is limited, often to ± 100 volts; in order to measure a temperature variation over a range greater than $\pm 100°F$, we would need a scale factor between potentials and temperatures. Similarly, the use of a time scale factor becomes necessary when the transient time of a problem is too short or too long; for short transients the components of the computer and of the recording equipment may not function adequately; for long transients amplifier drift and capacitor leakage may introduce appreciable errors. Furthermore, for reasons of economy, we often prefer to use the computer for a shorter time than the actual time. An extensive treatment of scale factors may be found in Chapter 3 of Reference 3.

Stability. When the gain through an amplifier is accompanied by a phase shift, the latter may lead to a condition of self-oscillation (instability) in the amplifier output. This important problem is beyond the scope of the text. A short discussion may be found in Section 7–8 of Reference 2.

High-speed computers. In a high-speed computer a problem is solved over and over again, many times in a second. Because of its limited component capability, this computer has a relatively low accuracy. Since no available graphical recording instrument is suitable to high-speed computers, the cathode-ray oscilloscope appears to be the only practical display device. Reference 7 is devoted to this type of computer.

Schematic and symbolic diagrams. In this chapter we have discussed circuit elements and corresponding circuits in terms of schematic and symbolic diagrams. The use of symbolic notation is desirable because of its simplicity; however, this notation often causes confusion for the beginner. Therefore, until enough experience has been gained in the use of symbolic diagrams, the use of schematic diagrams is suggested.

References

1. W. J. KARPLUS, *Analog Simulation.* New York: McGraw-Hill, 1958.
2. W. J. KARPLUS and W. W. SOROKA, *Analog Methods.* New York: McGraw-Hill, 1959.
3. A. S. JACKSON, *Analog Computation.* New York: McGraw-Hill, 1960.
4. C. L. JOHNSON, *Analog Computer Techniques.* New York: McGraw-Hill, 1956.
5. G. A. KORN and T. M. KORN, *Electronic Analog Computers.* New York: McGraw-Hill, 1956.
6. S. FIFER, *Analogue Computation I, II.* New York: McGraw-Hill, 1961.
7. R. TOMOVIC and W. J. KARPLUS, *High-Speed Analog Computers.* New York: Wiley, 1962.

Problems

10-1. Draw the passive circuit element corresponding to the conduction of heat governed by the equation

$$\frac{\partial T}{\partial t} + U \frac{\partial T}{\partial x} = a \frac{\partial^2 T}{\partial x^2}.$$

10-2. Draw the passive circuit elements corresponding to the boundary conditions

$$\pm k \left(\frac{\partial T}{\partial n}\right)_\sigma + q'' = h(T_\sigma - T_\infty)$$

and

$$\pm k_1 \left(\frac{\partial T_1}{\partial n}\right)_\sigma + \mu p V = \pm k_2 \left(\frac{\partial T_2}{\partial n}\right)_\sigma.$$

10-3. Find the effect of loading on the passive circuit element (adder) shown in Fig. 10-23.

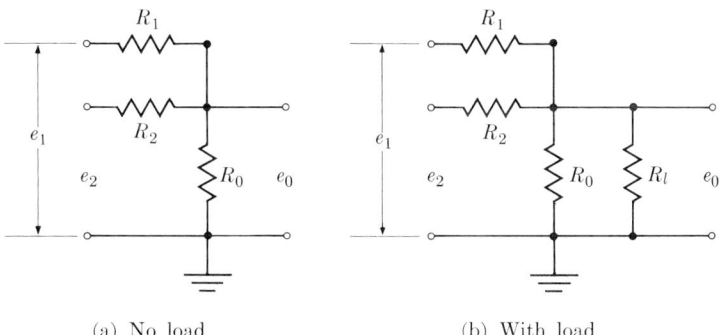

(a) No load (b) With load

FIG. 10-23

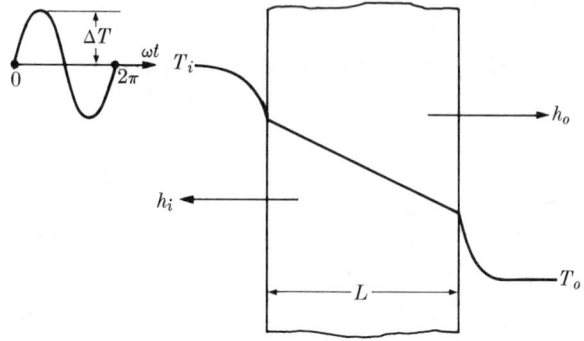

FIG. 10-24

10-4. Find the effect of loading on the passive circuit element (integrator) shown in Fig. 10-5.

10-5. Find the solution of Example 10-3 in terms of a passive circuit.

10-6. Find the solution of Example 2-2 in terms of an active circuit.

10-7. Draw the active circuit element for a typical inner nodal point of Problem 5-5. Connect three of these elements corresponding to three neighboring points.

10-8. Consider a wall through which heat is steadily transferred from an ambient of higher temperature to an ambient of lower temperature (Fig. 10-24). Assume that the higher ambient temperature starts to oscillate in the form

$$T_i(t) = T_i + \Delta T \sin \omega t.$$

(a) Using DC amplifiers, design a (function generator) circuit that generates sine wave fluctuations of the ambient temperature. (b) Give the solution of the problem in terms of an active circuit.

For the next three problems the reader is urged to use a locally available analog computer.

10-9. Solve Problem 9-9 by using an active circuit.

10-10. When axial conduction is included, Laplace transforms are no longer suitable to Problem 7-10. Investigate the effect of axial conduction on this problem by means of an active circuit.

10-11. Solve the general formulation of Example 2-5 in terms of an active circuit.

INDEX

Active circuit elements, 529
Ampere's circuit law, 17
Amplifiers, high-gain DC, 529
Amplitude and scale factors of analog solutions, 540
Analog solution, 524
  circuit elements of, active, 529
    passive, 527
  stability of, 540
  units of, 540
Analogy between conduction and electricity, 104, 524
Analytic functions, branch points of, 376
  definition of, 376
  line integral of, 378
  poles of, 378
  singularities of, 376
  Taylor series of, 383
Anisotropy, 38
Approximate profiles, see Profiles
Approximate solutions, see Solutions
Approximation in continuum (field) theory formulations, 316, 505

Backward difference, 514
Basic rule of relaxation method, 494
  modified, 496
    block relaxation, 496
    group relaxation, 496
    overrelaxation, 496
Bessel equation, 132, 134
Bessel functions, 103, 132, 133
  behavior of, 140
    large argument, 141
    small argument, 140
  derivatives of, 139
  graphical representation of, 141
  modified, of first and second kinds, of order ν, 136
  of first and second kinds, of order ν, 135
  of third kind, 139
  properties of, 139
  relation between, 140
    circular or hyperbolic functions and, 140

Binder-Schmidt method, graphical, 483, 508, 515
Biot modulus, 53, 207
Boundary conditions, 46
  heat transfer, by convection, 50
    by radiation, 53
  interface, 54
    in relative motion, 55
    moving, 56
  natural, of variational formulation, 440, 456
  nonhomogeneous, of separation of variables, 217
  prescribed heat flux, 46
    acting at a distance, 54
    insulation, 48
  prescribed temperature, 46
  time-dependent, 307
Boundary-value problems, 180
Boundaries, irregular, 492
Branch points of analytic functions, 376

Cauchy-Riemann conditions, 376
Cauchy's equation, 138
Cauchy's integral formula, 382, 383
Cauchy's integral theorem, 378, 380
Cauchy's residue theorem, 384, 385
Characteristic-value problems, 180
Characteristic values, 181
Charts, one-dimensional, 290
  cartesian, heat flux, 294
    semi-infinite geometry, infinite h, 302
    semi-infinite geometry, finite h, 303
    temperature, 292, 293
  cylindrical, heat flux, 298
    temperature, 296, 297
  spherical, heat flux, 301
    temperature, 299, 300
  use of, 290
Circuit law, Ampere's, 17
Coefficient of heat transfer, 50
  overall, 105

cartesian, 107
composite structures, 107
  cartesian, 108
  cylindrical, 108
  spherical, 108
cylindrical, 107
spherical, 107
Complex functions, elementary, 375
  single- and multiple-valued, 375
Complex line integral, 378
Complex temperature, method of, 324
Complex variable, argument of a, 374
  conjugate of a, 374
  functions of a, 374
  modulus of a, 374
Concepts, definition of, 17, 18
Condition, initial, 46
Conditions, boundary, see Boundary
    conditions
Conduction, 9
  equation of, 44
    for heterogeneous continua,
      anisotropic, 45
      isotropic, 44
    for homogeneous continua,
      anisotropic, 45
      isotropic, 44
  Fourier's law of, 17, 37
    for heterogenous continua,
      anisotropic, 40
      isotropic, 38
    for homogeneous continua,
      isotropic, 37
Conductive resistance, 104
Conductivity, thermal, 37
  variable, 129
Conservation laws, for control volume,
    see General laws
Continuum, definition of, 18
  description of, Eulerian, 18
    Lagrangian, 18
  (field) theory formulations,
    approximation in, 316, 505
Continuum approach, 3
  versus molecular approach, 9
Control volume, definition of, 18
  general laws for, see General laws
Convection, 9
  boundary condition, 50
Convective derivative, 33
Convective resistance, 104
Convolution integral, 341

relation between Duhamel's integral
    and, 342
Coordinate axes, selection of, 214
Coulomb's law, 525, 526
Critical radius, 120
Current law, Kirchhoff's, 526
Cycle, definition of, 18

Degree of irreversibility, 26
Depth, penetration, 76, 78, 159, 474
Derivative, convective, 33
  following motion, 33
  material, 33
  substantial, 33
Difference formulation, 483
  backward and forward differences, 514
  cylindrical, 490
  errors in, round-off, 504
    truncation, 488
  explicit, 514
  implicit, 514
  irregular boundaries, 492
  network of, 484
    finer, graded, hexagonal, and
      triangular, 489
    nodal points of, 484
    relation between differential and, 486
  spherical, 490
  steady, 483
  unsteady, 503
    stability of, 503
    unconditional, 514
Difference solution, graphical
    (Binder-Schmidt) method, 483,
    508, 515
  numerical (step-by-step) method, 483,
    508, 515
  relaxation method, 493
    basic rule of, 494
    modifications of (block, group,
      overrelaxations), 496
    residuals of, 494
    tables of (relaxation and operation),
      494
Differential control volume, general laws
    for, see General laws, for
    differential control volume
Differential-difference formulation, 524
  amplitude and scale factors of, 540
  circuit elements, active, 529
    high-gain DC amplifier, 529
    passive, 527

stability of, 540
units of, 540
Differential-difference solution (analog), 524
Differential formulation of general laws, see General laws, formulation of
Diffusion, 9
Diffusivity, thermal, 44
Duhamel's superposition integral, 307
  relation between convolution integral and, 342

Eigenfunctions, 182
Eigenvalue problems, 182
Eigenvalues, 182
Elasticity law, Hooke's, 4, 17
Electric charge, conservation of, 17, 524
Electricity law, Ohm's, 17, 104, 525
Energy, total, 23
  chemical, kinetic, nuclear, potential, thermal (internal), 23
  conservation of, for differential control volume, see General laws, for differential control volume
Energy generation, 24 31, 33
Enthalpy, stagnation, 25, 31
Entropy, conservation of, see General laws, formulation of
Entropy generation, 26, 31, 36
Equality of temperature, definition of, 19
Equation, Bessel, 132, 134
  Cauchy, 138
  equidimensional, 132, 138
  Euler, 138
    of variational formulation, 437, 457, 473
  Legendre, 240
Equidimensional equation, 132, 138
Euler equation, 138
Eulerian description of continuum, 18
Even functions, 187
Extended surfaces (fins, pins, or spines), 144
  constant cross section, 146
  efficiency of, 148
  solution of, general, 138
    approximate, 156
    higher-order, 161

Faraday's law of induction, 17
Field (continuum) theory formulations, approximation in, 316, 505

First law of thermodynamics, see General laws
Force law (electromagnetic), Lorentz's, 17
Formulation and solution, further methods of, 433
Formulation, five basic steps of, 61
  methods of, 59
  of general laws, see General laws, formulation of
Fourier-Bessel integral, 371
Fourier-Bessel series, 230
Fourier integral, 365
  complete, 370
  complex, 370
  cosine, 369
  sine, 368
Fourier-Legendre series, 240, 244
Fourier series, 186
  complete, 192
  cosine, 190
  double, 250, 290
  sine, 187
Fourier transform, 371
Fourier's law of conduction, 17, 37
  for heterogeneous continua, anisotropic, 40
    isotropic, 38
  for homogeneous continua, isotropic, 37
Functions, Bessel, see Bessel functions
  even, 187
  Gamma, 134
  Graetz, of first and second kinds, of order zero, 205
  Hankel, of first and second kinds, of order ν, 140
  Legendre, 240
  odd, 187
  of a complex variable, 374
    single- and multiple-valued, 375
  piecewise, continuous, 337
    differentiable, 186
Further methods of formulation and solution, 433

Gamma functions, 134
General laws, definition of, 7, 17
  for differential control volume,
    conservation of entropy, 36
    conservation of mass, 32
    conservation of mechanical energy, 34
    conservation of thermal energy, 34, 35

conservation of total energy, 34
first law of thermodynamics, 34
second law of thermodynamics, 35
for integral control volume,
conservation of entropy, 32
conservation of mass, 29
first law of thermodynamics, 31
second law of thermodynamics, 31
for lumped control volume,
conservation of entropy, 26
conservation of mass, 23
first law of thermodynamics, 25
second law of thermodynamics, 26
formulation of, differential,
conservation of mass, 32
first law of thermodynamics, 33
second law of thermodynamics, 35
integral, conservation of mass, 29
first law of thermodynamics, 29
second law of thermodynamics, 31
lumped, conservation of mass, 22
first law of thermodynamics, 23
second law of thermodynamics, 26
statement of, 19
distributed (integral, differential, variational, difference), 19
lumped, 19
Graphical (Binder-Schmidt) method, 483, 508, 515

Hankel functions of first and second kinds, of order ν, 140
Heat, definition of, 19
flow, 30
flux, 30
Heat transfer, coefficient of, 50
overall, 105
cartesian, 107
cylindrical, 107
spherical, 107
composite structures, 107
overall coefficient of, 107
cartesian, 108
cylindrical, 108
spherical, 108
continuum, 3
foundations of, 3
continuum, 10
modes of, 9
place of, 3
Heterogeneous continua, 38
equation of conduction for, anisotropic, 45

isotropic, 44
Fourier's law of conduction for,
anisotropic, 40
isotropic, 38
High-gain DC amplifier, 529
characteristics of, 530
functions of, 530
adder, 531
integrator, 532
multiplier, 530
sign changer, 530
Homogeneous boundary conditions and differential equations, definitions of, 126
Homogeneous continua, 38
equation of conduction for, anisotropic, 45
isotropic, 44
Fourier's law of conduction for, isotropic, 37
Hooke's law of elasticity, 4, 17

Ideal gas law, 8, 17
Implicit difference formulation, 514
Induction law, Faraday's, 17
Initial condition, 46
Initial-value problems, 180
Integral, complex line, 378
Integral control volume, general laws for, see General laws, for integral control volume
Integral formula, Cauchy's, 382, 383
Integral formulation of general laws, see General laws, formulation of, integral
Integral theorem, Cauchy's, 378, 380
Inversion, table of, 344
Inversion theorem, 339, 372, 373
Contour I for, 386
Contour II for, 388
solutions by, 390
Irregular boundaries, 492
Irreversibility, degree of, 26
Isotropy, 38

Kantorovich method, 460
approximate solutions by, see Solutions, steady two-dimensional and unsteady
extended, 466
Kantorovich profiles, selection of, 469, 473
Kirchhoff transformation, 130

Kirchhoff's current law, 526
Kirchhoff's method, 129

Lagrangian description of continuum, 18
Laplace transforms, definition of, 337
   inversion table of, 344
   inversion theorem for, 339, 372, 373
     Contour I of, 386
     Contour II of, 388
   properties of, 339
   table of, 343
   solutions, by expansion for large
     argument, 418
    by expansion for small argument, 413
    by inversion theorem, 390
    periodic, 421
    by tables, 343
Legendre equation, 240
Legendre function, 240
Legendre polynomial, 240
   of degree n, of first and second kinds, 241
    order m, of first and second kinds, associated, 242
Leibnitz's integral formula, 94
Line integral, of analytic functions, 378
Loading error, of passive circuit elements, 528
Lost work, 26
Lorentz's (electromagnetic) force law, 17
Lumped control volume, general laws for, see General laws, for lumped control volume
Lumped formulation of general laws, see General laws, formulation of, lumped

Material derivative, 33
Mechanical energy, conservation of, for differential control volume, 34
Method, analog, 524
   graphical (Binder-Schmidt), 483, 508, 515
   Kantorovich, steady two-dimensional, 460
     unsteady, 473
   Kirchhoff's, 129
   Laplace transforms, 336
   numerical (step-by-step), 483, 508, 515
   of complex temperature, 324
   relaxation, 493
   Ritz, steady one-dimensional, 444
   steady two-dimensional, 458
   separation of variables, steady, 180
     unsteady, 270
Methods of formulation, 59
   and solution, further, 433
Modified Bessel functions of first and second kinds, of order ν, 136
Modulus, Biot, 53, 207
   Fourier, 294
   Graetz, 204
   Nusselt, 207
   Péclèt, 117, 212
Molecular approach, 3
Multidimensional unsteady problems expressible in terms of one-dimensional unsteady problems, 290
   condition for, 301

Natural boundary conditions of variational calculus, 440, 456
Natural laws, statement of, 17
Network of difference formulation, 484
Newton's second law of motion, 17
Newton's viscosity law, 17
Nodal points of difference formulation, 484
Nonhomogeneity for separation of variables, 126
Nusselt modulus, 207

Odd functions, 187
Ohm's law of electricity, 17
One-dimensional charts, use of, 290
   heat flux, cartesian, 294
     cylindrical, 298
     spherical, 301
   temperature, cartesian, 292, 293
     semi-infinite geometry, infinite h, 302
     semi-infinite geometry, finite h, 303
     cylindrical, 296, 297
     spherical, 299, 300
Orthogonal functions, 183
   expansion of arbitrary functions in a series of, 185
Orthogonality of characteristic functions, 183
Overall coefficient of heat transfer, 105
   cartesian, 107
   composite structure, 107
     cartesian, 108

cylindrical, 108
spherical, 108
cylindrical, 107
spherical, 107

Particular laws, definition of, 7, 17
  statement of, 37
    Coulomb's law, 525, 526
    Fourier's law, 37
    Ohm's law, 525, 526
    Stefan-Boltzmann's law, 41
Passive circuit elements, 527
  loading error of, 528
Péclèt modulus, 117, 212
Penetration depth, 76, 78, 159, 474
Piecewise continuous functions, 337
Piecewise differentiable functions, 187
Poles of analytic functions, 378
Polynomial, Legendre, 240
  of degree n, of first and second kinds, 241
    order m, of first and second kinds, associated, 242
Power (rate of work), 23, 24, 29, 30
Power series, 132
Principle of superposition, 126, 218
Process, definition of, 18
Problem, Graetz, 203
Problems, boundary-value, 180
  characteristic-value, 180
  eigenvalue, 180
  mechanically determined or undetermined, 7, 17
  thermodynamically determined or undetermined, 7, 17
Profiles, Ritz, steady one-dimensional, 450
  steady two-dimensional, 469
  Kantorovich, steady two-dimensional, 469
  unsteady, 473
Property, definition of, 18

Radiation, 9
Radiation law, Stefan-Boltzmann's, 17, 37, 41
Radius, critical, 120
Relaxation, block, 496
  group, 496
  method, basic rule of, 494
    modifications of, 496
  operation table of, 494
  relaxation table of, 494
  residuals of, 494

solutions by, 493
overrelaxation, 496
Residue theorem, Cauchy's, 384, 385
Resistance, conductive, 104
Reynolds' transport theorem (transformation formula), 20
  for differential formulation, 33
  for integral formulation, 28
  for lumped formulation, 22
Ritz method, steady one-dimensional, 444
  steady two-dimensional, 458
Ritz profiles, steady one-dimensional, 450
  steady two-dimensional, 469
Round-off error, of difference formulation, 504

Second law, of motion, Newton's, 17
  of thermodynamics, 17
    for control volume, see General laws
    formulation of, see General laws, formulation of
Selection of coordinate axes, 214, 217
Separation parameter, 194
Separation of variables, 180
  nonhomogeneity, 217
  steady two-dimensional problems, cartesian, 193
    cylindrical, of type $\theta(r, \varphi)$, 224
    of type $\theta(r, z)$, 230
    spherical, 240
  steady three-dimensional problems, 248
  two conditions for use of, 194
  unsteady problems, 270
    disturbance, stepwise, 273
    time-dependent, 307
    one-dimensional charts for multidimensional, 290
    conditions for use of, 301
    two forms of superposition for, $\theta(x, t) = \psi(x, t) + \phi(x)$ and $\theta(x, t) = \psi(x, t) + \phi(x) + \varphi(t)$, 279
Series, Fourier, 186
  complete, 192
  cosine, 190
  double, 250, 290
  sine, 187
  Fourier-Bessel, 230
  Fourier-Legendre, 240, 244
  power, 132
  Taylor, 383, 384
Singularities of analytic functions, 376

branch points, 376
poles, 378
Solution, steady, one-dimensional problems, ordinary differential equations, 110
  extended surfaces, ordinary differential equations, 147
    Ritz-Integral, first-order, 156, 165
    Ritz-Integral, second-order, 163
    Ritz-Variational, first-order, 445, 449
    Ritz-Variational second-order, 446, 449
  two-dimensional problems, Kantorovich-Integral, first-order, 69
    Kantorovich-Integral, second-order, 222
    Kantorovich-Variational, first-order, 461
    Kantorovich-Variational, second-order, 462
    relaxation method, 493
    Ritz-Integral, first-order, 68
    Ritz-Integral, second-order, 221
    Ritz-Variational, first-order, 459
    Ritz-Variational, second-order, 459
    separation of variables, 194
  three-dimensional problems, separation of variables, 248
  unsteady, one-dimensional problems with stepwise disturbances, analog, 533
    graphical, 508, 515
    Kantorovich-Integral, first-order, 74, 79, 82, 159, 161, 359
    Kantorovich-Variational, first-order, 474, 475
    Kantorovich-Variational, second-order, 476
    Laplace transforms, by expansion for large argument, 418
    Laplace transforms, by expansion for small argument, 413
    Laplace transforms, by inversion, 390
    Laplace transforms, by tables, 343
    numerical, 515
    separation of variables, 273
  one-dimensional problems with time-dependent disturbances, Duhamel's superposition integral, 307
    Laplace transforms, 399
  one-dimensional problems with steady periodic disturbances, complex temperature, 324
    Laplace transforms, 343, 421, 422
  one-dimensional problems with unsteady periodic disturbances, Duhamel's superposition integral, 307
    Laplace transforms, 343, 421, 422
  two-dimensional problems, Laplace transforms, 409
    separation of variables, 288
Specific heat, constant pressure, 35
  constant volume, 34
Spherical problems, $\theta(r, t) = \psi(r, t)/r$ transformation of, 286
State, definition of, 18
Steady problems, *see* Solution, steady problems
Stefan-Boltzmann, constant, 43
  law of radiation, 17, 37, 41
Substantial derivative, 33
Superposition integral, Duhamel's, 307
Superposition principle, 126, 218
System, definition of, 18

Table, 2–1 property B, mass- or volume-dependent, 21
  2–2 heat transfer coefficient, range of, 51
  3–1 extended surfaces, general solution of, 138
  3–2 insulated finite fin, comparison of exact and first-order Integral-Ritz temperatures, tip, 158
  3–3 comparison of heat losses from problem of Table 3–2, 158
  3–4 comparison of exact, and first- and second-order Integral-Ritz heat losses from problem of Table 3–2, 164
  7–1 properties of Laplace transforms, 343
  7–2 inversion of Laplace transforms, 344
  8–1 comparison of exact heat loss with first- and second-order Integral- and Variational-Ritz heat losses from problem of Table 3–2, 448

8-2 thin disks in relative motion, comparison of first- and second-order Variational-Ritz temperatures, centerline, 450
8-3 two-dimensional heater, comparison of exact temperature with first-order Integral and Variational (Ritz and Kantorovich) temperatures, center, 465
operation, of relaxation method, 494
relaxation, 494
Taylor series of analytic functions, 383, 384
Temperature, definition of equality of, 19
method of complex, 324
Thermal conductivity, 37
variable, 129
Thermal diffusivity, 44
Thermal energy, conservation of, for differential control volume, 34, 35
Three-dimensional problems, steady, 248
Time-dependent boundary conditions, 307
Total energy, conservation of, for differential control volume, 34
Transformation, Kirchhoff, 130
$\theta(r, t) = \psi(r, t)/r$, for spherical problems 286
Transformation formula, see Reynolds' transport theorem
Transforms, Fourier, 371
Laplace, 336
Transport theorem, Reynolds', see Reynolds' transport theorem
Truncation error, 488
Two-dimensional problems, steady, see Solution, steady, two-dimensional problems
unsteady, see Solution, unsteady, two-dimensional problems

Unsteady problems, see Solution, unsteady

Variable, complex, see Complex variable
Variable thermal conductivity, 129
Variational calculus, 435
basic problem of, 435
Euler's equation of, 437
formulation, 435
functionals of, 436
Kantorovich method, see Kantorovich method
Kantorovich profiles, see Kantorovich profiles
meaning and rules of, 438
natural boundary conditions, 440, 456
Ritz method, see Ritz method
Ritz profiles, see Ritz profiles
Viscosity law, Newton's, 6, 17

Weighting function, 183
Weighting number, 183
Work, definition of, 18
lost, 26
rate of (power), 23, 24, 29, 30